Handbook of Systems Biology

Handbook of Systems Biology

Editor: Alexis White

R CALLISTO REFERENCE

www.callistoreference.com

Callisto Reference,
118-35 Queens Blvd., Suite 400,
Forest Hills, NY 11375, USA

Visit us on the World Wide Web at:
www.callistoreference.com

ISBN: 978-1-64116-143-5 (Hardback)

Trademark Notice: Registered trademark of products or corporate names are used only for explanation and identification without intent to infringe.

Cataloging-in-Publication Data

Handbook of systems biology / edited by Alexis White.
    p. cm.
Includes bibliographical references and index.
ISBN 978-1-64116-143-5
1. Systems biology. 2. Computational biology. 3. Biology. I. White, Alexis.
QH324.2 .H36 2019
570.28--dc23

Table of Contents

Preface

Every book is a source of knowledge and this one is no exception. The idea that led to the conceptualization of this book was the fact that the world is advancing rapidly; which makes it crucial to document the progress in every field. I am aware that a lot of data is already available, yet, there is a lot more to learn. Hence, I accepted the responsibility of editing this book and contributing my knowledge to the community.

Systems biology is an interdisciplinary field of science that is concerned with the computational and mathematical modeling of biological systems which are complex in nature. The field strives to model and discover properties of cells, tissues and organisms functioning as a system. These involve metabolic networks and cell signaling networks. Some associated disciplines of system biology are phenomics, genomics, epigenetics, proteomics, biomics, etc. Systems biology is an upcoming field of science that has undergone rapid development over the past few decades. This book covers in detail some existing theories and innovative concepts revolving around systems biology. It consists of contributions made by international experts. It is a vital tool for all researching and studying this field.

While editing this book, I had multiple visions for it. Then I finally narrowed down to make every chapter a sole standing text explaining a particular topic, so that they can be used independently. However, the umbrella subject sinews them into a common theme. This makes the book a unique platform of knowledge.

I would like to give the major credit of this book to the experts from every corner of the world, who took the time to share their expertise with us. Also, I owe the completion of this book to the never-ending support of my family, who supported me throughout the project.

Editor

Recurrent patterns of DNA copy number alterations in tumors reflect metabolic selection pressures

Nicholas A Graham[1,2,3,†] (iD), Aspram Minasyan[1,2,†] (iD), Anastasia Lomova[1,2] (iD), Ashley Cass[1,2], Nikolas G Balanis[1,2], Michael Friedman[1,2], Shawna Chan[1,2] (iD), Sophie Zhao[1,2], Adrian Delgado[1,2], James Go[1,2], Lillie Beck[1,2], Christian Hurtz[4,§] (iD), Carina Ng[4], Rong Qiao[2], Johanna ten Hoeve[1,2], Nicolaos Palaskas[1,2] (iD), Hong Wu[2,5,6], Markus Müschen[4,7], Asha S Multani[8], Elisa Port[9], Steven M Larson[10], Nikolaus Schultz[11,12], Daniel Braas[1,2,13], Heather R Christofk[2,5,13,‡], Ingo K Mellinghoff[12,14,15,16,‡] & Thomas G Graeber[1,2,5,13,17,*] (iD)

Abstract

Copy number alteration (CNA) profiling of human tumors has revealed recurrent patterns of DNA amplifications and deletions across diverse cancer types. These patterns are suggestive of conserved selection pressures during tumor evolution but cannot be fully explained by known oncogenes and tumor suppressor genes. Using a pan-cancer analysis of CNA data from patient tumors and experimental systems, here we show that principal component analysis-defined CNA signatures are predictive of glycolytic phenotypes, including ^{18}F-fluorodeoxy-glucose (FDG) avidity of patient tumors, and increased proliferation. The primary CNA signature is enriched for p53 mutations and is associated with glycolysis through coordinate amplification of glycolytic genes and other cancer-linked metabolic enzymes. A pan-cancer and cross-species comparison of CNAs highlighted 26 consistently altered DNA regions, containing 11 enzymes in the glycolysis pathway in addition to known cancer-driving genes. Furthermore, exogenous expression of hexokinase and enolase enzymes in an experimental immortalization system altered the subsequent copy number status of the corresponding endogenous loci, supporting the hypothesis that these metabolic genes act as drivers within the conserved CNA amplification regions. Taken together, these results demonstrate that metabolic stress acts as a selective pressure underlying the recurrent CNAs observed in human tumors, and further cast genomic instability as an enabling event in tumorigenesis and metabolic evolution.

Keywords aneuploidy; DNA copy number alterations; genomic instability; glycolysis; metabolism
Subject Categories Cancer; Genome-scale & Integrative Biology; Metabolism

Introduction

Cancer cells differ from normal cells in that they exhibit aberrant proliferation, resist apoptosis, and invade other tissues (Hanahan & Weinberg, 2011). Modern cancer classification relies on molecular

1 Crump Institute for Molecular Imaging, David Geffen School of Medicine, University of California, Los Angeles, CA, USA
2 Department of Molecular & Medical Pharmacology, David Geffen School of Medicine, University of California, Los Angeles, CA, USA
3 Mork Family Department of Chemical Engineering and Materials Science, University of Southern California, Los Angeles, CA, USA
4 Department of Laboratory Medicine, University of California, San Francisco, CA, USA
5 Jonsson Comprehensive Cancer Center, David Geffen School of Medicine, University of California, Los Angeles, CA, USA
6 School of Life Sciences & Peking-Tsinghua Center for Life Sciences, Peking University, Beijing, China
7 Department of Haematology, University of Cambridge, Cambridge, UK
8 Department of Genetics, M. D. Anderson Cancer Center, The University of Texas, Houston, TX, USA
9 Department of Surgery, Icahn School of Medicine at Mount Sinai, New York, NY, USA
10 Department of Radiology, Memorial Sloan Kettering Cancer Center, New York, NY, USA
11 Marie-Josée and Henry R. Kravis Center for Molecular Oncology, Memorial Sloan Kettering Cancer Center, New York, NY, USA
12 Human Oncology and Pathogenesis Program, Memorial Sloan Kettering Cancer Center, New York, NY, USA
13 UCLA Metabolomics Center, David Geffen School of Medicine, University of California, Los Angeles, CA, USA
14 Department of Neurology, Memorial Sloan Kettering Cancer Center, New York, NY, USA
15 Department of Pharmacology, Weill Cornell Medical College, New York, NY, USA
16 Department of Neurology, Weill Cornell Medical College, New York, NY, USA
17 California NanoSystems Institute, David Geffen School of Medicine, University of California, Los Angeles, CA, USA
*Corresponding author. E-mail: tgraeber@mednet.ucla.edu
†These authors contributed equally to this work
‡These authors contributed equally to this work
§Present address: Division of Hematology and Oncology, Department of Medicine, University of Pennsylvania, Philadelphia, PA, USA

characterization, including examination of genomic DNA mutations and copy number alterations (CNAs; Stuart & Sellers, 2009). Although individual oncogenes and tumor suppressor genes are preferential targets of DNA amplifications and deletions, respectively, the recurrent CNA patterns in tumors cannot be fully explained by canonical cancer genes (Beroukhim *et al*, 2010; Muller *et al*, 2012; Davoli *et al*, 2013). Thus, the unexplained recurrent CNA patterns observed in human cancer subtypes are suggestive of additional, not yet fully defined, selective pressures that are conserved across patients and tumor types (Cahill *et al*, 1999; Sheltzer, 2013; Cai *et al*, 2016). Reports that the cumulative phenotypic effects of many small gene dosage alterations across the genome can impact the resulting tumor copy number landscape (Solimini *et al*, 2012; Davoli *et al*, 2013) illustrate a need to consider more subtle and combinatorial effects to explain the remaining selective forces underlying recurrent CNA patterns observed in human cancers.

One of the fundamental and consequential differences between non-transformed and tumorigenic cells is the reprogramming of cellular metabolism (Hanahan & Weinberg, 2011). The altered metabolism of tumors is thought to benefit transformed cells in several ways. Upregulation of glucose metabolism allows proliferating cells to meet their energy demand through synthesis of adenosine triphosphate (ATP), while increased flux through glycolysis branch pathways provides dividing cells with intermediates necessary for biosynthesis of nucleotides and fatty acids, as well as reducing agents such as glutathione and NADPH (DeBerardinis *et al*, 2008; Cairns *et al*, 2011). Moreover, in addition to glucose, cancer cells frequently upregulate consumption of other metabolites for energy and biomass generation, including glutamine, serine, and glycine (Jain *et al*, 2012; Maddocks *et al*, 2013). Notably, several individual metabolic enzymes have been directly implicated in tumorigenesis (Kim *et al*, 2007; Dang *et al*, 2009; Locasale *et al*, 2011; Possemato *et al*, 2011; Patra *et al*, 2013; Li *et al*, 2014; Wang *et al*, 2014; Xie *et al*, 2014) and/or immortalization (Kondoh *et al*, 2005; Kondoh, 2009; Kaplon *et al*, 2013), suggesting that altered metabolism is not a passive bystander, but rather a driving force of oncogenesis (Yun *et al*, 2009; Zhang *et al*, 2012). Using an integrative analysis of CNA data from human tumors, mouse models of cancer, cancer cell lines, and a murine experimental immortalization system, here we show that the loci of metabolic genes impact the recurrent CNA changes observed in genomically unstable tumors. Our bioinformatic and experimental results support a tumorigenesis model in which copy number changes in metabolic genes contribute to an enhanced glycolytic and proliferative state (see Fig EV1 for a schematic of our overall approach).

Results

PCA-defined CNA signatures in human cancers

To develop an unbiased understanding of DNA copy number alterations (CNAs) in cancer, we performed principal component analysis (PCA) of gene-based CNA data derived from comparative genomic hybridization (CGH) microarrays from 15 tumor types available from The Cancer Genome Atlas (TCGA). This pan-cancer

PCA revealed a high degree of similarity in CNA profiles between basal breast invasive carcinoma (BRCA basal), lung squamous cell carcinoma (LUSC), ovarian serous cystadenocarcinoma (OV), and serous uterine corpus endometrial carcinoma (UCEC serous) (Fig 1A and Appendix Fig S1). In tumor type-specific PCA, analyzing the four tissue-defined tumor sets of BRCA, LU (lung cancer consisting of LUSC and LUAD [lung adenocarcinoma]), OV, and UCEC revealed two strong PCA-based CNA signatures, termed signatures A and B, in all four cases. Notably, signature A was highly consistent across each tumor type, and reflected the pattern of pan-cancer PC1 loadings (Figs 1B and C, and EV2A–C). In BRCA, signature A tumors were enriched for the basal subtype, p53 point mutations, high numbers of genomic breakpoints, and thus subchromosomal alterations (Figs 1B and D, and EV2D; $P < 0.001$, $P < 0.001$, $P = 2 \times 10^{-4}$, respectively). Signature B BRCA tumors, in contrast, were enriched for luminal type tumors ($P < 0.001$) and exhibited amplifications of the oncogenes *MYC* and *MDM2* and deletion of the tumor suppressor *CDKN2A* (Fig 1B). Amplification of *MDM2* and loss of *CDKN2A* were generally mutually exclusive in signature B tumors (Fig EV2E), reflecting alternate mechanisms for disabling the p53/ARF axis (Sherr & Weber, 2000). In the other tissues, signature A tumors were enriched for lung squamous cell carcinomas, the proliferative subtype of ovarian cancer (The Cancer Genome Atlas Research Network, 2011), and the serous subtype of uterine cancer (Fig EV2A–D). Overall, signature A tumors demonstrated enrichment of p53 mutations, more genomic breakpoints (BRCA, LU, and UCEC), and a higher degree of copy number alterations (LU, OV, and UCEC) than signature B tumors (Appendix Fig S2A and B, *P*-values indicated in figure). In signature A tumors, the per tumor average segment size is on the scale of 1×10^7 base pairs (approximately one-tenth of a chromosome, containing on the order of 100 genes). Overall, the segment sizes in signature A span from focal to arm-length/whole chromosome scale. The per tumor average segment size for signature B BRCA, LU, and UCEC tumors is statistically larger by 1.3-fold to sevenfold (Appendix Fig S2C–E; $P = 2 \times 10^{-4}$ or less). Unlike signature A tumors, the signature B CNA patterns were quite distinct between tumor types, although some commonalities were observed including point mutations in oncogenes such as *KRAS* (LU and UCEC) and amplification of *MYC* (BRCA, LU, and OV). An alternative approach using hierarchical clustering confirmed the existence of the shared pan-cancer CNA signatures across multiple tumor types (signature A), as well as distinct signature subtypes within each of the BRCA, OV, UCEC, and LU tumor types (signature A vs. signature B) (Fig EV3 and Appendix Fig S3, *P*-values of concordance between PCA and clustering approaches indicated in figure). In summary, PCA revealed the CNA signature A as a pattern shared across a subset of tumors in multiple tissue types, as well as several tumor type-specific cases of more distinct signature B patterns (Appendix Fig S1D).

Because altered metabolism is a hallmark of human tumors, we next tested whether the shared CNA signature A was enriched for genes from metabolic pathways. Using CNA-based gene set enrichment analysis over all metabolic pathways defined by KEGG (Kanehisa *et al*, 2014), we found that the conserved profile of core signature A tumors (i.e., OV, BRCA, UCEC, LU) (Fig 1C) was significantly enriched for DNA amplifications of core glycolysis pathway genes (Fig 1E and F, and Table EV1; $P = 0.024$). For example, BRCA

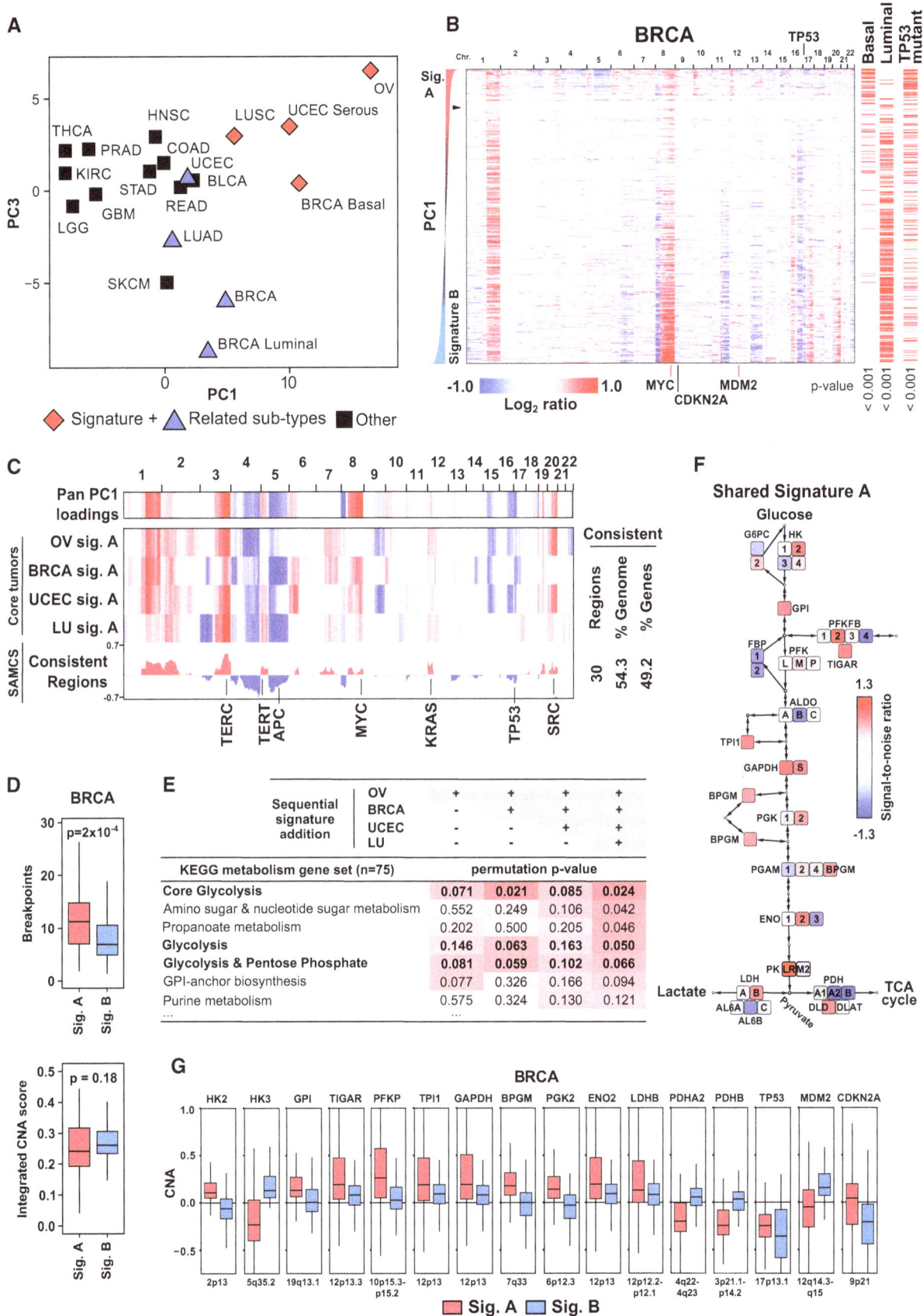

Figure 1.

◀ **Figure 1. Principal component analysis (PCA) reveals a shared CNA signature in breast, lung, ovarian, and uterine carcinomas.**

A   Pan-cancer PCA of copy number data from a balanced, random sampling of tumors of 15 tumor types from The Cancer Genome Atlas (TCGA). The average tumor PC scores for each tumor subtype are shown. Pan-cancer PC1 scores primarily separate diploid from highly aneuploid tumors, while PC2 distinguished GBM from the other tumor types (Appendix Fig S1A and B). Four tumor subtypes with similar PC scores are labeled as "Signature positive (+)". Tumor type abbreviations are as defined by TCGA: ovarian serous cystadenocarcinoma (OV), breast (BR), bladder urothelial (BL), and thyroid (TH) carcinoma (*CA), uterine corpus endometrial carcinomas (UCEC), lung (LU), and head and neck (HN) squamous carcinomas (*SC), skin cutaneous melanoma (SKCM), lung (LU), rectal (RE), colon (CO), stomach (ST), and prostate (PR) adenocarcinomas (*AD), glioblastoma (GBM), low-grade glioma (LGG), and kidney renal clear cell carcinoma (KIRC).

B   Copy number profiles of 873 breast invasive carcinomas (BRCA). Tumors (rows) sorted by tumor-specific principal component 1 (PC1) score with genomic locations listed across the top. PCA identified and distinguished two signatures with similar degrees of variance in the first component. The triangle marks the transition from signature A through diploid samples to signature B. The membership of each tumor with known molecular subtypes (e.g., basal/luminal) and mutant *TP53* status are indicated on the right with red horizontal bars and corresponding permutation-based enrichment *P*-values. Basal and luminal subtype classifications are from a published gene expression-based determination (The Cancer Genome Atlas Research Network, 2012b). Signature A was moderately enriched for claudin-low tumors, whereas the HER2-enriched subtype was not significantly associated with either signature A or B (Fig EV2D). Similar PCA-based distinctions were observed in the LU, OV, and UCEC datasets (Fig EV2A–C).

C   Pan-cancer PC1 loadings and signature A summary profiles for ovarian (OV), breast (BRCA), uterine (UCEC), and lung (LU) tumor types. Summary signatures are normalized gene loci signal-to-noise ratios (SNRs) of the top 10% of PC1-based core signature A tumors compared to normal (non-tumor) samples. Pan-cancer PC1-3 loadings are shown in Appendix Fig S1C. Consistency signatures of conserved amplification and deletions (consistent regions) are indicated by their signed absolute minimum consistency score (SAMCS). The SAMCS is non-zero when all CNA summary signatures have the same sign across all tumor types, and is derived from the absolute value-based minimum summary metric, and then re-signed positive for amplification or negative for deletion (see Materials and Methods). In the bar graph, red and blue denote consistently amplified and deleted regions, respectively. The number or percentage of consistent regions, genome coverage, and gene loci are indicated. The positions of canonical oncogenes, proto-oncogenes, tumor suppressor genes, and telomerase components (*TERC*, *TERT*) present in these regions are indicated.

D   Breast signature A tumors have more DNA breakpoints per chromosome than signature B tumors, but similar levels of copy number alteration (as measured by an integrated CNA score proportional to the extent and absolute value magnitude of amplifications and deletions). *P*-values are indicated (top, Mann–Whitney *U*-test; bottom, Student's *t*-test), and data presented in box (median, first and third quartiles) and whisker (extreme value) plots. The number of breakpoints is inversely proportional to the mean DNA segment length.

E   KEGG metabolism pathway enrichment analysis based on consistent CNA patterns in the core signature A tumors of (A). The OV, BRCA, UCEC, and LU signatures from (C) were sequentially added and only directionally consistent CNA changes were retained (see Materials and Methods). The combined glycolysis–gluconeogenesis (glycolysis) and pentose phosphate pathway was included based on our prior mRNA work identifying the predictive value of this gene set (Palaskas *et al*, 2011). Core glycolysis is a KEGG-defined gene subset (M00001). Table EV1 lists enrichment results for all KEGG metabolism pathways.

F   Schematic showing average signal-to-noise (SNR) metrics of core glycolysis pathway genes plus 6-phosphofructo-2-kinase/fructose-2,6-biphosphatase (PFKFB and TIGAR), lactate dehydrogenase (LDH), and pyruvate dehydrogenase (PDH) in core signature A tumors from (C) and (E). A more detailed listing of the gene names can be found in Materials and Methods.

G   Gene copy number alteration distributions of selected glycolysis genes, *TIGAR*, the tumor suppressor p53, and the p53-associated cancer genes *MDM2* and *CDKN2A* in BRCA tumors. Data are presented in box (median, first and third quartiles) and whisker (extreme value) plots.

Data information: In panels (C, E–G), the top 10% of tumors from each indicated signature based on PC1 scores were used in the analysis. See also Fig EV2 and Appendix Figs S1–S3.

signature A tumors exhibited DNA amplification of most genes from the glycolytic pathway, as well as amplification of lactate dehydrogenase B, deletion of pyruvate dehydrogenase subunits A and B, and amplification of the glycolysis-regulating metabolic enzyme TIGAR (human gene *C12orf5*; Fig 1G). Notably, this shared CNA signature was defined by genome-wide patterns, rather than by single gene loci, which were not consistently altered in all tumors with a strong signature (Appendix Fig S2F). Interestingly, signature A summary profiles from breast, ovarian, uterine, and lung tumors all exhibited hexokinase 2 (HK2) amplification (Fig 1F and G), whereas signature B profiles had primarily either HK3 (BRCA, LU, OV) or HK1 (UCEC) amplification (Fig 1G and Appendix Fig S2G). Thus, PCA identified a shared signature from breast, lung, ovarian, and uterine carcinomas that was enriched for p53 mutations, higher numbers of genomic breakpoints, and CNA of genes from the core glycolysis pathway.

Elimination of passenger genes via cross-species synteny mapping

While canonical oncogenes and tumor suppressor genes drive some recurrent DNA copy number alterations, many recurrent CNA regions cannot be fully explained by the presence of known cancer genes (Beroukhim *et al*, 2010; Muller *et al*, 2012; Davoli *et al*, 2013). The conservation of CNA signature A across breast, lung, ovarian, and uterine tumor subsets suggests the existence of one or more selective pressures that are potentially shared by additional tumor types. In that many of the human signature A conserved regions identified in Fig 1C still span large, chromosome-scale regions, we hypothesized that passenger genes could be diluting the enrichment signal. We subsequently reasoned that a previously reported approach of cross-species comparison of CNA data from human tumors and non-human tumorigenesis models would eliminate passenger genes via synteny mapping (Maser *et al*, 2007; Zhang *et al*, 2013; Tang *et al*, 2014).

For the cross-species analysis, we first empirically determined the full set of human signature A-like tumor tissue types that could be included based on CNA similarity as defined by the pan-cancer PC1 scores from Fig 1A. Namely, we sequentially added each human tumor type in order of decreasing pan-cancer PC1 score and tested whether inclusion improved the overall pathway enrichment. In a parallel analysis, at each sequential step we also added the corresponding tissue-matched mouse epithelial cancer model signature if it existed. Mouse CNA signatures obtained from the literature and public repositories included signatures for genetically engineered mouse models of mammary (breast) cancer (Brca) (Drost *et al*, 2011; Herschkowitz *et al*, 2012), melanoma (Skcm) (Viros *et al*, 2014), glioblastoma/high-grade astrocytoma (Gbm) (Chow *et al*, 2011), and prostate cancer (Prad) (Ding *et al*, 2012; Wanjala *et al*, 2015), as well as *in vitro* mouse epithelial cell models of bladder (Blca), colorectal (Coad), and kidney (Kirc) cancer (Padilla-Nash *et al*, 2013).

Using human tumors only, the sequential enrichment analysis showed a general improvement through the addition of the COAD tumor type (Fig 2A, dotted orange line; average enrichment signal of the top 10 metabolic pathways). Moreover, sequential inclusion of the corresponding tissue-matched mouse cancer model CNA signatures substantially enhanced the overall pathway enrichment results with the peak occurring at HNSC, a sequentially adjacent tumor type that has similar pan-cancer PC1 score with COAD (Fig 2A, solid blue line and arrow indicating the peak). Overall pathway enrichment then decreased as additional, less copy number similar (lower pan-cancer PC1 scores) tumor types were added.

The improvement upon including the mouse signatures supports the hypothesis that the cross-species analysis eliminates passenger genes and reveals pathways whose genes are enriched in the resulting cross-species consistent CNA regions. Taken together, this pan-cancer and cross-species analysis demonstrates that a combination of nine human tumor types (OV, BRCA, UCEC, LU, BLCA, READ, SKCM, COAD, and HNSC) and four corresponding mouse tumor models (Brca, Blca, Skcm, and Coad) gives the strongest overall enrichment signal strength across all metabolic pathways (Fig 2A).

Examining the highly ranked individual pathways from this optimized tumor type combination, we found that the carbohydrate metabolism pathway "amino sugar and nucleotide sugar metabolism" (hsa00520) was ranked first and glycolysis–gluconeogenesis (hsa00010, henceforth called glycolysis) was ranked second among KEGG metabolism gene sets (Fig 2B and Table EV1). As in the average of top 10 pathways (Fig 2A), the enrichment score for the glycolysis gene set improved as more human tumor and mouse models were sequentially included up through the additions of HNSC and Coad (Fig 2C). Compared to our previous enrichment analysis of the four core signature A tumors (Fig 1E), the permutation P-value of the glycolysis gene set improved from 0.05 to 0.001, reflecting that our pan-cancer and cross-species analysis has eliminated passenger genes in the non-consistent CNA regions. Thus, we hereafter refer to the consistent signature pattern from OV up through and including HNSC and Coad as "expanded signature A". Of particular note, upon expanding our analysis to include 1,321 gene sets from the MSigDB Canonical Pathways (CP) database, glycolysis remained a top result as the third ranked pathway, with the "glutathione-conjugation" pathway (involved in cell detoxification and oxidative stress responses) also scoring strongly (Fig 2B and Table EV2).

To visualize the pan-cancer and cross-species conserved genomic regions, we plotted the consistency profiles of the expanded signature A human and mouse tumors (Fig 2D). Examination of conserved regions revealed that *HK2*, *TPI1*, *GAPDH*, *PGAM2*, *ENO2*, and *LDHB* glycolysis genes contribute to the cross-species consistency signal. Among canonical oncogenes, *MYC* and *KRAS* were also present in the amplification regions of the expanded signature A human tumors and mouse models. Due to the well-documented and clinically relevant role of glycolysis and pentose phosphate pathways in tumorigenesis, we chose to further examine the recurrent amplification of the set of these gene loci in subsequent analyses. Equally important for functional validation of these candidate pathways, the activity of glycolysis can be directly and indirectly measured by many assays in both patients (e.g., FDG-PET imaging) and experimental systems.

CNA signatures are predictive of glycolysis

In that the CNA-consistent region-defined signature A is enriched for core glycolysis genes, we next tested whether signature A patient tumors *in vivo* were associated with increased tumor glycolysis. To assign a signature A score to a set of FDG-PET-imaged breast cancers (Palaskas *et al*, 2011), we projected CNA data from these tumors onto a PCA of the four core human tumor tissue types (OV, BRCA, UCEC, and LU, Fig 1C). We found a strong correlation between the strength of CNA signature A and the measured FDG-PET standardized uptake values (SUVs) (Fig 3A and B; Pearson rho = 0.94, $P = 5 \times 10^{-5}$). A similar analysis using all nine expanded signature A tumor types demonstrated equivalent results (Appendix Fig S4A; Pearson rho = 0.92, $P = 2 \times 10^{-4}$). Thus, the CNA-defined signature A is associated with increased FDG uptake in human primary tumors *in vivo*.

We next asked whether the CNA signature A-defined tumors associated with high FDG uptake also had RNA-based signatures of increased glycolysis. First, we compared signature A tumors to signature B tumors using RNA-based enrichment analysis (GSEA). We analyzed both BRCA and LU tumors because these tumor types show distinct signature A and signature B subtypes. (OV signature B is highly similar to the pan-cancer signature A pattern, and thus does not provide a differential test (Fig EV2B). UCEC tumors were not included due to a lack of sufficient paired RNA and DNA profiling data.) In the enrichment analysis, we included a gene set consisting of genes from the glycolysis and pentose phosphate pathways that were upregulated in FDG-high BRCA tumors, as defined by our previous work (Palaskas *et al*, 2011). In the GSEA, the empirically defined FDG-high gene set ranked number one overall (NES = 2.5, permutation P-value = 2×10^{-4}), confirming that Sig A tumors have glycolysis RNA expression profiles matching those of FDG-high tumors (Fig EV4A). Furthermore, BRCA and LU signature A tumors were significantly enriched for genes from the full glycolysis and pentose phosphate pathways (overall rank 4th of 76 pathways, NES = 2.0, permutation P-value = 6.8×10^{-3}), as well as other glycolysis-related pathways (Table EV3). This analysis of RNA expression data is consistent with the enrichment of glycolysis-associated pathways at the DNA copy number level (Fig 1E and 2B), and points to glycolysis-related pathways as selection targets for upregulation in signature A tumors.

To further explore the RNA expression data, we predicted the glycolytic phenotypes of the core signature A tumor types (BRCA, LU, and OV) using RNA-based weighted gene voting (WGV) (Golub *et al*, 1999) and our previously defined FDG prediction model (Palaskas *et al*, 2011). UCEC tumors were again excluded because there were not a sufficient number of samples with paired RNA and DNA data. We found that RNA-based predictions of high glycolysis were associated with the signature A end of tumor type-specific PC1 for BRCA and LU (Fig EV4B–D). Signature A tumors were predicted to be significantly more glycolytic than signature B tumors for BRCA and LU (P-values of 2.4×10^{-17} and 5.7×10^{-18}, respectively, Fig EV4E). There was no significant differential trend in OV tumors, potentially because almost all of these tumors are genomically unstable (integrated CNA scores > 0.2) and the signature B of OV is relatively signature A-like (Fig EV2B and Appendix Fig S1D). The OV predictions were consistent with predicted high glycolysis across all tumors (Fig EV4E). To control

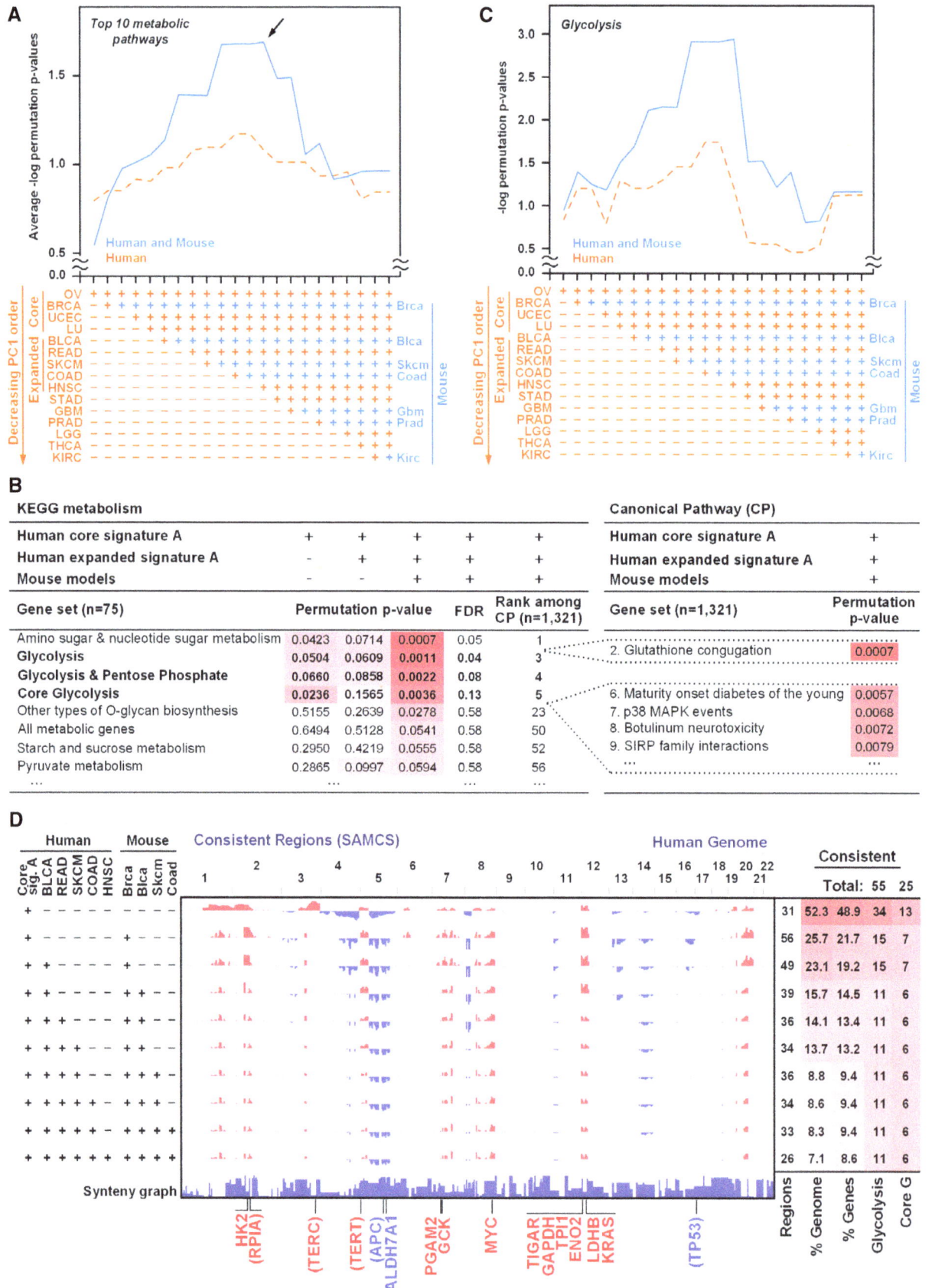

Figure 2.

Figure 2. Pan-cancer and cross-species analysis reveals enrichment of glycolytic genes in conserved amplification regions.

A   Average pathway enrichment signal strength for the top 10 enriched KEGG metabolism pathways based on the consistent CNA patterns across multiple tumor types. The four core CNA signature A patterns are defined as in Fig 1C. CNA signatures for other tissue tumor types are defined by the signal-to-noise ratios (SNRs) of all tumors compared to tissue-matched normal (non-tumor) samples. Starting with the four core signature A tumor types from Fig 1 (OV, BRCA, UCEC, and LU), additional human (orange) or human and mouse (blue) signatures were sequentially added based on their decreasing tissue type pan-cancer PC1 score from Fig 1A (see Materials and Methods). The resulting sequentially restricted consistency signatures are illustrated in (D). A maximal peak (indicated by an arrow) in the enrichment signal of this sequential analysis is seen after adding tumors through human HNSC and mouse Coad (no mouse Hnsc signature was available). Enrichment signal strength is based on the negative log frequency of equivalent enrichment by chance (i.e., permutation P-value). Mouse tumor type abbreviations are shown with lower case letters but otherwise match the human TCGA abbreviations.

B   Enrichment analysis of metabolism (75 KEGG pathways) and canonical pathways (CP; 1,321 MSigDB canonical pathways). Permutation P-values for the consistent CNA patterns from the four human core signature A tumors (defined in Fig 1C), the human expanded signature A (from A), and the human expanded signature A combined with corresponding mouse cancer models (from A). FDR calculations are described in Materials and Methods. Tables EV1 and EV2 list all sequential enrichment results for KEGG metabolism and canonical pathways, respectively.

C   Enrichment score for the glycolysis pathway upon sequential addition of human tumors and mouse models (as in A) showing a maximal peak in the enrichment signal corresponding to the expanded HNSC–Coad signature.

D   Genome view of the sequential consistency signatures used in the enrichment analysis of (A–C). The locations of consistently amplified glycolysis genes, canonical oncogenes, and tumor suppressors are indicated. Genes listed in parentheses (i.e., *RPIA*, *TERC*, *TERT*, *APC*, and *TP53*) were consistently altered across the human tumors but not consistently altered in the mouse models, due in part to the inclusion of p53 genetic knockout mouse models (*TP53*) and known differences in human and mouse telomere maintenance (*TERC*, *TERT*) (Sherr & DePinho, 2000). The synteny graph at the bottom indicates the syntenic mouse chromosome number and thus the broken synteny regions between human and mouse genomes. This genome view chronicles how the sequential pan-cancer and cross-species analysis of conserved amplification and deletion loci greatly reduces the percentage of consistent copy number alterations, thereby reducing the percentage of the genome implicated as candidate driver regions.

for the general increase in glycolysis predictions at increased levels of genomic instability, we also analyzed the correlation between glycolysis predictions and PCA scores for several range windows of integrated CNA scores. This analysis demonstrated that RNA-based predictions of glycolysis were significantly correlated with PC1 scores in BRCA and LU tumors with high integrated CNA scores (permutation P-value for BRCA and LU < 1 × 10^{-6} each, Fig EV4F). Taken together, the DNA-defined signature A tumors exhibit RNA expression patterns consistent with increased FDG uptake. These RNA-based results support the more downstream activity-based findings of elevated FDG uptake in signature A tumors (Fig 3A and B).

Returning to the result of the CNA-defined signature A being associated with increased FDG uptake in human primary tumors *in vivo*, we next asked which sets of metabolic gene loci copy number levels were most predictive of glycolytic phenotypes. We performed a CNA-based weighted gene voting (WGV) analysis to predict the glycolytic phenotypes of breast tumors and breast cancer cell lines (Neve *et al*, 2006) using individual gene sets from the KEGG metabolic pathways database (Kanehisa *et al*, 2014). Gene weights were calculated from each of our four tumor type CNA "training" signatures. Specifically, we tested the ability of individual metabolic pathways to predict (i) FDG uptake in patient primary breast tumors and (ii) the lactate secretion of a panel of 32 breast cancer cell lines (Hong *et al*, 2016). Averaging results across these two test cases revealed that genes from the glycolysis and pentose phosphate pathway were most predictive of these metabolic phenotypes (Fig 3C and Table EV4; P = 0.01). Moreover, signature A-based predictions were predictive of lactate secretion for basal cell lines more so than for luminal lines (Fig 3D and E, and Appendix Fig S4B), consistent with the observed basal-type tumor enrichment in signature A samples (Figs 1B and EV2D). Thus, consistent with the gene expression-based predictions above, the glycolysis and pentose phosphate pathway DNA copy number alterations from signature A are predictive of glycolytic phenotypes of primary human breast tumors and cancer cell lines.

Experimental recapitulation of tumor CNA signatures

Having demonstrated that the glycolytic pathway is statistically associated with genome-wide DNA copy number patterns, we sought an experimentally tractable system that would allow us to test the hypothesis that recurrent patterns of DNA amplification reflect metabolic selection pressures. We thus derived a panel of immortalized mouse embryonic fibroblasts (MEFs) using the classical 3T9 protocol (Todaro & Green, 1963). In this experimental system, under standard culture conditions (e.g., 21% O$_2$), diploid cells undergo a crisis-associated event that increases tolerance for genomic instability and allows them to escape senescence and evolve into cells with chromosomal instability and genomic aberrations (Fig 4A and Appendix Fig S5A; Sun & Taneja, 2007). This system has been used to study core cancer phenotypes such as proliferation, anti-apoptosis, and chromosomal instability (Lowe *et al*, 1993; Gupta *et al*, 2007; He *et al*, 2007; Sotillo *et al*, 2007; Sun *et al*, 2007; Weaver *et al*, 2007) and is one of the few experimentally tractable cancer models involving spontaneous genomic instability (Sherr & DePinho, 2000). In addition, because this system is not driven by strong oncogenes (e.g., KRAS mutation), it allows for complex CNA signatures to evolve from a combination of individual, presumably weaker, DNA alteration events.

We profiled the genome-wide copy number of 42 independent MEF sublines by CGH microarray (Fig 4A). Most samples were profiled after immortalization (post-senescence), with a few profiled before or during senescence. Analysis of this CNA data by PCA revealed that the MEF system recapitulated a two-signature pattern (signatures A and B). These two signatures were generally orthogonal, with the exception of a few "mixed" samples that had CNA characteristics of both signatures A and B (Appendix Fig S5B). As anticipated from prior MEF studies, immortalized MEF cells exhibited an increased number of genomic breakpoints and a higher degree of copy number alterations than the diploid, pre-senescent MEF cells (Appendix Fig S5C and D). Importantly, the MEF-derived signature A resembled the shared signature A pattern derived from

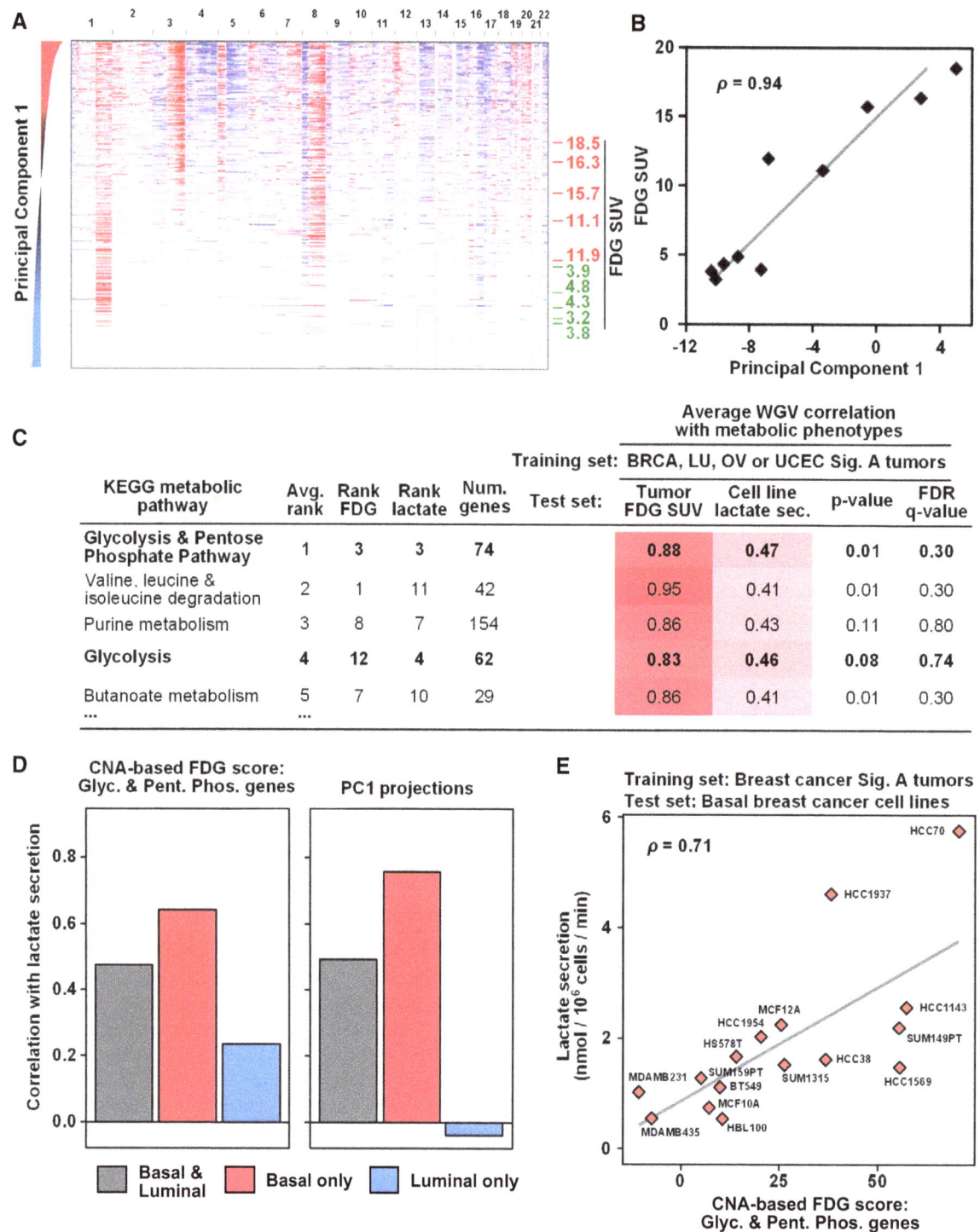

Figure 3.

human breast, lung, ovarian, and uterine tumors. In particular, the MEF-derived signature A was characterized by *p53* mutation and chromosome loss at the p53 locus, more genomic breakpoints, subchromosomal-sized alterations, and a higher degree of copy number alteration (Fig 4A and B, and Appendix Table S1). Additionally, this experimental system allowed us to profile the same MEF lines at subsequent passages, and a paired statistical analysis of five

evolving signature A lines revealed genomic regions changing from mid- to late passage (Fig 4A and Appendix Fig S6A and B). Similar to the shared human tumor signature A, both the MEF signature A and the evolving MEF signature demonstrated DNA amplifications of glycolysis and glycolysis-related genes (Table EV2). In particular, both human and mouse CNA signatures A included amplification of *Hk2, Bpgm, Rpia, Tigar* (mouse gene *9630033F20Rik*), *Eno2, Tpi1,*

◀ **Figure 3. PCA-based CNA signatures are predictive of breast cancer glycolytic metabolism *in vivo* and breast cancer cell line metabolism *in vitro*.**

A   PC1-sorted copy number profiles of a balanced, random sampling of tumors from the four core CNA-consistent tumor types (breast, lung squamous, ovarian, and uterine carcinomas, Fig 1A) along with copy number profiles from primary breast carcinoma tumors with glycolytic levels imaged *in vivo* by ^{18}F-fluorodeoxy-glucose positron emission tomography (FDG-PET) (Palaskas *et al*, 2011). On the right, red and green values indicate high and low FDG standardized uptake values (SUVs), respectively, for FDG-imaged tumors.

B   FDG uptake values in FDG-PET measured tumors are highly correlated with CNA signature A PC1 score (ρ, Pearson rho correlation; *P*-value = 5.3 × 10^{-5}).

C   CNA values for genes from glycolysis and pentose phosphate pathways have stronger predictive power of FDG-PET SUV in breast tumors and of lactate secretion (sec.) in breast cancer cell lines than other metabolism pathway-based sets of genes. The table indicates the correlation between weighted gene voting (WGV)-based predictions on the test sets and the measured metabolic phenotypes. WGV was performed with individual training on signature A tumors from each of the core four tumor types (breast (BRCA), lung (LU), ovarian (OV), and uterine (UCEC); top 10% signature A compared to normal (non-tumor) samples), the voting predictions were averaged, and compared to the measured metabolic phenotype. *P*-values were assessed by permutation analysis. See Table EV4 for all pathways.

D, E Signature A-based WGV predictions and signature A-based PC1 projections (as in B) have stronger correlation with measured lactate secretion for basal-like breast cancer cell lines than for luminal-like lines (D). See Appendix Fig S4B for luminal cell line data. Basal and luminal subtype classifications are from a published gene expression-based determination (Neve *et al*, 2006). The breast cancer-based glycolysis and pentose phosphate pathway (G & PP) WGV predictions are shown as a representative case (E).

Data information: See also Appendix Fig S4B.

Gapdh, Ldhb, Kras (mouse chr. 6), and *Pgk2* (mouse chr. 17) (Figs 1G and 4A, and Appendix Fig S6C).

In contrast, the MEF-derived signature B was characterized by amplification of *Mdm2* or deletion of *Cdkn2a* (Ink4a/Arf) and fewer overall copy number alterations. *Mdm2* amplification and *Cdkn2a* loss are alternative mechanisms for inactivating p53 function in human tumors and in MEFs (Sherr & DePinho, 2000), but as found here result in a distinct CNA signature as compared to the p53 mutation-associated signature A. As in human BRCA (Fig EV2E), amplification of *Mdm2* and loss of *Cdkn2a* were generally mutually exclusive in signature B MEFs (Appendix Fig S6D), reflecting alternate mechanisms for disabling the p53/ARF axis (Sherr & Weber, 2000). Additionally, *Mdm2* amplification tends to co-occur with *Hk1* amplification, both loci being located on mouse chr. 10. Thus, the signature B cases are associated with an alternate HK amplification, similar to our finding in signature B human tumors (HK1 or HK3 rather than HK2) (Fig 1G and Appendix Figs S2G and S6C). Immunoblotting confirmed that signature A MEF and signature B MEF lines generally had increased expression of Hk2 and Hk1 protein, respectively (Appendix Fig S6E and F).

To further characterize the association between p53 loss and CNA signatures, we profiled the CNA patterns of 29 independent p53$^{-/-}$ MEF sublines derived in standard 3T9 culture conditions. Comparison of the p53$^{-/-}$ MEF CNA patterns to wild-type MEFs by PCA demonstrated that the CNA patterns of p53$^{-/-}$ MEFs resemble the wild-type MEF signature A pattern, with no signature B-like sublines observed (Appendix Fig S7). p53$^{-/-}$ MEFs do not undergo senescence (Olive *et al*, 2004), and consistent with this, we observed that they tended to have less strong copy number alterations. In summary, strong p53 functional loss (p53 mutation or genetic loss) tends to lead to the CNA signature A pattern, which is associated with a higher degree of copy number alterations (higher integrated CNA score and more breakpoints) and Hk2 amplification, while weaker or less complete p53 functional loss (e.g., mediated by *Mdm2* amplification or *Cdkn2a* loss) is associated with an alternative signature (signature B).

Numerical and structural chromosomal abnormalities

Next we sought to understand how CNA signatures revealed by CGH relate to numerical and structural aneuploidy (i.e., whole chromosomal and subchromosomal gains or losses, respectively). Using propidium iodide-based DNA staining, we found that immortalized MEF cells had increased total DNA content compared to pre-senescent MEFs (Appendix Fig S8A–C). Similar results were observed in a comparison between a genome-stable, close to diploid, immortalized human mammary epithelial cell line (MCF10A) and a human breast cancer cell line with a high degree of DNA copy number alterations (Hs578T). In addition, we used spectral karyotyping (SKY) to assess the numerical and structural chromosomal aberrations in immortalized MEF cells. Profiling a representative signature A MEF subline just after immortalization (subline H1; passage 25, P25), SKY revealed increased total DNA content (105 ± 36 chromosomes) with nearly all chromosomes having experienced whole chromosome gains (Appendix Fig S8D, E and H). In addition, there was substantial cell-to-cell heterogeneity in the number and type of chromosomes at passage 25. Profiling the same MEF cell line 23 passages later (P48), SKY revealed that the average number of chromosomes had slightly decreased and stabilized (84 ± 12 chr), as has been reported previously (Hao & Greider, 2004), and clonal markers had begun to emerge (e.g., translocation (3:16)) (Fig 4C and Appendix Fig S8F, G and I). We also observed a substantial number of double minutes in some of the cells examined at P48. As expected, the most strongly amplified chromosomes by SKY were scored as gains in CGH data (e.g., chrs. 3 and 6 in subline H1 at P48/49). Since array CGH analysis is normalized by input DNA quantity, genomic regions scoring as negative on a log$_2$ CNA plot can be greater than diploid if most other chromosomes have amplified to an even greater extent (e.g., trisomy chrs. 1, 2, and 7 in subline H1 at P48/49). Similar results have been reported for karyotyping and CGH results of human cancer cell lines (Kytölä *et al*, 2000). Taken together, these data reveal that immortalized MEFs experience substantial numerical and structural chromosomal abnormalities, similar to what is observed in human tumors, and further support that selection for optimized rearranged genomes occurs.

Senescence-associated oxidative stress as a selective force for copy number alterations

The MEF system allowed us to investigate the selective pressures driving copy number changes during immortalization. Because MEFs cultured under physiological oxygen conditions (3% oxygen) undergo little to no senescence and exhibit less DNA damage

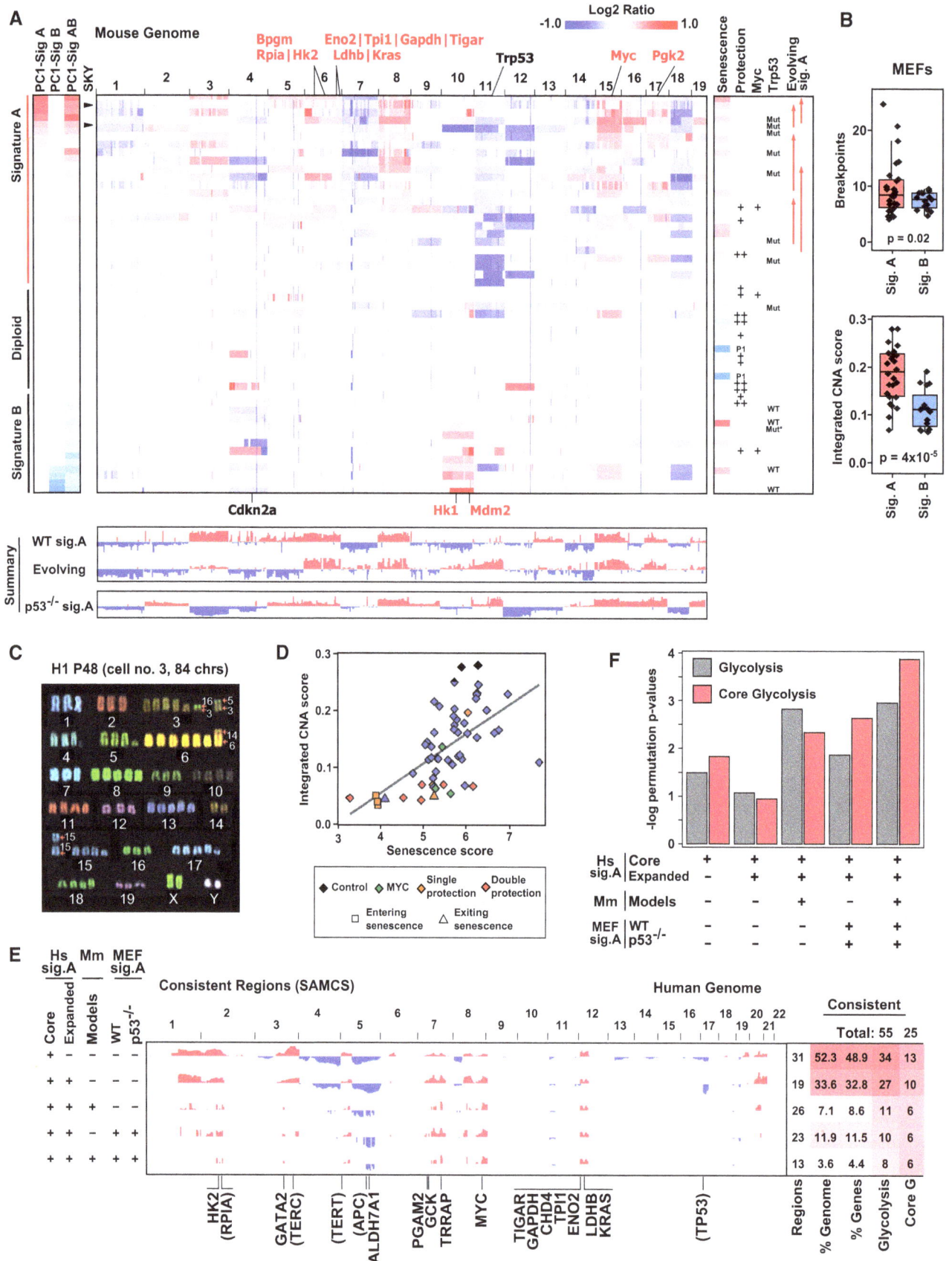

Figure 4.

Figure 4. A mouse embryonic fibroblast (MEF) immortalization system recapitulates the two-signature CNA patterns and glycolysis gene CNA enrichment observed in human tumors and mouse models.

A Top: Copy number profiles of 59 samples from 42 independent mouse embryonic fibroblasts (MEF) sublines before, during, and after senescence recapitulate the two-signature CNA patterns observed in human tumors. Twenty-two sublines evolved to Sig. A patterns, 13 sublines exhibited Sig. B, and seven sublines remained diploid-like due to redox protection. PC1 scores for analysis of only signature A MEFs (PC1-Sig A), only signature B MEFs (PC1-Sig B), or all MEFs (PC1-Sig AB) are indicated on the left. Also indicated are a metric of the degree of senescence observed during immortalization (senescence score), protection by 3% O_2-based hypoxia or catalase (+), or both (++) during immortalization, CNA profiling at passage 1 (P1), exogenous MYC expression (which enabled cells to bypass senescence), and *Trp53* sequencing status (* indicates a non-severe mutation p.183D > E, blank indicates not sequenced, additional p53 sequencing information in Appendix Table S1). Signature A MEF lines profiled at more than one passage number are indicated by the start (earlier passage) and end points (later passage) of the upward "evolving signature A" arrows. For comparison to (C), the early (25) and late (49) passages of the H1 subline that were characterized by spectral karyotyping (SKY) are indicated on the left by triangles. The indicated chromosome 6 region includes loci for *Bpgm, Hk2, Rpia, Eno2, Tpi1, Gapdh, Tigar, Ldhb,* and *Kras.* Bottom: Summary signatures of amplification and deletion loci in copy number profiles (normalized PC1 loadings) are shown for MEF WT and p53$^{-/-}$ signature A (Sig. A) samples. A paired t-test analysis of signature A MEF lines profiled at more than one passage number revealed genomic regions associated with mid- to late-passage CNA evolution (log$_{10}$ t-test P-value signed positive for amplifications, negative for deletions; labeled "Evolving"). Individual profiles of p53$^{-/-}$ signature A samples are not shown.

B Signature A MEFs have more DNA breakpoints and a larger degree of copy number alteration (integrated CNA score) than signature B MEFs. Signature A MEFs had a subset of cases with large numbers of breakpoints per chromosome (> 10) that were not observed in the signature B MEFs (hypergeometric P-value = 0.02, Student's t-test P-value = 4 × 10^{-5}). Data are presented in box (median, first and third quartiles) and whisker (extreme value) plots.

C Spectral karyotyping (SKY) of the H1 MEF subline at passage 48. Whole chromosomal gains of varying extent are observed for all chromosomes with some chromosomes experiencing translocation, for example, t(3:16), and deletion, for example, del(4). Additional chromosome spreads and a summary table for P25 and P48 karyotypes are shown in Appendix Fig S8. CGH profiles for H1 cells are in (A) (indicated by the triangles in the SKY column) and Appendix Fig S6A and B.

D The amount of senescence demonstrated during MEF line derivation (senescence score, see Materials and Methods) is highly correlated with the degree of copy number alterations obtained (integrated CNA score) (Spearman rho = 0.66, P-value = 3 × 10^{-8}). Single protection indicates protection by 3% O_2 culture conditions or media supplementation with 250 U/ml catalase. Double protection indicates that cells were cultured at 3% O_2 with catalase.

E Conserved amplification and deletion loci in copy number profiles across core and expanded signature A human tumor types (Hs Sig. A), corresponding mouse tumorigenesis model signatures (Mm models), and the MEF immortalization system (MEF Sig. A). The human and mouse signatures are as defined in Fig 2, and the wild-type MEF signatures are from (A). In the human consistent regions, the 52.3% consistently altered gene loci include *bona fide* tumor suppressor genes (p53), oncogenes (MYC, KRAS), telomere components (TERC, TERT), and core glycolysis genes (e.g., HK2). As in Fig 2D, genes listed in parentheses were consistently altered across the human tumors but not consistently altered in the mouse. The tissue-specific, non-lymphoid oncogenes from the COSMIC database that are in a conserved amplification or deletion region (TRRAP, CHD4) are listed for completeness (Forbes *et al*, 2015).

F Enrichment signal strength (negative of log permutation P-value) for the glycolysis and core glycolysis gene sets improve when MEF CNA signatures are added to the human and mouse CNA signatures from Fig 2. The degree of increased enrichment upon addition of the MEF signatures is quantitatively similar to adding mouse tumorigenesis model signatures. Adding all signatures into the analysis further strengthens the enrichment (last column). Table EV2 lists results for the enrichment analysis of the remaining MSigDB canonical pathways when human core and expanded signature A, mouse models, and MEF immortalization models are sequentially combined.

Data information: See also Appendix Figs S5–S9.

(Parrinello *et al*, 2003), we tested whether oxidative stress-induced senescence is a selective pressure for copy number alterations. To protect cells from oxidative stress, we cultured MEFs in 3% oxygen in media supplemented with or without catalase, an enzymatic scavenger of reactive oxygen species (ROS) (Halliwell, 2003; Graham *et al*, 2012). When doubly protected from oxidative stress, MEFs did not undergo senescence and maintained relatively diploid genomes (Fig 4A and Appendix Fig S5A). In addition, we observed a variation in the degree of senescence experienced by MEFs derived in atmospheric oxygen concentrations (21%) (Appendix Fig S5A). Upon calculating the degree of senescence encountered by each subline (senescence metric described in Materials and Methods), we found a correlation between senescence and the degree of copy number alterations (Fig 4D, Spearman rho = 0.66, $P = 3 \times 10^{-8}$; and Appendix Fig S9A). Furthermore, p53$^{-/-}$ cells did not undergo senescence (Olive *et al*, 2004) and exhibited less strong copy number alterations than wild-type signature A MEFs (Appendix Fig S7B; $P = 9 \times 10^{-4}$). Taken together, our results implicate senescence-associated redox stress as one of the selective forces driving the copy number alterations recurrently observed in human tumors.

MEF CNA signatures recapitulate recurrent patterns from human tumors and further implicate the glycolysis pathway in shaping the cancer genome

Our pan-cancer and cross-species analysis revealed that the glycolysis pathway is highly enriched in conserved CNA regions (Fig 2).

Because the MEF signature A is qualitatively similar to the human expanded signature A, we next asked how inclusion of our MEF signatures would affect metabolic and canonical pathway enrichment analysis. Although the MEF immortalization system utilizes fibroblast cells and the mouse tumors are epithelial in origin, both types of models share a similar cross-species consistency with the expanded signature A tumor types (Appendix Fig S9B). When the human tumors signatures were combined with the MEF signatures, consistent genome regions were reduced to 3.6% of the genome spread over 13 conserved regions (Fig 4E). In cross-species pathway enrichment analysis, MEF CNA signatures added a similar amount of enrichment signal to human tumor signatures as do the non-MEF mouse model signatures, and when combined together, the enrichment was even stronger (e.g., glycolysis and core glycolysis pathways, Fig 4F). When all signatures are used, glycolysis was the top-ranked enriched pathway out of 1,321 MSigDB canonical pathways (Table EV2). Thus, including the MEF signatures in the cross-species analysis further implicates the glycolysis pathway in shaping the cancer genome.

Examination of the cross-species genomic regions conserved in human tumors, mouse models of cancer, and the MEF immortalization system revealed consistent amplification of five regions containing the COSMIC database-enumerated oncogenes *GATA2* (human chr. 3; Fig 4E, last SAMCS line), *TRRAP* (chr. 7, region 4), *MYC* (chr. 8, region 1), *CHD4* (chr. 12, region 1), and *KRAS* (chr. 12, region 2) (Forbes *et al*, 2015). Notably, four of the cross-species conserved CNA genomic regions included genes from the glycolysis pathway: *HK2* (chr. 2), *GCK* and *PGAM2* (chr. 7, region 1), *ENO2*, *TPI1*, and *GAPDH* (chr. 12, region 1, which includes *TIGAR*), and *LDHB* (chr. 12, region 2, which also includes *KRAS*). In both human tumors and the MEF immortalization system, the genomic region harboring *TIGAR–GAPDH–TPI1–ENO2* was separated from the *LDHB–KRAS* region by a deletion-prone region that includes the tumor suppressor *CDKN1B* (Appendix Fig S9C–E).

To investigate whether amplification of the DNA cross-species conserved CNA regions results in upregulated RNA expression levels, we examined the correlation between DNA copy number and RNA expression levels in BRCA, LU, and OV tumors for glycolytic genes and, as a point of comparison, known oncogenes (Appendix Fig S10). We examined the average correlation across BRCA, LU, and OV signature A tumors for all genes in the 12 cross-species conserved regions. We found that three glycolytic genes (*LDHB*, *TPI1*, and *GAPDH*) and one oncogene (*KRAS*) exhibited strong DNA–RNA correlation ($r > 0.66$), three glycolytic genes (*HK2*, *ENO2*, and *PGAM2*) and three oncogenes (*MYC*, *CHD4*, and *TRRAP*) exhibited moderate correlation ($0.2 < r < 0.5$), and only one glycolytic gene (*GCK*) and one oncogene (*GATA2*) exhibited weak DNA–RNA correlation ($r < 0.2$) (Appendix Fig S10A and B). This analysis indicates that gene copy number alterations at the DNA level generally lead to increased RNA expression in signature A tumors in BRCA, LU, and OV tumors, with a similar degree of correlation observed for both glycolysis genes and oncogenes. Finally, we compared the upregulation of glycolytic genes and onco-genes on two cross-species conserved regions of human chr. 12 that contain both types of genes (Appendix Fig S10C). Within the centro-mere-distal region of chromosome 12, the glycolytic genes *TPI1* and *GAPDH* show strong correlation (Spearman rank correlation = 0.71 and 0.66, respectively) while the oncogene *CHD4* and the glycolytic gene *ENO2* exhibit moderate correlate (Spearman rank correlation = 0.42 and 0.32, respectively). Within the more centromere-proximal region, *LDHB* and *KRAS* both exhibit strong correlation (Spearman rank correlation = 0.68 and 0.66, respectively). Importantly, the correlation of RNA expression with DNA amplification is not stronger for oncogenes than for glycolytic genes within these cross-species regions. Taken together, these results support a model in which the selection pressures shared during immortalization and tumorigenesis result in cross-species conservation of the glycolysis gene loci copy number alterations (Fig 4E and F, and Table EV2).

Alteration of CNA signatures by exogenous expression of metabolic enzymes

The presence of core glycolysis genes in cross-species conserved amplification regions suggests that these metabolic gene loci drive the amplification of these regions. To test this hypothesis, we trans-duced pre-senescent MEFs with either wild-type HK2 or HK1, kinase-dead HK2 (D209A/D657A) (McCoy *et al*, 2014), or wild-type ENO2 and allowed the cells to senesce and immortalize in the pres-ence of these exogenously expressed proteins (Appendix Fig S11). Analyzing the signature A set of sublines, we found that the endoge-nous *Hk2* locus (chr. 6) was less amplified in cells expressing exoge-nous wild-type hexokinase ($P = 0.048$) (Fig 5A and B). As a control, a signature A MEF cell line expressing kinase-dead hexokinase did not show reduced amplification of the *Hk2* locus ($P = 2 \times 10^{-4}$). In that MEF lines exogenously expressing hexokinase still

demonstrated positive selection for the centromere-proximal half of chr. 6 (Fig 5A), we examined the ratio of *Hk2* gene locus copy number to the maximal amplification on chromosome 6 (Hk2:Chr6 max). In this analysis, we found that cells expressing exogenous hexokinase demonstrated significantly reduced Hk2:Chr6 max ratios ($P = 3 \times 10^{-3}$) (Fig 5A and C). Additionally, a MEF subline express-ing exogenous ENO2 exhibited deletion rather than amplification of the *Eno2* locus on chr. 6 ($P = 0.02$) (Fig 5A and D). Analyzing the signature B set of sublines, we found that the endogenous *Hk1* locus (chr. 10) was copy number neutral, rather than amplified, in a cell line expressing exogenous hexokinase ($P = 0.17$), whereas a signa-ture B MEF cell line expressing kinase-dead hexokinase did not show reduced amplification of the *Hk1* locus (Fig 5E and F). Fisher's combined statistical analysis of these results yielded P-values of 0.001 or less (Fig 5).

Taken together, these results demonstrate that exogenous expression of metabolic enzymes can alter the copy number status of the endogenous genomic loci, supporting these metabolic genes as drivers within the conserved amplification regions observed in human tumors and mouse models. In addition, these results support a model in which the net propensity for a chromosomal region to be amplified or deleted is in part related to the sum of the fitness effects of the genes present (Davoli *et al*, 2013). For example, when HK2 is exogenously expressed, the centromere-distal half of chromosome 6 had a low copy number value in early culture but after additional culture demonstrated increased copy number, supporting that the other gene loci of this region (such as Eno2, Gapdh, and other glycolytic genes) do have a remaining pro-fitness benefit (Fig 5A).

Metabolism and growth phenotypes of CNA signatures

To test whether there exist phenotypic differences between signa-ture A and signature B MEFs, we characterized 11 wild-type MEF lines representative of either signature A or B and one mixed signa-ture line. We found that signature A MEFs generally had higher rates of glucose consumption and lactate production than signature B MEF lines (Figs 6A and EV5F). Plotting the PC1 score versus glucose consumption, we found that CNA-based signature A was highly predictive of glucose consumption in the MEF signature A lines (Fig 6B). In contrast, signature B was only moderately predic-tive of the glucose consumption of signature B lines, and was not as accurate as signature A in predicting the glucose consumption of signature A lines (Figs 6B and EV5A). In addition, we noted that the signature A cell lines generally exhibited significantly higher rates of proliferation than the signature B cell lines (Fig 6C). Similar to glucose consumption, signature A was predictive of the growth rates of signature A MEFs and signature B was predictive of the growth rates of signature B MEFs, while cross-signature predictions had less power (Figs 6D and EV5B). Furthermore, we observed a general coevolution of higher growth rates and increased CNA signature strength in MEF lines that were profiled at different passage numbers (Fig 6D). As noted above, the evolving MEF CNA signature pattern was enriched for DNA amplifications of genes in the core glycolysis and glycolysis-associated pathways (Table EV2), particu-larly due to amplification of chromosome 6, which contains multiple metabolic gene loci including *Hk2* and *Eno2* (Fig 4A and E, and Appendix Fig S6B).

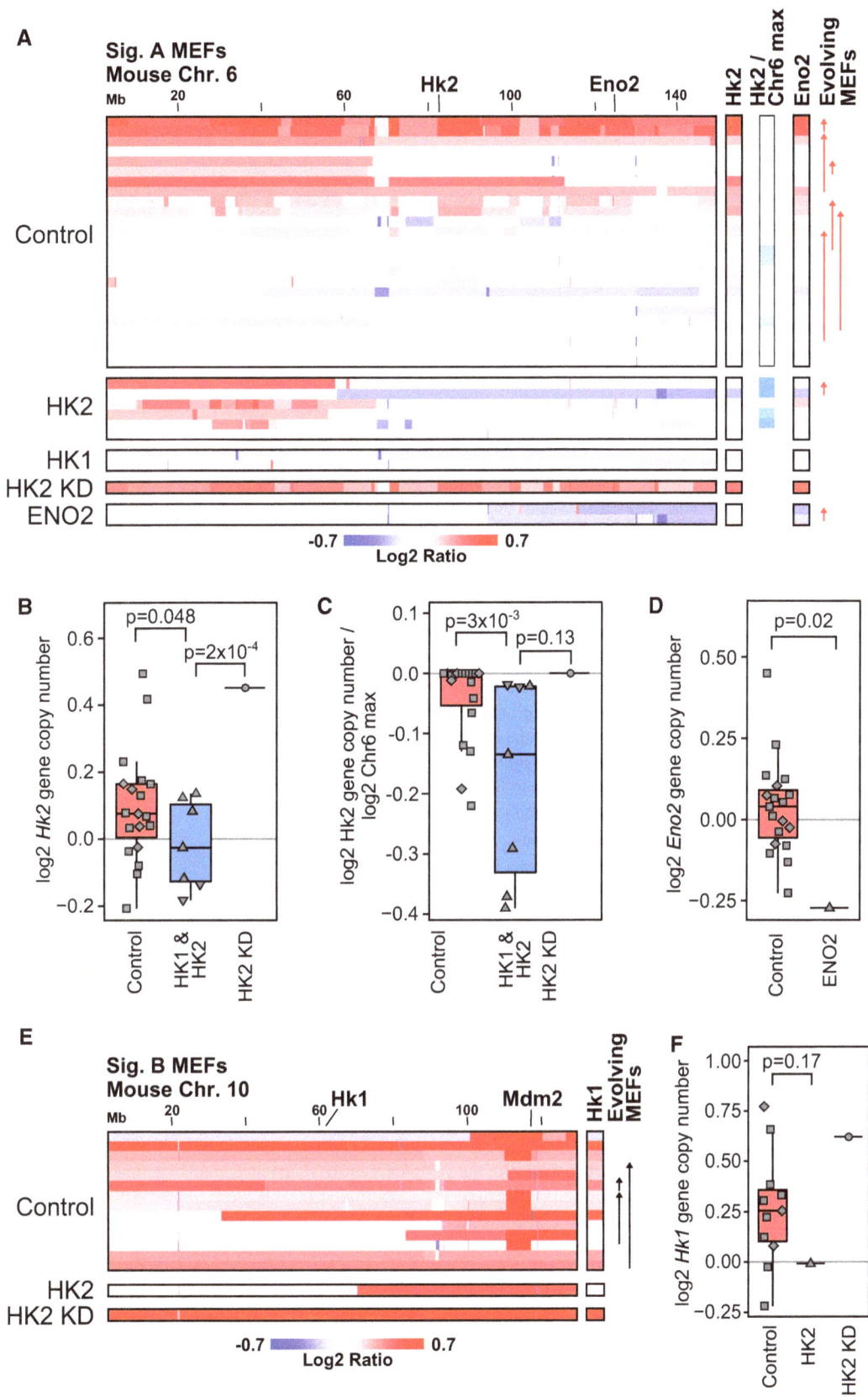

Figure 5.

◀ **Figure 5. Alteration of CNA signatures by exogenous expression of metabolic enzymes.**

   A   CD1 MEFs expressing exogenous HK1/HK2 or ENO2 exhibit reduced amplification of the endogenous *Hk2* or *Eno2* loci, respectively. Chr. 6 copy number profiles from control signature A MEFs (untransduced or expressing red fluorescent protein (RFP)) compared to signature A MEFs expressing either wild-type HK1 or HK2, kinase-dead HK2 (HK2 KD, D209A/D657A), or wild-type ENO2. The positions of endogenous *Hk2* and *Eno2* are indicated. MEF lines profiled at more than one passage number are indicated by the start (earlier passage) and end points (later passage) of the arrows under "Evolving MEFs".

   B–D Boxplots of *Hk2* copy number (B), ratio of *Hk2* copy number to maximum amplification of chromosome 6 (defined operationally as the 5th percentile CNA value across chr. 6 to avoid outlier effects) (C), and *Eno2* copy number (D). Control MEFs: untransduced (squares) or expressing RFP (diamonds); test MEFs expressing wild-type HK1 (upside-down triangles), HK2 (triangles in B and C), kinase-dead HK2 (HK2 KD, D209A/D657A, circles), or wild-type ENO2 (triangle in D). In this analysis, copy number data from samples profiled at multiple passages were averaged to prevent overrepresentation of these cell lines. *P*-values were calculated using Student's *t*-test (B), Mann–Whitney *U*-test (C), and *z*-score (B, C) based on criteria described in Materials and Methods. Data are presented in box (median, first and third quartiles) and whisker (extreme value) plots.

   E   Signature B MEFs expressing exogenous HK2 exhibit less amplification of the *Hk1* locus. Chr. 10 copy number profiles from control signature B MEFs (untransduced or expressing RFP) compared to signature B MEFs expressing wild-type HK2 or kinase-dead HK2 (HK2 KD). The positions of endogenous *Hk1* and *Mdm2* are indicated.

   F   Boxplot of the *Hk1* copy number with indicated *P*-value (*z*-score). Data are presented in box (median, first and third quartiles) and whisker (extreme value) plots.

   Data information: Fisher's combined *P*-values for the alteration of endogenous loci CNA by exogenous expression of metabolic enzymes is 3×10^{-5} (panels B, D, F) or 7×10^{-4} (panels C, D, F). For all comparisons in this figure, no significant differences were observed between untransduced and RFP-expressing MEFs. See also Appendix Fig S11.

To test whether signature A and signature B MEFs differentially use glucose, we cultured MEF cells with $[1,2\text{-}^{13}\text{C}]$-labeled glucose and conducted metabolomic profiling by mass spectrometry (Fig EV5C–G and Table EV5) (Metallo *et al*, 2009). In all MEF lines tested, we observed a low percentage of single heavy labeled carbon [M1 isotopomer compared to M2] in pyruvate, lactate, and alanine, indicating that the contribution of glucose-derived carbon from the oxidative arm of the pentose phosphate pathway to these metabolites was relatively low (Fig EV5D). Nonetheless, the patterns of heavy isotope labeling revealed differences in nutrient utilization between signature A and B MEF cell lines. On average, metabolites of early glycolysis, the pentose phosphate, and nucleotide synthesis pathways showed a higher percentage of glucose-derived heavy carbon labeling in signature A MEFs (Figs 6E and EV5E). In contrast, signature A cells had a lower percentage of glucose-derived heavy carbon labeling in metabolites of the serine synthesis pathway and the TCA cycle. When compared to signature B MEFs, signature A cells also tended to exhibit increased consumption of serine, glutamine, and other amino acids (Fig EV5F and G). The increased glutamine consumption in signature A MEFs reflects similarity with basal breast cancer cell lines (also signature A associated), which exhibit increased glutamine consumption relative to luminal breast cancer cell lines (Timmerman *et al*, 2013). In general, the percentage of individual metabolites from the early glycolysis, the pentose phosphate, and nucleotide synthesis pathways that incorporated heavy, glucose-derived carbon metabolites was positively correlated with glucose consumption rates (e.g., fructose-1,6-bisphosphate, Fig 6F). Conversely, metabolites from the serine synthesis pathway and the TCA cycle showed a negative correlation between glucose consumption rates and the percentage of each molecule containing a heavy, glucose-derived carbon (e.g., 3-phosphoserine, Fig 6F). While caution must be exercised when interpreting metabolite labeling results, we overall observed a consistent trend of differences between signature A and B lines. Of note, the differences between signature A and B lines were mainly in regard to scale, with the strongest signature B lines demonstrating similar glycolysis and proliferation rates, as well as similar metabolic profiles, as the weakest signature A lines.

Taken together, these results demonstrate that signature A MEFs, which resemble signature A human tumors, exhibit increased glycolysis and to a somewhat lesser extent increased proliferation, and have an increased relative proportion of glucose-derived carbon in metabolites of pentose phosphate-associated biosynthetic pathways such as nucleotide synthesis. These findings are consistent with published mouse model studies in which tumor cells that are channeling glucose toward nucleotide biogenesis achieve faster rates of proliferation (Boros *et al*, 1998; Ying *et al*, 2012).

Thus, our cross-species and pan-cancer CNA analysis revealed conserved amplification regions shared by the majority of tumor types studied that are enriched for genes involved in core glycolysis. To aid others in pan-cancer and cross-species CNA signature comparisons, we have created an interactive web-interface resource available at http://systems.crump.ucla.edu/cna_conservation/.

Discussion

Chromosomal instability and high glycolysis characterize some of the most aggressive tumors, and the complexity and plasticity of the genomes in aggressive tumors can hinder molecularly targeted therapies (Nakamura *et al*, 2011; McGranahan *et al*, 2012; Shi *et al*, 2012). While the glycolytic changes associated with tumorigenesis were one of the early defining phenotypes of cancer cells (Warburg, 1956), they have not previously been linked to recurrent DNA copy number patterns. Taken together, our experimental and computational data support a model in which glycolysis-linked selective pressures encountered during tumorigenesis (e.g., redox stress and senescence) shape the highly recurrent DNA copy number alterations found in aneuploid human tumors (Fig 6G). We found that CNAs in core glycolysis enzymes (e.g., HK2) and other cancer-linked metabolic enzymes such as TIGAR are coordinately enriched in tumors with distinct CNA signatures. These CNA signatures are predictive of glycolysis, including patient FDG-PET activity and cell line proliferation phenotypes. The strong correlation of CNA-based principle component scores to uptake of the glucose analogue FDG in breast cancer patients and the predictive power of CNA signatures for breast cancer cell metabolism (Fig 3) provide support that the CNAs affecting metabolic gene loci collectively act as a copy number-based driver of metabolic differences. Importantly, in that exogenous expression of hexokinase and enolase enzymes reduced

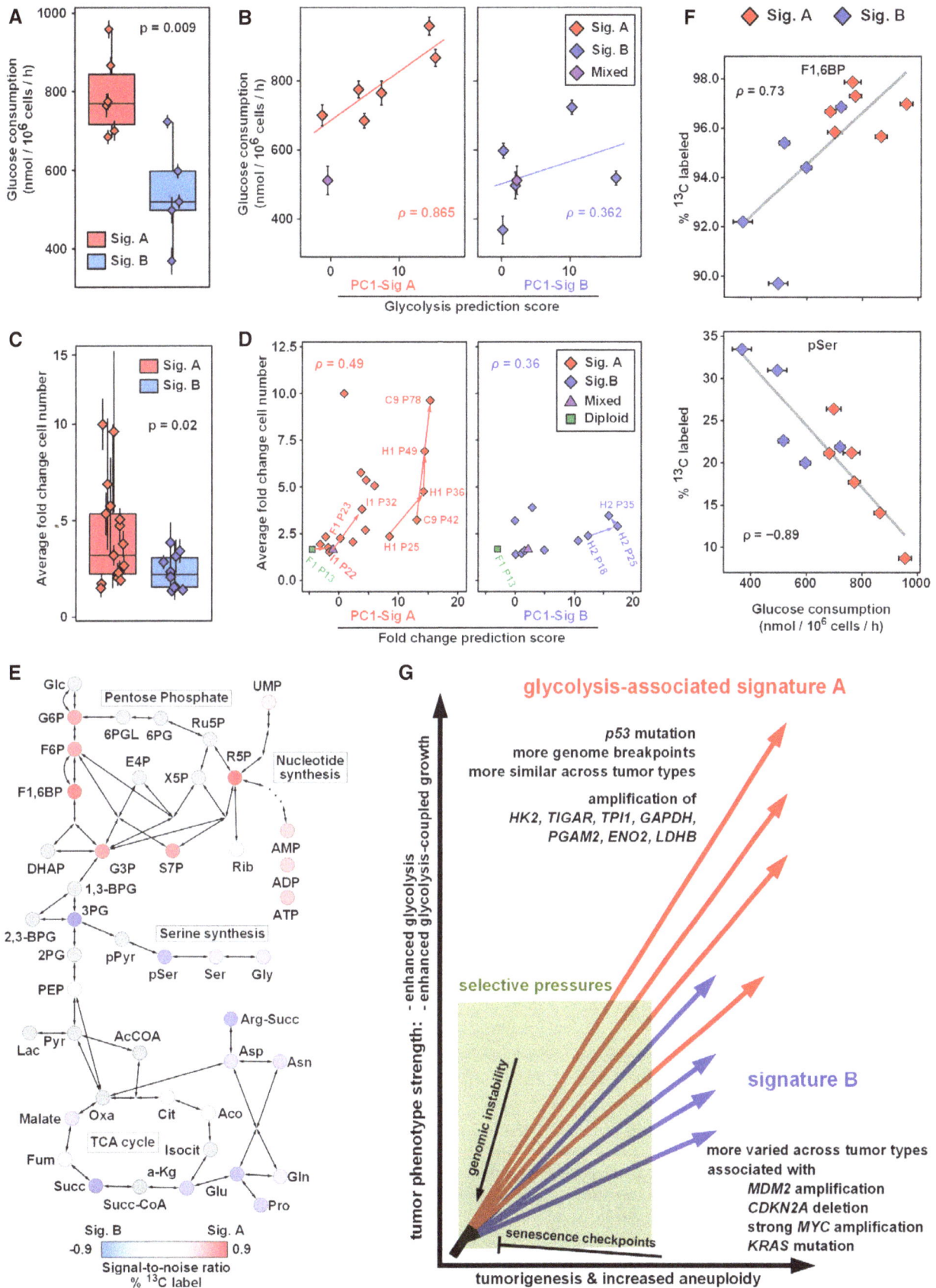

Figure 6.

◀ **Figure 6. Signature A and B MEFs exhibit differential metabolic and proliferation phenotype strengths.**

A   PCA-defined signature A MEFs exhibit higher glucose consumption rates than signature B MEFs (Student's t-test P-value = 0.009). Glucose consumption was measured using a bioanalyzer.

B   PC1 scores of signature A MEFs are predictive of glucose consumption rates in signature A MEFs, while scores from signature B MEFs have weaker predictive power in signature B MEFs. In cross-prediction tests, signature A-based predictions of signature A MEF line metabolic phenotypes perform the best (Fig EV5A). Glucose consumption was measured using a bioanalyzer.

C   PCA-defined signature A MEFs exhibit higher average growth fold change, as observed in 3T9 culture, compared to signature B MEFs. Signature A MEFs had a subset of cases with high growth rates (average fold change in cell number per passage > 4) that were not observed in the signature B MEFs (hypergeometric P-value = 0.02).

D   Correlation and a general coevolution of higher growth rates and increased CNA signature strength. Arrowed lines indicate progressing MEF lines profiled at more than one passage number (Appendix Figs S6A and B, and S9A). In cross-prediction tests, PC1 scores of signature A MEFs are more predictive of average growth fold change in signature A MEFs, while scores from signature B MEFs are more predictive in signature B MEFs (Fig EV5B).

E   Metabolomic profiling of 11 wild-type MEF lines representative of either signature A or B cultured for 24 h with [1,2-^{13}C]-labeled glucose. The metabolic pathway schematic is colored based on differences observed in the percent heavy label for intracellular metabolites (defined as percent of metabolite molecules with isotopomer mass greater than the monoisotopic molecular weight, M0) between signature A and B MEFs using the signal-to-noise (SNR) metric. Red indicates a higher heavy carbon-labeling percentage in signature A MEFs, and blue indicates a higher heavy carbon-labeling percentage in signature B MEFs. Signature A MEF lines incorporate more glucose-derived carbon in metabolites from the early glycolysis steps, the pentose phosphate pathway, and nucleotide synthesis, and a smaller fraction of glucose-derived carbon per molecule in metabolites from the later glycolysis steps, the serine synthesis pathway, and the TCA cycle. Full metabolomic data can be found in Table EV5.

F   Correlation of fructose-1,6-bisphosphate (F1,6BP) and negative correlation of 3-phosphoserine (pSer) with glucose consumption rates. Note, the high to low range for percent metabolite molecules with incorporated label varies for each metabolite, but generally does not extend from 0 to 100%. Signature A and B MEFs are colored red and blue, respectively.

G   Model of copy number selection and fitness gains during tumorigenesis. Genomic instability enables fitness gains in tumor metabolism. In human tumors, cancer cell lines, and an experimental MEF immortalization system, immortalization and tumorigenesis lead to multiple CNA signatures that are predictive of the tumor phenotypes of metabolism and proliferation. Senescence-associated redox stress and other tumorigenesis-related constraints select for stronger CNA signatures. A shared high glycolysis-associated signature A is observed in breast, lung, ovarian, and uterine tumors, additional tumor types, mouse models, and the MEF model system, and is linked with a higher range of glycolysis and proliferation phenotypes. Genetic manipulation of glycolysis enzymes leads to alteration of corresponding CNA signature propensities. Signature A and B genomes reflect two distinct trajectories from diploidy to tumor aneuploidy. Signature A tumors are enriched for mutations in *p53* and have smaller sized amplification and deletion genomic regions (i.e., have a higher number of genomic breakpoints), potentially providing increased alternative genome options. Signature A involves amplification of several genes in glycolysis-related pathways (such as *HK2, TIGAR, TPI1, GAPDH, ENO2, PGAM2,* and *LDHB*). Signature B CNA patterns occur in generally less glycolytic and proliferative samples and show more variation across different tumor types. In particular, signature B is enriched for *MDM2* amplification and *CDKN2A* loss, strong *MYC* amplification, and *KRAS* mutation and involves alternate hexokinase isoforms (*HK1, HK3*).

Data information: Data in (A, C) are presented in box (median, first and third quartiles) and whisker (extreme value) plots. Error bars indicate standard deviations of biological replicates in (A, B, C and F). See also Fig EV5.

the propensities for amplifications of the corresponding endogenous hexokinase and enolase loci, these metabolic genes empirically score as driver loci. However, we cannot exclude the possibility that the observed metabolic differences are in part due to other cancer-associated regulatory changes such as epigenetic events known to affect metabolism (Sebastián *et al*, 2012). Combined with the observation that metabolic genes can facilitate cellular immortalization (Kondoh *et al*, 2005; Kondoh, 2009; Kaplon *et al*, 2013), our results implicate tumor metabolism as an additional fitness measure linked to how genomic instability can enable tumorigenesis.

Chromosomal instability and aneuploidy—positive and negative impact on tumor cell fitness

Most solid human tumors exhibit both numerical and structural aneuploidy (Holland & Cleveland, 2012). Paradoxically, chromosomal instability can act either as an oncogene or as a tumor suppressor depending on the context (Weaver *et al*, 2007). Moreover, addition of a single chromosome in MEF cells induces a stress response that impairs proliferation and immortalization (Williams *et al*, 2008). However, numerical aneuploidy can lead to chromosomal instability (Nicholson *et al*, 2015) which results in subchromosomal gains and losses as observed in human tumors, in mouse models of cancer, and in immortalized MEF cells. Thus, aneuploidy can cause an initial fitness loss due to the costs of dealing with non-optimized chromosome numbers, gene copy numbers, and resulting proteomic imbalances. However, the associated state of

chromosomal instability enables further evolution of the genome toward fitness gains through refinement of gene copy number levels. In sum, these fitness gains either outweigh or offset the fitness losses due to aneuploidy. In this context, our data support a model (Fig 6G) in which metabolic selection forces and metabolic gene loci contribute to the recurrent patterns of DNA copy number alteration observed in human tumors.

Redox stress, biomass accumulation, and associated glycolytic changes in tumorigenesis

Tumorigenesis is a complex, multistage process during which cells must acquire the capability to maintain redox balance while accumulating the macromolecular precursors required for proliferation (DeBerardinis *et al*, 2008; Hanahan & Weinberg, 2011). Numerous stimuli, including RAS mutations, matrix detachment, altered metabolism, and hypoxia, induce the accumulation of intracellular ROS (Lee *et al*, 1999; Schafer *et al*, 2009; Weinberg *et al*, 2010; Anastasiou *et al*, 2011). Because increased ROS levels can trigger replicative senescence and subsequent cell cycle arrest (Lee *et al*, 1999; Takahashi *et al*, 2006), tumors must maintain pools of reduced glutathione using NADPH in part produced via the pentose phosphate pathway. Additionally, increased levels of ROS can divert glycolytic flux into the pentose phosphate pathway through, for example, oxidation and inhibition of the glycolytic enzyme PKM2, thereby supplying cells with the reducing power and precursors for anabolic processes (Boros *et al*, 1998; Anastasiou *et al*, 2011).

Consistent with this published knowledge on the role of metabolism in tumorigenesis, our study suggests that the metabolic stress associated with senescence (Fig 4A and D) and the metabolic demands of rapid proliferation (Fig 6) are components of the selective pressures underlying recurrent CNA changes.

Experimentally and computationally deciphering CNA patterns

Our experimental and bioinformatic approaches complement existing approaches for testing hypotheses for the selection pressures underlying recurrent CNA patterns observed in human tumors. Other CNA-analysis approaches target different resolutions of the genome. Statistical algorithms such as GISTIC (Genomic Identification of Significant Targets in Cancer) have identified many strong individual driver genes and candidate regions (Mermel et al, 2011). Integrating CNA data with RNA knockdown screens and gene expression data has further identified driver genes missed by statistical analysis of CNA data alone (Sanchez-Garcia et al, 2014). RNA knockdown-based analyses have also been used to support a more systems-level model in which the selection for amplification and deletion of a particular DNA region is based on the cumulative effects of many positive and negative fitness gains from multiple genes within that genomic region (Solimini et al, 2012; Cai et al, 2016). Subsequent computational extensions have incorporated somatic mutation patterns to infer the cumulative impact of co-localized genes on fitness, and to successfully predict whole chromosome and chromosome-arm resolution-level CNAs (Davoli et al, 2013).

Our approach using phenotypic data, functional gene sets, and cross-species syntenic mapping has yielded additional insight into the selective pressures shaping tumor CNA patterns, namely coordinated alteration of genes involved in glycolytic metabolism. Incorporation of copy number data from genetically engineered mouse models allowed synteny constraints to substantially reduce passenger amplifications and define a minimal collection of stringently conserved copy number regions. Additionally, to our knowledge, our experimental approach is the first reported system in which CNA and associated phenotypes have been followed and repeatedly sampled as non-immortalized cells undergo spontaneous genomic instability and proceed from a diploid state to an immortalized aneuploid state. Using this approach, we were able to validate genomic regions that are (i) associated with increased glycolysis and to a lesser but significant extent proliferation, (ii) enriched for genes from the core glycolysis pathway, and (iii) conserved in both human tumors and mouse models of cancer. Metabolic pathways are known to be coordinately regulated by modest changes in mRNA expression of functionally related genes (Mootha et al, 2003; Palaskas et al, 2011). The coordinated alterations of metabolic genes at the DNA level adds an additional mechanism, namely conserved sets of CNA changes, by which glycolysis is dysregulated to promote tumorigenesis.

A strength of our PCA-based approach is the ability to unbiasedly reveal distinct CNA subsignatures within a tumor type. Observed subsignatures, confirmed using an independent clustering analysis, were found to be associated with previously known pathology- or profiling-defined tumor subtypes (e.g., basal/luminal CNA signatures in BRCA) (Bergamaschi et al, 2006) (Figs 1B and EV2A–D). Across multiple tumor types, loss of p53 function through p53

mutation is associated with the high breakpoint signature A pattern. In contrast, loss of the p53/ARF axis via other mechanisms (MDM2 amplification or CDKN2A deletion) in BRCA human tumors is associated with a different CNA signature (signature B) that in general has fewer breakpoints. In our experimental follow-up, the MEF system recapitulated a two-signature pattern. Notably, the two mouse signatures and their associated phenotype strengths were defined by the initiating loss of tumor suppression event, namely Trp53 mutation versus either Mdm2 amplification or Cdkn2a loss. Thus, while the consequences of TP53 mutation and "MDM2 amplification/CDKN2A loss" are considered functionally similar and therefore mutually exclusive (Wade et al, 2013), our findings indicate they are not fully equivalent in terms of genomic instability and subsequent metabolic evolution. The tolerance of more highly disrupted and rearranged genomes upon p53 mutation appears to allow more flexibility in the evolution of aneuploid cancer genomes, thereby resulting in stronger glycolysis and somewhat enhanced proliferation. The specific combinations of CNA changes occurring in enzyme isoforms defining a metabolic pathway may be considered "onco-metabolic isoenzyme configurations" with differential potency, and the sets of combinations possible may be limited in part by the degree or specifics of tolerance to genomic rearrangements.

In summary, our work illustrates the value of cross-species comparisons in the analysis of DNA copy number data, much as recent pan-cancer and integrated genomic approaches have uncovered novel cancer subtypes and driver genes (Gatza et al, 2014; Hoadley et al, 2014). The broken chromosome synteny between human and mouse genomes reduced the size of potential driver regions by fivefold on average, and identified a relatively small number of amplification and deletion regions highly conserved between mouse and human (Figs 2D and 4E). These conserved regions are consistently found across the majority of tumor types examined, have subchromosomal scale with a median size on the order of 10 megabase pairs, and include the glycolysis pathway enzyme CNAs observed in our experimental work.

Therapeutic and diagnostic implications

The most copy number aberrant tumors tend to have fewer point mutations in canonical oncogenes (e.g., KRAS, Fig EV2A and C) and less canonical oncogene amplification (e.g., MYC, Fig 1B). Hence, genomic instability and subsequent coordinate alterations in multiple genes within a functional pathway may provide an alternate, more complex, pathway to acquisition of aggressive tumor phenotypes—with tumor evolution and selection guiding the trajectory (Ciriello et al, 2013). In that KRAS mutation and MYC amplification can drive glycolysis (Ying et al, 2012; Dang, 2013), the findings that signature A tumors are de-enriched in these events relative to signature B tumors and enriched for glycolysis gene loci CNA amplifications support that tumor cells can meet their metabolic demands through distinct mechanisms, or combinations thereof. Future models of the most aggressive cases of cancer will need to incorporate aspects of spontaneous genomic instability (mediated by distinct instability mechanisms) and resulting copy number alterations. Translationally, understanding how CNA patterns alter cancer genomes and impact cancer phenotypes will aid in the identification of metabolic or other "hard-wired" vulnerabilities that can be therapeutically targeted. Furthermore, the relative stability of DNA samples, combined with the growing linkage between highly recurrent copy number changes and phenotypes, supports the potential for molecular classification and diagnostic tests based on DNA copy number patterns (Hieronymus *et al*, 2014; Cai *et al*, 2016).

Materials and Methods

Cell culture and mouse strains

Tissue culture
CD1 mouse embryonic fibroblasts, E14.5, were purchased from Stem Cell Technologies. $p53^{fl/fl}$ MEFs were obtained at day E14.5 from $p53^{fl/fl}$ (FVB.129P2-Trp53^{tm1Brn}/Nci) crossed with C57BL/6-129/SV mice. MEF cells were maintained in Dulbecco's modified Eagle's medium without pyruvate and supplemented with 10% FBS and 1% SPF. Cells were lifted and re-plated at a density of 9×10^5 viable cells/60-mm dish every 3–4 days (i.e., 3T9 protocol; Todaro & Green, 1963). Cells were grown in atmospheric oxygen unless otherwise indicated. Breast cancer cell lines were obtained from the laboratory of Frank McCormick and extensively profiled both genomically and transcriptionally by the laboratory of Joe Gray (Neve *et al*, 2006; Hong *et al*, 2016).

Genetic engineering
Overexpression of HK1, HK2, kinase-dead HK2 (D209A/D657A), or ENO2 glycolysis enzymes and the MYC oncogene or the control protein RFP in CD1 MEFs was accomplished by transduction of non-immortalized cells with pDS-FB-neo retrovirus, followed by selection in 600 μg/ml G418. Deletion of p53 in $p53^{fl/fl}$ MEFs was induced by infection of non-immortalized cells with either retroviral Cre-GFP or Cre-ERT2 plus treatment with 1 μM 4-OHT.

ROS protection
To protect cells from reactive oxygen species (ROS), cells were cultured either at physiological oxygen concentrations (3% O_2), with 250 U/ml catalase from bovine liver (Sigma-Aldrich), or with both 3% O_2 and catalase.

Genome profiling

Array CGH profiling
Genomic DNA was harvested using the DNeasy kit (Qiagen). DNA from MEF lines and reference genomic DNA from C57BL/6J mouse tissue were hybridized to Agilent SurePrint G3 Mouse CGH 4×180 k CGH microarray chips at the UCLA Pathology Clinical Microarray Core. Bioconductor analysis tools were used for data processing: Moving minimum background correction and print-tip loess normalization were performed in snapCGH package (Smith *et al*, 2009); circular binary segmentation (with a minimum of three markers per segment) was performed on smoothed and log2-transformed copy number profiles using DNAcopy package (Seshan & Olshen, 2016); segmented data were converted into a matrix by genes for downstream analyses using *mus musculus 9* RefSeq reference genome from 2011.08.11 in CNTools (Zhang, 2016). Copy number profiles are presented using the Integrative Genomics Viewer (IGV) (Thorvaldsdóttir *et al*, 2013). CNA public datasets and TCGA tumor type abbreviations are available in the Data availability section.

Bioinformatic and statistical analysis

PCA
Principal component analysis (PCA) was performed using the mean-centered matrix of CNA values per gene locus. Genes with identical profiles across samples were collapsed to a single representative gene. Techniques such as PCA and the related singular value decomposition (SVD) have been applied to copy number data previously (Sankaranarayanan *et al*, 2015).

Mutation and tumor subtype enrichment analysis
To test for enrichment of mutations or tumor subtypes within CNA-defined PC scores, TCGA tumors were queried for mutations and tumor subtype designations using the cBioPortal for Cancer Genomics (Gao *et al*, 2013). For each tumor type, we included all genes with significant q-values as calculated by MutSigCV (Lawrence *et al*, 2013). We also included the gene *POLE*, a catalytic subunit of DNA polymerase epsilon, in our analysis because of its frequent mutation in UCEC (The Cancer Genome Atlas Research Network, 2013). Molecular subtypes tested were as follows: BRCA: basal, luminal, claudin-low, and HER2-enriched (The Cancer Genome Atlas Research Network, 2012b); LUAD: bronchioid, magnoid, and squamoid (The Cancer Genome Atlas Research Network, 2014); LUSC: basal, classical, primitive, and secretory (The Cancer Genome Atlas Research Network, 2012a); OV: proliferative, immunoreactive, differentiated, and mesenchymal (The Cancer Genome Atlas Research Network, 2011); and UCEC: *POLE* ultramutated, microsatellite instability hypermutated, copy number low, and copy number high (The Cancer Genome Atlas Research Network, 2013). Tumors were sorted by their PC1 score, and we calculated a Kolmogorov–Smirnov statistic against the expected distribution of mutations or tumor subtypes. The statistical significance of enrichment was determined by permutation analysis.

Hierarchical clustering
The TCGA tumors were hierarchically clustered using the pheatmap package in R (Kolde, 2015). Prior to clustering, the CNA values were filtered for gene loci that had zero values across all tumor samples. Non-centered and non-scaled tumor samples were then clustered using centered Pearson correlation distance and Ward's method (with dissimilarities squared before cluster updating). The strength of concordance between hierarchical clustering results and PCA-based signatures was assessed using the hypergeometric *P*-value.

Consistency signatures
Consistency signatures of conserved amplification and deletions regions were determined using stringent consistency criteria. The signed absolute minimum consistency score (SAMCS) was defined as non-zero when all CNA summary signatures have the same sign across all tumor types, and the score is derived from the absolute value-based minimum summary metric and then re-signed positive for amplification or negative for deletion. Consistent amplifications or deletions were combined into a "consistent region", when absolute SAMCS values greater than 0.05 spanned at least 1 Mbp. Any

two consistent regions separated by < 1 Mbp were combined into a single consistent region.

Enrichment analysis and weighted gene voting (WGV)

Metabolic pathway enrichment analysis (gene set enrichment analysis, GSEA; Subramanian *et al*, 2005) and pathway-specific weighted gene voting (WGV) prediction analysis (Golub *et al*, 1999) were performed using 75 metabolic pathways defined by the Kyoto Encyclopedia of Genes and Genomes (KEGG) database (Kanehisa *et al*, 2014), using pathways with seven or more measured genes. In the RNA-based enrichment analysis, we included a gene set consisting of genes from the glycolysis and pentose phosphate pathways that were upregulated in FDG-high BRCA tumors, as defined by our previous work (Palaskas *et al*, 2011). For CNA data, an expanded run of enrichment analysis was performed using 1,321 canonical pathways (CP) of seven or more measured genes as defined by the Broad Institute's Molecular Signatures Database (MSigDB). For KEGG-based metabolic pathway enrichment analysis of CNA data, we collapsed metabolic isoenzyme loci (genes with the same enzyme activity; Enzyme Commission [EC] numbers) that were within 100 kilobases from each other into a single representative locus. For mRNA expression analysis, metabolic isoenzymes were not collapsed. In CNA-based enrichment analysis using gene-based versions of the genome consistency signatures defined above, enrichment scores were calculated through the ranked set of consistently amplified genes since after this point the genes that are not consistent across signatures have a consistency value of zero and accordingly have tied ranks. Consistency regions were stepwise restricted by sequentially adding human, and corresponding tissue-matched mouse model of cancer, signatures based on their decreasing tissue type PC1 value from Fig 1A. To combine BRCA and LU into one enrichment analysis for mRNA data, genes were ordered by their average rank in BRCA and LU tumor types. For WGV predictions of metabolic phenotypes, *t*-scores were used as gene weights. To calculate permutation *P*-values, we calculated the fraction of 1,000 randomly chosen gene sets of equal size that gave average gene set rankings (column 2 of Fig 3C) better than the true gene set. False discovery rate (FDR) *q*-values were calculated using the Benjamini–Hochberg procedure. To increase statistical stringency, FDR values for each individual glycolysis–gluconeogenesis gene set (KEGG Glycolysis-Gluconeogenesis, KEGG Core Glycolysis, KEGG Glycolysis-Gluconeogenesis & Pentose Phosphate Pathway, Reactome Gluconeogenesis, Reactome Glucose Metabolism, Reactome Glycolysis, Biocarta Glycolysis Pathway) were calculated while removing the other glycolysis–gluconeogenesis gene sets. CNA and metabolite changes were visualized in the context of metabolic pathway structure using Cytoscape (Smoot *et al*, 2011).

Core glycolysis and glycolysis-associated gene list (from Fig 1F)

HK—hexokinase (HK1, HK2, HK3, HK4/glucokinase/GCK); GPI—glucose-6-phosphate isomerase; G6PC—glucose-6-phosphatase catalytic subunit (G6PC, G6PC2); FBP—fructose-bisphosphatase (FBP1, FBP2); PFK—phosphofructokinase (PFKL—liver type, PFKM—muscle, PFKP—platelet); PFKFB—6-phosphofructo-2-kinase/fructose-2,6-biphosphatase (PFKFB1, PFKFB2, PFKFB3, PFKFB4); TIGAR—TP53 induced glycolysis regulatory phosphatase; ALDO—aldolase, fructose-bisphosphate (ALDOA, ALDOB, ALDOC); TPI1—triosephosphate isomerase 1; GAPDH—glyceraldehyde-3-phosphate

dehydrogenase (GAPDH, GAPDHS—spermatogenic); PGK—phosphoglycerate kinase (PGK1, PGK2); PGAM—phosphoglycerate mutase (PGAM1, PGAM2, PGAM4, BPGM—bisphosphoglycerate mutase); ENO—enolase (ENO1, ENO2, ENO3); PK—pyruvate kinase (PKLR—liver and red blood cell, PKM2—muscle); LDH—lactate dehydrogenase (LDHA, LDHB, LDHC, LDHAL6A—LDH A-like 6A, LDHAL6B—LDH A-like 6B); and PDH—pyruvate dehydrogenase (PDHA1, PDHA2, PDHB, DLD—dihydrolipoamide dehydrogenase, DLAT—dihydrolipoamide S-acetyltransferase).

Metrics

Signal-to-noise ratio (SNR) = $(\mu_1 - \mu_2)/(\sigma_1 + \sigma_2)$, *t*-score = $(\mu_1 - \mu_2)/\mathrm{sqrt}(\sigma_1^2/n_1 + \sigma_2^2/n_2)$, where μ = mean, σ = standard deviation, n = number of samples.

Correlation of mRNA and CNA

Copy number alteration and mRNA expression levels for genes found in the cross-species conserved regions in Fig 4E were compared by calculating the Spearman rank correlation coefficient. BRCA, LU, and OV samples with paired mRNA and CNA data were included. UCEC tumors were excluded because there were not sufficient UCEC samples with paired RNA expression data (26% of all samples with CNA data). Known oncogenes were identified by comparison with the Catalogue Of Somatic Mutations In Cancer database (COSMIC, http://cancer.sanger.ac.uk/cosmic; Forbes *et al*, 2015).

Senescence score

A summary senescence score for each MEF subline was calculated by subtracting the area under the MEF's growth curve, Z(x), in $\log_2$ scale, from the area under an ideal growth curve, Y(x) (i.e., consistent growth at the fastest observed rate). The area difference was then averaged by dividing by the passage number, p, and $\log_2$-transformed, resulting in a normally distributed score.

$$Senescence\,Score = log_2 \frac{\sum_{x=2}^{p}\left(\frac{(Y_{(x)}+Y_{(x-1)})}{2} - \frac{(Z_{(x)}+Z_{(x-1)})}{2}\right)}{p}$$

Integrated CNA

A genomic instability score termed "integrated CNA" was calculated by summation of the Circular Binary Segmentation algorithm-inferred absolute mean copy number of segments multiplied by the length of each segment.

$$Int.CNA_{sample} = \frac{\sum_{segments} |segment\,end - segment\,start| \times |segment\,mean|}{\#base\,pairs\,in\,sample}$$

Box and whisker plots

In box and whisker plots, the box represents the median, as well as the first and third quartiles, and the whisker indicates the extreme values within 1.5 times the inter-quartile range. In cases where the number of samples permitted, individual values are superimposed as jitter plots.

Statistical tests

Indicated *P*-values were calculated using (i) Student's *t*-test for normally distributed data (with normality confirmed by the *P*-value of the Shapiro–Wilk test being > 0.05 for both datasets under comparison); (ii) Mann–Whitney U-test for non-normally distributed data; (iii) hypergeometric distribution P-values; and (iv) permutation-based approaches, as described in the figure legends. For data with a single data point in one comparison group, the z-score was used.

Propidium iodide staining

Mouse embryonic fibroblasts cells were washed in cold PBS and then fixed in ice-cold 70% ethanol. Fixed cells were washed in PBS and then incubated for 15 min in PBS with 20 µg/ml propidium iodide and 0.1% (v/v) Triton X-100. Data were acquired using a FACSCalibur (Becton Dickinson) analytic flow cytometer in the UCLA Jonsson Comprehensive Cancer Center and Center for AIDS Research Flow Cytometry Core Facility. Cells were gated using forward scatter and side scatter to remove debris and dead cells, and 10,000 cell events were recorded.

Spectral karyotyping

Exponentially growing MEF cells were exposed to colcemid (0.04 µg/ml) for 1 h at 37°C and to hypotonic treatment (0.075 M KCl) for 20 min at room temperature. Cells were fixed in a mixture of methanol and acetic acid (3:1 by volume) for 15 min, and then washed three times in the fixative. Slides were prepared by dropping the cell suspension onto wet slides followed by air-drying. Slides were processed for spectral karyotyping (SKY) according to the manufacturer's protocol with slight modifications using mouse paint probes (ASI, Vista, CA). Images were captured using Nikon 80i microscope equipped with spectral karyotyping software from ASI, Vista, CA; 12–18 metaphases were karyotyped from each cell line.

Immunoblot analysis

Cells were lysed in modified RIPA buffer (50 mM Tris–HCl (pH 7.5), 150 NaCl, 10 mM β-glycerophosphate, 1% NP-40, 0.25% sodium deoxycholate, 10 mM sodium pyrophosphate, 30 mM sodium fluoride, 1 mM EDTA, 1 mM vanadate, 20 µg/ml aprotinin, 20 µg/ml leupeptin, and 1 mM phenylmethylsulfonyl fluoride). Whole-cell lysates were resolved by SDS–PAGE on 4–15% gradient gels and blotted onto nitrocellulose membranes (Bio-Rad). Membranes were blocked overnight and then incubated sequentially with primary and either HRP-conjugated (Pierce) or IRDye-conjugated secondary antibodies (Li-Cor). Blots were imaged using the Odyssey Infrared Imaging System (Li-Cor). Protein levels were quantitated using ImageJ (http://imagej.nih.gov/ij/). Primary antibodies used for Western blot analysis included hexokinase 1 (2024, Cell Signaling Technology), hexokinase 2 (2867, Cell Signaling Technology), p53 (NB200-103, Novus Biologicals), and enolase 2 (8171, Cell Signaling Technology).

Glucose consumption and lactate secretion measurements

BRCA cell lines
Lactate secretion rates of breast cancer cell lines were measured from the culture media using a colorimetric assay kit (BioVision) (Hong *et al*, 2016).

MEFs
Glucose consumption and lactate secretion rates of MEFs were measured using a BioProfile Basic bioanalyzer (NOVA Biomedical). Data were normalized to the integrated cell number, which was calculated based on cell counts at the start and end of the time course and an exponential growth equation. Because the proliferation rates of MEF sublines vary, each cell line was seeded at the appropriate density so as to give an integrated cell number of approximately 6.5×10^5 cells in a 6-well plate. All samples were run as biological triplicates, and consistent results were seen in multiple independent experiments.

Mass spectrometry-based metabolomic analyses

Sample preparation
Mouse embryonic fibroblasts sublines were seeded onto 6-well plates, and after 24 h, media was replaced with media containing 4.5 g/l [1,2-^{13}C]-labeled glucose. Sample collection occurred after 24 h of culture in the labeled glucose media. For intracellular metabolite analysis, cells were washed with ice-cold 150 mM ammonium acetate (NH_4AcO) pH 7.3 and metabolites extracted in 1 ml ice-cold 80% MeOH. The cells were quickly transferred into a microfuge tube, and 10 nmol norvaline was added to the cell suspension for use as an internal standard. The suspension was subsequently vortexed three times over 15 min and then spun down at 4°C for 5 min. The supernatant was transferred into a glass vial, the cell pellet was re-extracted with 200 µl ice-cold 80% MeOH and spun down and the supernatants were combined. Metabolites were dried at 30°C under vacuum and re-suspended in 50 µl of 70% acetonitrile (ACN). For cell culture media metabolite analysis (footprint profiling), 20 µl of cell-free media samples was collected. Metabolites were extracted by adding 300 µl ice-cold 80% methanol, followed by vortexing three times over 15 min, and centrifugation for 10 min at 13,000 rpm at 4°C. The supernatant was transferred to a fresh tube, dried using a vacuum evaporator, and re-suspended in 50 µl of 70% acetonitrile (ACN); 5 µl was used for mass spectrometry-based analysis.

Mass spectrometry runs
Samples were run on a Q-Exactive mass spectrometer coupled to an UltiMate 3000RSLC UHPLC system (Thermo Scientific). The mass spectrometer was run in polarity switching mode (+3.00 kV/ −2.25 kV) with an m/z window ranging from 65 to 975. Mobile phase A was 5 mM NH_4AcO, pH 9.9, and mobile phase B was ACN. Metabolites were separated on a Luna 3 µm NH2 100 Å (150 × 2.0 mm) (Phenomenex) column. The flow was kept at 200 µl/min, and the gradient was from 15% A to 95% A in 18 min, followed by an isocratic step for 9 min and re-equilibration for 7 min.

Data analysis
Metabolites were detected and quantified as area under the curve (AUC) based on retention time and accurate mass ($\leq$ 3 ppm) using the TraceFinder 3.1 (Thermo Scientific) software. Relative amounts of metabolites between various conditions, percentage of metabolite isotopomers (relative to all isotopomers of that metabolite), and percentage of labeled metabolite molecules (isotopomer M1 and greater, relative to all isotopomers) were calculated and corrected

for naturally occurring ^{13}C abundance (Yuan *et al*, 2008). Footprinting data were normalized to the integrated cell number as described above, and intracellular metabolite concentrations were normalized to the number of cells present at the time of extraction. All samples were run as biological triplicates, and consistent results were seen in independent experiments. Our analysis focused on metabolite level, percent isotopomer, and percent labeled metabolite measurements with ANOVA *P*-values across the sample panel of < 0.05 in individual experiments, and Pearson correlation coefficients across all samples and between independent experiments of > 0.5.

Patient tumor samples and quantitative FDG-PET imaging

Breast cancer patient samples with imaged FDG uptake within 4 weeks prior to surgery, excluding patients with secondary breast cancers and recurrent disease, were collected surgically and processed as previously described (Palaskas *et al*, 2011). Of eighteen tumors collected in the original study for RNA microarrays (Palaskas *et al*, 2011), ten samples had sufficient remaining frozen tissue for array CGH profiling. None of the patients received systemic therapy or radiation prior to imaging. ^{18}FDG tumor uptake was quantified as standardized uptake values (SUVs) and showed the expected wide dynamic range (3.8–18.5). There was no significant difference in patient age, tumor volume, and lymph node involvement between the groups of FDG-high and FDG-low breast cancers. Breast cancers with high ^{18}FDG-PET SUVs frequently lacked expression of the estrogen receptor (ER) and the progesterone receptor (PR), but hormone receptor-negative tumors were also represented among the tumors with the lowest FDG uptake (Palaskas *et al*, 2011). We excluded lobular breast carcinomas, because they have been shown to take up less FDG than ductal carcinomas (Avril *et al*, 2001) We excluded large breast carcinomas (> 5 cm) and breast carcinomas with multifocal FDG uptake because our protocol did not include tissue autoradiography to direct the molecular tissue analysis to areas of distinct radiotracer retention. This study was approved by the Institutional Review Board (IRB) of Memorial Sloan-Kettering Cancer Center, and all participating patients signed the informed consent.

CNA dataset
Copy number profiling data for wild-type and genetically modified MEF samples and FDG-PET-imaged human breast tumors are available through Gene Expression Omnibus (GEO) accession GSE63306 (https://www.ncbi.nlm.nih.gov/geo/query/acc.cgi?acc = GSE63306).

CNA conservation web resource
An interactive website for user-defined pan-cancer and cross-species CNA conservation analysis to perform analysis analogous to that in Figs 1C, 2D, and 4E using any combination of tens of available CNA signatures from human tumors and mouse models (and additional signatures as they become available) and/or the inclusion of uploaded CNA signatures (http://systems.crump.ucla.edu/cna_con servation/). Signatures and genome reference files used in the interactive website are additionally available through the Biostudies repository, accession number S-BSST7.

Metabolomics dataset
Provided in Table EV5.

TCGA CNA dataset
The Cancer Genome Atlas (TCGA) CNA profiles were downloaded from the TCGA portal in September 2012 (https://cancergenome. nih.gov/). Copy number profiles obtained were pre-processed level 3 data based on human genome 19, with copy number variations (CNVs) removed. TCGA tumor type abbreviations and number of samples analyzed: bladder urothelial carcinoma (BLCA, 97 samples), brain lower grade glioma (LGG, 181), breast invasive carcinoma (BRCA, 873), colon adenocarcinoma (COAD, 447), glioblastoma multiforme (GBM, 593), head and neck squamous cell carcinoma (HNSC, 308), kidney renal clear cell carcinoma (KIRC, 539), lung adenocarcinoma (LUAD, 368), lung squamous cell carcinoma (LUSC, 359), ovarian serous cystadenocarcinoma (OV, 584), prostate adenocarcinoma (PRAD, 171), rectum adenocarcinoma (READ, 168), skin cutaneous melanoma (SKCM, 256), stomach adenocarcinoma (STAD, 162), thyroid carcinoma (THCA, 333), and uterine corpus endometrial carcinoma (UCEC, 492).

TCGA mRNA expression dataset
The Cancer Genome Atlas mRNA expression data were downloaded from the TCGA portal in May 2014. Gene-based mRNA expression levels were pre-processed, normalized Level 3 RNA Seq V2 RSEM values. TCGA tumor type abbreviations and number of samples analyzed: BRCA (865), LUAD (357), LUSC (358), OV (263).

Mouse tumor model CNA datasets
The genetically engineered mouse models with characterized CNA were obtained from public datasets: mammary (breast) tumors (Brca, 57 samples, GSE30710; 62 samples, GSE43997; 44 samples, GSE27101) (Drost *et al*, 2011; Herschkowitz *et al*, 2012); melanoma (Skcm, 30 samples, GSE58265) (Viros *et al*, 2014); glioblastoma/ high-grade astrocytoma (Gbm, 72 samples, GSE22927) (Chow *et al*, 2011); and prostate tumors (Prad, 18 samples GSE35247; 55 samples, GSE61382) (Ding *et al*, 2012; Wanjala *et al*, 2015). Additionally, *in vitro* epithelial murine cell lines modeling human carcinomas were obtained from public datasets: transformed colon cells (Coad, seven samples, GSE70790) and transformed bladder and kidney cells (Blca and Kirc, 6 and 7 samples, GSE45128; Padilla-Nash *et al*, 2013). Abbreviated mouse tumor names match to the corresponding tissue-based human tumor abbreviations from the TCGA datasets. The data were obtained from GEO and segmented using the algorithm described above. Datasets for which no mm9 genome annotation was available on the repository were lifted over to mm9 using UCSC web tools (Rosenbloom *et al*, 2015).

Acknowledgements
We thank Harvey Herschman, Judith Campisi, Edward Driggers, Michael Phelps, and Steven Bensinger for critical discussion of the project and experimental approaches. We thank Johannes Czernin for providing the FDG-PET image used in the graphical abstract. Flow cytometry was performed in the UCLA Jonsson Comprehensive Cancer Center (JCCC) and Center for AIDS Research Flow Cytometry Core Facility that is supported by National Institutes of Health awards P30 CA016042 and 5P30 AI028697, and by the JCCC, the

UCLA AIDS Institute, the David Geffen School of Medicine at UCLA, the UCLA Chancellor's Office, and the UCLA Vice Chancellor's Office of Research. We acknowledge the Molecular Cytogenetics Facility, Center for Genetics and Genomics, University of Texas, M.D. Anderson Cancer Center, Houston, Texas. N.A.G. is a postdoctoral trainee supported by the UCLA Scholars in Oncologic Molecular Imagining Program (NCI/NIH grant R25T CA098010). A.M. is supported by a UCLA Eugene V. Cota-Robles Fellowship and a UCLA Dissertation Year Fellowship. S.M.L. is supported by the NCI/NIH (P50 CA086438). T.G.G. is supported by the NCI/NIH (P01 CA168585, P50 CA086306, U19 AI067769), an American Cancer Society Research Scholar Award (RSG-12-257-01-TBE), a Melanoma Research Alliance Established Investigator Award (20120279), the Norton Simon Research Foundation, the UCLA Jonsson Cancer Center Foundation, the National Center for Advancing Translational Sciences UCLA CTSI Grant UL1TR000124, the UC Cancer Research Coordinating Committee, a Concern Foundation CONquer CanCER Now Award, and the UCLA Stein/Oppenheimer Endowment.

Author contributions

NAG, AM, AL, AC, DB, HRC, IKM, and TGG designed the study. NAG, AM, AL, AC, MF, SC, SZ, AD, LB, JtH, DB, and HRC performed experiments. NAG, AM, AL, AC, NGB, JG, NP, NS, DB, and TGG performed bioinformatic analysis. NAG, AM, AL, CH, CN, RQ, HW, MM, HRC, and TGG provided or derived reagents. NAG, AM, AL, DB, HRC, and TGG performed and analyzed metabolomics. AM and ASM performed spectral karyotyping. EP, SML, and IKM generated patient data. NAG, AM, DB, HRC, IKM, and TGG wrote the manuscript. The laboratories of HRC and IKM contributed equally. Experiments were performed at UCLA.

References

Anastasiou D, Poulogiannis G, Asara JM, Boxer MB, Jiang J, Shen M, Bellinger G, Sasaki AT, Locasale JW, Auld DS, Thomas CJ, Heiden MGV, Cantley LC (2011) Inhibition of pyruvate kinase M2 by reactive oxygen species contributes to cellular antioxidant responses. *Science* 334: 1278−1283

Avril N, Menzel M, Dose J, Schelling M, Weber W, Jänicke F, Nathrath W, Schwaiger M (2001) Glucose metabolism of breast cancer assessed by 18F-FDG PET: histologic and immunohistochemical tissue analysis. *J Nucl Med* 42: 9−16

Bergamaschi A, Kim YH, Wang P, Sørlie T, Hernandez-Boussard T, Lonning PE, Tibshirani R, Børresen-Dale A-L, Pollack JR (2006) Distinct patterns of DNA copy number alteration are associated with different clinicopathological features and gene-expression subtypes of breast cancer. *Genes Chromosom Cancer* 45: 1033−1040

Beroukhim R, Mermel CH, Porter D, Wei G, Raychaudhuri S, Donovan J, Barretina J, Boehm JS, Dobson J, Urashima M, Mc Henry KT, Pinchback RM, Ligon AH, Cho Y-J, Haery L, Greulich H, Reich M, Winckler W, Lawrence MS, Weir BA et al (2010) The landscape of somatic copy-number alteration across human cancers. *Nature* 463: 899−905

Boros LG, Lee PWN, Brandes JL, Cascante M, Muscarella P, Schirmer WJ, Melvin WS, Ellison EC (1998) Nonoxidative pentose phosphate pathways and their direct role in ribose synthesis in tumors: is cancer a disease of cellular glucose metabolism?. *Med Hypotheses* 50: 55−59

Cahill DP, Kinzler KW, Vogelstein B, Lengauer C (1999) Genetic instability and darwinian selection in tumours. *Trends Cell Biol* 9: M57−M60

Cai Y, Crowther J, Pastor T, Abbasi Asbagh L, Baietti MF, De Troyer M, Vazquez I, Talebi A, Renzi F, Dehairs J, Swinnen JV, Sablina AA (2016) Loss of chromosome 8p governs tumor progression and drug response by altering lipid metabolism. *Cancer Cell* 29: 751−766

Cairns RA, Harris IS, Mak TW (2011) Regulation of cancer cell metabolism. *Nat Rev Cancer* 11: 85−95

Chow LML, Endersby R, Zhu X, Rankin S, Qu C, Zhang J, Broniscer A, Ellison DW, Baker SJ (2011) Cooperativity within and among Pten, p53, and Rb pathways induces high-grade astrocytoma in adult brain. *Cancer Cell* 19: 305−316

Ciriello G, Miller ML, Aksoy BA, Senbabaoglu Y, Schultz N, Sander C (2013) Emerging landscape of oncogenic signatures across human cancers. *Nat Genet* 45: 1127−1133

Dang L, White DW, Gross S, Bennett BD, Bittinger MA, Driggers EM, Fantin VR, Jang HG, Jin S, Keenan MC, Marks KM, Prins RM, Ward PS, Yen KE, Liau LM, Rabinowitz JD, Cantley LC, Thompson CB, Vander Heiden MG, Su SM (2009) Cancer-associated IDH1 mutations produce 2-hydroxyglutarate. *Nature* 462: 739−744

Dang CV (2013) MYC, metabolism, cell growth, and tumorigenesis. *Cold Spring Harb Perspect Med* 3: a014217

Davoli T, Xu AW, Mengwasser KE, Sack LM, Yoon JC, Park PJ, Elledge SJ (2013) Cumulative haploinsufficiency and triplosensitivity drive aneuploidy patterns and shape the cancer genome. *Cell* 155: 948−962

DeBerardinis RJ, Lum JJ, Hatzivassiliou G, Thompson CB (2008) The biology of cancer: metabolic reprogramming fuels cell growth and proliferation. *Cell Metab* 7: 11−20

Ding Z, Wu C-J, Jaskelioff M, Ivanova E, Kost-Alimova M, Protopopov A, Chu GC, Wang G, Lu X, Labrot ES, Hu J, Wang W, Xiao Y, Zhang H, Zhang J, Zhang J, Gan B, Perry SR, Jiang S, Li L et al (2012) Telomerase reactivation following telomere dysfunction yields murine prostate tumors with bone metastases. *Cell* 148: 896−907

Drost R, Bouwman P, Rottenberg S, Boon U, Schut E, Klarenbeek S, Klijn C, van der Heijden I, van der Gulden H, Wientjens E, Pieterse M, Catteau A, Green P, Solomon E, Morris JR, Jonkers J (2011) BRCA1 RING function is essential for tumor suppression but dispensable for therapy resistance. *Cancer Cell* 20: 797−809

Forbes SA, Beare D, Gunasekaran P, Leung K, Bindal N, Boutselakis H, Ding M, Bamford S, Cole C, Ward S, Kok CY, Jia M, De T, Teague JW, Stratton MR, McDermott U, Campbell PJ (2015) COSMIC: exploring the world's knowledge of somatic mutations in human cancer. *Nucleic Acids Res* 43: D805−D811

Gao J, Aksoy BA, Dogrusoz U, Dresdner G, Gross B, Sumer SO, Sun Y, Jacobsen A, Sinha R, Larsson E, Cerami E, Sander C, Schultz N (2013) Integrative analysis of complex cancer genomics and clinical profiles using the cBioPortal. *Sci Signal* 6: pl1

Gatza ML, Silva GO, Parker JS, Fan C, Perou CM (2014) An integrated genomics approach identifies drivers of proliferation in luminal-subtype human breast cancer. *Nat Genet* 46: 1051−1059

Golub TR, Slonim DK, Tamayo P, Huard C, Gaasenbeek M, Mesirov JP, Coller H, Loh ML, Downing JR, Caligiuri MA, Bloomfield CD, Lander ES (1999) Molecular classification of cancer: class discovery and class prediction by gene expression monitoring. *Science* 286: 531−537

Graham NA, Tahmasian M, Kohli B, Komisopoulou E, Zhu M, Vivanco I, Teitell MA, Wu H, Ribas A, Lo RS, Mellinghoff IK, Mischel PS, Graeber TG (2012) Glucose deprivation activates a metabolic and signaling amplification loop leading to cell death. *Mol Syst Biol* 8: 589

Gupta S, Ramjaun AR, Haiko P, Wang Y, Warne PH, Nicke B, Nye E, Stamp G, Alitalo K, Downward J (2007) Binding of ras to phosphoinositide 3-Kinase p110α is required for Ras- driven tumorigenesis in mice. *Cell* 129: 957–968

Halliwell B (2003) Oxidative stress in cell culture: an under-appreciated problem? *FEBS Lett* 540: 3–6

Hanahan D, Weinberg RA (2011) Hallmarks of cancer: the next generation. *Cell* 144: 646–674

Hao L-Y, Greider CW (2004) Genomic instability in both wild-type and telomerase null MEFs. *Chromosoma* 113: 62–68

He L, He X, Lim LP, de Stanchina E, Xuan Z, Liang Y, Xue W, Zender L, Magnus J, Ridzon D, Jackson AL, Linsley PS, Chen C, Lowe SW, Cleary MA, Hannon GJ (2007) A microRNA component of the p53 tumour suppressor network. *Nature* 447: 1130–1134

Herschkowitz JI, Zhao W, Zhang M, Usary J, Murrow G, Edwards D, Knezevic J, Greene SB, Darr D, Troester MA, Hilsenbeck SG, Medina D, Perou CM, Rosen JM (2012) Comparative oncogenomics identifies breast tumors enriched in functional tumor-initiating cells. *Proc Natl Acad Sci USA* 109: 2778–2783

Hieronymus H, Schultz N, Gopalan A, Carver BS, Chang MT, Xiao Y, Heguy A, Huberman K, Bernstein M, Assel M, Murali R, Vickers A, Scardino PT, Sander C, Reuter V, Taylor BS, Sawyers CL (2014) Copy number alteration burden predicts prostate cancer relapse. *Proc Natl Acad Sci USA* 111: 11139–11144

Hoadley KA, Yau C, Wolf DM, Cherniack AD, Tamborero D, Ng S, Leiserson MDM, Niu B, McLellan MD, Uzunangelov V, Zhang J, Kandoth C, Akbani R, Shen H, Omberg L, Chu A, Margolin AA, van't Veer LJ, Lopez-Bigas N, Laird PW et al (2014) Multiplatform analysis of 12 cancer types reveals molecular classification within and across tissues of origin. *Cell* 158: 929–944

Holland AJ, Cleveland DW (2012) Losing balance: the origin and impact of aneuploidy in cancer. *EMBO Rep* 13: 501–514

Hong CS, Graham NA, Gu W, Espindola Camacho C, Mah V, Maresh EL, Alavi M, Bagryanova L, Krotee PAL, Gardner BK, Behbahan IS, Horvath S, Chia D, Mellinghoff IK, Hurvitz SA, Dubinett SM, Critchlow SE, Kurdistani SK, Goodglick L, Braas D et al (2016) MCT1 modulates cancer cell pyruvate export and growth of tumors that Co-express MCT1 and MCT4. *Cell Rep* 14: 1590–1601

Jain M, Nilsson R, Sharma S, Madhusudhan N, Kitami T, Souza AL, Kafri R, Kirschner MW, Clish CB, Mootha VK (2012) Metabolite profiling identifies a key role for glycine in rapid cancer cell proliferation. *Science* 336: 1040–1044

Kanehisa M, Goto S, Sato Y, Kawashima M, Furumichi M, Tanabe M (2014) Data, information, knowledge and principle: back to metabolism in KEGG. *Nucleic Acids Res* 42: D199–D205

Kaplon J, Zheng L, Meissl K, Chaneton B, Selivanov VA, Mackay G, van der Burg SH, Verdegaal EME, Cascante M, Shlomi T, Gottlieb E, Peeper DS (2013) A key role for mitochondrial gatekeeper pyruvate dehydrogenase in oncogene-induced senescence. *Nature* 498: 109–112

Kim J, Gao P, Liu Y-C, Semenza GL, Dang CV (2007) Hypoxia-inducible factor 1 and dysregulated c-Myc cooperatively induce vascular endothelial growth factor and metabolic switches hexokinase 2 and pyruvate dehydrogenase kinase 1. *Mol Cell Biol* 27: 7381–7393

Kolde R (2015) pheatmap: Pretty Heatmaps. R Package Version 1.0.8. https://CRAN.R-project.org/package=pheatmap

Kondoh H, Lleonart ME, Gil J, Wang J, Degan P, Peters G, Martinez D, Carnero A, Beach D (2005) Glycolytic enzymes can modulate cellular life span. *Cancer Res* 65: 177–185

Kondoh H (2009) The role of glycolysis in cellular immortalization. In *Cellular respiration and carcinogenesis*, Sarangarajan R, Apte S (eds) pp 91–102. Totowa, NJ: Humana Press

Kytölä S, Rummukainen J, Nordgren A, Karhu R, Farnebo F, Isola J, Larsson C (2000) Chromosomal alterations in 15 breast cancer cell lines by comparative genomic hybridization and spectral karyotyping. *Genes Chromosom Cancer* 28: 308–317

Lawrence MS, Stojanov P, Polak P, Kryukov GV, Cibulskis K, Sivachenko A, Carter SL, Stewart C, Mermel CH, Roberts SA, Kiezun A, Hammerman PS, McKenna A, Drier Y, Zou L, Ramos AH, Pugh TJ, Stransky N, Helman E, Kim J et al (2013) Mutational heterogeneity in cancer and the search for new cancer-associated genes. *Nature* 499: 214–218

Lee AC, Fenster BE, Ito H, Takeda K, Bae NS, Hirai T, Yu Z-X, Ferrans VJ, Howard BH, Finkel T (1999) Ras proteins induce senescence by altering the intracellular levels of reactive oxygen species. *J Biol Chem* 274: 7936–7940

Li B, Qiu B, Lee DSM, Walton ZE, Ochocki JD, Mathew LK, Mancuso A, Gade TPF, Keith B, Nissim I, Simon MC (2014) Fructose-1,6-bisphosphatase opposes renal carcinoma progression. *Nature* 513: 251–255

Locasale JW, Grassian AR, Melman T, Lyssiotis CA, Mattaini KR, Bass AJ, Heffron G, Metallo CM, Muranen T, Sharfi H, Sasaki AT, Anastasiou D, Mullarky E, Vokes NI, Sasaki M, Beroukhim R, Stephanopoulos G, Ligon AH, Meyerson M, Richardson AL et al (2011) Phosphoglycerate dehydrogenase diverts glycolytic flux and contributes to oncogenesis. *Nat Genet* 43: 869–874

Lowe SW, Ruley HE, Jacks T, Housman DE (1993) p53-dependent apoptosis modulates the cytotoxicity of anticancer agents. *Cell* 74: 957–967

Maddocks ODK, Berkers CR, Mason SM, Zheng L, Blyth K, Gottlieb E, Vousden KH (2013) Serine starvation induces stress and p53-dependent metabolic remodelling in cancer cells. *Nature* 493: 542–546

Maser RS, Choudhury B, Campbell PJ, Feng B, Wong K-K, Protopopov A, O'Neil J, Gutierrez A, Ivanova E, Perna I, Lin E, Mani V, Jiang S, McNamara K, Zaghlul S, Edkins S, Stevens C, Brennan C, Martin ES, Wiedemeyer R et al (2007) Chromosomally unstable mouse tumours have genomic alterations similar to diverse human cancers. *Nature* 447: 966–971

McCoy MK, Kaganovich A, Rudenko IN, Ding J, Cookson MR (2014) Hexokinase activity is required for recruitment of parkin to depolarized mitochondria. *Hum Mol Genet* 23: 145–156

McGranahan N, Burrell RA, Endesfelder D, Novelli MR, Swanton C (2012) Cancer chromosomal instability: therapeutic and diagnostic challenges. *EMBO Rep* 13: 528–538

Mermel CH, Schumacher SE, Hill B, Meyerson ML, Beroukhim R, Getz G (2011) GISTIC2.0 facilitates sensitive and confident localization of the targets of focal somatic copy-number alteration in human cancers. *Genome Biol* 12: R41

Metallo CM, Walther JL, Stephanopoulos G (2009) Evaluation of 13C isotopic tracers for metabolic flux analysis in mammalian cells. *J Biotechnol* 144: 167–174

Mootha VK, Lindgren CM, Eriksson K-F, Subramanian A, Sihag S, Lehar J, Puigserver P, Carlsson E, Ridderstråle M, Laurila E, Houstis N, Daly MJ, Patterson N, Mesirov JP, Golub TR, Tamayo P, Spiegelman B, Lander ES, Hirschhorn JN, Altshuler D et al (2003) PGC-1α-responsive genes involved in oxidative phosphorylation are coordinately downregulated in human diabetes. *Nat Genet* 34: 267–273

Muller FL, Colla S, Aquilanti E, Manzo VE, Genovese G, Lee J, Eisenson D, Narurkar R, Deng P, Nezi L, Lee MA, Hu B, Hu J, Sahin E, Ong D, Fletcher-Sananikone E, Ho D, Kwong L, Brennan C, Wang YA et al (2012) Passenger deletions generate therapeutic vulnerabilities in cancer. *Nature* 488: 337–342

Nakamura K, Hongo A, Kodama J, Hiramatsu Y (2011) The measurement of SUVmax of the primary tumor is predictive of prognosis for patients with endometrial cancer. *Gynecol Oncol* 123: 82–87

Neve RM, Chin K, Fridlyand J, Yeh J, Baehner FL, Fevr T, Clark L, Bayani N, Coppe J-P, Tong F, Speed T, Spellman PT, DeVries S, Lapuk A, Wang NJ, Kuo W-L, Stilwell JL, Pinkel D, Albertson DG, Waldman FM *et al* (2006) A collection of breast cancer cell lines for the study of functionally distinct cancer subtypes. *Cancer Cell* 10: 515–527

Nicholson JM, Macedo JC, Mattingly AJ, Wangsa D, Camps J, Lima V, Gomes AM, Dória S, Ried T, Logarinho E, Cimini D (2015) Chromosome mis-segregation and cytokinesis failure in trisomic human cells. *eLife* 4: e05068

Olive KP, Tuveson DA, Ruhe ZC, Yin B, Willis NA, Bronson RT, Crowley D, Jacks T (2004) Mutant p53 gain of function in two mouse models of li-fraumeni syndrome. *Cell* 119: 847–860

Padilla-Nash HM, McNeil NE, Yi M, Nguyen Q-T, Hu Y, Wangsa D, Mack DL, Hummon AB, Case C, Cardin E, Stephens R, Difilippantonio MJ, Ried T (2013) Aneuploidy, oncogene amplification and epithelial to mesenchymal transition define spontaneous transformation of murine epithelial cells. *Carcinogenesis* 34: 1929–1939

Palaskas N, Larson SM, Schultz N, Komisopoulou E, Wong J, Rohle D, Campos C, Yannuzzi N, Osborne JR, Linkov I, Kastenhuber ER, Taschereau R, Plaisier SB, Tran C, Heguy A, Wu H, Sander C, Phelps ME, Brennan C, Port E *et al* (2011) 18F-fluorodeoxy-glucose positron emission tomography marks MYC-overexpressing human basal-like breast cancers. *Cancer Res* 71: 5164–5174

Parrinello S, Samper E, Krtolica A, Goldstein J, Melov S, Campisi J (2003) Oxygen sensitivity severely limits the replicative lifespan of murine fibroblasts. *Nat Cell Biol* 5: 741–747

Patra KC, Wang Q, Bhaskar PT, Miller L, Wang Z, Wheaton W, Chandel N, Laakso M, Muller WJ, Allen EL, Jha AK, Smolen GA, Clasquin MF, Robey RB, Hay N (2013) Hexokinase 2 is required for tumor initiation and maintenance and its systemic deletion is therapeutic in mouse models of cancer. *Cancer Cell* 24: 213–228

Possemato R, Marks KM, Shaul YD, Pacold ME, Kim D, Birsoy K, Sethumadhavan S, Woo H-K, Jang HG, Jha AK, Chen WW, Barrett FG, Stransky N, Tsun Z-Y, Cowley GS, Barretina J, Kalaany NY, Hsu PP, Ottina K, Chan AM *et al* (2011) Functional genomics reveal that the serine synthesis pathway is essential in breast cancer. *Nature* 476: 346–350

Rosenbloom KR, Armstrong J, Barber GP, Casper J, Clawson H, Diekhans M, Dreszer TR, Fujita PA, Guruvadoo L, Haeussler M, Harte RA, Heitner S, Hickey G, Hinrichs AS, Hubley R, Karolchik D, Learned K, Lee BT, Li CH, Miga KH *et al* (2015) The UCSC genome browser database: 2015 update. *Nucleic Acids Res* 43: D670–D681

Sanchez-Garcia F, Villagrasa P, Matsui J, Kotliar D, Castro V, Akavia U-D, Chen B-J, Saucedo-Cuevas L, Rodriguez Barrueco R, Llobet-Navas D, Silva JM, Pe'er D (2014) Integration of genomic data enables selective discovery of breast cancer drivers. *Cell* 159: 1461–1475

Sankaranarayanan P, Schomay TE, Aiello KA, Alter O (2015) Tensor GSVD of patient- and platform-matched tumor and normal DNA copy-number profiles uncovers chromosome arm-wide patterns of tumor-exclusive platform-consistent alterations encoding for cell transformation and predicting ovarian cancer survival. *PLoS One* 10: e0121396

Schafer ZT, Grassian AR, Song L, Jiang Z, Gerhart-Hines Z, Irie HY, Gao S, Puigserver P, Brugge JS (2009) Antioxidant and oncogene rescue of metabolic defects caused by loss of matrix attachment. *Nature* 461: 109–113

Sebastián C, Zwaans BMM, Silberman DM, Gymrek M, Goren A, Zhong L, Ram O, Truelove J, Guimaraes AR, Toiber D, Cosentino C, Greenson JK, MacDonald AI, McGlynn L, Maxwell F, Edwards J, Giacosa S, Guccione E, Weissleder R, Bernstein BE *et al* (2012) The histone deacetylase SIRT6 is a tumor suppressor that controls cancer metabolism. *Cell* 151: 1185–1199

Seshan VE, Olshen A (2016) DNAcopy: DNA copy number data analysis. R Package Version 1.48.0

Sheltzer JM (2013) A transcriptional and metabolic signature of primary aneuploidy is present in chromosomally unstable cancer cells and informs clinical prognosis. *Cancer Res* 73: 6401–6412

Sherr CJ, DePinho RA (2000) Cellular senescence: minireview mitotic clock or culture shock? *Cell* 102: 407–410

Sherr CJ, Weber JD (2000) The ARF/p53 pathway. *Curr Opin Genet Dev* 10: 94–99

Shi H, Moriceau G, Kong X, Lee M-K, Lee H, Koya RC, Ng C, Chodon T, Scolyer RA, Dahlman KB, Sosman JA, Kefford RF, Long GV, Nelson SF, Ribas A, Lo RS (2012) Melanoma whole-exome sequencing identifies (V600E)B-RAF amplification-mediated acquired B-RAF inhibitor resistance. *Nat Commun* 3: 724

Smith ML, Marioni JC, McKinney S, Hardcastle T, Thorne NP (2009) snapCGH: Segmentation, normalisation and processing of aCGH data. R package version 1.44.0

Smoot ME, Ono K, Ruscheinski J, Wang P-L, Ideker T (2011) Cytoscape 2.8: new features for data integration and network visualization. *Bioinformatics* 27: 431–432

Solimini NL, Xu Q, Mermel CH, Liang AC, Schlabach MR, Luo J, Burrows AE, Anselmo AN, Bredemeyer AL, Li MZ, Beroukhim R, Meyerson M, Elledge SJ (2012) Recurrent hemizygous deletions in cancers may optimize proliferative potential. *Science* 337: 104–109

Sotillo R, Hernando E, Díaz-Rodríguez E, Teruya-Feldstein J, Cordón-Cardo C, Lowe SW, Benezra R (2007) Mad2 overexpression promotes aneuploidy and tumorigenesis in mice. *Cancer Cell* 11: 9–23

Stuart D, Sellers WR (2009) Linking somatic genetic alterations in cancer to therapeutics. *Curr Opin Cell Biol* 21: 304–310

Subramanian A, Tamayo P, Mootha VK, Mukherjee S, Ebert BL, Gillette MA, Paulovich A, Pomeroy SL, Golub TR, Lander ES, Mesirov JP (2005) Gene set enrichment analysis: a knowledge-based approach for interpreting genome-wide expression profiles. *Proc Natl Acad Sci* 102: 15545–15550

Sun H, Gulbagci NT, Taneja R (2007) Analysis of growth properties and cell cycle regulation using mouse embryonic fibroblast cells. *Methods Mol Biol* 383: 311–319

Sun H, Taneja R (2007) Analysis of transformation and tumorigenicity using mouse embryonic fibroblast cells. In *Cancer genomics and proteomics*, Fisher P (ed) pp 303–310. Totowa, NJ: Humana Press

Takahashi A, Ohtani N, Yamakoshi K, Iida S, Tahara H, Nakayama K, Nakayama KI, Ide T, Saya H, Hara E (2006) Mitogenic signalling and the p16INK4a–Rb pathway cooperate to enforce irreversible cellular senescence. *Nat Cell Biol* 8: 1291–1297

Tang J, Li Y, Lyon K, Camps J, Dalton S, Ried T, Zhao S (2014) Cancer driver-passenger distinction via sporadic human and dog cancer comparison: a proof-of-principle study with colorectal cancer. *Oncogene* 33: 814–822

The Cancer Genome Atlas Research Network (2011) Integrated genomic analyses of ovarian carcinoma. *Nature* 474: 609–615

The Cancer Genome Atlas Research Network (2012a) Comprehensive genomic characterization of squamous cell lung cancers. *Nature* 489: 519–525

The Cancer Genome Atlas Research Network (2012b) Comprehensive molecular portraits of human breast tumours. *Nature* 490: 61–70

The Cancer Genome Atlas Research Network (2013) Integrated genomic characterization of endometrial carcinoma. *Nature* 497: 67–73

The Cancer Genome Atlas Research Network (2014) Comprehensive molecular profiling of lung adenocarcinoma. *Nature* 511: 543–550

Thorvaldsdóttir H, Robinson JT, Mesirov JP (2013) Integrative genomics viewer (IGV): high-performance genomics data visualization and exploration. *Brief Bioinform* 14: 178–192

Timmerman LA, Holton T, Yuneva M, Louie RJ, Padró M, Daemen A, Hu M, Chan DA, Ethier SP, van't Veer LJ, Polyak K, McCormick F, Gray JW (2013) Glutamine sensitivity analysis identifies the xCT antiporter as a common triple-negative breast tumor therapeutic target. *Cancer Cell* 24: 450–465

Todaro GJ, Green H (1963) Quantitative studies of the growth of mouse embryo cells in culture and their development into established lines. *J Cell Biol* 17: 299–313

Viros A, Sanchez-Laorden B, Pedersen M, Furney SJ, Rae J, Hogan K, Ejiama S, Girotti MR, Cook M, Dhomen N, Marais R (2014) Ultraviolet radiation accelerates BRAF-driven melanomagenesis by targeting TP53. *Nature* 511: 478–482

Wade M, Li Y-C, Wahl GM (2013) MDM2, MDMX and p53 in oncogenesis and cancer therapy. *Nat Rev Cancer* 13: 83–96

Wang L, Xiong H, Wu F, Zhang Y, Wang J, Zhao L, Guo X, Chang L-J, Zhang Y, You MJ, Koochekpour S, Saleem M, Huang H, Lu J, Deng Y (2014) Hexokinase 2-mediated warburg effect is required for PTEN- and p53-deficiency-driven prostate cancer growth. *Cell Rep* 8: 1461–1474

Wanjala J, Taylor BS, Chapinski C, Hieronymus H, Wongvipat J, Chen Y, Nanjangud GJ, Schultz N, Xie Y, Liu S, Lu W, Yang Q, Sander C, Chen Z, Sawyers CL, Carver BS (2015) Identifying actionable targets through integrative analyses of GEM model and human prostate cancer genomic profiling. *Mol Cancer Ther* 14: 278–288

Warburg O (1956) On the origin of cancer cells. *Nature* 123: 309–314

Weaver BAA, Silk AD, Montagna C, Verdier-Pinard P, Cleveland DW (2007) Aneuploidy acts both oncogenically and as a tumor suppressor. *Cancer Cell* 11: 25–36

Weinberg F, Hamanaka R, Wheaton WW, Weinberg S, Joseph J, Lopez M, Kalyanaraman B, Mutlu GM, Budinger GRS, Chandel NS (2010) Mitochondrial metabolism and ROS generation are essential for Kras-mediated tumorigenicity. *Proc Natl Acad Sci USA* 107: 8788–8793

Williams BR, Prabhu VR, Hunter KE, Glazier CM, Whittaker CA, Housman DE, Amon A (2008) Aneuploidy affects proliferation and spontaneous immortalization in mammalian cells. *Science* 322: 703–709

Xie H, Hanai J, Ren J-G, Kats L, Burgess K, Bhargava P, Signoretti S, Billiard J, Duffy KJ, Grant A, Wang X, Lorkiewicz PK, Schatzman S, Bousamra M II, Lane AN, Higashi RM, Fan TWM, Pandolfi PP, Sukhatme VP, Seth P (2014) Targeting lactate dehydrogenase-a inhibits tumorigenesis and tumor progression in mouse models of lung cancer and impacts tumor-initiating cells. *Cell Metab* 19: 795–809

Ying H, Kimmelman AC, Lyssiotis CA, Hua S, Chu GC, Fletcher-Sananikone E, Locasale JW, Son J, Zhang H, Coloff JL, Yan H, Wang W, Chen S, Viale A, Zheng H, Paik J, Lim C, Guimaraes AR, Martin ES, Chang J *et al* (2012) Oncogenic kras maintains pancreatic tumors through regulation of anabolic glucose metabolism. *Cell* 149: 656–670

Yuan J, Bennett BD, Rabinowitz JD (2008) Kinetic flux profiling for quantitation of cellular metabolic fluxes. *Nat Protoc* 3: 1328–1340

Yun J, Rago C, Cheong I, Pagliarini R, Angenendt P, Rajagopalan H, Schmidt K, Willson JKV, Markowitz S, Zhou S, Diaz LA, Velculescu VE, Lengauer C, Kinzler KW, Vogelstein B, Papadopoulos N (2009) Glucose deprivation contributes to the development of KRAS pathway mutations in tumor cells. *Science* 325: 1555–1559

Zhang WC, Shyh-Chang N, Yang H, Rai A, Umashankar S, Ma S, Soh BS, Sun LL, Tai BC, Nga ME, Bhakoo KK, Jayapal SR, Nichane M, Yu Q, Ahmed DA, Tan C, Sing WP, Tam J, Thirugananam A, Noghabi MS *et al* (2012) Glycine decarboxylase activity drives non-small cell lung cancer tumor-initiating cells and tumorigenesis. *Cell* 148: 259–272

Zhang G, Hoersch S, Amsterdam A, Whittaker CA, Beert E, Catchen JM, Farrington S, Postlethwait JH, Legius E, Hopkins N, Lees JA (2013) Comparative oncogenomic analysis of copy number alterations in human and zebrafish tumors enables cancer driver discovery. *PLoS Genet* 9: e1003734

Zhang J (2016) CNTools: Convert segment data into a region by sample matrix to allow for other high level computational analyses. R package version 1.30.0

2

Genetic circuit characterization and debugging using RNA-seq

Thomas E Gorochowski[1,†] (iD), Amin Espah Borujeni[1,†] (iD), Yongjin Park[1,†], Alec AK Nielsen[1], Jing Zhang[1], Bryan S Der[1], D Benjamin Gordon[1,2] & Christopher A Voigt[1,2,*] (iD)

Abstract

Genetic circuits implement computational operations within a cell. Debugging them is difficult because their function is defined by multiple states (e.g., combinations of inputs) that vary in time. Here, we develop RNA-seq methods that enable the simultaneous measurement of: (i) the states of internal gates, (ii) part performance (promoters, insulators, terminators), and (iii) impact on host gene expression. This is applied to a three-input one-output circuit consisting of three sensors, five NOR/NOT gates, and 46 genetic parts. Transcription profiles are obtained for all eight combinations of inputs, from which biophysical models can extract part activities and the response functions of sensors and gates. Various unexpected failure modes are identified, including cryptic antisense promoters, terminator failure, and a sensor malfunction due to media-induced changes in host gene expression. This can guide the selection of new parts to fix these problems, which we demonstrate by using a bidirectional terminator to disrupt observed antisense transcription. This work introduces RNA-seq as a powerful method for circuit characterization and debugging that overcomes the limitations of fluorescent reporters and scales to large systems composed of many parts.

Keywords biofab; combinatorial logic; omics; synthetic biology; systems biology

Subject Categories Genome-Scale & Integrative Biology; Synthetic Biology & Biotechnology

Introduction

Natural regulatory networks control the timing and conditions for gene expression. An ability to construct synthetic networks would enable the spatiotemporal control of biological processes (Basu et al, 2004). These could be used to react to environmental conditions (e.g., different phases of growth in a bioreactor; Anderson et al, 2006; Gupta et al, 2017) or implement a dynamic response (e.g., avoiding the accumulation of toxic intermediates; Zhang et al, 2012). However, there are many challenges when building synthetic regulatory networks. Obtaining a desired response requires numerous interacting genes and precise control over their expression. This results in large systems that contain many genetic parts, all of which must function correctly in concert. While fluorescent reporters have been critical for quantifying the response of such systems to date (Kelly et al, 2009), they are only capable of probing a single gene at a time (usually the output) and require repetition of the assay for each state or time point of interest. Mapping the fluorescence data of the output back to the specific internal failure can be difficult or impossible.

Analogies to electronic circuits are often made when describing the computational operations performed by a regulatory network. Such "genetic circuits" have been built that function as logic gates (Anderson et al, 2006; Moon et al, 2012; Qi et al, 2012; Siuti et al, 2013; Nielsen & Voigt, 2014) as well as dynamic (Elowitz & Leibler, 2000; Zhang et al, 2012) and analog (Daniel et al, 2013) circuits. Larger circuits can be built by connecting simpler gates (Moon et al, 2012; Kiani et al, 2014; Nielsen & Voigt, 2014). This process is facilitated by defining the signal between gates as the RNA polymerase (RNAP) flux (Canton et al, 2008). In practice, this is achieved by designing gates such that their inputs and outputs are both promoters. The response function can then be defined as how the output promoter activity changes as a function of the input promoter activity at steady-state (Weiss, 2001; Nielsen et al, 2016). Gate response functions can be used to computationally predict how to build a circuit (Hooshangi et al, 2005; Nielsen et al, 2016). However, the genetic context of the gates in the circuit differs from that used to measure them in isolation and this can impact their function and, in turn, lead to circuit failures (Brophy & Voigt, 2014). Therefore, it is valuable to be able to directly measure the performance of individual gates in the final context of a circuit.

Systems biology has led to new –omics tools that offer the potential to take a snapshot of the entire internal workings of a circuit with a single experiment. Transcriptomic methods, such as RNA sequencing (RNA-seq), enable the measurement of genomewide mRNA levels with nucleotide resolution (Zhong et al, 2009). This

---

1 Synthetic Biology Center, Department of Biological Engineering, Massachusetts Institute of Technology, Cambridge, MA, USA
2 Broad Institute of MIT and Harvard, Cambridge, MA, USA
  *Corresponding author. E-mail: cavoigt@gmail.com
  †These authors contributed equally to this work

can be used to calculate promoter and terminator strengths (Smanski *et al*, 2014; Li *et al*, 2015; Srikumar *et al*, 2015), which are closely related to the RNAP fluxes used to connect transcriptional gates. RNA-seq takes advantage of next-generation sequencing to quantify genomewide RNA levels at a moment in time. It has been used to address problems in strain and metabolic engineering (Yuan *et al*, 2011; Kim *et al*, 2012; Zhang *et al*, 2012; Woodruff *et al*, 2013), but has not been applied to the characterization of genetic circuits. This stems in part from the cost of RNA-seq and the large numbers of states and time points required to fully characterize a circuit. Another problem is that sequencing generates a deluge of data and a lack of software tools for synthetic biology, and especially genetic circuit design, hinders its processing and interpretation.

Several advances have been made that reduce the cost of RNA-seq and enable multiple circuit states to be assayed in a single sequencing run. RNAtag-seq uses nucleotide barcodes to tag total fragmented RNA before depletion of ribosomal RNA (rRNA) to allow for many samples to be efficiently pooled and sequenced together (Shishkin *et al*, 2015). The tags denote which sample a fragment originates and allows for data from each sample to be separated post-sequencing. This leads to significant reductions in the cost of reagents, accelerates preparation time, and decreases biases due to the amplification of individual sample libraries. Here, we apply this method to barcode samples associated with different states of a genetic circuit. Specifically, we characterize the eight states of a three-input one-output combinatorial logic circuit. This approach can be scaled-up: a single flow cell on an Illumina HiSeq 2500 machine generates ~4 billion paired-end reads and is thus capable of characterizing up to 1,000 samples (Haas *et al*, 2012), which could be used to simultaneously assay many different circuits and states.

Another advantage of RNA-seq is that data is captured for the entire host genome. This enables the direct observation of how differing circuit states impact host gene expression and the burden imposed by the circuit (Ceroni *et al*, 2015). It has been shown that availability and sequestering of shared cellular resources can significantly impact circuit function (Cardinale *et al*, 2013; Jayanthi *et al*, 2013; Gorochowski *et al*, 2016). Therefore, as circuits become larger, accounting for these effects will become increasingly important.

In this manuscript, we present methodologies for the application of RNA-seq to characterize genetic circuits (Fig 1). A combinatorial logic circuit is chosen as a demonstration, and data are collected for all permutations of the inputs. Cells with circuits in these different states are sequenced using RNAtag-seq and new algorithms and software are used to automate data processing. Biophysical models are developed that connect the functions of promoters, terminators, and insulators to the expected transcriptional profiles. This is used to algorithmically quantify the performance of genetic parts. Furthermore, the data are used to quantify the response functions of the three sensors and five NOT/NOR gates in the context of the circuit. This analysis reveals several mechanistic causes of circuit failures. The ability to observe the internal workings of genetic circuits will lead to a better understanding of the mechanisms that lead to part failure, how this propagates to impact system function, and ultimately will support the construction of larger genetic systems.

Results

Data collection

The first step in characterizing a genetic circuit is to gather data covering the range of states (Fig 1A). This differs depending on the type of circuit; for logic, this corresponds to steady-state measurements for each combination of inputs, whereas for a dynamic circuit, it would involve sampling time points. Once reaching either steady-state or a specific time point, aliquots of cells harboring the circuit are taken and flash-frozen in liquid nitrogen to minimize RNA degradation (Materials and Methods). The total RNA is then harvested, purified, and concentrated.

Samples are next converted to a pooled sequencing library using RNAtag-seq (Shishkin *et al*, 2015). First, they are separately fragmented before short DNA adaptors containing unique barcode sequences are ligated to the 3′-end of the RNAs. These barcodes uniquely "tag" every molecule such that the originating sample is known. Due to end specificity of ligation, they also capture strand-specific information. To ensure later sequencing is not affected by barcode choice, we use a set where minimal sequencing bias has been observed (Shishkin *et al*, 2015). Tagged samples are pooled to simplify the remaining preparatory steps. Unwanted rRNA is depleted, and cDNA generated by reverse transcription. Then, any remaining RNA is degraded and 3′ DNA adaptors are ligated such that a final library can be produced by amplification with indexed sequencing primers. Finally, the library is sequenced to generate strand-specific reads and the barcodes are used to associate each read to its original sample. Files containing these data are then used as inputs for circuit characterization.

Conversion of raw RNA-seq reads to transcription profiles

We developed a suite of algorithms to process RNA-seq data and characterize part performance, sensor/gate function and host response (Fig 1A). This requires the conversion of the raw sequencing data into a transcription profile for each input state or time point (Box 1). The profiles capture the observed number of transcripts at every position along the DNA encoding the circuit. To perform this conversion, the second step in our pipeline takes as input the raw reads generated from sequencing and maps these to user-provided reference sequences containing both the host genome and synthetic circuit in a multi-FASTA format using BWA (Li & Durbin, 2009; Fig 1A). Strand-specific transcription profiles are then generated by separately extracting reads mapping to the sense and antisense strand and their start and end position using SAMtools (Li *et al*, 2009). A mathematical model is then applied to correct the transcription profiles for the localized drops in sequencing depth at the ends of transcripts, using the mapped fragment length distribution as an input variable (Box 1). This correction is required to be able to characterize parts that occur near transcript start and end sites (e.g., promoters and terminators). To provide further gene-level expression estimates for the host and circuit, a user-provided sequence annotation in GFF format containing the region of each gene is used by HTSeq (Anders *et al*, 2015) to count the reads mapping to each gene.

Because RNA-seq provides relative measurements of transcript abundance (Robinson & Oshlack, 2010; *i.e.*, fractional abundance of

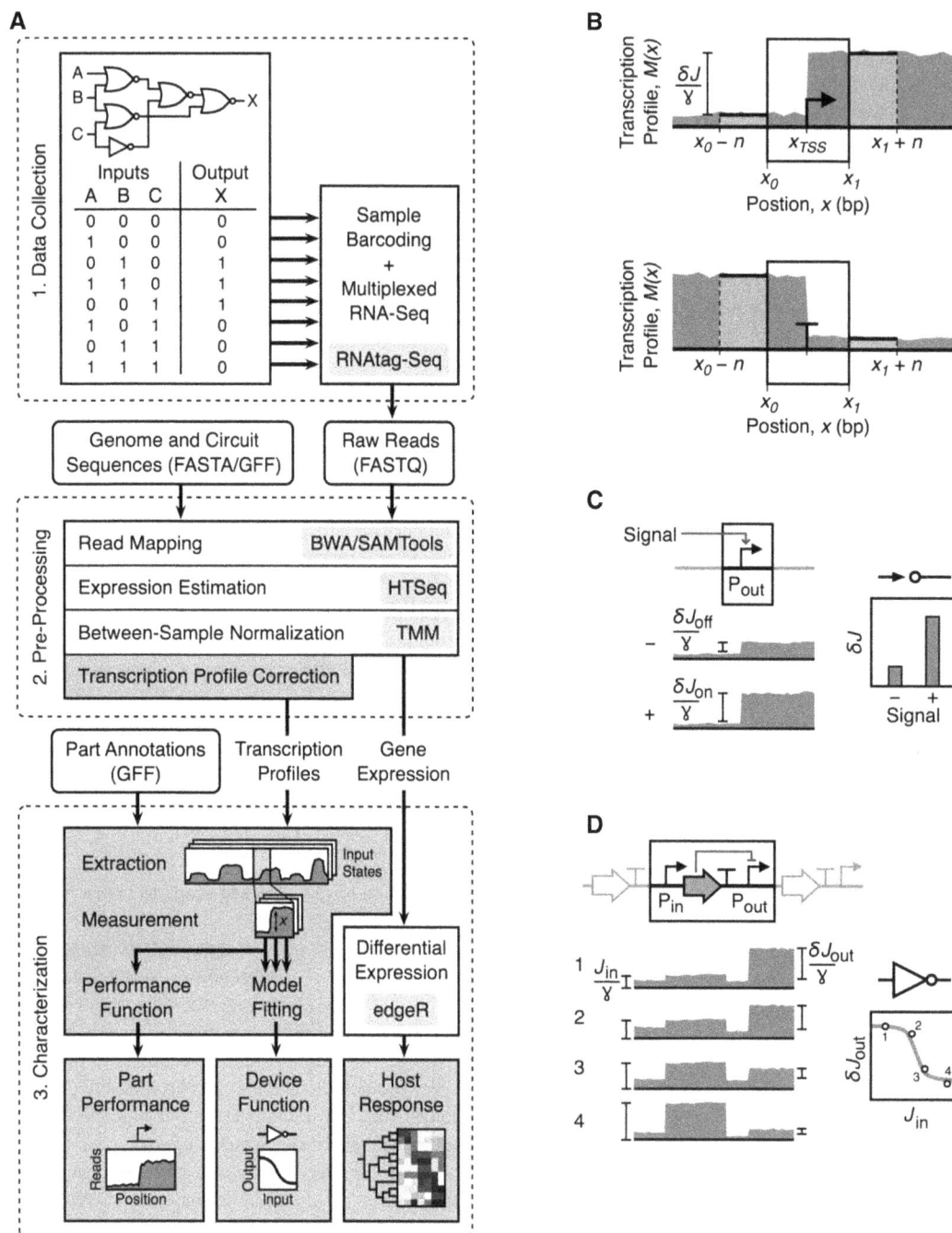

Figure 1. Overview of RNA-seq circuit characterization.

A Circuit characterization pipeline. Square boxes are the major steps in the process, and rounded boxes are input/output files. Light gray boxes denote experimental protocols or computational tools used during that process. Dark gray boxes correspond to the algorithms developed in this work and the major outputs from the pipeline. Details regarding the software to process sequencing data are provided in the Materials and Methods.

B Quantification of the performance of promoters and terminators from transcription profiles. Transcription profiles are shown in dark gray, with the location and extent of the promoter and terminator shown by a box. Parameters correspond to equations 2 and 3.

C Characterization of sensors. The activity of the output promoter is measured in the absence and presence of the associated signal.

D Characterization of gates. Measurements of the total input RNAP flux and the change due to the promoter are measured for each state, and these data are used to parameterize the response function. Parameters correspond to equation 4.

the total sequenced fragments), potential differences introduced during sample preparation or sequencing require correction to allow for comparison between samples. This is performed by calculating between-sample normalization factors using a trimmed mean of M-values (TMM) approach (Robinson & Oshlack, 2010). These factors are applied to produce normalized gene expression levels in

Box 1: Generation of transcription profiles for part characterization

A widely used approach to generate transcription profiles from RNA-seq data is to count the number of mapped fragments that cover each nucleotide position along the DNA sequence (Nagalakshmi *et al*, 2008; Zhong *et al*, 2009). The resulting profiles often exhibit a gradual but significant increase and decrease at the start and end of transcription units, respectively. These non-uniformities (shown by red circles) are problematic for characterizing the performance of genetic parts such as promoters and terminators, where their function is defined by the regions near the transcription start and end points.

To address this problem, a probabilistic method was developed to correct for curvature at the ends of each transcription unit, which utilizes the fragment length distribution as the only input variable. We begin by generating a fragment length distribution by directly analyzing the sequenced fragments. Using a Monte Carlo approach, large numbers of fragment lengths are drawn from this distribution. Fragments of these lengths are randomly mapped to positions falling within the boundaries of a hypothetical transcription unit 2,000 nt long. By counting the number of mapped fragments covering each nucleotide x, a hypothetical profile $T(x)$ is produced that defines the expected curvature at each end of a transcription unit. Because the curvature is localized and fully captured by the first 500 nt of the hypothetical profile, this region is extracted and normalized by its maximum value to generate a correction factor profile $C(x)$.

Next, the RNA-seq data are used to generate the transcription profile for each transcription unit within the circuit. Fragments mapping exclusively within transcription unit boundaries are selected and a transcription profile $P(x)$ generated by counting the number of mapped fragments covering each nucleotide. Unwanted curvature is corrected for by dividing the value of $P(x)$ for the first and last 500 nt of each transcription unit by $C(x_n)$, where x_n is the distance in nucleotides to the nearest end of the transcript. Specifically, the corrected transcription profile is given by

$$P_c(x) = \begin{cases} \frac{P(x)}{C(x_n)}, & 0 < x_n \le 500, \\ P(x), & \text{otherwise} \end{cases}$$

Because the correction factor is only based on the length of the fragments and does not consider their sequence, the same correction

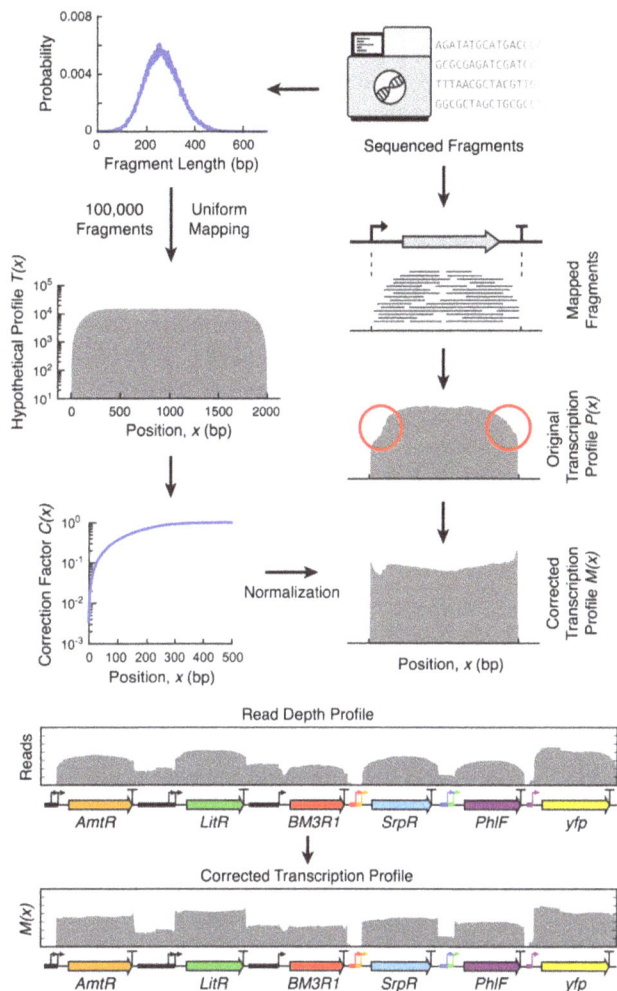

factor profile is used for both the 5′- and 3′-end, while the middle of a transcription unit, where no curvature is present, remains unmodified. The correction factor is only applied to known transcripts internal to the genetic circuit. It is not applied to other regions where part function does not have to be calculated; for example, read-through between transcriptional units, transcripts from internal promoters, antisense transcription, and genomic transcription. Further details regarding the correction method are provided in Appendix Text S1.

---

fragments per kilobase of exon per million fragments mapped (FPKM) units (Trapnell *et al*, 2010) and to the transcription profiles to enable comparisons between samples (Appendix Text S1).

Genetic part characterization from transcription profiles

Genetic parts, including promoters and terminators, impact the shape of the transcription profile by altering the flux of RNAP and mRNA transcripts produced. Techniques have been developed in bioinformatics and systems biology to naively scan transcription profiles for natural regulatory features, such as promoters (Conway *et al*, 2014). Analyzing a synthetic system has different objectives that require a different computational approach. First, the parts are modular and defined with clear start and end points. Second, the function of a part needs to be quantified (even if it is non-functional) and doing so requires a biophysical model that can

process RNA-seq data. Here, we develop models to characterize those parts that are the most critical in the design of transcriptional genetic circuits. Regions of the transcription profiles corresponding to each part are extracted, and measurements of localized changes in the profile depth are taken. These are interpreted in the context of biophysical models of each part type to infer their performance.

Promoters cause sharp increases in the transcription profile (Fig 1B). The activity of a promoter can be quantified as the increase in the flux of RNAPs that occurs between the beginning x_0 and end x_1 of a part. The RNAP flux $J(x)$ is the number of RNAPs passing nucleotide position x per second. Here, we assume that all RNAPs that pass a nucleotide lead to an mRNA transcript and that all transcripts within the circuit degrade at the same rate. With these assumptions, the flux at a position x is given by the steady-state number of transcripts $M(x)$ at that position (in effect, counting the number of RNAPs passing that position that occur on the timescale of degradation).

The transcription profile provides the steady-state number of transcripts $M(x)$ at each position x. However, the profiles cannot be quantified in units of transcripts and are thus presented in arbitrary units (au) and the fluxes are in au/s. The change in the number of transcripts at position x is given by

$$\frac{dM(x)}{dt} = J(x) - \gamma M(x), \tag{1}$$

where $\gamma = 0.0067 \text{ s}^{-1}$ is the degradation rate of mRNA (Chen et al, 2015). At steady state, $dM(x)/dt = 0$ and the flux of RNAP $J(x) = \gamma M(x)$. The activity of a promoter can be quantified as the change in flux δJ that occurs over the length of the part (note that a promoter part could have multiple transcription start sites, x_{TSS}). To reduce the effect of fluctuations in the profile, an averaging window is applied immediately before and after the part boundaries (Fig 1B). The promoter strength is then given by

$$\delta J = \frac{\gamma}{n} \left[\sum_{i=x_1+1}^{x_1+n} M(i) - \sum_{i=x_0-1}^{x_0-n} M(i) \right], \tag{2}$$

where $n = 10$ is the window length (Fig 1B). The background RNAP flux originating upstream of the promoter is subtracted to ensure that only flux originating from the promoter is measured.

Terminators cause drops in the transcription profile at the 3'-end of the poly-A region (Fig 1B). The terminator strength T_S has been previously defined as the fold decrease in gene expression before and after the terminator (Chen et al, 2013). Based on the profile, it can be calculated as

$$T_s = \frac{\sum_{i=x_1+1}^{x_1+n} M(i)}{\sum_{i=x_0-1}^{x_0-n} M(i)}, \tag{3}$$

where x_0 and x_1 are the beginning and end positions of the terminator part. Following the approach for promoters described above, the activity of a terminator can also be calculated as a change in flux δJ as RNAPs either dissociate from the DNA or read-through.

Characterization of genetic devices from transcription profiles

Sensors and gates are examples of genetic devices, where a set of parts collectively performs a function. RNA-seq is particularly suitable for characterizing transcriptional devices, where the inputs and/or outputs are defined as RNAP fluxes. For example, the input to a sensor is a stimulus (e.g., inducer or environmental signal) and the output is the control of a promoter (turning RNAP flux on or off). For gates, the inputs and outputs are both promoters and the response function captures how the output changes as a function of the input at steady-state. Unlike genetic parts, whose function can be extracted from a single profile, characterizing a sensor or circuit requires sampling the device in different states, extracting the activities of the input/output promoters, and then fitting these data to a mathematical model of device performance.

The response of a sensor is given by the activity of the output promoter in the presence and absence of signal, δJ_{on} and δJ_{off}, respectively (Fig 1C). This can be calculated by performing RNA-seq

experiments under these conditions and then calculating the promoter activity according to equation 2. More states with intermediate levels of inducer are required to calculate the full dose-dependent response function. In this manuscript, the circuit is characterized in multiple states, a subset of which may have the sensor in the on or off state. In these cases, we simply average the promoter activities across those states where the sensor is on and those where the sensor is off and this is presented as the response.

A transcriptional gate has one or more input promoters and a single output promoter. The response function captures how the activity of the output promoter changes as a function of the input flux (the input promoters and upstream transcriptional read-through) at steady-state. For example, a NOT gate has one input promoter that drives the expression of a repressor that turns off an output promoter. The response function of this gate is

$$\delta J_{out} = \delta J_{out}^{min} + \left(\delta J_{out}^{max} - \delta J_{out}^{min} \right) \left(\frac{K^n}{K^n + J_{in}^n} \right) \tag{4}$$

where J_{in} is total input flux, δJ_{out}^{min} and δJ_{out}^{max} are the minimal and maximal output promoter activities, K is threshold, and n is the cooperativity. When there is no transcriptional read-through from upstream of the input promoters, then $J_{in} = \delta J_{in}$. NOR gates have a similar structure as a NOT gate, but include multiple input promoters. For two-input NOR gates, $J_{in} = \delta J_{in,1} + \delta J_{in,2} + J_0$, where the 1 and 2 subscripts indicate the activity of the two input promoters and J_0 is the read-through from upstream of these promoters.

RNA-seq experiments could be designed to characterize the response function of individual gates by taking samples where the inputs are varied, calculating the promoter activities from the profiles and then fitting them to a mathematical form of a response function. Here, we wanted to be able to quantify multiple gates within the context of a circuit. For example, when characterizing combinatorial logic, the sensors are induced in all combinations (e.g., a three-input logic gate has eight combinations of inputs). Under these different conditions, the magnitude of the input promoter activity to the gate varies because of changes to the remainder of the circuit. We utilize those changes to plot data points for J_{in} and δJ_{out} (Fig 1D) and this can be fitted to a response function (equation 4) to extract the parameters δJ_{out}^{min}, δJ_{out}^{max}, K, and n.

Characterization of a combinatorial logic circuit

We applied our characterization method to a three-input one-output combinatorial logic circuit (Fig 2A). Three sensors respond to IPTG, aTc, and arabinose, and their activities are processed by five layered NOR/NOT gates. The complete circuit consists of 46 genetic parts (Fig 2B), including promoters, genes, terminators, and ribozyme insulators (Lou et al, 2012). The output of the circuit is yellow fluorescent protein (YFP), which allows for the use of flow cytometry to measure the response in single cells. Cello was used to simulate circuit performance based on the sensor and gate functions measured in isolation (Materials and Methods) (Nielsen et al, 2016). This circuit was selected because overall it functioned as predicted in terms of producing the correct pattern of on and off outputs, but several of the responses (+/−/+, −/+/+, +/+/+) had wide distributions indicating that some of the cells were responding improperly (Fig 2C).

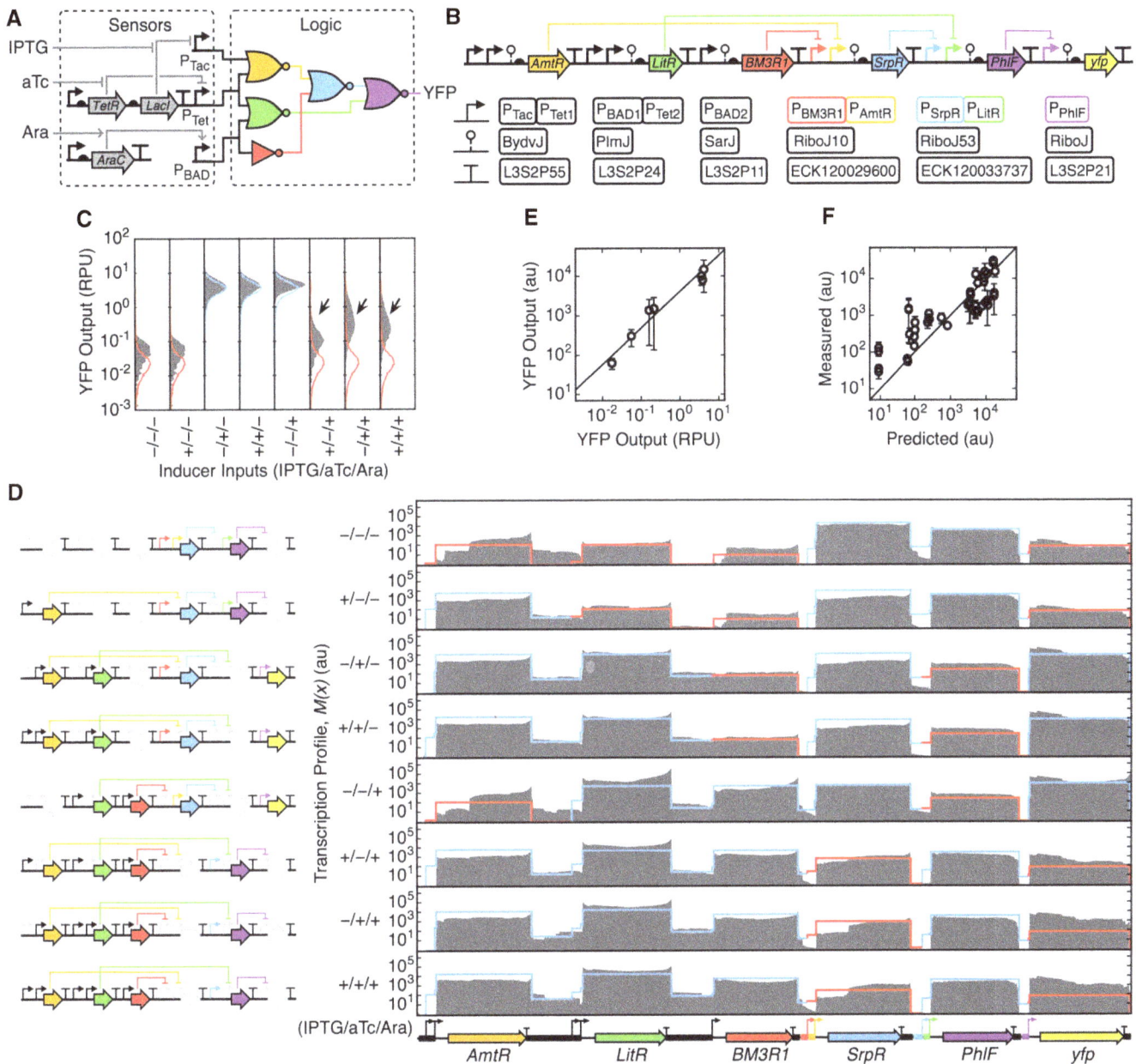

Figure 2. Characterization of a genetic circuit.

A The sensors and wiring diagram for the three-input combinatorial logic circuit are shown. Colors correspond to the repressors used for each gate.

B Genetic implementation of the logic circuit with annotated part names. Genetic parts are shown using SBOLv notation.

C Flow cytometry data of the YFP output for all inducer input combinations (filled gray distributions) and the predicted distributions from Cello (blue are on; red are off). Arrows highlight cells responding improperly.

D Transcription profiles for the circuit are shown for all combinations of inducers (0.5 mM IPTG/22 nM aTc/5 mM arabinose). The transcription profiles are calculated as the average of three biological replicates measured on different days. Predicted transcription profiles are shown by red and blue lines corresponding to when the gate should be off and on, respectively, as calculated using Cello (Materials and Methods). To the left of the profiles, the genes that are expressed in each state are highlighted.

E Determination of the conversion factor between RPU measured via cytometry and the expression of the *yfp* gene measured by RNA-seq. The black line shows the linear fit. The averages and standard deviations were calculated from three replicates measured on different days.

F Comparison of the expression of circuit genes predicted by Cello and measured experimentally from the transcription profile (Materials and Methods). Black line shows x = y. The averages and standard deviations were calculated from three replicates measured on different days.

The states of a combinatorial logic circuit are defined as the steady-state responses to all combinations of inputs. Cells containing the circuit were grown for 5 h in media supplemented with the eight combinations of inducers, and RNA samples were collected and processed (Materials and Methods). Sequencing of these samples generated between 2.0×10^5 and 1.8×10^7 mapped fragments (Dataset EV1). These were pre-processed to generate transcription profiles and normalized gene expression levels. The resultant transcription profiles displayed punctuated forms with expression of the transcriptional units for each gene clearly separated in most cases (Fig 2D). Two additional replicates were performed for all input states on different days and processed in the same way, and the resulting profiles are consistent (Appendix Fig S2).

Comparing simulation predictions to the measured transcription profiles is complicated by the fact that Cello reports relative promoter units (RPUs). To convert RPUs to arbitrary units that are compatible with the transcription profiles, the activity of the YFP output was measured in RPUs and compared to the average transcription profile across the *yfp* gene. These were linearly correlated (Fig 2E) with a conversion factor of 1 RPU = 2,895 au. Cello predictions for all promoter activities were converted using this factor. These were then used to trace a predicted profile along the length of the construct (Fig 2D; Materials and Methods). A correlation was found between the predicted and experimental transcription profiles for each gene (Fig 2F). Note that this is the first time we have been able to compare the levels of the repressors to that predicted; previously, the correlation could only be quantified for the output (YFP).

Next, every promoter and terminator in the circuit was characterized across each of the eight states. An idiosyncrasy in our gate designs required a modification to the approach to characterize promoters. Normally, the rise in the transcription profile would be observed just after the promoter part. However, we use ribozymes as part of our gate design to insulate against contextual effects caused by changing the upstream promoters (Lou *et al*, 2012). The ribozymes cleave the 5'-UTR sequences, releasing a small RNA that is filtered during sequencing preparation (Appendix Text S2; Appendix Fig S3). This causes the increase in the transcription profile

to occur at the cleavage site of the ribozyme (Fig 3A; Appendix Fig S4). To calculate the promoter activities when there is a ribozyme, equation 2 is changed so that the fluxes are calculated downstream of the cleavage site. A mathematical model of ribozyme efficiency was also constructed that can be used to quantify imperfect cleavage efficiencies (Appendix Text S2; Appendix Fig S4; Appendix Table S1).

The use of ribozymes also makes it impossible to resolve the individual activities of multiple promoters in series (because the cleaved 5'-RNA from transcripts from either promoter are lost during processing). Due to the use of multiple NOR gates, there are many examples of this in the circuit (e.g., $P_{Tac}-P_{Tet1}$). Therefore, we calculate the dual promoter as a single promoter part using equation 2. If sufficient data are generated across states, the contributions of the individual promoters to the total can be deduced (see next section).

The transcriptional profiles for all of the promoters in the circuit are shown in Fig 3A, and the activities are provided in Table 1. The activities of the promoters for all eight input states are shown, which includes cases where the promoters should be off and on. The values calculated from the profiles are compared to the activities measured in isolation using fluorescent proteins (Table 1). Terminator strengths were calculated from the drops in the profile at the end of transcripts (Fig 3B) and compared to strengths measured in isolation using fluorescent reporters (Table 1).

Characterization of devices internal to the circuit

The circuit contains eight genetic devices: three sensors, four NOR gates, and a NOT gate. Each was characterized in isolation by empirically measuring the response function using a fluorescent reporter (Nielsen *et al*, 2016). This information was then used by Cello to predict how to connect them to build the larger circuit.

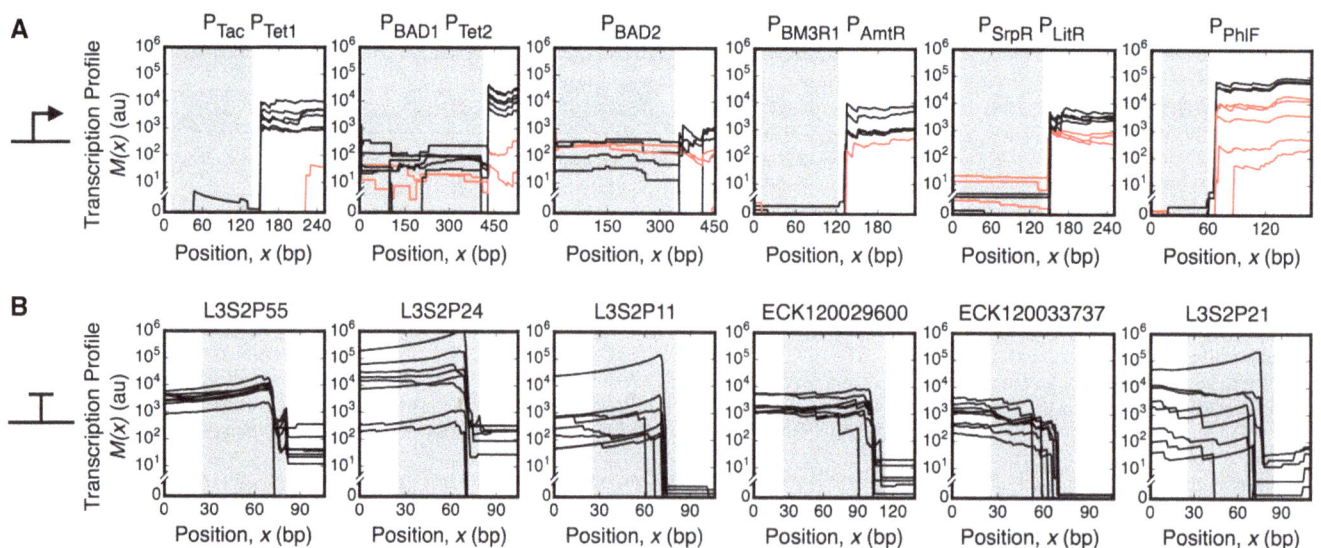

Figure 3. Quantifying part function.

A, B Alignment of transcription profiles for promoters (A) and terminators (B). Lines show the transcription profile for each of the eight combinations of inducers. Shaded regions denote the boundaries of the part. The lines are black for states where the promoter should be on and red when it should be off. The data are shown for the profiles derived from a single experiment (part parameterization from three replicates is provided in Table 1).

Here, we calculated the performance of the devices in the context of the circuit using RNA-seq and compare the values to their measurement in isolation using a fluorescent protein and cytometry.

The sensor response function is specified by the activity of the output promoter in response to the presence and absence of an inducer. When a sensor occurs alone (e.g., P_{BAD2}), we used the measured promoter strengths (Table 1) to calculate δJ_{off} and δJ_{on} directly. When the output promoters appear in pairs (e.g., $P_{Tac}-P_{Tet1}$ and $P_{BAD1}-P_{Tet2}$), the promoters cannot be individually resolved directly from the RNA-seq data. To separate the activities of the two promoters, the individual outputs of the first and second promoters are defined as $\delta J_{1,on}$ and $\delta J_{2,on}$, or $\delta J_{1,off}$ and $\delta J_{2,off}$, where on and off refers to the presence or absence of the inducer. Then, for each state, the measured combined activity of the promoter pair δJ_{1+2} is equated to the expected activities for different combinations of inducers. For example, when the inducers that activate both sensors are present, $\delta J_{1+2} = \delta J_{1,on} + \delta J_{2,on}$. This yields a set of algebraic equations to which $\delta J_{1,on}$, $\delta J_{1,off}$, $\delta J_{2,on}$, and $\delta J_{2,off}$ can be fitted (Materials and Methods).

Sensor responses in the context of the circuit can then be compared to the values measured for the sensor in isolation (Fig 4A; Table 2). Strong context effects were observed for P_{BAD} (Table 2). In one location in the circuit, it performs as expected (P_{BAD1}). However, P_{BAD2} produces a lower than expected on state, which manifests as a nearly flat transcription profile (Fig 3A). This coincided with read-through from the highly expressed *LitR* gene upstream (Fig 2D).

The internal gates were characterized by calculating the full response functions from the transcription profiles for all eight states. NOR gates are characterized as NOT gates by having the multiple upstream input promoters serve as the combined input to the gate. The promoter activities were calculated from the profiles and fit to equation 4. The total flux into the gate serves as the input J_{in} (either one or two promoters and upstream read-through), and the output is the activity of the output promoter δJ_{out}. It is simple to calculate when this appears as a single promoter. If it is part of two promoters in series, the individual promoter activities are calculated as described for sensors, above (Materials and Methods).

The calculated J_{in} and δJ_{out} values for the eight states are fit to equation 4 to obtain the response functions for each gate in the circuit (Fig 4B). The response functions (solid lines, Fig 4B) and parameter values (Table 2) are compared to those obtained from the gate measurements performed in isolation using fluorescent reporters (dashed lines, Fig 4B). The performance of the five gates is similar with several notable exceptions. The LitR gate displayed lower output flux for both on and off states, while the PhlF gate saw elevated output flux and a > 2-fold shift in the input flux required to switch the gate into an off output state.

Part substitution to correct antisense transcription

All genes in the circuit are organized on the sense strand (Fig 2B), but the RNA-seq data also report transcription in the antisense direction (Fig 5A; Appendix Fig S5). There are several mechanisms by which antisense transcription can interfere with the function of the circuit (Shearwin et al, 2005; Brantl, 2007; Brophy & Voigt, 2015). Most reads corresponding to antisense transcription cluster within the AmtR, LitR, and BM3R1 genes, implying the existence of reverse promoters internal to these genes. The P_{BAD} promoters we used have known antisense transcription start sites (Schleif, 2003), from which antisense transcription could be observed (Fig 5A).

A part substitution was made to correct for the observed antisense transcription. In the original circuit design, the terminators only stop RNAP coming from the sense direction. One mechanism to stop antisense termination is to use a bidirectional terminator (Chen et al, 2013) or to follow a terminator with a terminator oriented in the opposite direction. To this end, we replaced the terminator after the LitR gene with two terminators, each of which blocks transcription in opposite directions. This completely blocked antisense transcription (Fig 5B), demonstrating the use of RNA-seq to rationally correct an observed error in a circuit.

Impact of circuit state on host gene expression

Different combinations of inputs will cause different genes in the circuit to be expressed (Fig 2D), and this can change the burden on the host cell (Ceroni et al, 2015). For the three-input logic circuit, an increase in cell doubling times is observed for input states where four genes (including repressors and *yfp*) were expressed (gray bars, Fig 6A). For the input states expressing only two and three repressors ($-/-/-$ and $+/-/-$), the growth rates were similar. Input states with the slowest growth rates ($+/-/+, -/+/+, +/+/+$) also corresponded to those with the broadest flow cytometry distributions, deviating from the Cello predictions (Fig 2C).

Table 1. Promoter and terminator part characterization.

| | Genetic context | | | |
| | Isolation (Cytometry)[b] | | Circuit (RNA-seq)[c] | |
Promoters[a]	1st	2nd	1st	2nd
P_{Tac}-P_{Tet1}	54	85	8 ± 2	17 ± 5
P_{BAD1}-P_{Tet2}	49	85	242 ± 118	283 ± 237
P_{BAD2}	49		26 ± 29	
P_{BM3R1}-P_{AmtR}	10	75	9 ± 4	55 ± 32
P_{SrpR}-P_{LitR}	41	85	37 ± 20	14 ± 4
P_{PhlF}	80		323 ± 25	
Terminators				
L3S2P55	418		24 ± 6	
L3S2P24	212		295 ± 176	
L3S2P11	384		110 ± 89	
ECK120029600	374		380 ± 98	
ECK120033737	391		870 ± 344	
L3S2P21	505		187 ± 127	

[a]The strength of the left promoter is 1st; right is 2nd.
[b]Previously reported promoter strengths (in au/s) based on a fluorescent reporter (Nielsen et al, 2016), converted to au/s as described in text. Previously reported termination strengths (Chen et al, 2013).
[c]Average and standard deviations are calculated from three replicates performed on different days. For promoters, all states where the promoter is predicted to be on are included and the units are au/s. For double promoters, separate strengths for each promoter are calculated as described in the text. Median terminator strengths are calculated for states where the upstream gene is on. For terminators L3S2P24 and L3S2P11 in one replicate, the data for input state $-/-/+$ were excluded due to a mapping bias (Appendix Fig S1).

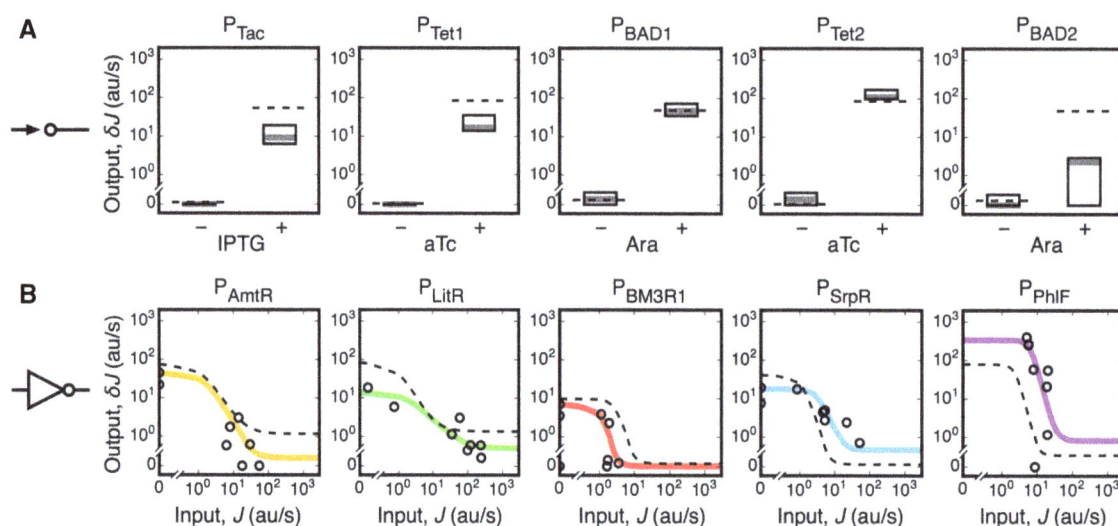

Figure 4. Extraction of sensor and gate response functions from the transcription profiles.

A The responses of the output promoters of the sensors are shown in the presence and absence of each inducer. The dashed lines show the sensor outputs measured in isolation (Nielsen *et al*, 2016). The boxes show the median (gray line) and range of promoter activities measured for the four states where it is off (δJ_{off}) and four where it is on (δJ_{on}).

B Solid colored lines show the response functions of the gates obtained by fitting the promoter activities to the RNA-seq data (circles denote the measured values for the eight input states). The dashed lines show the output of the gate measured in isolation (Nielsen *et al*, 2016). The data are shown for the profiles derived from a single experiment (device parameterization from three replicates is provided in Table 2).

Table 2. Sensor and gate response function parameters.

	Genetic context						
	Isolation (Cytometry)[a]				**Circuit (RNA-seq)**[b]		
Sensor	δJ_{off}	δJ_{on}			δJ_{off}	δJ_{on}	
P_{Tac}	0.1	54			0.0 ± 0.0	9 ± 0.9	
P_{Tet1}	0.0	85			0.0 ± 0.0	16.0 ± 4.8	
P_{Tet2}	0.0	85			0.8 ± 0.6	274 ± 165	
P_{BAD1}	0.2	49			0.8 ± 0.6	165 ± 114	
P_{BAD2}	0.2	49			0.0 ± 0.0	13.0 ± 9.2	
Gate	δJ_{out}^{min}	δJ_{out}^{max}	K	n	δJ_{out}^{min}	δJ_{out}^{max}	K	n
P_{AmtR}	1.2	75	1.4	1.6	0.3 ± 0.1	80 ± 48	1.0 ± 0.7	1.7 ± 0.2
P_{LitR}	1.4	85	1.0	1.7	1.9 ± 1.2	53 ± 36	2.7 ± 0.5	1.4 ± 0.2
P_{BM3R1}	0.1	10	2.9	2.9	0.0 ± 0.0	12 ± 8	1.6 ± 0.6	2.9 ± 0.2
P_{SrpR}	0.1	41	1.2	2.8	0.3 ± 0.2	16 ± 2	3.0 ± 1.6	2.5 ± 0.2
P_{PhlF}	0.4	80	2.5	3.9	0.7 ± 0.1	337 ± 2	8.9 ± 1.9	3.6 ± 0.2

[a]Previously reported values (in au/s) based on a fluorescent reporter (Nielsen *et al*, 2016), converted to au/s as described in the text. The units of δJ_{off}, δJ_{on}, δJ_{out}^{min}, δJ_{out}^{max}, and K are au/s.
[b]The units of δJ_{off}, δJ_{on}, δJ_{out}^{min}, δJ_{out}^{max}, and K are au/s. Average and standard deviations are calculated from three replicates measured on different days.

The impact on the host of carrying a circuit can be observed as changes in the expression of native genes. To generate a baseline for comparison, we preformed duplicate RNA-seq experiments using cells harboring the circuit backbone (pAN1201), but without the remainder of the circuit. These data were used for differential gene analysis to search for potential differences in host gene expression. Expression of endogenous genes in cells containing the circuit for input states expressing only two or three genes ($-/-/-$ and $+/-/-$) was highly correlated with the baseline ($R^2 = 0.83$; Fig 6B). For input states expressing four genes, we found a lower correlation to the baseline ($R^2 = 0.55$), and 125 significantly differentially expressed genes ($P < 0.01$; Dataset EV2). Most of these were downregulated ($N = 106$) with enrichment for functions broadly related to energy generation, anaerobic respiration, and

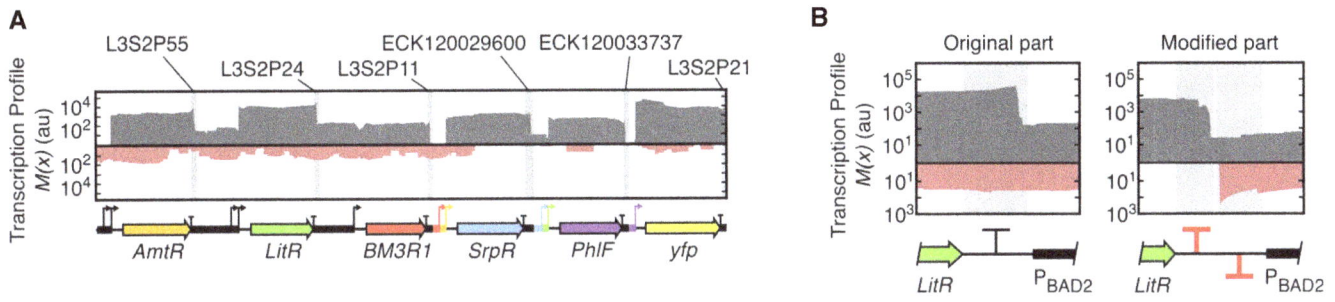

Figure 5. Antisense transcription, measurement, and correction.

A   Transcription profiles for both sense (gray) and antisense (red) strands for circuit grown in culture tube conditions for the −/+/− combination of inducers (22 nM aTc). Terminators are shown by light gray-labeled regions. The antisense profiles for all combinations of inducers are shown in Appendix Fig S5. The profiles correspond to a single experiment that is representative of three replicates.

B   The change in the transcription profile that occurs due the addition of a reverse terminator is shown. Shaded regions denote terminator part boundaries. The original terminator is L3S2P24, and this is replaced by the terminators ECK120033736 (forward) and ECK120010818 (reverse). The profile normalization (Box 1) has been applied to the 3′-end of the antisense profile in the modified part.

fermentation (Appendix Table S2). Of those genes upregulated, there was significant enrichment for DNA replication and repair, iron assimilation and homeostasis, and functions linked to colanic acid biosynthesis, the production of which occurs in response to low temperature, osmotic shock, and desiccation (Navasa et al, 2009).

Environmental robustness

Growth conditions impact host physiology (Kram & Finkel, 2014), and this can influence the performance of genetic parts and devices (Moser et al, 2012; Gorochowski et al, 2014). The experiments described so far were performed in 14-ml culture tubes. However, we observed serendipitously that the circuit failed when cells were grown in the same media in Erlenmeyer flasks (Materials and Methods). Under these conditions, three of the input combinations (+/−/+, −/+/+ and +/+/+) resulted in much broader flow cytometry distributions of the YFP output with an incorrect on output state for most cells (Fig 6C). This coincided with increases in the doubling times when arabinose was present (Fig 6A).

Transcription profiles were compared between cells grown in Erlenmeyer flasks and culture tubes (Fig 6D). For the working states, there are only minor differences, but for the three broken states, there are major differences. Using these data, we re-quantified part and device performance (Appendix Figs S6 and S7; Appendix Tables S3 and S4). Few changes were observed for the gates and the IPTG (P_{Tac}) and aTc (P_{Tet}) sensors, with similar response functions across conditions (Appendix Fig S7). However, large changes in performance were found for both the arabinose sensors that displayed ~2-fold increases in their induced activities. This change propagates through the circuit and impacts the levels of the repressors, culminating in a > 2-fold decrease in PhlF, which increases YFP expression (Fig 6D), thus causing the average of the cytometry populations to appear to be on for these states (Fig 6C). The level of antisense transcription is also higher from the reverse promoter within P_{BAD} under these conditions (Appendix Fig S5).

We analyzed the host transcriptome under both conditions to ascertain whether a shift in cellular physiology might be the cause for the failure of the arabinose sensors (P_{BAD1} and P_{BAD2}).

Differential expression analysis of the broken input states highlighted significant changes for 179 genes ($P < 0.01$), with enrichment of transport-related functions for arabinose, xylose, and maltose ($P < 5.69 \times 10^{-3}$; Fig 6E; Dataset EV2). Upregulation of these genes coincided with the presence of arabinose, suggesting its role in their regulation (Fig 6E). Notably, host genes involved in arabinose transport (araEFGH) saw significant eight- to 56-fold upregulation in culture tubes for the three broken input states ($P < 5.75 \times 10^{-4}$; Dataset EV2). Such a difference would facilitate greater intracellular accumulation of arabinose due to an increased uptake and is consistent with the large measured increases in activity of both P_{BAD} promoters when induced (Khlebnikov et al, 2001).

Discussion

DNA synthesis and assembly methods enable the construction of large genetic circuits that can implement complex functions through the layering of simple transcriptional gates. Gates can be built based on many classes of biochemistry (e.g., DNA-binding proteins (Moon et al, 2012; Stanton et al, 2014), recombinases (Bonnet et al, 2013; Siuti et al, 2013; Fernandez-Rodriguez et al, 2015), and CRISPRi (Gilbert et al, 2013; Larson et al, 2013; Qi et al, 2013; Kiani et al, 2014; Nielsen & Voigt, 2014; Gander et al, 2017)). Orthogonal libraries of these parts enable many to be reliably used in a single cell without fear of interference. These advances have led to circuits that can consist of 10+ regulators and > 40 genetic parts and the size is growing. However, the ability to debug systems of this size has lagged, particularly when the function is defined by many states. Here, we have developed methodologies to characterize the inner workings of a circuit using RNA-seq data.

Circuit characterization has been limited by the use of fluorescent reporters to provide a single datum for the output of the circuit as a whole. One way to measure the response of internal parts and gates is to separate them onto characterization plasmids that can be assayed in isolation of the remainder of the circuit (Kelly et al, 2009; Stanton et al, 2014; Nielsen et al, 2016). In contrast, RNA-seq enables the function of multiple parts to be simultaneously measured in situ within a circuit. Using this approach, we have

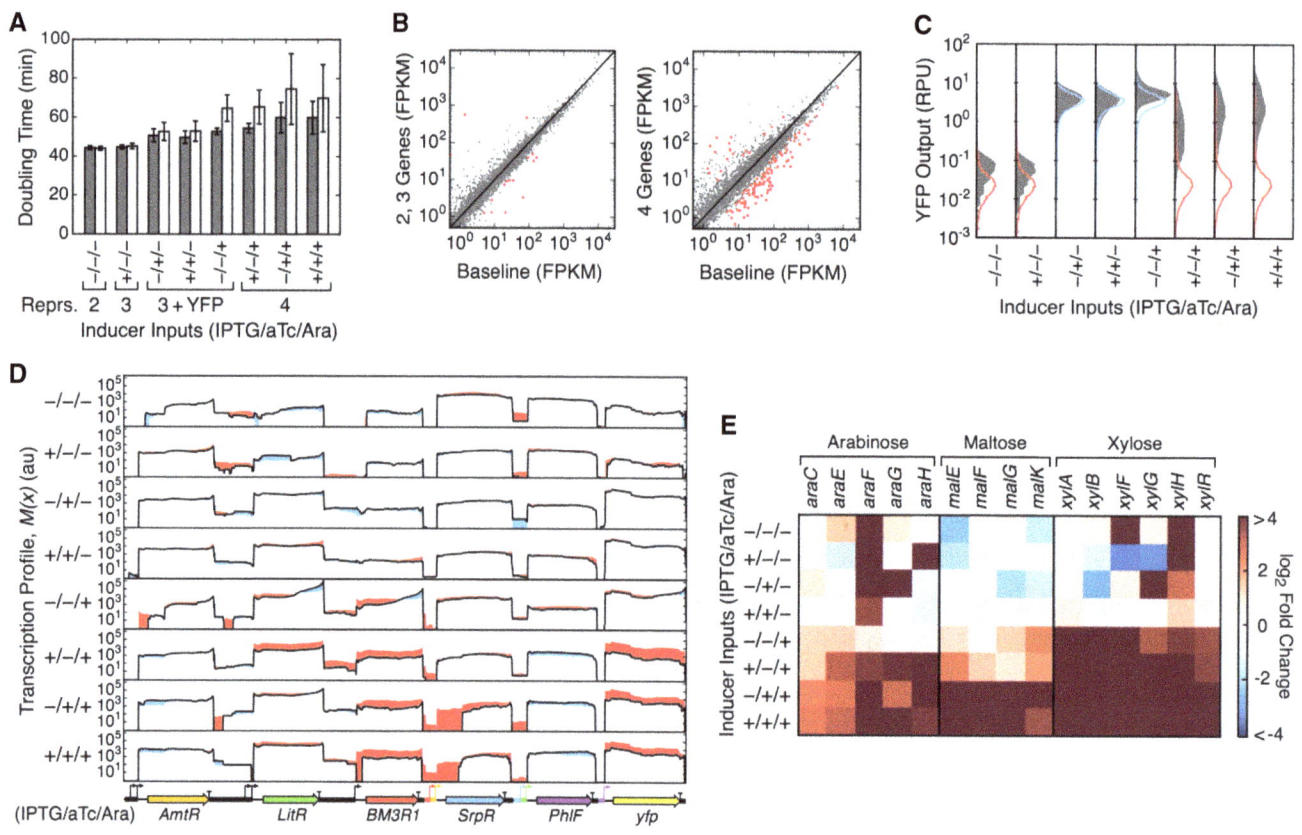

Figure 6. Changes to the growth conditions break the circuit by changing host gene expression.

A   Doubling times for cells carrying the circuit grown in culture tubes (gray) and Erlenmeyer flasks (white). Error bars are calculated as the standard deviation from three biological replicates performed on different days.

B   Comparison of host gene expression under culture tube conditions for sets of input states with differing numbers of expressed circuit genes (including *yfp*): two = −/−/−, three = +/−/−, and four = −/+/−, +/+/−, −/−/+, +/−/+, −/+/+ and +/+/+ (0.5 mM IPTG/22 nM aTc/5 mM arabinose). The baseline is calculated from RNA-seq data collected from cells harboring an empty circuit plasmid backbone (pAN1201). Points show the mean expression level of each gene. Red points denote genes with statistically significantly differential expression in comparison with the baseline ($P < 0.01$; Dataset EV2).

C   Flow cytometry data of the YFP output for the circuit grown in Erlenmeyer flasks (filled gray distributions) and the predicted output distributions from Cello (blue is on and red is off).

D   Comparison of transcription profiles for the circuit when cells are grown in culture tubes (black line; data for biological replicate 1) and Erlenmeyer flasks. If the profile obtained from Erlenmeyer flasks is higher the difference is shown in red and if it is lower the difference is shown in blue.

E   Fold-change analysis of differentially expressed sugar transport-related genes ($P < 0.003$) compared between Erlenmeyer flask and culture tube growth conditions (Dataset EV2).

revealed several failure modes and showed that some parts functioned unreliably. For example, the P_{BAD} promoter was particularly context dependent and sensitive to shifts in culturing conditions.

One limitation of RNA-seq is that it only measures mRNA levels. This is useful when characterizing transcriptional circuits. However, some parts such as ribosome binding sites (RBSs) and bi-cistronic insulators (Mutalik *et al*, 2013) operate at the level of translation. It has been shown that ribosome profiling (Ingolia *et al*, 2009; Ingolia, 2014) can accurately estimate protein synthesis rates (Li *et al*, 2014) and therefore could be used to characterize translational parts such as ribosome binding site strengths, if they are the rate-limiting step during protein synthesis (Li *et al*, 2014). This offers a powerful complementary technique that could, along with RNA-seq, fully characterize all of the parts in a circuit. Another limitation is that the RNA-seq method that we use is based on population measurements. Expression can differ between cells, and it is useful to have

the population data provided by cytometry or microscopy (Rosenfeld *et al*, 2005). RNA-seq can be used to measure expression in single cells, but current techniques could not be used to extract part performance data (Tang *et al*, 2009; Shapiro *et al*, 2013; Grün *et al*, 2014; Lasken & McLean, 2014).

Much of the mystery of genetic engineering comes from a lack of being able to see what you are doing; that is, how design choices impact the system and cell. Problems that can be seen are simple to correct. For example, once unwanted antisense transcription was detected, it could be blocked easily by adding a reverse terminator. If a problem cannot be seen, then this necessitates the creation of a large library of designs where many potential fixes are tried randomly until one blindly solves the problem (Smanski *et al*, 2014). New techniques to quantify mRNA, protein, and metabolites —driven by plummeting DNA sequencing costs—are giving a fuller picture of the cell. Furthermore, the routine sequencing of laboratory strains can show how genomic mutations that emerge can impact the system (Lamb *et al*, 2006; Fernandez-Rodriguez *et al*, 2015; Song *et al*, 2017). Collectively, these are quickly clearing the fog and hand-waving underlying cellular design. However, this is also leading to a deluge of data that are increasingly perplexing to the designer. Fully utilizing these datasets will require new software that simplifies the process of collection, processing, merging data from diverse techniques, and learning. This will help reveal the design principles that allow for robust part function and support the effective development of increasingly complex genetic systems.

Materials and Methods

Strain, media, and inducers

The *Escherichia coli* DH10B derivative NEB 10-beta Δ(ara-leu) 7697 araD139 fhuA ΔlacX74 galK16 galE15 e14-ϕ80dlacZΔM15 recA1 relA1 endA1 nupG rpsL (StrR) rph spoT1 Δ(mrr-hsdRMS-mcrBC) was used for cloning and measurements (New England Biolabs, MA, C3019). Cells were grown in LB Miller broth (Difco, MI, 90003-350) for harvesting plasmids. Cells were grown in MOPS EZ Rich (Teknova, M2105) defined medium with 0.2% glycerol for circuit performance measurements. To select for the presence of plasmids, 50 µg/ml kanamycin (Gold Biotechnology, MO, K-120-5) and 50 µg/ml spectinomycin (Gold Biotechnology, MO, S-140-5) antibiotics were used.

Circuit induction

Chemicals used to induce input promoters were isopropyl β-D-1-thiogalactopyranoside (IPTG; Sigma-Aldrich, MO, I6758), anhydrotetracycline hydrochloride (aTc; Sigma-Aldrich, MO, 37919), and L-arabinose (Sigma-Aldrich, MO, A3256). Individual colonies were inoculated into MOPS EZ Rich Defined Medium (Teknova, CA, M2105) with 0.2% glycerol carbon source and 50 µg/ml kanamycin (Gold Biotechnology, MO, K-120-5) and grown overnight for 16 h at 37°C and 1,000 rpm in V-bottom 96-well plates (Nunc, Roskilde, Denmark, 249952) in an ELMI Digital Thermos Microplates shaker incubator (Elmi Ltd, Riga, Latvia). The following day, cultures were diluted 178-fold (two serial dilutions of 15 µl into 185 µl) into EZ Rich glycerol with kanamycin and grown under the same ELMI shaker incubator conditions for 3 h. For culture tube assays (Falcon 14 ml round-bottom polypropylene tubes; Corning, MA, 352059), cells were diluted 658-fold (4.56 µl into 3 ml) into EZ Rich glycerol with kanamycin and inducers. For Erlenmeyer flask assays (Pyrex 250 ml; Cole-Palmer, IL, 4980-250), cells were diluted 658-fold (76 µl into 50 ml) into EZ Rich glycerol with kanamycin and inducers. Eight inducer combinations were used that cover the presence or absence of 0.5 mM IPTG, 10 ng/ml aTc, and 5 mM L-arabinose. Culture tubes and Erlenmeyer flasks were then grown in an Innova 44 shaker (Eppendorf, CT) at 37°C and 250 rpm for 5 h. Finally, 40 µl of cell culture was placed into 160 µl of phosphate-buffered saline (PBS) containing 2 mg/ml kanamycin to arrest translation and cell growth. These growth-arrested cells were incubated in the PBS with kanamycin for one hour before fluorescence was measured using flow cytometry.

Flow cytometry analysis

Fluorescence of individual cells was measured using an LSRII Fortessa flow cytometer (BD Biosciences, San Jose, CA) controlled by the BD FACSDiva software. More than 20,000 gated events were collected for each state, and analysis of flow cytometry data was performed using FlowJo (TreeStar, Inc., Ashland, OR).

RNA-seq library preparation and sequencing

Total RNA was harvested from *E. coli* DH10B cells harboring the genetic circuit plasmid and cultured under the circuit induction assay condition described above. Cultures were spun down at 4°C, 15,000 *g* for 3 min. Supernatants were discarded after centrifugation, and cell pellets were flash-frozen in liquid nitrogen for storage at −80°C. Cells were lysed with 1 mg of lysozyme (Sigma-Aldrich, MO, L6871) in 10 mM Tris–HCl (pH 8.0) (USB 75825) supplemented with 0.1 mM EDTA (USB 15694). RNA was extracted with PureLink RNA Mini Kit (Life Technologies, CA, 12183020) and further purified and concentrated with RNA Clean & Concentrator-5 (Zymo Research, R1015) to ensure sufficient RNA concentrations (> 280 ng per sample). The purified RNA samples were analyzed using a Bioanalyzer (Agilent, CA), and Ribo-Zero rRNA Removal Kit for bacteria (Illumina, CA, MRZMB126) was used to deplete rRNA from the samples. We also checked the quality of the RNA extracted by calculating the RNA integrity number (RIN), which ranges from a value of 10 if all RNA is intact to 1 if the RNA is totally degraded (Schroeder *et al*, 2006). We only consider highly intact samples with a RIN > 8.5 (Imbeaud *et al*, 2005). Strand-specific RNAtag-seq (Shishkin *et al*, 2015) libraries were created by the Broad Technology Labs specialized service facility (SSF). A total of 16 samples (one for each of the eight inducer combinations under the two different culturing conditions) were pooled, split, and run on two separate lanes of an Illumina HiSeq 2500 as technical replicates. Both lanes were checked for quality and re-pooled before reads were demultiplexed into the original samples. Barcode sequences were trimmed from reads before further analysis was performed.

Processing of sequencing data

Raw reads were mapped to the host genome (NCBI RefSeq: NC_01473.1) and circuit reference sequences using BWA (Li & Durbin, 2009) version 0.7.4 with default settings. The "view" command of the SAMtools (Li *et al*, 2009) suite was then used with default settings to convert the generated SAM files into a BAM format for downstream analyses. Read counts for each host and circuit gene were carried out using the "htseq-count" command of the HTSeq toolkit (Anders *et al*, 2015) with user-defined GFF annotations of the reference sequences and the options "-s reverse -a 10 -m union". Transcription profiles were generated by first splitting each BAM file into two separate BAM files that contained reads from either the sense or antisense strands. This was achieved by filtering the complete BAM file using the "view" command of SAMtools and the filter codes 83 and 163 for sense reads, and 99 and 147 for antisense reads. Normalized FPKM values were generated from the raw gene counts by custom scripts that calculated and applied a trimmed mean of M-values (TMM) factor using edgeR (Robinson *et al*, 2010) version 3.8.6. The BAM files were also separately processed by custom Python scripts to extract the position of the mapped reads and generate the transcription profiles. Gene expression in arbitrary units (au) that are compatible with the transcription profiles was calculated as the average of the transcription profile height along the length of a gene. Differential gene expression was performed using edgeR to calculate adjusted P-values using the built-in false discovery rate (FDR) correction. Characterization of promoters, ribozymes, and terminators was performed using custom Python scripts that took a GFF reference of the circuit defining all part locations, their types, and any further information (e.g., predicted cutsite for ribozymes). To ensure termination strength and ribozyme cleavage were not underreported, measurements were filtered if there was a low level of RNAP flux ($<$ 1 au/s) entering the terminator or leaving the ribozyme. Characterization of genetic parts was performed using a window size of $n = 10$ bp in equations 2 and 3, and S1. All scripts were executed using either Python version 2.7.9 or R version 3.2.1.

Genetic circuit design and simulations

Cello (Nielsen *et al*, 2016) version 1.0 (http://www.cellocad.org) was used for all circuit simulations. For inputs, "low RPU" values of $P_{Tac} = 0.0034$, $P_{Tet} = 0.0013$, and $P_{BAD} = 0.0082$, and "high RPU" values of $P_{Tac} = 2.8$, $P_{Tet} = 4.4$, and $P_{BAD} = 2.5$ were used and the Eco1C1G1T1.UCF.json UCF file describing the gate response functions. Genetic circuit visualizations in SBOLv format (Myers *et al*, 2017) were produced using DNAplotlib (Der *et al*, 2017) version 1.0.

Numerical fitting

To fit the response function of a gate when the output promoter was in isolation, we defined a least squares error function $E_1 = \Sigma_{i \in S}[\log(\delta J_i) - \log(\delta J_{out,i})]^2$. Here, S is the set of states (e.g., combinations of inducer), δJ_i is the measured activity of the output promoter for state i, and $\delta J_{out,i}$ is the expected output promoter activity calculated according to the gate's response function. For NOT gates, the response function is given by equation 4 with parameters δJ_{out}^{min}, δJ_{out}^{max}, K, and n. For each state, to calculate the expected output promoter activity of the gate, the total input RNAP flux J_{in} was extracted from the transcription profile and used as input to equation 4 (Fig 1D). The "minimize" function of SciPy version 0.15.0 and the built-in sequential least squares programming algorithm were then used to fit the parameters such that the error function was minimized.

Fitting of promoters in series and estimating their individual activities

When the output promoters of two sensors or gates were found in series, we characterized their individual behaviors by employing an adapted error function $E_2 = \Sigma_{i \in S}[\log(\delta J_{1+2,i}) - \log(\delta J_{1,i} + \delta J_{2,i})]^2$. Here, $\delta J_{1+2,i}$ is the measured combined activity of both output promoters for state i, and $\delta J_{1,i}$ and $\delta J_{2,i}$ are the expected individual output promoter activities calculated using the response functions of the first and second sensor/gate, respectively. For sensors, the response function was equated to either δJ_{on} or δJ_{off} depending on the inducers that were present for that state. For example, given all

combinations of inducers for two sensors in series: $-/-$, $-/+$, $+/-$, and $+/+$, then the expected combined output activity $\delta J_1 + \delta J_2$ would be given by: $\delta J_{1,off} + \delta J_{2,off}$, $\delta J_{1,off} + \delta J_{2,on}$, $\delta J_{1,on} + \delta J_{2,off}$, and $\delta J_{1,on} + \delta J_{2,on}$, respectively. For gates, the combined activity of two output promoters in series $\delta J_1 + \delta J_2$ was calculated as $\delta J_{1,out} + \delta J_{2,out}$, where the output promoter activity of each gate ($\delta J_{1,out}$ and $\delta J_{2,out}$) was given by equation 4 with parameters δJ_{out}^{min}, δJ_{out}^{max}, K, and n. For each state, to calculate the expected output promoter activity of a gate, the total input RNAP flux J_{in} was extracted from the transcription profile and used as input to equation 4 (Fig 1D). Parameters for sensors (δJ_{on} and δJ_{off}) and gates (δJ_{out}^{min}, δJ_{out}^{max}, K and n) were then fitted to minimize the adapted error function. To extract the individual contributions of each promoter to the combined measured activity of the output promoters $\delta J_{1+2,i}$ (Fig 4B and Appendix Fig S7B), we calculated the fractional contribution and split the measured combined activity according to this. Specifically, the estimated measured activity of output promoter 1 for state i was calculated as $\delta J_{1+2,i} \times [\delta J_{1,i}/(\delta J_{1,i} + \delta J_{2,i})]$, and for promoter 2 as $\delta J_{1+2,i} \times [\delta J_{2,i}/(\delta J_{1,i} + \delta J_{2,i})]$.

Calculation of the predicted transcription profiles

Predicted promoter activities in RPUs were calculated by Cello and converted to arbitrary units (au) compatible with the experimentally measured transcription profiles by using the conversion factor 1 RPU = 2,895 au. The circuit DNA is scanned from the 5'- to 3'-end to trace out the predicted profile. Upon reaching a promoter, the height of the profile is raised at the transcription start site by the predicted promoter strength (from Cello in au) multiplied by $(1-p_c)$, where p_c is the cleavage efficiency (Nielsen *et al*, 2016) of the next downstream ribozyme. Upon reaching a ribozyme, the profile height it set to the combined strength of upstream promoters directly driving expression (from Cello in au) plus any read-through from the nearest upstream terminator. This height is traced along the circuit until the end of the next downstream terminator is reached. Then, to capture RNAP read-through, the profile height is multiplied by $1/T_s$, where T_s is the terminator strength from Chen *et al* (2013).

Measurement of doubling times

Individual colonies were inoculated into MOPS EZ Rich Defined Medium (Teknova, CA, M2105) with 0.2% glycerol carbon source and 50 µg/ml kanamycin (Gold Biotechnology, MO, K-120-5) and grown overnight for 16 h at 37°C and 1,000 rpm in V-bottom 96-well plates (Nunc, Roskilde, Denmark, 249952) in an ELMI Digital Thermos Microplates shaker incubator (Elmi Ltd, Riga, Latvia). The following day, cultures were diluted 178-fold (16.9 µl into 3 ml) into EZ Rich glycerol with kanamycin and grown in an Innova 44 shaker (Eppendorf, CT) at 37°C and 250 rpm for 3 h. Next, 1 ml of culture was added to a VWR disposable cuvette (VWR, PA, 97000-586) and the optical density at 600 nm (OD_{600}) was measured using a Cary 50 Bio spectrophotometer (Agilent, CA, 10068900). For culture tube assays (Falcon 14-ml round-bottom polypropylene tubes; Corning, MA, 352059), cells were diluted 658-fold (4.56 µl into 3 ml) into EZ Rich glycerol with kanamycin and inducers. For Erlenmeyer flask assays (Pyrex 250 ml; Cole-Palmer, IL, 4980-250), cells were diluted 658-fold (76 µl into 50 ml) into EZ Rich glycerol with kanamycin and inducers. To calculate the initial OD_{600} post-dilution, the aforementioned OD_{600} value was divided by 658. Culture tubes and Erlenmeyer flasks were then grown in an Innova 44 shaker at 37°C and 250 rpm for five hours. The final OD_{600} for each vessel/inducer condition was measured for 1 ml of culture in the manner described above. The doubling time was calculated as $(5 \text{ h})/\log_2(OD_{Final}/OD_{Initial})$.

Characterization of the modified circuit

New genetic parts were synthesized as gBlock gene fragments (Integrated DNA Technologies, CA) and assembled into full genetic circuit using Gibson assembly (New England Biolabs, MA, 2611L). The modified circuit was sequence verified and transformed into *E. coli* DH10-beta (New England Biolabs, MA, C3019). Cultures were grown in culture tubes under different induction conditions, and RNA-seq samples were prepared by following the protocols described above.

Acknowledgements

This work was supported by US Defense Advanced Research Projects Agency (DARPA) Living Foundries awards HR0011-13-1-0001, HR0011-12-C-0067 and HR0011-15-C-0084 (C.A.V., T.E.G. and D.B.G.); US Department of Commerce – NIST (National Institute of Standards and Technology) award 70NANB16H164 (C.A.V.); Office of Naval Research, Multidisciplinary University Research Initiative grant N00014-13-1-0074 (C.A.V. and A.A.K.N.); Samsung Scholarship (Y.P.).

Author contributions

CAV, TEG, YP, and AAKN conceived of the study. CAV, TEG, AEB, and YP wrote the manuscript. TEG, AEB, BSD, and DBG developed algorithms and performed simulations. AAKN, YP, and JZ carried out the experiments.

References

Anders S, Pyl PT, Huber W (2015) HTSeq – a Python framework to work with high-throughput sequencing data. *Bioinformatics* 31: 166–169

Anderson JC, Clarke EJ, Arkin AP, Voigt CA (2006) Environmentally controlled invasion of cancer cells by engineered bacteria. *J Mol Biol* 355: 619–627

Basu S, Mehreja R, Thiberge S, Chen M-T, Weiss R (2004) Spatiotemporal control of gene expression with pulse-generating networks. *Proc Natl Acad Sci USA* 101: 6355–6360

Bonnet J, Yin P, Ortiz ME, Subsoontorn P, Endy D (2013) Amplifying genetic logic gates. *Science* 340: 599–603

Brantl S (2007) Regulatory mechanisms employed by *cis*-encoded antisense RNAs. *Curr Opin Microbiol* 10: 102–109

Brophy JA, Voigt CA (2014) Principles of genetic circuit design. *Nat Methods* 11: 508–520

Brophy JAN, Voigt CA (2015) Antisense transcription as a tool to tune gene expression. *Mol Syst Biol* 12: 854

Canton B, Labno A, Endy D (2008) Refinement and standardization of synthetic biological parts and devices. *Nat Biotechnol* 26: 787–793

Cardinale S, Joachimiak MP, Arkin AP (2013) Effects of genetic variation on the *E. coli* host-circuit interface. *Cell Rep* 4: 231–237

Ceroni F, Algar R, Stan GB, Ellis T (2015) Quantifying cellular capacity identifies gene expression designs with reduced burden. *Nat Methods* 12: 415–418

Chen YJ, Liu P, Nielsen AAK, Brophy JA, Clancy K, Peterson T, Voigt CA (2013) Characterization of 582 natural and synthetic terminators and quantification of their design constraints. *Nat Methods* 10: 659–664

Chen H, Shiroguchi K, Ge H, Xie XS (2015) Genome-wide study of mRNA degradation and transcript elongation in *Escherichia coli*. *Mol Syst Biol* 11: 781

Conway T, Creecy JP, Maddox SM, Grissom JE, Conkle TL, Shadid TM, Teramoto J, San Miguel P, Shimada T, Ishihama A, Mori H, Wanner BL (2014) Unprecedented high-resolution view of bacterial operon architecture revealed by RNA sequencing. *MBio* 5: e01442-14

Daniel R, Rubens JR, Sarpeshkar R, Lu TK (2013) Synthetic analog computation in living cells. *Nature* 497: 619–623

Der BS, Glassey E, Bartley BA, Enghuus C, Goodman DB, Gordon DB, Voigt CA, Gorochowski TE (2017) DNAplotlib: programmable visualization of genetic designs and associated data. *ACS Synth Biol* 6: 1115–1119

Elowitz MB, Leibler S (2000) A synthetic oscillatory network of transcriptional regulators. *Nature* 403: 335–338

Fernandez-Rodriguez J, Yang L, Gorochowski TE, Gordon DB, Voigt CA (2015) Memory and combinatorial logic based on DNA inversions: dynamics and evolutionary stability. *ACS Synth Biol* 4: 1361–1372

Gander MW, Vrana JD, Voje WE, Carothers JM, Klavins E (2017) Digital logic circuits in yeast with CRISPR-dCas9 NOR gates. *Nat Commun* 8: 15459

Gilbert LA, Larson MH, Morsut L, Liu Z, Brar GA, Torres SE, Stern-Ginossar N, Brandman O, Whitehead EH, Doudna JA, Lim WA, Weissman JS, Qi LS (2013) CRISPR-mediated modular RNA-guided regulation of transcription in eukaryotes. *Cell* 154: 442–451

Gorochowski TE, van den Berg E, Kerkman R, Roubos JA, Bovenberg RA (2014) Using synthetic biological parts and microbioreactors to explore the protein expression characteristics of *Escherichia coli*. *ACS Synth Biol* 3: 129–139

Gorochowski TE, Avcilar-Kucukgoze I, Bovenberg RAL, Roubos JA, Ignatova Z (2016) A minimal model of ribosome allocation dynamics captures trade-offs in expression between endogenous and synthetic genes. *ACS Synth Biol* 5: 710–720

Grün D, Kester L, van Oudenaarden A (2014) Validation of noise models for single-cell transcriptomics. *Nat Methods* 11: 637–640

Gupta A, Reizman IM, Reisch CR, Prather KL (2017) Dynamic regulation of metabolic flux in engineered bacteria using a pathway-independent quorum-sensing circuit. *Nat Biotechnol* 35: 273–279

Haas BJ, Chin M, Nusbaum C, Birren BW, Livny J (2012) How deep is deep enough for RNA-Seq profiling of bacterial transcriptomes. *BMC Genom* 13: 734

Hooshangi S, Thiberge S, Weiss R (2005) Ultrasensitivity and noise propagation in a synthetic transcriptional cascade. *Proc Natl Acad Sci USA* 102: 3581–3586

Imbeaud S, Graudens E, Boulanger V, Barlet X, Zaborski P, Eveno E, Mueller O, Schroeder A, Auffray C (2005) Towards standardization of RNA quality assessment using user-independent classifiers of microcapillary electrophoresis traces. *Nucleic Acids Res* 33: e56

Ingolia NT, Ghaemmaghami S, Newman JRS, Weissman JS (2009) Genome-wide analysis *in vivo* of translation with nucleotide resolution using ribosome profiling. *Science* 324: 218–223

Ingolia NT (2014) Ribosome profiling: new views of translation, from single codons to genome scale. *Nat Rev Genet* 15: 205–213

Jayanthi S, Nilgiriwala KS, Del Vecchio D (2013) Retroactivity controls the temporal dynamics of gene transcription. *ACS Synth Biol* 2: 431–441

Kelly JR, Rubin AJ, Davis JH, Ajo-Franklin CM, Cumbers J, Czar MJ, de Mora K, Glieberman AL, Monie DD, Endy D (2009) Measuring the activity of BioBrick promoters using an *in vivo* reference standard. *J Biol Eng* 3: 4

Khlebnikov A, Datsenko KA, Skaug T, Wanner BL, Keasling JD (2001) Homogeneous expression of the PBAD promoter in *Escherichia coli* by constitutive expression of the low-affinity high-capacity AraE transporter. *Microbiology* 147: 3241–3247

Kiani S, Beal J, Ebrahimkhani MR, Huh J, Hall RN, Xie Z, Li Y, Weiss R (2014) CRISPR transcriptional repression devices and layered circuits in mammalian cells. *Nat Methods* 11: 723–726

Kim IK, Roldao A, Siewers V, Nielsen J (2012) A systems-level approach for metabolic engineering of yeast cell factories. *FEMS Yeast Res* 12: 228–248

Kram KE, Finkel SE (2014) Culture volume and vessel affect long-term survival, mutation frequency, and oxidative stress of *Escherichia coli*. *Appl Environ Microbiol* 80: 1732–1738

Lamb J, Crawford ED, Peck D, Modell JW, Blat IC, Wrobel MJ, Lerner J, Brunet J-P, Subramanian SA, Haggarty SJ, Clemons PA, Wei R, Carr SA, Lander ES, Golub TR (2006) The connectivity map: using gene-expression signatures to connect small molecules, genes, and disease. *Science* 313: 1929–1935

Larson MH, Gilbert LA, Wang X, Lim WA, Weissman JS, Qi LS (2013) CRISPR interference (CRISPRi) for sequence-specific control of gene expression. *Nat Protoc* 8: 2180–2196

Lasken RS, McLean JS (2014) Recent advances in genomic DNA sequencing of microbial species from single cells. *Nat Rev Genet* 15: 577–584

Li H, Durbin R (2009) Fast and accurate short read alignment with Burrows-Wheeler transform. *Bioinformatics* 25: 1754–1760

Li H, Handsaker B, Wysoker A, Fennell T, Ruan J, Homer N, Marth G, Abecasis G, Durbin R, Genome Project Data Processing S (2009) The sequence alignment/Map format and SAMtools. *Bioinformatics* 25: 2078–2079

Li GW, Burkhardt D, Gross C, Weissman JS (2014) Quantifying absolute protein synthesis rates reveals principles underlying allocation of cellular resources. *Cell* 157: 624–635

Li S, Wang J, Li X, Yin S, Wang W, Yang K (2015) Genome-wide identification and evaluation of constitutive promoters in streptomycetes. *Microb Cell Fact* 14: 172

Lou C, Stanton B, Chen YJ, Munsky B, Voigt CA (2012) Ribozyme-based insulator parts buffer synthetic circuits from genetic context. *Nat Biotechnol* 30: 1137–1142

Moon TS, Lou C, Tamsir A, Stanton BC, Voigt CA (2012) Genetic programs constructed from layered logic gates in single cells. *Nature* 491: 249–253

Moser F, Broers NJ, Hartmans S, Tamsir A, Kerkman R, Roubos JA, Bovenberg R, Voigt CA (2012) Genetic circuit performance under conditions relevant for industrial bioreactors. *ACS Synth Biol* 1: 555–564

Mutalik VK, Guimaraes JC, Cambray G, Lam C, Christoffersen MJ, Mai QA, Tran AB, Paull M, Keasling JD, Arkin AP, Endy D (2013) Precise and reliable gene expression via standard transcription and translation initiation elements. *Nat Methods* 10: 354–360

Myers CJ, Beal J, Gorochowski TE, Kuwahara H, Madsen C, McLaughlin JA, Misirli G, Nguyen T, Oberortner E, Samineni M, Wipat A, Zhang M, Zundel Z (2017) A standard-enabled workflow for synthetic biology. *Biochem Soc Trans* 45: 793–803

Nagalakshmi U, Wang Z, Waern K, Shou C, Raha D, Gerstein M, Synder M (2008) The transcriptional landscape of the yeast genome defined by RNA sequencing. *Science* 320: 1344–1349

Navasa N, Rodriguez-Aparicio L, Martinez-Blanco H, Arcos M, Ferrero MA (2009) Temperature has reciprocal effects on colanic acid and polysialic acid biosynthesis in *E. coli* K92. *Appl Microbiol Biotechnol* 82: 721–729

Nielsen AAK, Voigt CA (2014) Multi-input CRISPR/Cas genetic circuits that interface host regulatory networks. *Mol Syst Biol* 10: 763

Nielsen AAK, Der B, Shin J, Vaidyanathan P, Paralanov V, Strychalski EA, Ross D, Densmore D, Voigt CA (2016) Genetic circuit design automation. *Science* 352: aac7341

Qi L, Haurwitz RE, Shao W, Doudna JA, Arkin AP (2012) RNA processing enables predictable programming of gene expression. *Nat Biotechnol* 30: 1002–1006

Qi LS, Larson MH, Gilbert LA, Doudna JA, Weissman JS, Arkin AP, Lim WA (2013) Repurposing CRISPR as an RNA-guided platform for sequence-specific control of gene expression. *Cell* 152: 1173–1183

Robinson MD, McCarthy DJ, Smyth GK (2010) edgeR: a Bioconductor package for differential expression analysis of digital gene expression data. *Bioinformatics* 26: 139–140

Robinson MD, Oshlack A (2010) A scaling normalization method for differential expression analysis of RNA-seq data. *Genome Biol* 11: R25

Rosenfeld N, Young JW, Alon U, Swain PS, Elowitz MB (2005) Gene regulation at the single-cell level. *Science* 307: 1962–1965

Schleif R (2003) AraC protein: a love-hate relationship. *BioEssays* 25: 274–282

Schroeder A, Mueller O, Stocker S, Salowsky R, Leiber M, Gassmann M, Lightfoot S, Menzel W, Granzow M, Ragg T (2006) The RIN: an RNA integrity number for assigning integrity values to RNA measurements. *BMC Mol Biol* 7: 3

Shapiro E, Biezuner T, Linnarsson S (2013) Single-cell sequencing-based technologies will revolutionize whole-organism science. *Nat Rev Genet* 14: 618–630

Shearwin KE, Callen BP, Egan JB (2005) Transcriptional interference – a crash course. *Trends Genet* 21: 339–345

Shishkin AA, Giannoukos G, Kucukural A, Ciulla D, Busby M, Surka C, Chen J, Bhattacharyya RP, Rudy RF, Patel MM, Novod N, Hung DT, Gnirke A, Garber M, Guttman M, Livny J (2015) Simultaneous generation of many RNA-seq libraries in a single reaction. *Nat Methods* 12: 323–325

Siuti P, Yazbek J, Lu TK (2013) Synthetic circuits integrating logic and memory in living cells. *Nat Biotechnol* 31: 448–452

Smanski MJ, Bhatia S, Zhao D, Park Y, Woodruff LBA, Giannoukos G, Ciulla D, Busby M, Calderon J, Nicol R, Gordon DB, Densmore D, Voigt CA (2014) Functional optimization of gene clusters by combinatorial design and assembly. *Nat Biotechnol* 32: 1241–1249

Song M, Sukovich DJ, Ciccarelli L, Mayr J, Fernandez-Rodriguez J, Mirsky EA, Tucker AC, Gordon DB, Marlovits TC, Voigt CA (2017) Control of type III protein secretion using a minimal genetic system. *Nat Commun* 8: 14737

Srikumar S, Kröger C, Hébrand M, Colgan A, Owen SV, Sivasankaran SK, Cameron ADS, HHokamp K, Hinton JCD (2015) RNA-seq brings new insights to the intra- macrophage Transcriptome of *Salmonella* Typhimurium. *PLoS Pathog* 11: e1005262

Stanton BC, Nielsen AA, Tamsir A, Clancy K, Peterson T, Voigt CA (2014) Genomic mining of prokaryotic repressors for orthogonal logic gates. *Nat Chem Biol* 10: 99–105

Tang F, Barbacioru C, Wang Y, Nordman E, Lee C, Xu N, Wang X, Bodeau J, Tuch BB, Siddiqui A, Lao K, Surani MA (2009) mRNA-Seq whole-transcriptome analysis of a single cell. *Nat Methods* 6: 377–382

Trapnell C, Williams BA, Pertea G, Mortazavi A, Kwan G, van Baren MJ, Salzberg SL, Wold BJ, Pachter L (2010) Transcript assembly and quantification by RNA-Seq reveals unannotated transcripts and isoform switching during cell differentiation. *Nat Biotechnol* 28: 511–515

Weiss R (2001) *Cellular computation and communications using engineered genetic regulatory networks*. Cambridge, MA: Massachusetts Institute of Technology

Woodruff LB, Pandhal J, Ow SY, Karimpour-Fard A, Weiss SJ, Wright PC, Gill RT (2013) Genome-scale identification and characterization of ethanol tolerance genes in *Escherichia coli*. *Metab Eng* 15: 124−133

Yuan T, Ren Y, Meng K, Feng Y, Yang P, Wang S, Shi P, Wang L, Xie D, Yao B (2011) RNA-Seq of the xylose-fermenting yeast *Scheffersomyces stipitis* cultivated in glucose or xylose. *Appl Microbiol Biotechnol* 92: 1237−1249

Zhang F, Carothers JM, Keasling JD (2012) Design of a dynamic sensor-regulator system for production of chemicals and fuels derived from fatty acids. *Nat Biotechnol* 30: 354−359

Zhong W, Gerstein M, Snyder M (2009) RNA-Seq: a revolutionary tool for transcriptomics. *Nat Rev Genet* 10: 57−63

3

Network analyses identify liver-specific targets for treating liver diseases

Sunjae Lee[1,†] , Cheng Zhang[1,†] , Zhengtao Liu[1,†], Martina Klevstig[2], Bani Mukhopadhyay[3],
Mattias Bergentall[2], Resat Cinar[3] , Marcus Ståhlman[2], Natasha Sikanic[1], Joshua K Park[3],
Sumit Deshmukh[1], Azadeh M Harzandi[1], Tim Kuijpers[1], Morten Grøtli[4], Simon J Elsässer[5],
Brian D Piening[6], Michael Snyder[6], Ulf Smith[2], Jens Nielsen[1,7] , Fredrik Bäckhed[2], George Kunos[3],
Mathias Uhlen[1] , Jan Boren[2,*] & Adil Mardinoglu[1,7,**]

Abstract

We performed integrative network analyses to identify targets
that can be used for effectively treating liver diseases with minimal side effects. We first generated co-expression networks (CNs)
for 46 human tissues and liver cancer to explore the functional
relationships between genes and examined the overlap between
functional and physical interactions. Since increased *de novo* lipogenesis is a characteristic of nonalcoholic fatty liver disease
(NAFLD) and hepatocellular carcinoma (HCC), we investigated the
liver-specific genes co-expressed with fatty acid synthase (FASN).
CN analyses predicted that inhibition of these liver-specific genes
decreases FASN expression. Experiments in human cancer cell
lines, mouse liver samples, and primary human hepatocytes validated our predictions by demonstrating functional relationships
between these liver genes, and showing that their inhibition
decreases cell growth and liver fat content. In conclusion, we
identified liver-specific genes linked to NAFLD pathogenesis, such
as pyruvate kinase liver and red blood cell (PKLR), or to HCC
pathogenesis, such as PKLR, patatin-like phospholipase domain
containing 3 (PNPLA3), and proprotein convertase subtilisin/kexin
type 9 (PCSK9), all of which are potential targets for drug
development.

Keywords co-expression; co-regulation; HCC; metabolism; NAFLD
Subject Categories Genome-Scale & Integrative Biology; Molecular Biology
of Disease; Network Biology

Introduction

Nonalcoholic fatty liver disease (NAFLD) is characterized by the
accumulation of excess fat in the liver and is associated with
obesity, insulin resistance (IR), and type 2 diabetes (T2D). NAFLD
includes a spectrum of diseases ranging from simple steatosis to
nonalcoholic steatohepatitis (NASH) and plays a major role in the
progression of cirrhosis and hepatocellular carcinoma (HCC), a
cancer with one of the highest mortality rates worldwide (Kew,
2010). Although NAFLD is the most common cause of chronic liver
disease in developed countries, and its worldwide prevalence
continues to increase along with the growing obesity epidemic,
there is no approved pharmacological treatment for NAFLD. NAFLD
is projected to become the most common indication leading to liver
transplantation in the United States by 2030 (Shaker *et al*, 2014).
The incidence of HCC has also increased significantly in the United
States over the past few decades, in parallel with the epidemic of
NAFLD (Petrick *et al*, 2016). Hence, there is an urgent need to
develop new strategies for preventing and treating such chronic
hepatic diseases.

Biological networks can be used to uncover complex systems-
level properties. Systems biology combining experimental and
computational biology to decipher the complexity of biological
systems can be used for the development of effective treatment
strategies for NAFLD, HCC, and other complex diseases (Mardinoglu
& Nielsen, 2015; Yizhak *et al*, 2015; Mardinoglu *et al*, 2017b;
Nielsen, 2017). To date, several metabolic processes that are altered
in NAFLD (Mardinoglu *et al*, 2014a, 2017a; Hyötyläinen *et al*, 2016)
and HCC (Agren *et al*, 2012, 2014; Björnson *et al*, 2015; Elsemman
et al, 2016) have been revealed through the use of genome-scale

1   Science for Life Laboratory, KTH – Royal Institute of Technology, Stockholm, Sweden
2   Department of Molecular and Clinical Medicine, University of Gothenburg and Sahlgrenska University Hospital, Gothenburg, Sweden
3   Laboratory of Physiologic Studies, National Institute on Alcohol Abuse and Alcoholism, National Institutes of Health, Bethesda, MD, USA
4   Department of Chemistry and Molecular Biology, University of Gothenburg, Gothenburg, Sweden
5   Department of Medical Biochemistry and Biophysics, Karolinska Institutet, Stockholm, Sweden
6   Department of Genetics, Stanford University, Stanford, CA, USA
7   Department of Biology and Biological Engineering, Chalmers University of Technology, Gothenburg, Sweden
    *Corresponding author.E-mail: jan.boren@wlab.gu.se
    **Corresponding author. E-mail: adilm@scilifelab.se
    †These authors contributed equally to this work

metabolic models (GEMs), a popular tool in systems biology. GEMs are reconstructed on the basis of detailed biochemical information and have been widely used to determine the underlying molecular mechanisms of metabolism-related disorders (Mardinoglu & Nielsen, 2012, 2015, 2016; Mardinoglu *et al*, 2013b, 2015; Yizhak *et al*, 2013, 2014a,b; Bordbar *et al*, 2014; Björnson *et al*, 2015; O'Brien *et al*, 2015; Shoaie *et al*, 2015; Zhang *et al*, 2015; Uhlen *et al*, 2016).

Recently, we have integrated GEMs for hepatocytes (*iHepato-cytes2322*; Mardinoglu *et al*, 2014a), myocytes (*iMyocytes2419*; Varemo *et al*, 2015), and adipocytes (*iAdipocytes1850*; Mardinoglu *et al*, 2013a, 2014b) with transcriptional regulatory networks (TRNs) and protein–protein interaction networks (PPINs) to generate tissue-specific integrated networks (INs) for liver, muscle, and adipose tissues (Lee *et al*, 2016). The INs allowed us to comprehensively explore the tissue biological processes altered in the liver and adipose tissues of obese subjects, thus accounting for the effects of transcriptional regulators and their interacting proteins and enzymes. Although INs provide physical interactions between pairs or groups of enzymes, transcription factors (TFs), and other proteins, these physical interactions may not necessarily have close functional connections.

Co-expression connections are enriched for functionally related genes, and co-expression networks (CNs) allow the simultaneous investigation of multiple gene co-expression patterns across a wide range of clinical and environmental conditions. In this study, we constructed CNs for major human tissues including liver, muscle, and adipose tissues and studied the overlap between functional connections and physical interactions defined by CNs and INs, respectively. We also constructed CNs for HCC to investigate the functional relationship between genes. Finally, we used these comprehensive biological networks to explore the altered biological processes in NAFLD and HCC and identified liver-specific gene targets that may be used in the development of effective treatment strategies for NAFLD and HCC with likely minimal negative side effects.

Results

Generation of CNs for human tissues

A common observation in gene expression analysis performed for different clinical conditions is that many genes known to be functionally related often show similar expression patterns, thus potentially indicating shared biological functions under common regulatory control. Thus, identifying co-expression patterns instead of only differential expression patterns may be informative for understanding altered biological functions. To identify genes with similar gene expression profiles, we retrieved RNA-seq data comprising 51 tissues, including liver, muscle, and subcutaneous and omental adipose, along with other major human tissue samples (Dataset EV1), from the Genotype-Tissue Expression (GTEx) database (GTEx, 2013). To measure the tendency of gene expression correlation, we calculated the Pearson correlation coefficients (r) between all gene pairs in 46 human tissues (Dataset EV1) with more than 50 samples, and we ranked all genes according to the calculated r. We used the top 1% correlation value of each tissue as a cutoff indicating that two genes were co-expressed (average

$r = 0.576$ for 46 tissues) and therefore had the same number of co-expression links for all tissues, thus yielding 1,498,790 links, and combined them to construct the tissue-specific CNs. The resulting CNs were as follows: Liver tissue contained 11,580 co-expressed genes, muscle tissue contained 10,728 co-expressed genes, and subcutaneous and omental adipose tissue contained 12,120 and 11,117 co-expressed genes (Dataset EV1).

Given the connectivity of tissue-specific co-expression links, we found groups of highly co-expressed genes, termed co-expression clusters, by using the random walk community detection algorithm (Pons & Latapy, 2005; Fig 1A). Among these clusters, we selected the most highly co-expressed key clusters on the basis of their clustering coefficients (average clustering coefficient = 0.522). We found that genes associated with key co-expression clusters in tissues significantly overlapped with tissue-specific genes presented in the Human Protein Atlas (HPA; Uhlen *et al*, 2015), when global proteomics and transcriptomics data were available for more than 30 human tissues. Our analysis indicated that 75.6% of 41 HPA-matched tissues had key co-expression clusters significantly enriched in tissue-specific genes (hypergeometric test $P < 0.01$), thus suggesting the tissue-specific roles of the genes associated with the key clusters (Dataset EV2).

To investigate tissue-specific functions of the genes associated with the key co-expression clusters, we performed gene ontology (GO) enrichment analysis by using the GO biological processes (BP) terms in MSigDB (Subramanian *et al*, 2005; hypergeometric test $P < 1.0 \times 10^{-4}$; Fig EV1). For example, in liver tissue, we found that genes associated with key co-expression clusters were enriched in terms comprising the immune response, hemopoiesis, and fatty acid metabolic processes (Fig 1B), whereas in testis tissue, co-expression clusters were enriched in cell cycle, metabolism, and reproduction (Fig EV1C and Dataset EV3). Likewise, we investigated the enrichment of the genes associated with key co-expression clusters in other metabolically active tissues, including skeletal muscle and subcutaneous and omental adipose tissues, compared with the significantly enriched GO BP terms in liver tissue, and found that the GO BP term "hemopoiesis" was enriched only in liver tissue, whereas "muscle development" was enriched only in skeletal muscle (Fig EV1A). In both subcutaneous adipose tissue and liver tissue, genes associated with the key clusters were significantly enriched in fatty acid metabolic processes (Fig EV1B). Hence, our analysis indicated that genes involved in fatty acid metabolism are significantly co-expressed in liver and adipose tissues.

Increased co-regulation results in increased co-expression

We have previously presented INs for metabolic tissues, including liver, muscle, and adipose tissues, and have identified the physical links between TFs and target proteins and enzymes (Lee *et al*, 2016). Here, we investigated the overlap between the INs and CNs and analyzed the potential deregulation of metabolism in metabolically active human tissues by using the topology on the basis of the regulatory interactions and protein–protein interactions of INs by comparing the mean co-expression of actual networks with that of networks randomly permuted by 1,000 repetitions (Fig 1C). We found that most of the actual tissue regulatory networks (RNs; Fig 1D) and PPINs (Fig 1E) showed higher co-expression than randomly permuted networks in liver and other metabolic tissues

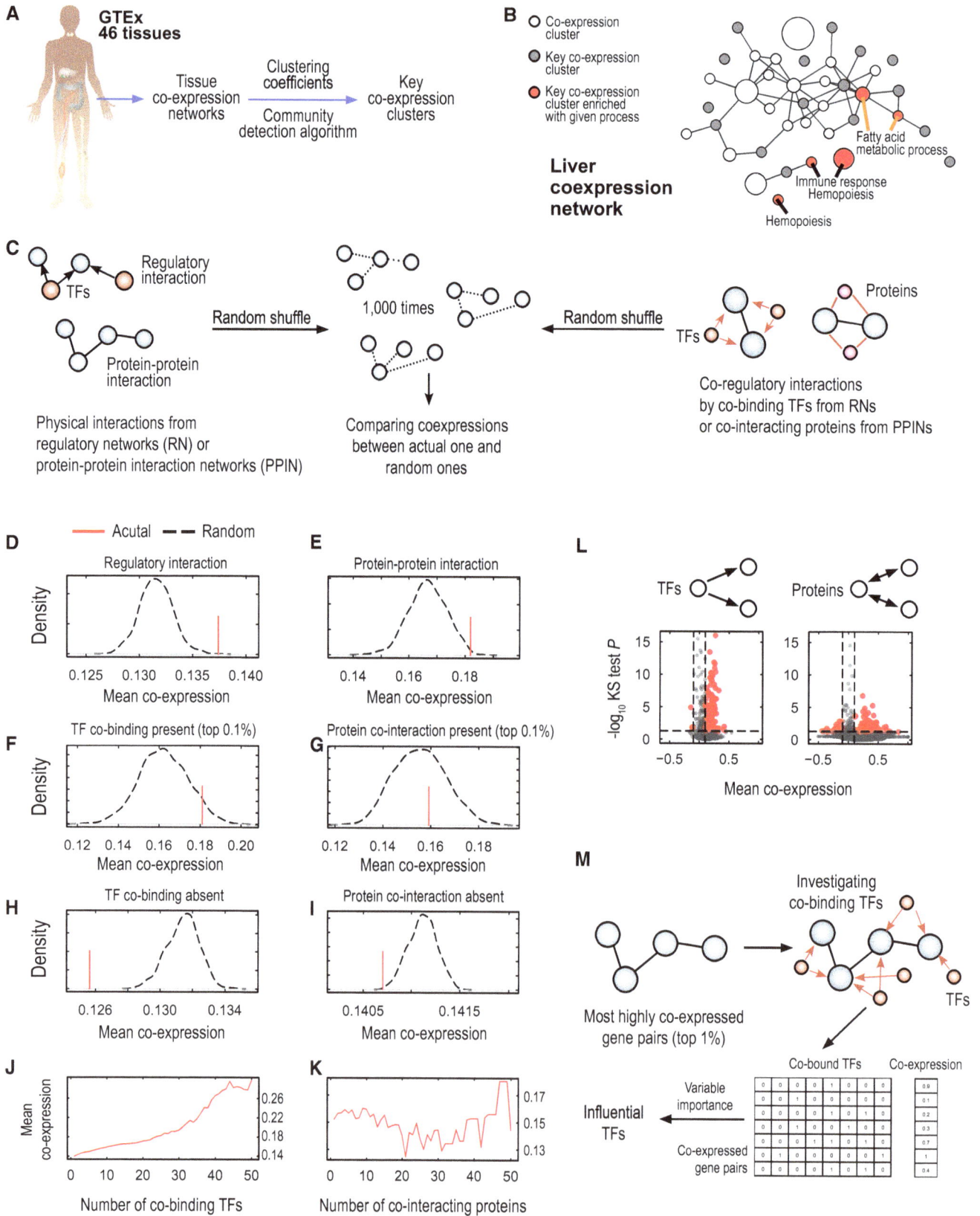

Figure 1.

Figure 1. Characteristics of tissue-specific co-expression networks (CNs).

A We generated co-expression networks for 46 tissues that had more than 50 samples from GTEx RNA-seq data. In each tissue, we found groups of highly co-expressed genes, called co-expression clusters, by using a community detection algorithm. Among these groups, we selected key co-expression clusters on the basis of their clustering coefficients.

B For example, in liver tissue, we found genes from key co-expression clusters significantly enriched in biological processes required for liver function, such as fatty acid metabolism, hemopoiesis, and immune response. All nodes in the network stand for co-expression clusters, and edges are connected when genes belonging to those clusters were highly connected more frequently than at random. Node sizes are proportional to the number of genes involved in the respective clusters.

C–K In liver tissue, we investigated co-expression of physical interactions from the liver regulatory network (RN) or liver protein–protein interaction network (PPIN) (red lines). We also generated randomly permutated gene pairs from those physical interactions by 1,000× (blue dashed lines) and compared them with actual gene pairs from RN or PPIN (D and E, respectively). We identified co-regulatory interactions from RN or PPIN (F and G, respectively), which indicate gene pairs co-bound by the same TFs or co-interacting with the same proteins, and compared them with randomly permutated gene pairs; here, we examined only the co-regulated gene pairs with the highest numbers of co-bound TFs or co-interacting proteins (top 0.1%). In addition, we identified gene pairs that had no co-regulation by TF binding or protein interactions (H and I, respectively) and compared them with randomly permutated gene pairs. We found that gene pairs from actual physical interactions (D and E) or co-regulatory interactions (F and G) had higher co-expression than at random; however, gene pairs with no co-regulations (H and I) had lower co-expression than at random. We examined the co-expression profiles of co-regulated gene pairs according to their co-bound TFs (J) or co-interacting proteins (K). Here, we found that only co-regulations from RN were associated with increased co-expression, whereas co-regulations from PPIN were not, thus suggesting that RN has more specificity for increasing co-expression.

L We examined which TFs or proteins were highly co-expressed with their bound target genes or interacting proteins by comparing their gene pairs with the overall expression by Kolmogorov–Smirnov two-sided test ($P < 0.05$) and the absolute value of mean co-expression ($|r| > 0.1$).

M To find the most influential TFs in co-expression, we established a feature matrix between highly co-expressed gene pairs (top 1%) and their co-bound TFs and fitted it to a Random Forest model; we considered co-bound TFs as predictor variables and co-expression values as response variables. From this model, we calculated variable importance scores of TFs, and on the basis of these scores, we identified the most influential TFs in co-expression in liver tissue. The top 1% most influential TFs included YY1, CTCF, RAD21, SREBF1, and SREBF2 and were enriched in liver development, interphase of the mitotic cell cycle, and response to lipids.

(Figs EV2A and B, and EV3A and B). We also observed that regulatory interactions built from *in vitro* differentiated adipocytes (Lee *et al*, 2016) had less specificity than co-expression calculated by using subcutaneous adipose tissue RNA-seq data demonstrating higher tissue specificity (Fig EV2B).

Target proteins regulated by the same TFs or interacting with the same proteins may have similar gene expression patterns (Zhang *et al*, 2016). Hence, we examined the co-expression of co-regulated gene pairs in RNs and found that their mean co-expression was higher in actual RNs than in randomly permutated RNs in the liver (Fig 1F) and other metabolic tissues (Fig EV2C and D). Moreover, we examined the co-expression of gene pairs that were not regulated by the same TFs and found that their mean co-expression was lower in actual liver RNs than in randomly permutated RNs in liver tissue (Fig 1H) and other metabolic tissues (Fig EV2E and F). Similarly, we examined the co-expression of co-interacting gene pairs for the same proteins in PPINs and found relatively lower mean co-expression in actual liver PPINs than in randomly permutated PPINs in liver tissue (Fig 1G) and other metabolic tissues (Fig EV3C and D), as compared with the mean co-expression of the actual RNs. We also found that mean co-expression in PPINs was lower in actual PPINs than in randomly permutated PPINs in liver tissue (Fig 1I) and other metabolic tissues (Fig EV3E and F) when we compared proteins that did not interact with the same proteins. Our analysis indicated that physical interactions defined by the RNs and PPINs can be used to explain protein co-expression. Moreover, we observed that the two target proteins regulated by the same TFs may have similar expression patterns, whereas two target proteins interacting with the same protein may have less similar expression patterns.

Target proteins may be regulated by more than one TF in RNs, thus potentially affecting their expression patterns. In this context, we determined the co-expression of two target proteins regulated by the same TFs and found that an increased number of co-bound TFs were likely to be associated with increased mean

co-expression levels in liver (Fig 1J), skeletal muscle, and adipose tissues (Fig EV2G and H). We also repeated a similar analysis in PPINs on the basis of the two target proteins that co-interacted with the same proteins and found that the increased number of co-interacting proteins was not directly proportional to increased mean co-expression although the mean co-expression levels were high in the liver (Fig 1K) and other tissues analyzed (Fig EV3G and H). Our analysis indicates that protein interactions provide evidence for increased co-expression by RNs compared with PPINs in metabolic tissues.

We also identified the TFs and proteins that were highly co-expressed with their target genes or proteins in RNs and PPINs, respectively ($|r| > 0.1$ and Kolmogorov–Smirnov (KS) two-sided test, $P < 0.05$) (Fig 1L, Datasets EV4 and EV5). We investigated the biological functions associated with these TFs and proteins using DAVID (Huang *et al*, 2009) and found that the TFs in liver RN were enriched in the regulation of transcription and cell differentiation (false discovery rate [FDR] < 0.01) and that the proteins in liver PPIN were enriched in RNA splicing (FDR < 0.01), which is involved in post-transcriptional regulation. To identify the most influential TFs in the co-expression of gene pairs, we used the most highly co-expressed gene pairs (top 1%), established the co-binding matrix of bound TFs on the basis of RNs, and calculated the variable importance score by using the Random Forest model in the liver (Fig 1M and Dataset EV6). Among the top 1% most influential TFs (76 TFs), for example, YY1, CTCF, RAD21, SREBF1, and SREBF2, we found TFs enriched in liver development, interphase of the mitotic cell cycle, and response to lipids (FDR < 0.001), thus suggesting that many influential TFs are involved in liver-specific functions.

Highly co-expressed metabolic pathways in the liver

On the basis of physical and functional links provided by the networks, we examined which metabolic pathways are regulated

Figure 2. Highly co-expressed metabolic pathways in liver tissue with their co-regulating TFs.

A   Among human metabolic reactions with known enzymes (HMR2), we calculated the co-expression of respective enzymes in liver tissue and established a co-expression matrix of those metabolic reactions for liver tissue. Performing hierarchical clustering on the matrix, we found 100 reaction clusters of highly co-expressed enzymes in liver tissue. We compared the mean co-expression profiles of given reaction clusters in liver tissue with the co-expression profiles of clusters in other tissues, such as skeletal muscle and adipose subcutaneous tissues. On the basis of their differential co-expression levels, we identified liver-specific reaction clusters (B) and their co-regulating TFs (C and D).

B   Among the 100 reaction clusters for liver tissue, we selected ten reaction clusters with the most differential co-expression between liver tissue and skeletal muscle tissue (left) or between liver tissue and adipose subcutaneous tissue (right), regarding them as liver-specific reaction clusters, colored red and blue, respectively (purple for those in both cases).

C   We examined co-regulating TFs of liver-specific reaction clusters found in (B). Using the hypergeometric test, we identified co-regulating TFs significantly enriched in given reaction clusters ($P < 0.05$). Here, we found that reaction cluster 14 was enriched in binding of metabolic nuclear receptors, such as PXR, FXR, and RXR, whereas reaction cluster 9 was enriched in binding of SREBF2, a regulator of lipid homeostasis.

D   We found additional evidence of strong regulation in some liver-specific reaction clusters, on the basis of variable TF importance scores. We compared the variable importance scores (Dataset EV6) of enriched TFs in given reaction clusters with the overall score by using Kolmogorov–Smirnov tests and selected significant clusters ($P < 0.25$) as highly regulated reaction clusters, including reaction clusters 9, 14, 40, and 58.

E, F   We identified HCC-deregulated reaction clusters by comparing co-expression levels between liver tissue and HCC tumor tissue similarly to (B). Yellow-colored are HCC-deregulated liver reaction clusters; green-colored are liver-specific clusters found in (B); and purple-colored are reaction clusters shown in both cases. Here, we found that reaction clusters 9 and 14 were regulated in a liver-specific manner at the levels of co-expression (B) and co-regulation (D) but were deregulated in HCC tumor tissues (E). Reaction clusters 9 and 14 included reactions associated with fatty acid synthesis, including glucose uptake, pyruvate synthesis, and citrate transport (F).

specifically in the liver. We first identified the group of metabolic reactions catalyzed by highly co-expressed enzymes in liver tissue by using the Human Metabolic Reaction database (HMR2; Mardinoglu et al, 2014a). Taking the maximal co-expression of enzymes between two metabolic reactions, we established a reaction co-expression matrix among all reactions (Fig 2A). From this matrix, we identified hundreds of clusters of reactions with highly co-expressed enzymes by using hierarchical clustering (Dataset EV7). By comparing their co-expression in liver tissue with the co-expression in adipose subcutaneous and skeletal muscle tissues, we identified metabolic reaction clusters that were co-expressed only in liver tissue (Fig 2B). On the basis of Fisher Z-transformed differences in tissue co-expression, we identified ten reaction clusters of the most differential co-expression levels between liver and muscle tissue (left, Fig 2B) or between liver and adipose tissue (right, Fig 2B); four reaction clusters overlapped in both cases.

Among those clusters, we determined the TFs that were highly co-bound with enzymes of those reaction clusters by using the hypergeometric test ($P < 0.05$) and found that each reaction cluster was governed by different sets of TFs (Fig 2C and Dataset EV8). For example, metabolic nuclear receptors, such as the farnesoid X receptor (FXR or NR1H4), pregnane X receptor (PXR or NR1I2), and RXRB, were highly co-bound with enzymes catalyzing the reaction in cluster 14, whereas SREBF2, a regulator of lipid homeostasis, was highly co-bound with enzymes catalyzing reactions in cluster 9. These findings indicated that enzymes catalyzing reactions in clusters 14 and 9 would share regulatory controls in response to metabolic alterations. Next, we determined which reaction clusters were enriched in influential TFs that we had identified. Using variable importance scores of TFs (Dataset EV6), we determined whether highly co-bound TFs of reaction clusters (Dataset EV8) had variable importance scores higher than the overall scores (KS one-sided test, $P < 0.25$; Fig 2D). We found that reaction clusters including 9, 14, 40, and 58 had significantly enriched TFs with highly variable importance scores, thus providing strong evidence of liver-specific regulation from physical and co-expression links. Through metabolic subsystem annotation from HMR2, we observed that reaction

cluster 9 was enriched in mitochondrial transport, pyruvate metabolism, and lipid metabolism, cluster 14 was enriched in fatty acid synthesis, cluster 40 was enriched in cholesterol metabolism, and cluster 58 was enriched in amino acid metabolism (hypergeometric test $P < 0.01$; Dataset EV9); these results indicated reactions involved in pathways that are regulated in only the liver and are primarily associated with lipid metabolism.

Next, we determined which liver reaction clusters were deregulated in HCC (Dataset EV7), on the basis of co-expression of those clusters in HCC tumor tissue from The Cancer Genome Atlas (TCGA; Fig 2E). We first retrieved RNA-seq data for 371 HCC tumors from TCGA and calculated Pearson's correlation coefficients between the gene pairs. As we identified liver-specific co-expression clusters (Fig 2B), we identified 10 reaction clusters that were deregulated in HCC tumors on the basis of Fisher Z-transformed differences in co-expression between GTEx liver data and TCGA HCC data (yellow and purple points in Fig 2E). Among those deregulated clusters, five reaction clusters were identified as liver-specific clusters as opposed to adipose and/or muscle tissue clusters (purple points in Fig 2E): reaction clusters 9, 14, 56, 62, and 65. Of note, reaction clusters 9 and 14 were identified on the basis of their regulation with strong evidence regarding co-expression (Fig 2B) and physical (Fig 2D) clues. In HCC, deregulation of these liver-specific reaction clusters primarily associated with fatty acid synthesis may be linked to HCC pathogenesis (Fig 2F).

Identification of liver-specific FASN inhibitors for the treatment of NAFLD and HCC

The expression of fatty acid synthase (FASN), which catalyzes the last step in *de novo* lipogenesis (DNL), is significantly upregulated in NAFLD (Dorn et al, 2010) and HCC (Björnson et al, 2015). We have recently shown that short-term intervention with an isocaloric carbohydrate-restricted diet causes a large decrease in liver fat accompanied by striking rapid metabolic improvements (Mardinoglu et al, in preparation). We measured clinical characteristics, body composition, liver fat, hepatic DNL, and hepatic

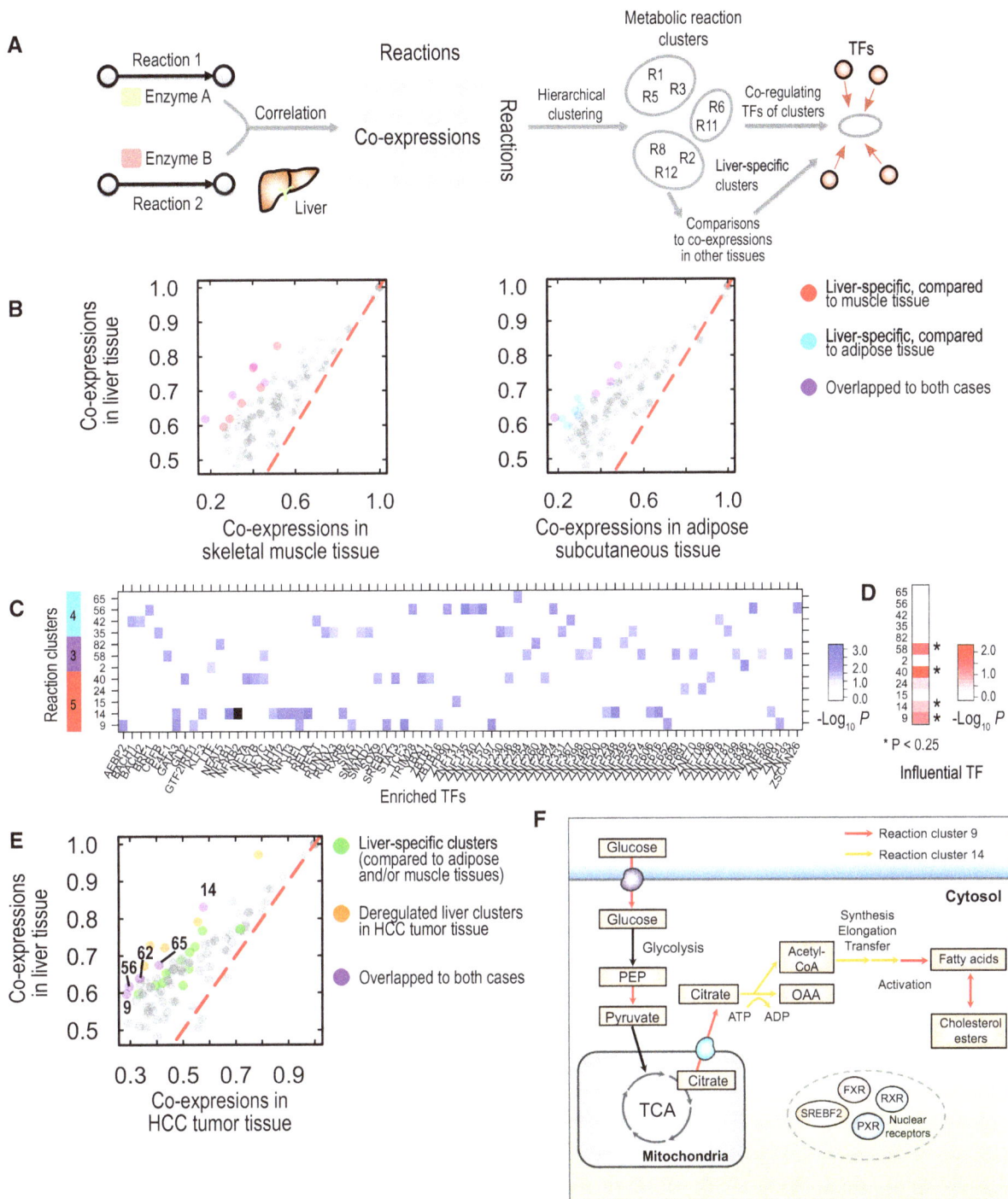

Figure 2.

β-oxidation and found that the diet caused rapid and significant sustained decreases in liver fat, DNL, plasma triglycerides, and very low-density lipoprotein triglycerides and a parallel increase in β-hydroxybutyrate, an indicator of increased hepatic β-oxidation. Hence, DNL may be targeted for the development of effective treatment strategies for NAFLD and other chronic liver diseases, for example, HCC.

However, small-molecule FASN inhibitors (e.g., C75, cerulenin) suffer from pharmacological limitations that prevent their

development as systemic drugs (Pandey *et al*, 2012). These side effects can also be explained by the high expression of FASN in almost all major human tissues (Uhlen *et al*, 2015, 2016). Thus, we hypothesized that the identification of liver-specific FASN inhibitors might allow for the development of effective treatment strategies for NAFLD and HCC. We identified highly co-expressed genes with FASN on the basis of CNs generated using GTEx and TCGA data, which have been used as representative datasets for NAFLD (Dataset EV10) and HCC (Dataset EV11), respectively. Our CN

A

B

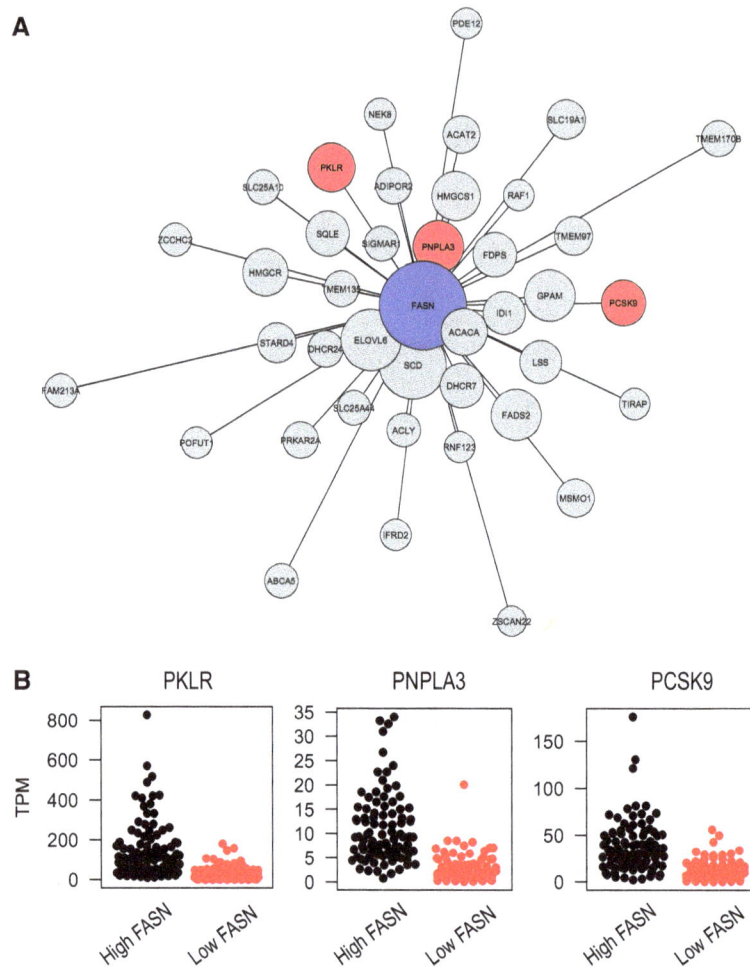

Figure 3. FASN CN in HCC tumor tissue.

A   We present the 40 genes with the highest co-expression with FASN in HCC tumor tissue based on both log-transformed and raw expression values (Datasets EV11 and EV12). The lengths of edges were inversely proportional to the co-expression of corresponding gene pairs; thus, genes close to FASN were more co-expressed than others. Node sizes were inversely proportional to adjusted p-values of differential expressions between patients between high (upper quartile) and low (lower quartile) FASN expressions (Dataset EV13). We colored co-expressed genes on the basis of their liver specificity; red-colored genes were liver tissue-enriched (based on HPA ver. 16 annotation) and co-expressed with FASN in fewer than three human tissues (Dataset EV10); FASN alone was colored blue.

B   We show expressions of PKLR, PNPLA3, and PCSK9 in HCC patients with high and low FASN expressions. We found that liver-specific genes were significantly (adjusted $P < 1.0 \times 10^{-10}$) upregulated in patients high FASN expression compared to those with low FASN expression (Dataset EV13).

analysis allowed the identification of genes functionally related to FASN in liver and other major human tissues.

Using DAVID, we found that the top 100 genes co-expressed with FASN in liver CN (Dataset EV10) were significantly enriched in GO BP terms, including carboxylic acid, oxoacid, hexose, monosaccharide, acyl-CoA and fatty acid metabolic processes, protein transport, secretion, regulation of secretion, and negative regulation of cell communication (Huang *et al*, 2009). We also compared the top 100 co-expressed genes with FASN in liver CN (Dataset EV10) with the genes in 45 other tissue CNs (Dataset EV10) and identified pyruvate kinase liver and red blood cell (PKLR), an enzyme phosphorylating pyruvate from glycolysis to the TCA (citric acid) cycle and also in fatty acid synthesis; PKLR is also shown in Fig 2F, as a liver-specific gene co-expressed with FASN.

In HCC CN, we found that the top 100 genes co-expressed with FASN (Dataset EV11) were significantly enriched in GO BP

terms, including animal organ development, carboxylic acid, oxoacid, cholesterol, secondary alcohol, sterol and fatty acid metabolic processes, along with steroid, sterol, and alcohol biosynthetic processes. We also found ELOVL6, ACACA, and SCD, involved in fatty acid biosynthesis, as the top genes co-expressed with FASN in HCC CN (Dataset EV11). In addition, we calculated Pearson's correlations from log-transformed gene expressions in HCC in order to check for robustly co-expressed genes (Dataset EV12). We found that 40 of the genes in top-100 co-expressed genes were significantly co-expressed with FASN before (Dataset EV11) and after log transformation (Dataset EV12). We compared those robustly co-expressed genes with FASN in HCC CN (Dataset EV12) with the top 100 co-expressed genes in CNs of 46 human tissues (Dataset EV10) and found PKLR, patatin-like phospholipase domain containing 3 (PNPLA3), and proprotein convertase subtilisin/kexin type 9 (PCSK9), referred to as liver-specific genes, as the only genes co-expressed with FASN in less than three human tissues (Fig 3A).

Next, we compared the global gene expression profiling of the HCC tumors in the upper quartile with highest expression of FASN ($n = 93$) with the lower quartile with the lowest expression of FASN ($n = 93$) using DESeq package (Anders & Huber, 2010) to analyze the expression profile of liver-specific genes and their key role in cancer progression. We found that the expression of PKLR, PNPLA3, and PCSK9 was significantly increased in patients with high FASN expression compared to those with low FASN expression (Fig 3B and Dataset EV13).

We also determined the expression patterns of PKLR, PCSK9 and PNPLA3 in the Human Protein Atlas, in which the expression of all human protein coding genes have been measured in 32 major human tissues, and these genes were identified as liver-specific genes based on protein and mRNA expression (Kampf et al, 2014; Uhlen et al, 2015). Hence, these liver-specific genes, including PKLR, PNPLA3, and PCSK9, may potentially be targeted for the treatment of HCC, and PKLR may potentially be targeted for the treatment of NAFLD. Due to the direct involvement of PKLR, PNPLA3, and PCSK9 in lipid metabolism, we focused on the relationship between FASN and these three liver-specific genes in the rest of our studies.

Validation of physical interactions by using human cancer cell lines

We hypothesized that inhibition of liver-specific targets that are co-expressed with FASN would inhibit FASN expression and decrease fat synthesis. This inhibition would also inhibit tumor growth in the case of HCC, because fatty acids play key roles in HCC progression and development. To validate our hypothesis and to demonstrate the physical interactions between FASN and liver-specific genes, we first screened the cell lines with the highest PKLR mRNA expression levels by using the data generated in the Human Protein Atlas (Uhlen et al, 2015). We identified the K562 leukemia cell line as having the highest PKLR expression using our recently published data in Cell Atlas (Thul et al, 2017) and treated the cell line with C75, a FASN inhibitor, at different concentrations for 24 h. The purpose of this experiment was to test whether FASN inhibition may affect the co-expressed genes with FASN (i.e., PKLR).

We found that FASN and PKLR expression levels were significantly decreased (Fig 4A). Moreover, we determined that decreased FASN and PKLR expression resulted in a significant decrease in cell growth (Fig 4B). Next, we treated the HepG2 human cancer cell line with C75 and found that FASN expression and the expression levels of liver-specific genes, including PKLR, PNPLA3, and PCSK9, were significantly decreased after 24 h (Fig 4C). Similarly, we found that the growth of HepG2 cells was significantly decreased after treatment of the cells with different concentrations of C75 (Fig 4D).

However, during the treatment of the cells with C75, other metabolic pathways may also have been regulated, and gene expression may have been affected by nonspecific binding of C75. In this context, we used siRNA to decrease PKLR expression in the K562 cell line and observed that FASN expression (Fig 4E) and cell growth (Fig 4F) were significantly decreased, similarly to our observations after treating the cells with C75. Moreover, we inhibited PKLR expression in the HepG2 cell line and found that FASN expression (Fig 4G) and cell growth (Fig 4H) were significantly decreased, as expected.

Validation of physical interactions in mice and human

To show the functional relationship between FASN and the liver-specific targets identified here in vivo, we fed mice (C57Bl/6N) a zero-fat high-sucrose diet (HSD) for 2 weeks that induced the development of fatty livers in mice (Fig 5A). We collected liver tissue samples from the mice fed a HSD and compared the expression profiles of the genes in the liver of these mice with those in the liver of mice fed a chow diet (CD). We observed that liver TG content in the mice was significantly increased in the HSD fed group compared with the control (Fig 5A). Next, we determined the expression profiles of FASN and our liver-specific gene targets and found that their expression levels were significantly increased in mice fed the HSD compared with mice fed the CD in parallel with the increase in liver fat (Fig 5B).

It has been reported that circulating PCSK9 levels increase with the severity of hepatic fat accumulation in patients at risk of NASH and PCSK9 mRNA levels in liver have been linked with steatosis severity (Ruscica et al, 2016). In this context, we fed the wild-type (WT) and PCSK9 knockout (KO) mice a CD for 10 weeks. We collected liver tissue samples from the WT and PCSK9 KO mice and compared the expression levels of FASN and the other identified gene targets. We determined the expression levels of these genes in PCSK9 KO mice and found that they were significantly downregulated (Fig 5C). Hence, our analysis suggested to targeting of PCSK9 for the development of efficient treatment strategies for HCC patients.

Endocannabinoids acting on the hepatic cannabinoid-1 receptor (CB_1R) promote DNL by increasing the expression of genes involved in lipid metabolism including FASN, SREBF1, and acetyl-CoA carboxylase-1 (ACACA) (Osei-Hyiaman et al, 2005). CB_1R has been implicated in the pathology of different liver diseases with various etiologies including NAFLD (Osei-Hyiaman et al, 2008), AFLD (Jeong et al, 2008), viral hepatitis (Hezode et al, 2005), liver fibrosis (Teixeira-Clerc et al, 2006), cirrhosis (Giannone et al, 2012), and liver cancer (Mukhopadhyay et al, 2015). Activation of the endocannabinoid/CB_1R system inhibits fatty acid β-oxidation in the liver (Osei-Hyiaman et al, 2008), interrupts hepatic carbohydrate and cholesterol metabolism (Jourdan et al, 2012), and contributes to diet-induced obesity and NAFLD. It has been observed that activation of hepatic CB_1R promoted the initiation and progression of chemically induced HCC in mice (Mukhopadhyay et al, 2015). Considering the associations between CB_1R, liver fibrosis, and HCC even in the absence of obesity, we analyzed the expression of Fasn as well as liver-specific gene targets in animal models of liver cancer. We measured the expression of the Fasn and the liver-specific gene targets including Pklr, Pcsk9, and Pnpla3 and found that their expression was significantly ($P < 0.05$) increased in HCC tumor compared to adjacent noncancerous samples obtained from $CB_1R^{+/+}$ mice (Fig 5D), with much smaller increases noted in corresponding samples $CB_1R^{-/-}$ mice. When we measured the expression of these genes in HCC tumor and noncancerous samples obtained from $CB_1R^{-/-}$ mice, we found that the increase in the expression of Fasn as well as liver-specific gene targets is attenuated in conjunction with the decrease in tumor growth $CB_1R^{-/-}$ mice (Fig 5E).

We finally validated our predictions by treating primary human hepatocytes with C75 and found that the expression levels of FASN PKLR, PCSK9 and PNPLA3 were significantly decreased (Fig 5F). Liver diseases connected to FASN can thus be treated (by silencing

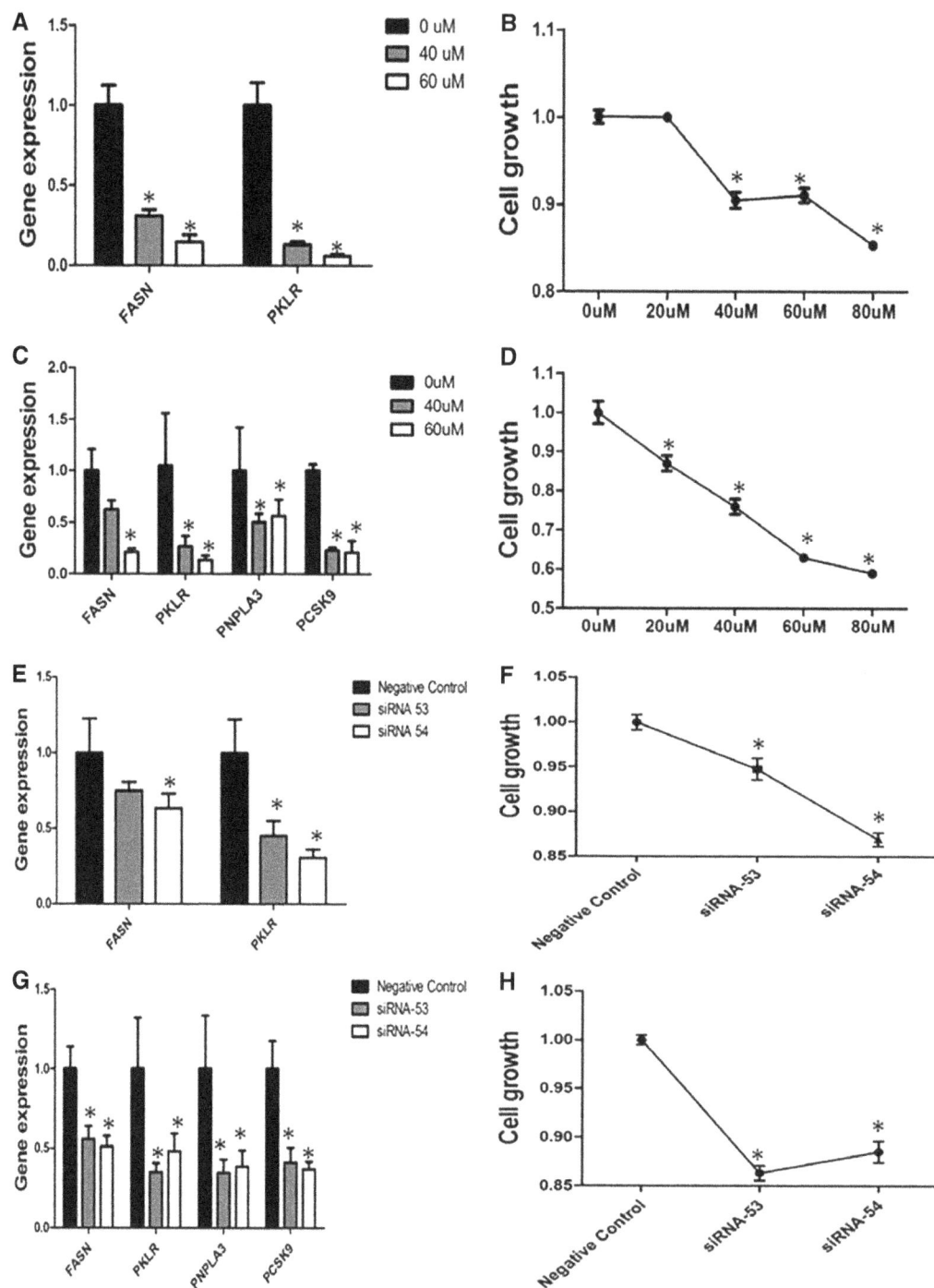

Figure 4. Gene expression and proliferation of K562 and HepG2 cells after interference by C75 and PKLR-specific siRNA.

A   FASN and PKLR expression levels in K562 cells after interference by different doses (0, 40, and 60 μM) of C75.

B   Cell growth in K562 cells after interference by different doses (0, 20, 40, 60, and 80 μM) of C75.

C   FASN, PKLR, PNPLA3, and PCSK9 expression levels in HepG2 cells after interference by different doses (0, 40, and 60 μM) of C75.

D   Cell growth in HepG2 cells after interference by different doses (0, 20, 40, 60, and 80 μM) of C75.

E   FASN and PKLR expression levels in K562 cells after interference by PKLR-specific siRNA (siRNA 53, siRNA 54).

F   Cell growth in K562 cells after interference by PKLR-specific siRNA (siRNA 53, siRNA 54).

G   FASN, PKLR, PNPLA3, and PCSK9 expression levels in HepG2 cells after interference by PKLR-specific siRNA (siRNA 53, siRNA 54).

H   Cell growth in HepG2 cells after interference by PKLR-specific siRNA (siRNA 53, siRNA 54).

Data information: RNA was isolated for RT–PCR after interference for 24 h; GAPDH was set as the internal reference. Cell counting was performed after interference for 72 h. Data are presented as the means ± standard errors of five independent experiments. Comparisons were performed by one-way ANOVA. Samples without any interference were assigned as controls. * represents a significant difference compared with the value in the control group ($P < 0.05$).

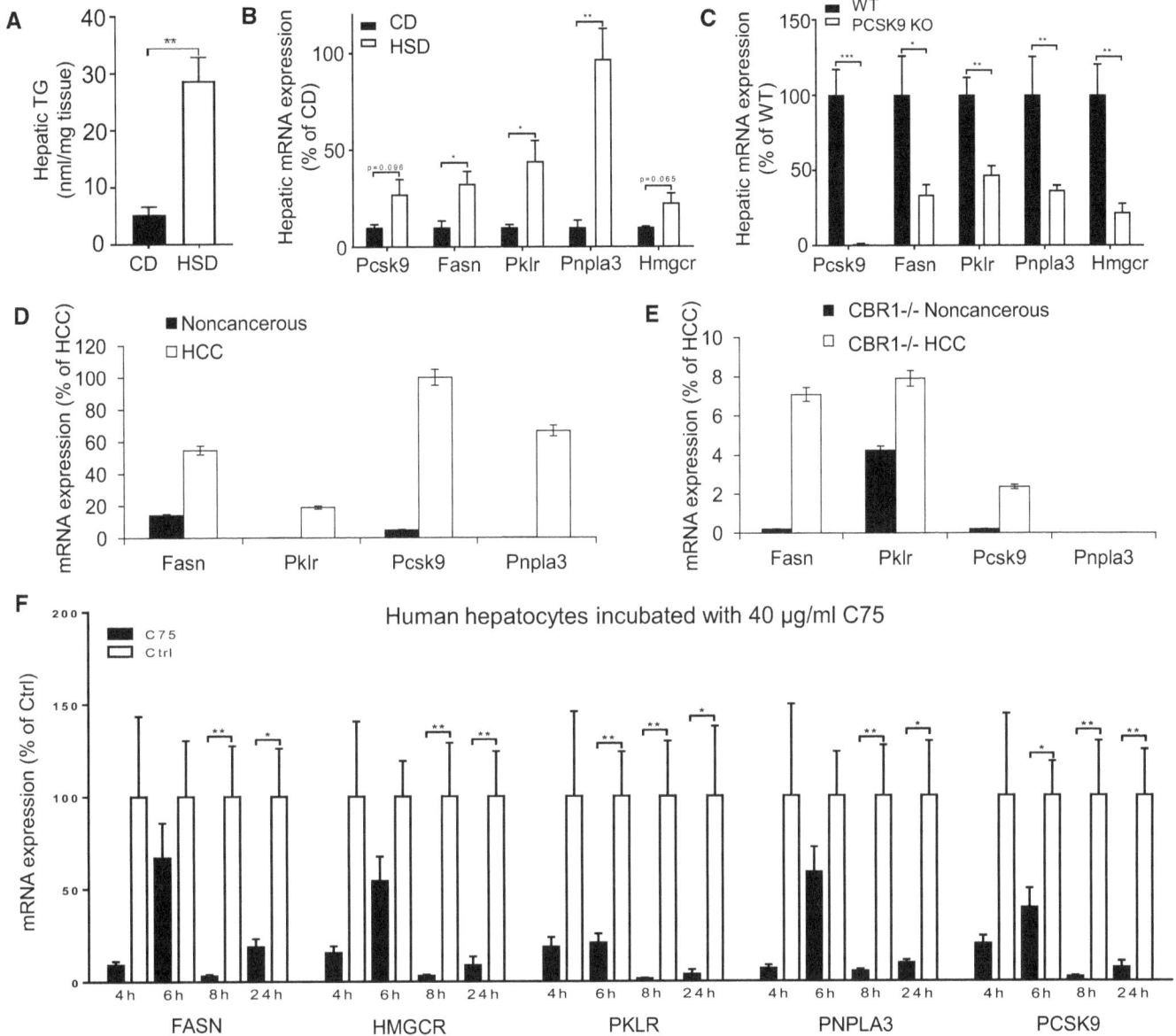

Figure 5. The relationship between the genes in mice and human samples.

A   TG content in the liver tissue of mice fed a zero-fat high-sucrose diet (HSD) and chow diet (CD) for 2 weeks ($n = 10$).

B–F The hepatic mRNA expressions of the Fasn, Pklr, Pcsk9, Pnpla3, and Hmgcr is measured in (B) mice fed a HSD and CD for 2 weeks, (C) Pcsk9 knockout and its littermates (WT) fed a CD for 10 weeks, (D) wild-type HCC tumor and noncancerous samples, (E) CB$_1$R knockout HCC tumor and noncancerous samples, and (F) primary human hepatocytes treated with C75 for 4, 6, 8, and 24 h.

Data information: Data are presented as the means ± standard errors of independent experiments. Student's t-test; *$P < 0.05$, **$P < 0.01$, ***$P < 0.001$ represents a significant difference compared with the value in the control group.

PKLR, PCSK9 or PNPLA3) without experiencing the side effects associated with direct FASN inhibition.

Discussion

Rapid advances in omics technologies along with the adoption of large shared public databases have allowed for the generation and aggregation of massive sample datasets that can be used to construct comprehensive biological networks. These networks may provide a

scaffold for the integration of omics data, thereby revealing the underlying molecular mechanisms involved in disease appearance and providing a better understanding of the variations in healthy and diseased tissue that may be used in the development of effective treatment strategies. In this study, we generated tissue-specific CNs for 46 major human tissues and human liver cancer and explored the tissue-specific functions by using the topologies provided by these networks. An important aspect of a gene CN is modularity: Genes that are highly interconnected within the network are usually involved in the same biological modules or pathways. We compared the tissue-specific CNs with the recently generated INs and found that physical interactions revealed meaningful functional relationships between genes. Next, using CNs, we investigated the emergent properties and behaviors of the affected genes in response to NAFLD and HCC at the system level rather than focusing on their individual functions and clinical utilities. The use of CNs also allowed us to obtain detailed information about the systems-level properties of such complex liver diseases.

Correlation analysis is used to identify co-expressions between different genes based on mRNA expression data. Although co-expression does not necessarily indicate a relationship among transcript levels, functional relationships between encoded protein-coding genes have been shown in our study. Co-expression analysis is applied to identify important alterations related to lipid metabolism, and promotion of cell survival and cell growth. In order to avoid possible false positives from co-expressed genes, we applied strict cutoff of Pearson's correlation coefficients, but it would be necessary to develop advanced method to identify co-expressed genes for smaller number of false positives. Here, we calculated correlations of gene expressions using normal tissue samples obtained from general population (i.e., GTEx) without selection bias, thus expected to identify general functional relationships, not only associated with a specific clinical indication. However, it is noteworthy that RNA-seq data generated for large and extensively phenotyped cohorts would improve our understanding of functional relationships between genes in a specific clinical indication.

Considering the increased prevalence of NAFLD in the worldwide as a hidden epidemic (Younossi et al, 2016), it is not implausible to predict that NAFLD may become responsible for the future clinical and economical burden of HCC (Younossi et al, 2015). It is also reported that HCC patients with NAFLD have poor outcome for disease progression (Younossi et al, 2015). Therefore, it is important to identify common biological pathways and gene targets as driving forces in the pathologies for developing effective therapeutic modalities. In this context, system biology may contribute to identification of novel targets that can be used in the development of efficient treatment strategies. Our recent analysis described in Mardinoglu et al (in preparation) and previous analysis (Mardinoglu et al, 2014a) indicated that DNL is a key pathway involved in the progression of the NAFLD. Previous analysis has also revealed DNL as one of the most tightly regulated pathways in HCC compared with noncancerous liver tissue (Björnson et al, 2015). FASN (rate-limiting enzyme in DNL) is believed to play key roles in NAFLD progression and development and HCC; therefore, inhibiting FASN may downregulate the DNL, decrease the accumulated fat within the cells, and decrease tumor growth. To improve the success rate of NAFLD and HCC treatment and to improve the survival prognosis of HCC patients, identification of nontoxic FASN inhibitors is required. To identify possible inhibitors, correlation analysis was applied to mRNA expression data derived from liver tissue with varying degree of fat and tumor samples. Co-expression analysis in these samples revealed that FASN is co-expressed with a number of genes that play roles in crucial biological processes involved in fat accumulation and cancer cell metabolism.

We analyzed the majority of publicly available liver tissue gene expression data, performed correlation analysis, and ultimately generated co-expression networks for FASN. We identified a number of liver-specific gene targets that can be inhibited with chemical compounds or monoclonal antibodies including PKLR, PCSK9, and PNPLA3, for effective treatment of chronic liver diseases. We finally validated our predictions by demonstrating the functional relationships among the expression of these genes and FASN, liver fat, and cell growth, by using human cancer cell lines, mouse liver samples (four different mouse studies), and human hepatocytes.

Wang et al (2002) have provided evidence that variants in the PKLR gene are associated with an increased risk of T2D, which has a pathogenesis similar to that of NAFLD. Ruscica et al (2016) have associated circulating PCSK9 levels with accumulated liver fat. In 201 consecutive patients biopsied for suspected NASH, liver damage has been quantified by NAFLD activity score, circulating PCSK9 by ELISA, and hepatic mRNA by qRT–PCR in 76 of the patients. Circulating PCSK9 has been found to be significantly associated with hepatic steatosis grade, necroinflammation, ballooning, and fibrosis stage (Ruscica et al, 2016). Circulating PCSK9 has also been found to be significantly associated with hepatic expression of SREBP-1c and FASN, whereas PCSK9 mRNA levels have been found to be significantly correlated with steatosis severity and hepatic APOB, SREBP-1c, and FASN expression (Ruscica et al, 2016). Aragones et al (2016) have evaluated the association between liver PNPLA3 expression, key genes in lipid metabolism, and the presence of NAFLD in morbidly obese women and have reported that PNPLA3 expression was related to HS in these subjects. Their analysis indicates that PNPLA3 may be related to lipid accumulation in the liver, mainly in the development and progression of simple steatosis. PNPLA3 was also emphasized as a genetic determinant of risk factor for the severity of NAFLD (Salameh et al, 2016). Furthermore, higher prevalence for HCC development and poorer prognosis was reported to be associated with PNPLA3 polymorphism in viral and nonviral chronic liver diseases (Khlaiphuengsin et al, 2015). A potential unifying factor upstream of these genes is the cannabinoid-1 receptor, stimulation of which was found to upregulate several of the above-listed target proteins, including Fasn, Pklr, Pnpla3, and Pcsk9 in mouse models of obesity/metabolic syndrome and HCC, as documented and detailed.

Moreover, we found a number of genes, for example, ACACA, were significantly co-expressed with FASN. It has been suggested that inhibition of ACACA may be useful in treating a variety of metabolic disorders, including metabolic syndrome, type 2 diabetes mellitus, and fatty liver disease (Harriman et al, 2016). However, our analysis indicated that potential inhibition of ACACA may have severe side effects in other human tissues as the inhibitors of FASN.

In conclusion, we demonstrated a strategy whereby tissue-specific CNs can be used to identify deregulations of biological functions in response to disease and reveal the effects on relevant expression of genes in liver. Eventually, we identified liver-specific drug targets that can be used in effective treatment of liver diseases including NAFLD and HCC.

Materials and Methods

Tissue-specific CNs

RNA-seq data from human tissues were downloaded from the GTEx database, and their reads per kilobase per million (RPKM) values were transformed into transcripts per kilobase per million (TPM) values. From each RNA-seq dataset, we excluded one-third of the genes with the lowest expression levels from calculating Pearson's correlation coefficients of gene expression and combined the highest correlated gene pairs into a respective tissue CN.

On the basis of the network connectivity of each tissue CN, we clustered co-expressed genes by using the modularity-based random walk method from the cluster walktrap function of the igraph package in R (Pons & Latapy, 2005). Among those gene groups, we selected the half of the highest connected groups as key co-expression clusters, on the basis of their clustering coefficients. We produced a liver CN with those co-expression clusters as nodes and their significant connections by edges (Fig 1B). For example, the edges of two co-expression clusters, A and B, were identified if their observed connections (O_{AB}) were twice as high as the expected connections (E_{AB}). Expected connections were defined by the normalized sum of multiplications of the degree of connectivity of genes in two clusters (i.e., $E_{AB} = \Sigma\ k_a \times k_b/2N$; $a \in A$, $b \in$, N = all edges in the network). Next, we collected genes belonging to key co-expression clusters in each tissue and tested whether they were enriched in biological process GO terms in MSigDB (Subramanian et al, 2005) by using hypergeometric tests.

Comparing physical networks with CNs

We used RNs and PPINs of liver, skeletal muscle, and adipose tissues from our prior published data (Lee et al, 2016) as sources of physical interaction network data. We identified co-regulated gene pairs from the physical interactions by selecting genes sharing TF binding (from RNs) or genes sharing protein interactions (from PPINs); on the basis of the number of co-bound TFs or co-interacting proteins, called co-regulators, we selected the highest co-regulated gene pairs (i.e., gene pairs with the highest numbers of co-bound TFs or co-interacting proteins; top 0.1%) and gene pairs with no co-regulation. Using Pearson's correlation coefficients of the gene expression level of each tissue from GTEx, we calculated the mean Pearson's correlation coefficients of gene pairs of interest, such as physically linked gene pairs or co-regulated gene pairs. The edges of physical networks or their co-regulatory networks were randomly permutated among genes in respective actual networks by 1,000×, and their mean correlation coefficients were compared with those from original networks. We also identified TFs or proteins highly co-expressed with respective bound genes by comparing the co-expression levels of given gene pairs to overall levels (Datasets EV4 and EV5). We selected TFs or proteins if co-expression levels of given linked gene pairs were higher than the overall levels by using Kolmogorov–Smirnov (KS) tests ($P < 0.05$) and absolute values of mean Pearson's correlation coefficients (> 0.1). Finally, we examined mean co-expression levels of co-regulated gene pairs according to their number of co-regulators. Increasing the cutoff number of co-regulators, we selected co-regulated gene pairs exceeding the cutoff and calculated mean Pearson's correlation coefficients of the corresponding gene pairs.

Finding the most influential TFs for co-expression on the basis of variable importance score

For liver tissue, we constructed a feature matrix between the most highly co-expressed gene pairs (top 1%) and their co-bound TFs. In this matrix, we assigned a value of 1 for given co-expressed gene pairs that were co-bound by TFs and zero otherwise. For each co-expressed gene pair, we considered co-bound TFs as predictor variables and the co-expression value as the response variable and fitted them to the Random Forest model. From the model, we calculated variable importance scores of all TFs and used them as a metric to show the most influential TFs for co-expression.

Identifying liver-specific reaction clusters on the basis of tissue co-expression of enzymes

From the HMR2 database, we collected human metabolic reactions with known enzymes. Between the two reactions, we calculated Pearson's correlation coefficients of enzyme gene expression in liver tissue, and if there were multiple enzymes for a single reaction, we took the maximum value among possible co-expressions. On the basis of co-expression values among metabolic reactions, we performed hierarchical clustering and classified reactions into 100 clusters for liver tissue. To compare the co-expression values of metabolic reactions in different tissues and tumor tissue, we calculated the co-expression values in not only liver tissue but also adipose subcutaneous, skeletal muscle, and HCC tumor tissues. Subsequently, we identified the differences in mean co-expression values of given reaction clusters between liver tissue and other tissues after transforming co-expression values into Fisher Z-values. On the basis of differential Fisher Z-values, we took the top 1% of reaction clusters of the highest in each comparison (adipose or muscle tissues) and identified them as liver-specific reaction clusters. Likewise, we identified an HCC-deregulated reaction cluster on the basis of differential Fisher Z-values (top 1%).

Finally, using liver RN, we examined TFs highly bound at genes encoding enzymes in given liver reaction clusters and identified TFs whose binding was significantly enriched in given clusters by hypergeometric tests. From those enriched TFs, we selected reaction clusters in which the enriched TFs had higher variable importance scores than overall scores according to KS tests ($P < 0.25$) and denoted them as highly regulated reaction clusters in liver tissue.

FASN co-expressed genes in human tissues and HCC tumors

In each normal tissue (GTEx) or HCC tumor tissue (TCGA), we calculated Pearson's correlation coefficients of gene expression between FASN and other protein-coding genes expressing more than 1 TPM and selected the top 100 most correlated genes (Datasets EV10 and EV11). In addition, we calculated correlations of log-transformed expression values in HCC tumors (Dataset EV12). To select tissue-specific genes, we examined the top 100 most correlated genes to FASN in HCC tumor tissue by the RNA tissue category in the Human Protein Atlas (ver. 16); in particular, genes "tissue-enhanced" or "tissue-enriched" in liver tissue were selected (Fig 3). We also examined those genes with tissues that were most correlated (Dataset EV10) and selected genes that were present in fewer than three human tissues.

Differential expression analysis of HCC patients stratified based on FASN expressions

Using raw count data from HCC patients, we stratified patients into two groups: patients having FASN expression above the upper quartile and patients having FASN expression under the lower quartile. Between the two groups, we examined differentially expressed genes by negative binomial test using DESeq (Anders & Huber, 2010; Dataset EV13).

Cell line experiments

For subsequent experiments, we selected K562 and HepG2, human immortalized myelogenous leukemia and hepatic cell lines, respectively. Both cell lines were cultured in RPMI-1640 medium (R2405; Sigma-Aldrich) supplemented with 10% fetal bovine serum (FBS, F2442; Sigma-Aldrich) and incubated in 5% CO_2 humidity at 37°C.

To confirm the speculated CN, we chose chemical inhibitors and RNA interference (RNAi) assays to interfere with the immortalized cell lines (including K562 and HepG2) and observed the subsequent candidate gene expression and cell growth patterns. More specifically, the experimental protocol was as follows: (i) C75 (C5490; Sigma-Aldrich), as a well-known fatty acid synthase [FAS] inhibitor, was added to cells at 80% confluence with a final concentration of 20, 40, 60, or 80 μM (taking cells without C75 interference as the control); and (ii) cells at 80% confluence were separately transfected with three pairs of Silencer® pre-designed PKLR-targeted siRNAs (clone ID: 53, 54; Life Technologies; Dataset EV14) at 15 nM by using Lipofectamine® RNAiMAX (13778075; Life Technologies). Cells incubated in medium with nontarget negative control siRNA at 15 nM (4390843; Life Technologies) were assigned as the control.

Total RNA was isolated with TRIzol reagent (15596026, Thermo Fisher Scientific) after treatment with C75 or siRNA for 24 h. The expression profiles of key genes (FASN, PKLR in K562/HepG2, and PNPLA3, PCSK9 only in HepG2 cells) in the co-expression network were measured and analyzed via quantitative real-time PCR with iTaq Universal SYBR Green One-Step Kit (1725151; Bio-Rad), using anchored oligo (dT) primer based on CFX96™ detection system (Bio-Rad). GAPDH was set as the internal control for normalization, and the primer sequences are listed in Dataset EV15. Variation in cell proliferation was detected with a Cell Counting Kit-8 (CCK-8, CK04; Dojindo) after interference by C75/siRNA for 72 h. All experiments were performed strictly according to the manufacturer's instructions and were repeated at least in triplicate for three samples and yielded similar results.

Mouse experiments

Twenty male C57BL/6N mice were fed a standard mouse chow diet (Purina 7012; Harlan Teklad) and housed under a 12-h light–dark cycle. From 8 weeks of age, the mice were fed either a HSD diet (TD.88137; Harlan Laboratories, WI, USA) or CD for 2 weeks. The mice were housed at the University of Gothenburg animal facility (Lab of Exp Biomed) and supervised by university veterinarians and professional staff. The health status of our mice is constantly monitored according to the rules established by the Federation of European Laboratory Animal Science Associations. The experiments were approved by the Gothenburg Ethical Committee on Animal Experiments.

In liver cancer mouse model, we injected 25 mg/kg of DEN (Sigma) to $CB_1R^{+/+}$ and $CB_1R^{-/-}$ littermates in C57BL/6J background after 2 weeks of birth, verified the presence of the HCC tumor, and measured its size with magnetic resonance imaging (MRI) 8 months after the DEN administration (Mukhopadhyay et al, 2015). It has been observed that activation of hepatic CB_1R promoted the initiation and progression of chemically induced HCC in mice (Mukhopadhyay et al, 2015). Total RNA was isolated from tumor area and noncancerous area of liver tissue samples obtained from six DEN-treated (HCC) $CB_1R^{+/+}$ and six $CB_1R^{-/-}$ mice (Mukhopadhyay et al, 2015). rRNA-depleted RNA, 100 ng for each sample, was treated with RNase III to generate 100- to 200-nt fragments, which were pooled and processed for RNA sequencing. All data were normalized based on housekeeping genes used by CLC Genomics Workbench program (version 5.1; CLC Bio, Boston, USA). These absolute numbers were extracted from the reads, and the data were adjusted to non-HCC biopsies for each gene.

Human hepatocytes

Human primary hepatocytes were purchased from Biopredic International. Twenty-four hours after arrival (48 h after isolation), human hepatocytes were treated with 40 μg/ml of C75 or dimethyl sulfoxide (DMSO) for 4, 6, 8, or 24 h.

mRNA expression in mouse liver and human primary hepatocytes

Total RNA was isolated from human hepatocytes and snap-frozen mouse liver with an RNeasy Mini Kit (Qiagen). cDNA was synthesized with a high-capacity cDNA Reverse Transcription Kit (Applied Biosystems) and random primers. The mRNA expression levels of genes of interest were analyzed via TaqMan real-time PCR in a ViiATM7 System (Applied Biosystems). The TaqMan Gene Expression assays used were Mm01263610_m1 (for mouse Pcsk9), Mm00662319_m1 (mouse Fasn), Mm00443090_m1 (mouse Pklr), Mm00504420_m1 (mouse Pnpla3), Mm01282499_m1 (mouse Hmgcr), Hs00545399_m1 (human PCSK9), Hs01005622_m1 (human FASN), Hs00176075_m1 (human PKLR), Hs00228747_m1 (human PNPLA3), and Hs00168352_m1 (human HMGCR) (all from Applied Biosystems). Hprt (mouse Mm03024075_m1) and GADPH (human Hs02758991_g1) (Applied Biosystems) were used as internal controls.

Acknowledgements

This work was financially supported by the Knut and Alice Wallenberg Foundation, Swedish Research Foundation, and EU Seventh Framework Programme RESOLVE. The research leading to these results received support from the Innovative Medicines Initiative Joint Undertaking under EMIF grant agreement no. 115372. The computations of network generations were performed using resources provided by the Swedish National Infrastructure for Computing (SNIC) at C3SE and UPPMAX.

Author contributions

SL, CZ, and AM generated the co-expression networks and analyzed the clinical data, together with TK, MG, BDP, MSn, JN, and MU. ZL, MK, BM, MB, RC, NS, SD, SJE, FB, GK, MSt, JKP, AMH, US, and JB measured the expression levels of the genes in the cell lines, mouse liver samples, and human hepatocytes. SL, CZ, and AM wrote the manuscript, and all authors were involved in editing the manuscript.

References

Agren R, Bordel S, Mardinoglu A, Pornputtapong N, Nookaew I, Nielsen J (2012) Reconstruction of genome-scale active metabolic networks for 69 human cell types and 16 cancer types using INIT. *PLoS Comput Biol* 8: e1002518

Agren R, Mardinoglu A, Asplund A, Kampf C, Uhlen M, Nielsen J (2014) Identification of anticancer drugs for hepatocellular carcinoma through personalized genome-scale metabolic modeling. *Mol Syst Biol* 10: 721

Anders S, Huber W (2010) Differential expression analysis for sequence count data. *Genome Biol* 11: R106

Aragones G, Auguet T, Armengol S, Berlanga A, Guiu-Jurado E, Aguilar C, Martinez S, Sabench F, Porras JA, Ruiz MD, Hernandez M, Sirvent JJ, Del Castillo D, Richart C (2016) PNPLA3 expression is related to liver steatosis in morbidly obese women with non-alcoholic fatty liver disease. *Int J Mol Sci* 17: 630

Björnson E, Mukhopadhyay B, Asplund A, Pristovsek N, Cinar R, Romeo S, Uhlen M, Kunos G, Nielsen J, Mardinoglu A (2015) Stratification of hepatocellular carcinoma patients based on acetate utilization. *Cell Rep* 13: 2014–2026

Bordbar A, Monk JM, King ZA, Palsson BO (2014) Constraint-based models predict metabolic and associated cellular functions. *Nat Rev Genet* 15: 107–120

Dorn C, Riener MO, Kirovski G, Saugspier M, Steib K, Weiss TS, Gabele E, Kristiansen G, Hartmann A, Hellerbrand C (2010) Expression of fatty acid synthase in nonalcoholic fatty liver disease. *Int J Clin Exp Pathol* 3: 505–514

Elsemman IE, Mardinoglu A, Shoaie S, Soliman TH, Nielsen J (2016) Systems biology analysis of hepatitis C virus infection reveals the role of copy number increases in regions of chromosome 1q in hepatocellular carcinoma metabolism. *Mol BioSyst* 12: 1496–1506

Giannone FA, Baldassarre M, Domenicali M, Zaccherini G, Trevisani F, Bernardi M, Caraceni P (2012) Reversal of liver fibrosis by the antagonism of endocannabinoid CB1 receptor in a rat model of CCl(4)-induced advanced cirrhosis. *Lab Invest* 92: 384–395

GTEx (2013) The genotype-tissue expression (GTEx) project. *Nat Genet* 45: 580–585

Harriman G, Greenwood J, Bhat S, Huang X, Wang R, Paul D, Tong L, Saha AK, Westlin WF, Kapeller R, Harwood HJ Jr (2016) Acetyl-CoA carboxylase inhibition by ND-630 reduces hepatic steatosis, improves insulin sensitivity, and modulates dyslipidemia in rats. *Proc Natl Acad Sci USA* 113: E1796–E1805

Hezode C, Roudot-Thoraval F, Nguyen S, Grenard P, Julien B, Zafrani ES, Pawlotsky JM, Dhumeaux D, Lotersztajn S, Mallat A (2005) Daily cannabis smoking as a risk factor for progression of fibrosis in chronic hepatitis C. *Hepatology* 42: 63–71

Huang DW, Sherman BT, Lempicki RA (2009) Systematic and integrative analysis of large gene lists using DAVID bioinformatics resources. *Nat Protoc* 4: 44–57

Hyötyläinen T, Jerby L, Petäjä EM, Mattila I., Jäntti S, Auvinen P, Gastaldelli A, Yki-Järvinen H, Ruppin E, Orešič M (2016) Genome-scale study reveals reduced metabolic adaptability in patients with non-alcoholic fatty liver disease. *Nat Commun* 7: 8994

Jeong WI, Osei-Hyiaman D, Park O, Liu J, Batkai S, Mukhopadhyay P, Horiguchi N, Harvey-White J, Marsicano G, Lutz B, Gao B, Kunos G (2008) Paracrine activation of hepatic CB1 receptors by stellate cell-derived endocannabinoids mediates alcoholic fatty liver. *Cell Metab* 7: 227–235

Jourdan T, Demizieux L, Gresti J, Djaouti L, Gaba L, Verges B, Degrace P (2012) Antagonism of peripheral hepatic cannabinoid receptor-1 improves liver lipid metabolism in mice: evidence from cultured explants. *Hepatology* 55: 790–799

Kampf C, Mardinoglu A, Fagerberg L, Hallström B, Edlund K, Nielsen J, Uhlen M (2014) The human liver-specific proteome defined by transcriptomics and antibody-based profiling. *Faseb J* 28: 2901–2914

Kew MC (2010) Epidemiology of chronic hepatitis B virus infection, hepatocellular carcinoma, and hepatitis B virus-induced hepatocellular carcinoma. *Pathol Biol (Paris)* 58: 273–277

Khlaiphuengsin A, Kiatbumrung R, Payungporn S, Pinjaroen N, Tangkijvanich P (2015) Association of PNPLA3 polymorphism with hepatocellular carcinoma development and prognosis in viral and non-viral chronic liver diseases. *Asian Pac J Cancer Prev* 16: 8377–8382

Lee S, Zhang C, Kilicarslan M, Bluher M, Uhlen M, Nielsen J, Smith U, Serlie MJ, Boren J, Mardinoglu A (2016) Integrated network analysis reveals an association between plasma mannose levels and insulin resistance. *Cell Metab* 24: 172–184

Mardinoglu A, Nielsen J (2012) Systems medicine and metabolic modelling. *J Intern Med* 271: 142–154

Mardinoglu A, Agren R, Kampf C, Asplund A, Nookaew I, Jacobson P, Walley AJ, Froguel P, Carlsson LM, Uhlen M, Nielsen J (2013a) Integration of clinical data with a genome-scale metabolic model of the human adipocyte. *Mol Syst Biol* 9: 649

Mardinoglu A, Gatto F, Nielsen J (2013b) Genome-scale modeling of human metabolism – a systems biology approach. *Biotechnol J* 8: 985–996

Mardinoglu A, Agren R, Kampf C, Asplund A, Uhlen M, Nielsen J (2014a) Genome-scale metabolic modelling of hepatocytes reveals serine deficiency in patients with non-alcoholic fatty liver disease. *Nat Commun* 5: 3083

Mardinoglu A, Kampf C, Asplund A, Fagerberg L, Hallstrom BM, Edlund K, Bluher M, Ponten F, Uhlen M, Nielsen J (2014b) Defining the human adipose tissue proteome to reveal metabolic alterations in obesity. *J Proteome Res* 13: 5106–5119

Mardinoglu A, Nielsen J (2015) New paradigms for metabolic modeling of human cells. *Curr Opin Biotech* 34: 91–97

Mardinoglu A, Shoaie S, Bergentall M, Ghaffari P, Zhang C, Larsson E, Backhed F, Nielsen J (2015) The gut microbiota modulates host amino acid and glutathione metabolism in mice. *Mol Syst Biol* 11: 834

Mardinoglu A, Nielsen J (2016) Editorial: the impact of systems medicine on human health and disease. *Front Physiol* 7: 552

Mardinoglu A, Bjornson E, Zhang C, Klevstig M, Soderlund S, Stahlman M, Adiels M, Hakkarainen A, Lundbom N, Kilicarslan M, Hallstrom BM, Lundbom J, Verges B, Barrett PH, Watts GF, Serlie MJ, Nielsen J, Uhlen M, Smith U, Marschall HU et al (2017a) Personal model-assisted identification of NAD$^+$ and glutathione metabolism as intervention target in NAFLD. *Mol Syst Biol* 13: 916

Mardinoglu A, Boren J, Smith U, Uhlen M, Nielsen J (2017b) The employment of systems biology in gastroenterology and hepatology. *Nat Rev Gastroenterol Hepatol* In press

Mukhopadhyay B, Schuebel K, Mukhopadhyay P, Cinar R, Godlewski G, Xiong K, Mackie K, Lizak M, Yuan Q, Goldman D, Kunos G (2015) Cannabinoid receptor 1 promotes hepatocellular carcinoma initiation and progression through multiple mechanisms. *Hepatology* 61: 1615–1626

Nielsen J (2017) Systems biology of metabolism: a driver for developing personalized and precision medicine. *Cell Metab* 25: 572–579

O'Brien EJ, Monk JM, Palsson BO (2015) Using genome-scale models to predict biological capabilities. *Cell* 161: 971–987

Osei-Hyiaman D, DePetrillo M, Pacher P, Liu J, Radaeva S, Batkai S, Harvey-White J, Mackie K, Offertaler L, Wang L, Kunos G (2005) Endocannabinoid activation at hepatic CB1 receptors stimulates fatty acid synthesis and contributes to diet-induced obesity. *J Clin Investig* 115: 1298–1305

Osei-Hyiaman D, Liu J, Zhou L, Godlewski G, Harvey-White J, Jeong WI, Batkai S, Marsicano G, Lutz B, Buettner C, Kunos G (2008) Hepatic CB1 receptor is required for development of diet-induced steatosis, dyslipidemia, and insulin and leptin resistance in mice. *J Clin Investig* 118: 3160–3169

Pandey PR, Liu W, Xing F, Fukuda K, Watabe K (2012) Anti-cancer drugs targeting fatty acid synthase (FAS). *Recent Pat Anticancer Drug Discov* 7: 185–197

Petrick JL, Kelly SP, Altekruse SF, McGlynn KA, Rosenberg PS (2016) Future of hepatocellular carcinoma incidence in the United States forecast through 2030. *J Clin Oncol* 34: 1787–1794

Pons P, Latapy M (2005) Computing communities in large networks using random walks. *ISCIS* 3733: 284–293

Ruscica M, Ferri N, Macchi C, Meroni M, Lanti C, Ricci C, Maggioni M, Fracanzani AL, Badiali S, Fargion S, Magni P, Valenti L, Dongiovanni P (2016) Liver fat accumulation is associated with circulating PCSK9. *Ann Med* 48: 384–391

Salameh H, Hanayneh MA, Masadeh M, Naseemuddin M, Matin T, Erwin A, Singal AK (2016) PNPLA3 as a genetic determinant of risk for and severity of non-alcoholic fatty liver disease spectrum. *J Clin Transl Hepatol* 4: 175–191

Shaker M, Tabbaa A, Albeldawi M, Alkhouri N (2014) Liver transplantation for nonalcoholic fatty liver disease: new challenges and new opportunities. *World J Gastroenterol* 20: 5320–5330

Shoaie S, Ghaffari P, Kovatcheva-Datchary P, Mardinoglu A, Sen P, Pujos-Guillot E, de Wouters T, Juste C, Rizkalla S, Chilloux J, Hoyles L, Nicholson JK, Dore J, Dumas ME, Clement K, Backhed F, Nielsen J (2015) Quantifying diet-induced metabolic changes of the human gut microbiome. *Cell Metab* 22: 320–331

Subramanian A, Tamayo P, Mootha VK, Mukherjee S, Ebert BL, Gillette MA, Paulovich A, Pomeroy SL, Golub TR, Lander ES, Mesirov JP (2005) Gene set enrichment analysis: a knowledge-based approach for interpreting genome-wide expression profiles. *Proc Natl Acad Sci USA* 102: 15545–15550

Teixeira-Clerc F, Julien B, Grenard P, Tran Van Nhieu J, Deveaux V, Li L, Serriere-Lanneau V, Ledent C, Mallat A, Lotersztajn S (2006) CB1 cannabinoid receptor antagonism: a new strategy for the treatment of liver fibrosis. *Nat Med* 12: 671–676

Thul PJ, Akesson L, Wiking M, Mahdessian D, Geladaki A, Ait Blal H, Alm T, Asplund A, Bjork L, Breckels LM, Backstrom A, Danielsson F, Fagerberg L, Fall J, Gatto L, Gnann C, Hober S, Hjelmare M, Johansson F, Lee S *et al* (2017) A subcellular map of the human proteome. *Science* 356: eaal3321

Uhlen M, Fagerberg L, Hallstrom BM, Lindskog C, Oksvold P, Mardinoglu A, Sivertsson A, Kampf C, Sjöstedt E, Asplund A, Lundberg E, Djureinovic D, Odeberg J, Habuka M, Tahmasebpoor S, Danielsson A, Edlund K, Szigyarto CA, Skogs M, Takanen JO *et al* (2015) Tissue-based map of the human proteome. *Science* 347: 1260419

Uhlen M, Hallstrom BM, Lindskog C, Mardinoglu A, Ponten F, Nielsen J (2016) Transcriptomics resources of human tissues and organs. *Mol Syst Biol* 12: 862

Varemo L, Scheele C, Broholm C, Mardinoglu A, Kampf C, Asplund A, Nookaew I, Uhlén M, Pedersen BK, Nielsen J (2015) Transcriptome and proteome driven reconstruction of the human myocyte metabolic model and its use for identification of metabolic markers for type 2 diabetes. *Cell Rep* 11: 921–933

Wang H, Chu W, Das SK, Ren Q, Hasstedt SJ, Elbein SC (2002) Liver pyruvate kinase polymorphisms are associated with type 2 diabetes in northern European Caucasians. *Diabetes* 51: 2861–2865

Yizhak K, Gabay O, Cohen H, Ruppin E (2013) Model-based identification of drug targets that revert disrupted metabolism and its application to ageing. *Nat Commun* 4: 2632

Yizhak K, Gaude E, Le Devedec S, Waldman YY, Stein GY, van de Water B, Frezza C, Ruppin E (2014a) Phenotype-based cell-specific metabolic modeling reveals metabolic liabilities of cancer. *Elife* 3: e03641

Yizhak K, Le Devedec SE, Rogkoti VM, Baenke F, de Boer VC, Frezza C, Schulze A, van de Water B, Ruppin E (2014b) A computational study of the Warburg effect identifies metabolic targets inhibiting cancer migration. *Mol Syst Biol* 10: 744

Yizhak K, Chaneton B, Gottlieb E, Ruppin E (2015) Modeling cancer metabolism on a genome scale. *Mol Syst Biol* 11: 817

Younossi ZM, Otgonsuren M, Henry L, Venkatesan C, Mishra A, Erario M, Hunt S (2015) Association of nonalcoholic fatty liver disease (NAFLD) with hepatocellular carcinoma (HCC) in the United States from 2004 to 2009. *Hepatology* 62: 1723–1730

Younossi ZM, Koenig AB, Abdelatif D, Fazel Y, Henry L, Wymer M (2016) Global epidemiology of nonalcoholic fatty liver disease-Meta-analytic assessment of prevalence, incidence, and outcomes. *Hepatology* 64: 73–84

Zhang C, Ji B, Mardinoglu A, Nielsen J, Hua Q (2015) Logical transformation of genome-scale metabolic models for gene level applications and analysis. *Bioinformatics* 31: 2324–2331

Zhang C, Lee S, Mardinoglu A, Hua Q (2016) Investigating the combinatory effects of biological networks on gene co-expression. *Front Physiol* 7: 160

4

Mathematical modeling links Wnt signaling to emergent patterns of metabolism in colon cancer

Mary Lee[1,†], George T Chen[2,†], Eric Puttock[1], Kehui Wang[3,4], Robert A Edwards[3,4], Marian L Waterman[2,4,5,*] & John Lowengrub[1,4,5,6,**] (iD)

Abstract

Cell-intrinsic metabolic reprogramming is a hallmark of cancer that provides anabolic support to cell proliferation. How reprogramming influences tumor heterogeneity or drug sensitivities is not well understood. Here, we report a self-organizing spatial pattern of glycolysis in xenograft colon tumors where pyruvate dehydrogenase kinase (PDK1), a negative regulator of oxidative phosphorylation, is highly active in clusters of cells arranged in a spotted array. To understand this pattern, we developed a reaction–diffusion model that incorporates Wnt signaling, a pathway known to upregulate PDK1 and Warburg metabolism. Partial interference with Wnt alters the size and intensity of the spotted pattern in tumors and in the model. The model predicts that Wnt inhibition should trigger an increase in proteins that enhance the range of Wnt ligand diffusion. Not only was this prediction validated in xenograft tumors but similar patterns also emerge in radiochemotherapy-treated colorectal cancer. The model also predicts that inhibitors that target glycolysis or Wnt signaling in combination should synergize and be more effective than each treatment individually. We validated this prediction in 3D colon tumor spheroids.

Keywords glycolysis; spatial pattern; tumor metabolism; Warburg effect; Wnt signaling

Subject Categories Cancer; Quantitative Biology & Dynamical Systems; Signal Transduction

Introduction

A hallmark feature of many cancers is "aerobic glycolysis", or the Warburg effect, a form of metabolism whereby cells skew their balance of cellular metabolism away from oxidative phosphorylation (OXPHOS) to favor glycolysis, despite the availability of sufficient levels of oxygen (Warburg, 1956). Cellular emphasis on Warburg metabolism is intriguing since it is much less efficient than OXPHOS in producing energy (four molecules of ATP produced by glycolysis for each molecule of glucose consumed versus 36 molecules by OXPHOS). Warburg metabolism has been hypothesized to be beneficial because glycolytic intermediates can be used as biosynthetic building blocks for cell growth and proliferation, suggesting that this mode of glucose utilization is essential for actively expanding tumors (Vander Heiden et al, 2009; Pavlova & Thompson, 2016). There are other effects as well: The production of lactate acidifies the tumor microenvironment, an environmental condition that can enhance tumor invasiveness (Gatenby & Gillies, 2004), and induces angiogenic responses for increased delivery of glucose, oxygen, and other nutrients (Végran et al, 2011), effects that are growth promoting and provide cancer cells with a fitness advantage.

Oncogenic, overactive Wnt signaling has been recently linked to metabolic and nutrient programming in tumors. For example, in colon cancer, Wnt signaling is proposed to increase expression of key glycolytic factors that enhance Warburg metabolism and angiogenesis (Pate et al, 2014). Oncogenic Wnt signaling most commonly derives from genetic inactivation of one or more signaling components (e.g., adenomatous polyposis coli, APC), inactivating mutations that cause the pathway to become chronically activated and to trigger overexpression of Wnt target genes. One such target gene is pyruvate dehydrogenase kinase 1 (PDK1), a mitochondrial kinase that inhibits the pyruvate dehydrogenase complex (PDC) via phosphorylation of the component pyruvate dehydrogenase (PDH) (Pate et al, 2014). Since PDC converts pyruvate to acetyl CoA for mitochondrial respiration, phosphorylation/inhibition of PDH by PDK1 suppresses OXPHOS modes of metabolism (ATP and CO_2

1 Department of Mathematics, University of California, Irvine, Irvine, CA, USA
2 Department of Microbiology and Molecular Genetics, University of California, Irvine, Irvine, CA, USA
3 Department of Pathology, School of Medicine, University of California, Irvine, Irvine, CA, USA
4 Chao Family Comprehensive Cancer Center, University of California, Irvine, Irvine, CA, USA
5 Center for Complex Biological Systems, University of California, Irvine,Irvine, CA, USA
6 Department of Biomedical Engineering, University of California, Irvine, Irvine, CA, USA
   *Corresponding author. E-mail: marian.waterman@uci.edu
   **Corresponding author. E-mail: jlowengr@uci.edu
   †These authors contributed equally to this work

production) to favor glycolytic modes that produce lactate (Roche et al, 2001). Thus, at least in some tissues such as colon, Wnt signaling elevates PDK1 to suppress OXPHOS and to encourage glycolysis and the production of lactate.

Our previous study of xenograft colon tumors established that oncogenic Wnt signals directly activate PDK1 gene transcription as well as other glycolysis-connected gene targets including the lactate transporter MCT1 (SLC16A1) (Pate et al, 2014; Sprowl-Tanio et al, 2016). That Wnt signals might be directly responsible for shaping the metabolic profile of cells is a discovery from multiple studies focused on diseased [e.g., melanoma, breast (Sherwood, 2015)] and normal tissues (Esen et al, 2013). At least two types of Wnt signals have been defined. One signal utilizes canonical signaling and β-catenin-regulated transcription to drive sustained expression of glycolysis regulators. A second signal utilizes a novel Rac-mTORC2 pathway to increase the protein levels of glycolytic enzymes in the cytoplasm (Esen et al, 2013). Both signals can be triggered by secreted Wnt ligands, and these, in addition to oncogenic Wnt pathway activities created by genetic mutations, can direct the metabolic and proliferative capacity of colon tumors. However, because metabolism is shaped by the collective activity of multiple pathways and environmental influences—including those that enhance or diminish Wnt signaling—there is still much to learn about how signatures of metabolism are established.

Metabolic symbiosis has emerged as a powerful model to explain tumor heterogeneity and survival. As a concept, metabolic symbiosis means that glycolysis is not a singular metabolic choice for cells in a tumor; OXPHOS modes of metabolism may be dominant in subpopulations. The proposed outcome of this heterogeneity is that cooperation between two groups of cancer cells can maximize delivery and consumption of nutrients and minimize the environmental stresses that are imposed on a tumor. Glycolytic cells are likely the dominant consumers of glucose, and their fermentation of this carbon source produces an acidic by-product (lactate) that must be exported to the tumor microenvironment. Lactate can be angiogenic, and thus, the activities of glycolytic cells can be important for delivery of nutrients and growth factors to the tumor microenvironment (Murray & Wilson, 2001; Sonveaux et al, 2012). In turn, cancer cells with prominent modes of OXPHOS metabolism can uptake and utilize lactate (and other metabolic by-products) from neighboring glycolytic cells and metabolize it as a stable source of energy over long time scales (De Saedeleer et al, 2012; Epstein et al, 2014). An important example of this is the "reverse Warburg" effect observed in breast cancer (Martinez-Outschoorn et al, 2014). Thus, not all cancer cells show a preference for glycolysis at all times because microenvironmental, spatial, and temporal factors may direct them to emphasize OXPHOS modes of metabolism (Sonveaux et al, 2008; Pavlides et al, 2009; Obre & Rossignol, 2015). Such back-and-forth influences on glycolysis and OXPHOS create nongenetic tumor heterogeneity, meaning that genetically identical cancer cells might adopt different modes of metabolism depending on cell-intrinsic and cell-extrinsic influences. Identifying these influences and signals, and understanding the spatial and temporal forces that direct their cooperation is important, as metabolic symbiosis is not just a manifestation of tumor heterogeneity, but it is likely a fundamental aspect of tumor survival.

In the course of our study of Wnt signaling and glycolysis in xenograft colon tumors, we observed heterogeneous patterns of metabolism. Heterogeneity was observed via immunohistochemical stain of PDK1 activity, a major inhibitor of mitochondrial activity, and immunohistochemical stains of Wnt signaling. In particular, these stains revealed a pattern of discrete clusters of cells, or "spots", indicating groups of cells with different levels of glycolysis relative to OXPHOS, and differences in Wnt activity. We refer to these groups of cells as "glycolytic (P_g)" and "OXPHOS (P_o)" to indicate that they differ in the relative balance between these two modes of metabolism. As the spots of PDK activity and Wnt signaling appeared as a regular array in space, we hypothesized that metabolism was subject to rules of pattern formation, and we therefore developed a mathematical model with spatial features to study the organization of this pattern. Using reaction–diffusion equations to describe the dynamics of Wnt signaling, nutrients, cell substrates, and the populations of the different metabolic cell types, we elucidate the mechanisms that underlie this spatial pattern and find good agreement between the model and experiments. We lastly exploit this knowledge to identify promising therapeutic strategies.

Results

A spotted pattern of PDK activity and LEF-1 expression in xenograft tumors

Xenograft tumors of (human) colon cancer cell line SW480 (containing homozygous loss-of-function mutations in APC and intrinsically activated Wnt signaling) were produced by subcutaneous injection of cells in immunocompromised mice. To investigate metabolic changes within the tumor, 5.0- to 6.0-μm serial sections of formalin-fixed, paraffin-embedded tumor were probed with antisera specific for phosphorylated PDH (pPDH) as an indicator of PDK activity, and lymphoid enhancer factor-1 (LEF-1), a Wnt signaling transcription factor and Wnt target gene (Hovanes et al, 2001). Both stains revealed a general, high level of pPDH and LEF-1, but also heterogeneity in the form of a prominent spotted pattern (Fig 1A and B, where "mock" refers to tumors from parental SW480 cells). The pattern appeared as discrete localized clusters of cells with increased levels of pPDH, and these clusters, or spots, were detected at seemingly regular intervals. Since pPDH staining is an indicator of PDK activity, the darker stained cell clusters indicate increased rates of glycolysis relative to neighboring cells. The lighter staining, neighboring cells are likely utilizing greater levels of OXPHOS since PDH is less inhibited (less phosphorylated). Since it is known that lactate, the secreted by-product of glycolytic cells, can be imported into neighboring cells for use as an OXPHOS metabolic fuel, this pattern points to a potential symbiotic spatial relationship between these two cell populations, a metabolic relationship proposed by other groups studying cancer metabolism (Sonveaux et al, 2008; Pavlides et al, 2009)—glycolytic cells that are localized in distinct regions uptake glucose and produce metabolic fuel such as lactate for surrounding oxidative cells, a mode of sharing and metabolic distribution. In addition to the spotted metabolic pattern, an overlying gradient in pPDH staining level was observed wherein the spots were more densely arrayed toward the outer edges of the tumor, decreasing in frequency toward the center of the tumor, suggesting that more glycolysis occurs at the outer regions of the tumor where there is more vasculature (Pate et al, 2014; Appendix A1.4).

SW480 Mock Xenograft Tumor
A Phosphorylated PDH

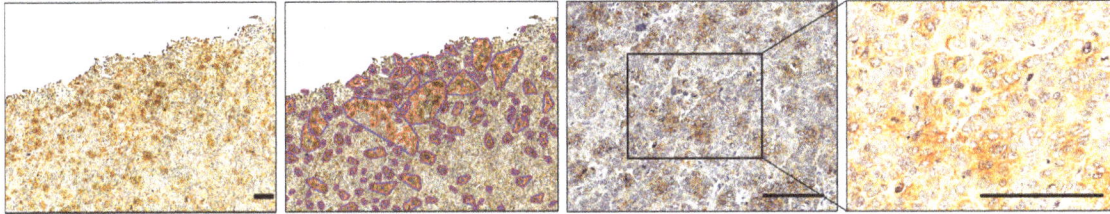

B LEF1

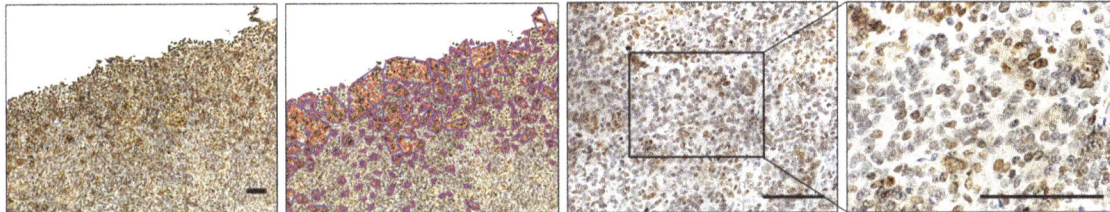

C

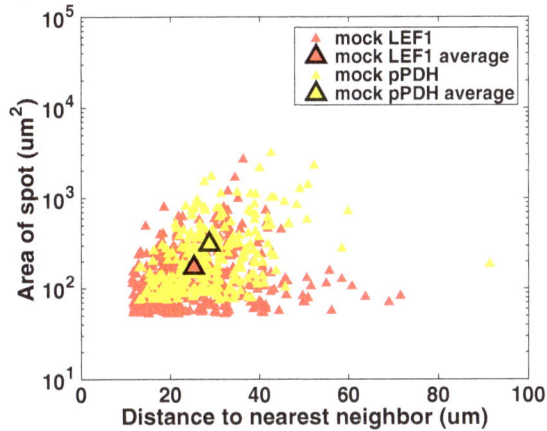

D Phosphorylated PDH   LEF1

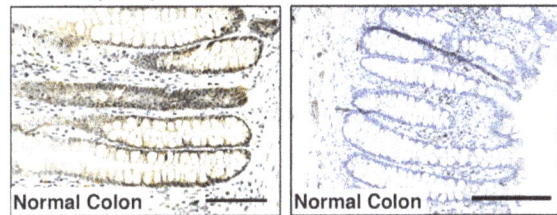

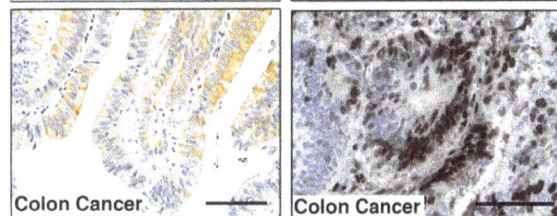

E Phosphorylated PDH   LEF1

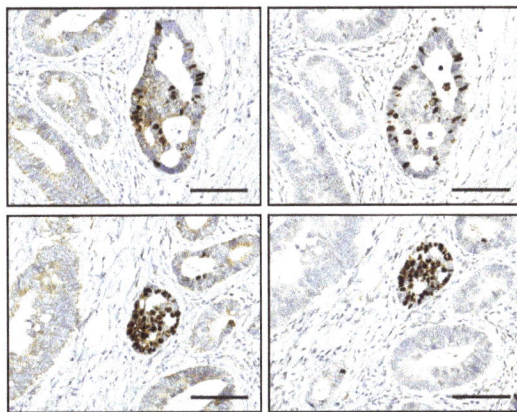

Figure 1. SW480 xenograft tumors reveal a spotted pattern of metabolic heterogeneity.

A, B   SW480 cells lentivirally transduced with empty pCDH vector (mock) were subcutaneously injected into immunocompromised mice. The resulting tumors were stained for (A) phosphorylated pyruvate dehydrogenase (pPDH) and counterstained with hematoxylin or (B) lymphoid enhancer factor-1 (LEF-1). Scale bars indicate 100 μm in the series of 4×, 20×, and 40× images. The red curves denote spot contours and the blue curves denote convex hulls, which group together spots that are sufficiently close to one another (see Appendix A1).

C   Image analysis of spot size versus distance of spot to nearest neighbor, using analyzed 20× images (third panels of A and B). The outlined data points indicate the average distance and area for pPDH and LEF-1 spots. Results show that quantifiable features of the spotted patterns in pPDH and LEF-1 are similar.

D   Colorectal carcinoma patient samples (tumors 1, 2, and 3) stained for pPDH (top) and LEF-1 (bottom) show spatial heterogeneity in expression levels. Scale bars are 100 μm (LEF-1 samples from Uhlén et al, 2015).

E   Serial section of human colorectal carcinoma stained with pPDH and LEF-1 antisera. Scale bars are 100 μm.

As we previously identified a link between Wnt signaling and glycolysis, we used immunohistochemistry to assess the activity of the canonical pathway. Interestingly, a spotted pattern was also evident in immunohistochemical stains of the Wnt target gene and effector, LEF-1 (Fig 1B), indicating that the spotted array might be linked to a pattern of heterogeneity in Wnt signaling. Automated image analysis was used to quantify the spatial parameters of each of the spotted patterns (see Appendix A1.1–A1.3 and Appendix Figs S1–S5). Figure 1C shows the quantification of each spot area and distances to each nearest neighbor, showing that the parameters of the spots for pPDH and LEF-1 are very similar (data on the number of cells per spot are given in Appendix A1.15). We found that the total area fraction of tumor covered by each set of spots in Fig 1A and B was nearly the same (pPDH: 21.2%; LEF-1: 20.2%). To assess the overlap between the pPDH and LEF-1 spots in the serial sections in Fig 1A and B (see Appendix A1.2), we counted spots that partially overlap and found that there was a significant overlap of 65–77% in the spatial arrangement of the pPDH and LEF-1 spots (Fig EV1). We also found that the area fraction of tumor covered by the overlapping region (pPDH spots that are contained in LEF-1 spots and LEF-1 spots that are contained in pPDH spots) is 7.4%.

To determine the significance of the association between the spots (see Appendix A1.2 for details), we analyzed staining in pairs of pixels, assuming that each pixel location in one section corresponds to the same pixel location in the other section. We performed a Cochran–Mantel–Haenszel test (Cochran, 1954; Mantel & Haenszel, 1959) and found that the pPDH and LEF-1 spots are significantly associated with one another ($P < 0.0001$). While this analysis is not definitive because it does not guarantee that the paired pixels are in the same cell (we found it difficult to directly match cells in the serial sections) and also does not take into account spatial variation in spot densities, it suggests that the patterned heterogeneity of metabolism and Wnt signaling are linked.

Xenograft tumors from colon cancer cell lines are different from primary human colon cancers, the latter of which develop in immunocompetent patients and contain a greater variety of cell types and stromal involvement. We asked whether PDK activity and Wnt signaling were uniform or heterogeneous in primary human colon tumors. In Fig 1D, pPDH and LEF-1 stains in primary human colon tumors compared to normal colon tissue demonstrate that there is indeed significant spatial heterogeneity in human tumors. In addition, serial sections of a primary human colon tumor stained with pPDH and LEF-1 antisera show a striking concordance in expression pattern (Fig 1E). While a regular spotted array is not apparent in primary tumors like it is in xenografts, the heterogeneous pattern of clusters of cells with high glycolysis and high LEF-1 in the epithelial portion of the tumor suggests that although xenograft tumors are artificial and have a different microenvironment, understanding the mechanisms underlying the observed spatial patterning in xenograft tumors can provide insight into the forces that create nongenetic heterogeneity in primary human colon tumors.

Reaction–diffusion modeling mimics the self-organizing patterns of PDK activity and Wnt signaling in xenograft tumors

The regular spotted pattern in the xenograft IHC stains suggests the development of a mathematical model consisting of reaction–diffusion equations similar to those first described by Alan Turing (Turing, 1952). Turing's equations describe how an initial perturbation in the concentrations of chemicals, or morphogens, can grow in the presence of diffusion (the Turing instability) and self-organize into a spatial pattern. Because diffusion is normally a stabilizing process, diffusion-driven instabilities occur only under certain conditions (Murray, 2003; Kondo & Miura, 2010). Recently, Marcon et al (2016) performed an automated analysis of Turing-type reaction–diffusion equations and identified general conditions for which instabilities could occur. When two species are considered (e.g., activator–inhibitor models), the species need to diffuse at sufficiently different rates as observed previously (e.g., short-range activator, long-range inhibitor). However, when multiple diffusing species are present, instabilities can be obtained even for arbitrary diffusivities. Here, we focus on reaction–diffusion models that link cell metabolic phenotypes with Wnt signaling and argue that conditions for instability are met in colon cancer.

Despite the fact that colon cancers are most often driven by genetically activated Wnt signaling, a cell-autonomous condition, there are numerous studies that highlight that secreted Wnt ligands and their bona fide signaling through Frizzled receptors on the plasma membrane are abundantly active in human colon cancer and that they influence colon cancer biology (Holcombe et al, 2002; Seshagiri et al, 2012; Voloshanenko et al, 2013; Giannakis et al, 2014). Importantly, Wnt ligands are highly constrained in their diffusion, traveling only one to two cells from the origin of their secretion, meaning that the range of their influence is highly localized (Farin et al, 2016). This is in contrast to the longer-range diffusion properties of known, secreted inhibitors that bind to Wnt ligands and/or interfere with receptor binding (i.e., DKK, SFRPs) (Mii & Taira, 2009, 2011). Some of these inhibitors are Wnt target genes, for example, DKK4, an inhibitor that is expressed in human colon cancer, and SFRP2, a secreted Wnt inhibitor induced by Wnt4 in the developing kidney (Lescher et al, 1998). Thus, a Turing-type model, wherein short-range nonlinear activation by Wnt ligands and long-range inhibition of their activities, fits well with the known physical and regulatory properties of Wnts and their inhibitors. Moreover, this type of model is capable of forming patterns (Murray, 2003; Kondo & Miura, 2010). Previously, Turing models have been used to describe Wnt-directed patterns in a variety of contexts including hair follicles (Sick et al, 2006; Kondo & Miura, 2010), colon crypts (Zhang et al, 2012), limb development (Raspopovic et al, 2014), and stem cell-driven cancers (Youssefpour et al, 2012; Yan et al, 2016). Additionally, the BMP family, known to be Wnt signaling antagonists, has been recently described to direct murine intestinal patterning (Walton et al, 2015).

We therefore developed a Turing-type model for simulating the spatial and temporal dynamics of different metabolic phenotypes, nutrients, and Wnt signaling activity through a system of reaction–diffusion equations (Fig 2A and B; Appendix A2 and A3). We included populations of cells that perform less glycolysis, which we refer to as oxidative (P_o) cells, and those that perform more glycolysis, which are termed glycolytic (P_g) cells. Both types of cells may divide, die, and undergo random movement. Depending on local environmental conditions, the cells may switch from one phenotype to the other. A diffusible substrate (N), which accounts for concentrations of nutrients such as glucose and growth factors, regulates cell division and death (χ_N), the switching function χ_W^* χ_N^* from

OXPHOS to glycolysis, and the ability of cells to generate Wnt (W) and Wnt inhibitor (W_I) activities. The Wnt and Wnt inhibitor equations are based on the Gierer–Meinhardt activator–inhibitor model (Gierer & Meinhardt, 1972), where Wnt is the short-range activator which produces a long-range factor that inhibits Wnt activity (e.g., SFRP2). Because Wnt signaling is assumed to be constitutively active, both OXPHOS and glycolytic cells are assumed to upregulate Wnt activity at the rate S_W. In the model shown in Fig 2A and B, the glycolytic cell proliferation rates and the metabolic switching rates (χ_W and χ_W^*) also depend on Wnt activity where a higher activity level increases cell propensities for glycolysis over OXPHOS, if sufficient nutrients are available (χ_N^*). To model the angiogenic response of the mouse vasculature to the lactate produced by the glycolytic cells and the accompanying increased delivery of nutrients, we introduced sources N_S that increase the amount of nutrient in the system proportionally to the amount of glycolytic activity of the cells. We also assumed that the vascular density was largest at the domain boundary and thus, we modified the boundary conditions for nutrients analogously. See Appendix A2 for the precise functional relationships.

We also considered a more general *in vivo* model, which accounted for PDK activity, hypoxia-inducible transcription factor concentrations (HIF1α), lactate concentration, and cross-feeding between glycolytic and OXPHOS cells (Appendix A3). Assuming that Wnt and HIFs promote PDK expression/activity (Kim *et al*, 2006; Pate *et al*, 2014; Prigione *et al*, 2014), that PDK activity promotes lactate production (Pate *et al*, 2014), and that lactate increases HIF1α expression levels and provides a source of fuel for OXPHOS cells (De Saedeleer *et al*, 2012; Epstein *et al*, 2014), we obtained results that were qualitatively similar to the simpler model shown in Fig 2A and B where these additional processes were not considered directly. In particular, the effects of Wnt signaling dominate those of cross-feeding between the cell types, and the positive feedback loop between Wnt and PDK (high PDK implies more P_g cells; higher numbers of P_g cells imply more Wnt activity; more Wnt activity means increased PDK) has been distilled in the simpler model so that Wnt activity level, rather than PDK levels, provides an effective metabolic switch between relative amounts of OXPHOS and glycolysis. Because PDK drives the switch in metabolism in SW480 cells, we use the P_g and P_o spatial distributions to compare to our xenograft stains.

The model equations were solved in nondimensional form using a characteristic proliferation time T of 1 day to rescale time and a characteristic diffusion length l of the Wnt inhibitor to rescale space.

Since we did not know l (in fact, there may be many factors that contribute to Wnt inhibition), we varied l and found good agreement between the experimental and numerical patterns when $l \approx 40$ μm. A full description of the both models, boundary conditions, and the nondimensionalization can be found in Materials and Methods and in the Appendix. The parameter values can be found in Table 1 (Jiang *et al*, 2005; Rockne *et al*, 2010; Mendoza-Juez *et al*, 2012).

Figure 2C presents the numerical results for the fractions of glycolytic and oxidative cells and the concentrations of Wnt and Wnt inhibitor, where each two-dimensional plot is a horizontal slice through the center of the three-dimensional spatial domain (nutrient distributions can be found in Appendix Fig S15 and Appendix A4). The cells were initially seeded randomly near the boundary of the domain to reflect the fact that the cells that survive the initial implantation are likely close to nutrient sources (alternative initial distributions of cells give similar results). The cells then proliferate and grow inwards toward the center of the domain with angiogenesis-induced nutrient sources fueling the growth. Consistent with the xenograft data, a distinct spotted pattern in the population of glycolytic cells is produced by the model over time. Over the entire domain, there is a high level of glycolytic-dominant cells with localized areas of highly active glycolytic cells (dark red spots). Similar to the xenograft tumors, the spots are denser toward the boundary of the tumor space, where there is a higher density of vasculature, a spatial pattern that agrees with the overall pattern of pPDH staining of the mock tumors in Fig 1A. The oxidative cell fractions are close to 0 in the same spots where glycolysis is high, and their levels are relatively higher in regions surrounding these spots. Wnt and Wnt inhibitor activity show a similar pattern, with high levels distributed in a spotted array throughout the domain, surrounded by lower levels in the neighboring regions. Like the pattern of glycolysis, the frequency of spots is higher near the boundary relative to the interior. The square symmetries in the simulated spot distributions are due to the use of a cubic spatial domain in the simulations. Quantitative and comparative analysis of the patterns in the xenograft tumors to the simulated pattern generated by the model indicates that the model predicts similar dimensions for the size of the spots and distance between the spots (see Fig 3D and E), although there is significant scatter in the data.

Because the model parameters were largely unknown, we investigated their influence on the results through a parameter study (Appendix A5). Using a diffusive stability analysis to determine the ranges of values for which patterns were predicted to occur (see

Figure 2. A mathematical model for Wnt signaling regulation of metabolism.

This set of reaction–diffusion equations describes the change over time of oxidative (P_o) and glycolytic (P_g) cell populations, Wnt signaling activity (W), and Wnt inhibitor activity (WI).

A The cells can diffuse, proliferate, and "switch" metabolism programs depending on Wnt signaling activity and nutrient levels and die from lack of nutrient (N).

B Wnt and Wnt inhibitor activity equations are based on the Gierer–Meinhardt activator–inhibitor model. The Wnt signal diffuses short range relative to the longer-range diffusion of the Wnt inhibitor. Wnt also auto-upregulates its activity in glycolytic cells at a rate proportional to nutrient level, is inhibited by a Wnt inhibitor, is constitutively upregulated in both cell types, and decays (downregulation term). The Wnt inhibitor diffuses long range, is nonlinearly upregulated by Wnt, and decays. Equations for nutrient and dead cells (P_d) are not shown; their descriptions are in the main text.

C Three-dimensional numerical simulations that model the spatial distribution and level of glycolytic and oxidative cells, Wnt, and Wnt inhibitor reveal an emergent self-organizing pattern of metabolic heterogeneity (spots). The simulations shown depict the heterogeneity in a 3D and 2D representation. The 3D representation includes a portion of the "tumor" removed to visualize the interior of the domain. The 2D representation is a horizontal slice of the respective 3D simulation in the center of the domain. Color bars refer to unitless concentrations.

D Summary of parameter effects on the spotted pattern.

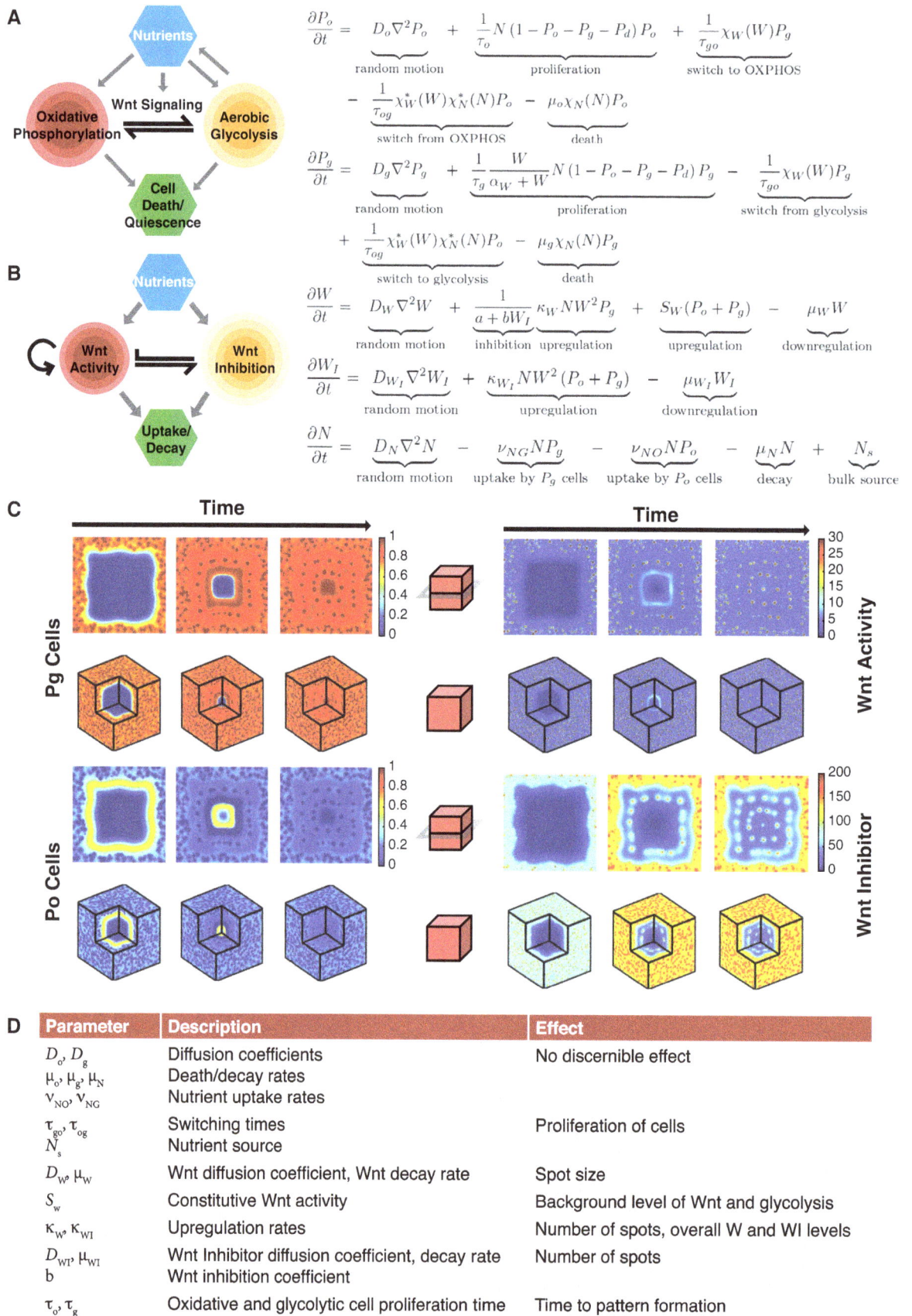

A

$$\frac{\partial P_o}{\partial t} = \underbrace{D_o \nabla^2 P_o}_{\text{random motion}} + \underbrace{\frac{1}{\tau_o} N \left(1 - P_o - P_g - P_d\right) P_o}_{\text{proliferation}} + \underbrace{\frac{1}{\tau_{go}} \chi_W(W) P_g}_{\text{switch to OXPHOS}}$$

$$- \underbrace{\frac{1}{\tau_{og}} \chi_W^*(W) \chi_N^*(N) P_o}_{\text{switch from OXPHOS}} - \underbrace{\mu_o \chi_N(N) P_o}_{\text{death}}$$

$$\frac{\partial P_g}{\partial t} = \underbrace{D_g \nabla^2 P_g}_{\text{random motion}} + \underbrace{\frac{1}{\tau_g} \frac{W}{\alpha_W + W} N \left(1 - P_o - P_g - P_d\right) P_g}_{\text{proliferation}} - \underbrace{\frac{1}{\tau_{go}} \chi_W(W) P_g}_{\text{switch from glycolysis}}$$

$$+ \underbrace{\frac{1}{\tau_{og}} \chi_W^*(W) \chi_N^*(N) P_o}_{\text{switch to glycolysis}} - \underbrace{\mu_g \chi_N(N) P_g}_{\text{death}}$$

B

$$\frac{\partial W}{\partial t} = \underbrace{D_W \nabla^2 W}_{\text{random motion}} + \underbrace{\frac{1}{a + b W_I}}_{\text{inhibition}} \underbrace{\kappa_W N W^2 P_g}_{\text{upregulation}} + \underbrace{S_W (P_o + P_g)}_{\text{upregulation}} - \underbrace{\mu_W W}_{\text{downregulation}}$$

$$\frac{\partial W_I}{\partial t} = \underbrace{D_{W_I} \nabla^2 W_I}_{\text{random motion}} + \underbrace{\kappa_{W_I} N W^2 (P_o + P_g)}_{\text{upregulation}} - \underbrace{\mu_{W_I} W_I}_{\text{downregulation}}$$

$$\frac{\partial N}{\partial t} = \underbrace{D_N \nabla^2 N}_{\text{random motion}} - \underbrace{\nu_{NG} N P_g}_{\text{uptake by } P_g \text{ cells}} - \underbrace{\nu_{NO} N P_o}_{\text{uptake by } P_o \text{ cells}} - \underbrace{\mu_N N}_{\text{decay}} + \underbrace{N_s}_{\text{bulk source}}$$

C

D

Parameter	Description	Effect
D_o, D_g	Diffusion coefficients	No discernible effect
μ_o, μ_g, μ_N	Death/decay rates	
ν_{NO}, ν_{NG}	Nutrient uptake rates	
τ_{go}, τ_{og}	Switching times	Proliferation of cells
N_s	Nutrient source	
D_W, μ_W	Wnt diffusion coefficient, Wnt decay rate	Spot size
S_w	Constitutive Wnt activity	Background level of Wnt and glycolysis
κ_W, κ_{WI}	Upregulation rates	Number of spots, overall W and WI levels
D_{WI}, μ_{WI}	Wnt Inhibitor diffusion coefficient, decay rate	Number of spots
b	Wnt inhibition coefficient	
τ_o, τ_g	Oxidative and glycolytic cell proliferation time	Time to pattern formation

Figure 2.

Table 1. Parameters.

Parameter	Description	Mock value	dnLEF value	References
D_o	Diffusion coefficient of oxidative cells	0.01	0.01	Rockne *et al* (2010)
D_g	Diffusion coefficient of glycolytic cells	0.01	0.01	Rockne *et al* (2010)
D_W	Diffusion coefficient of Wnt	0.004	0.008	Chosen to be small
D_{W_I}	Diffusion coefficient of Wnt inhibitor	1	1.5	Chosen to be large
D_N	Diffusion coefficient of nutrient	100	100	Jiang *et al* (2005)
τ_o	Oxidative cell proliferation time	1	1	Non-dimensionalization (time scale)
τ_g	Glycolytic cell proliferation time	1	1	Non-dimensionalization (time scale)
τ_{og}	Switch time from OXPHOS to glycolysis	1/24	1/24	Mendoza-Juez *et al* (2012)
τ_{go}	Switch time from glycolysis to OXPHOS	1	1	Mendoza-Juez *et al* (2012)
α_W	Constant for Michaelis-Menten dynamics	1	1	Parameter estimation
κ_W	Rate of nonlinear Wnt production	5	5	Parameter estimation
κ_{WI}	Rate of Wnt inhibitor production	1	1	Parameter estimation
μ_o	Decay rate of P_o cells	1	1	Parameter estimation
μ_g	Decay rate of P_g cells	1	1	Parameter estimation
μ_d	Decay rate of P_d cells	1	1	Parameter estimation
μ_W	Decay rate of Wnt	2	2	Parameter estimation
μ_{W_I}	Decay rate of Wnt inhibitor	3	3	Parameter estimation
μ_N	Decay rate of nutrient	0.1	0.1	Parameter estimation
S_W	Rate of Wnt production through cells	7.5	6.5	Parameter estimation
a	Constant of inhibition	10^{-8}	10^{-8}	Parameter estimation
b	Constant of inhibition by W_I	1	1	Parameter estimation
γ_W	Sensitivity level of Wnt switch functions	1	1	Parameter estimation
γ_N	Sensitivity level of nutrient switch function	100	100	Assumed to be high
v_{NG}	Uptake of nutrient by P_g cells	10	10	Parameter estimation
v_{NO}	Uptake of nutrient by P_o cells	10	10	Parameter estimation
N_s	Parameter for nutrient source	2	2	Parameter estimation
W^*	Wnt level at which 50% of cells switch metabolism	5	5	Parameter estimation
N^*	Nutrient level below which cells die	0.07	0.07	Parameter estimation
N_g^*	Nutrient level below which P_o cells cannot switch to glycolysis	0.1	0.1	Parameter estimation
α_N	Value of scaling function when $\int P_g = 0$	0.025	0.025	Parameter estimation
S_x	Horizontal length of spatial domain	12	12	
S_y	Vertical length of spatial domain	12	12	

Model parameters for mock and dnLEF/dnTCF simulations.

Appendix A6), we modified the parameters one by one within the pattern-forming range and tested for phenotype changes in metabolism and patterning. The results are summarized in the table in Fig 2D. Increasing the Wnt diffusion coefficient or decreasing the Wnt decay rate increased the extent of Wnt activity, so that the spots of glycolysis increased in size. Increasing the Wnt inhibitor diffusion coefficient or decreasing the decay of the Wnt inhibitor caused the inhibitor to stay within the system for longer times and resulted in fewer spots. Modifying the switching times between the phenotypes changed the proportion of P_o and P_g. Increasing or decreasing the Wnt switch changed the background levels of P_g cells and spot sizes without affecting the number of spots. Small reductions in S_W, which can be thought of as reducing overall Wnt signaling, reduced the background levels of glycolytic cells without much effect on the sizes or numbers of spots. Sufficiently reducing S_W resulted in all terms (P_o, P_g, W, and W_I) decreasing to 0. Decreasing κ_W (nonlinear Wnt activity) or increasing b (Wnt response to inhibition) paradoxically increases the number of glycolytic cells because nonlinear interactions actually result in a decreased amount of W_I. Analogously, when κ_{WI} (nonlinear Wnt inhibitor activity) decreases, the number of glycolytic cells decreases. Modifying the cell diffusion coefficients, death and decay rates, and the nutrient uptake rates did not significantly influence the self-organization of a spotted array. Similarly, varying the proliferation times only changed the time it took to reach a steady state but otherwise had no effect on pattern formation.

SW480 dnLEF1 Xenograft Tumor

A Phosphorylated PDH

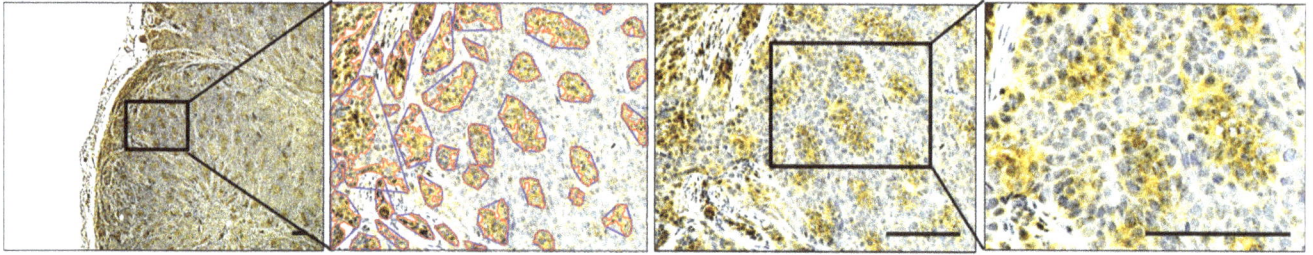

B β-catenin

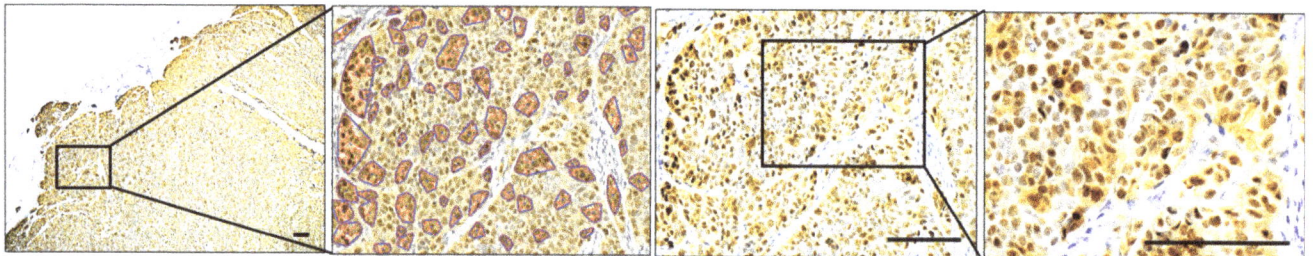

C

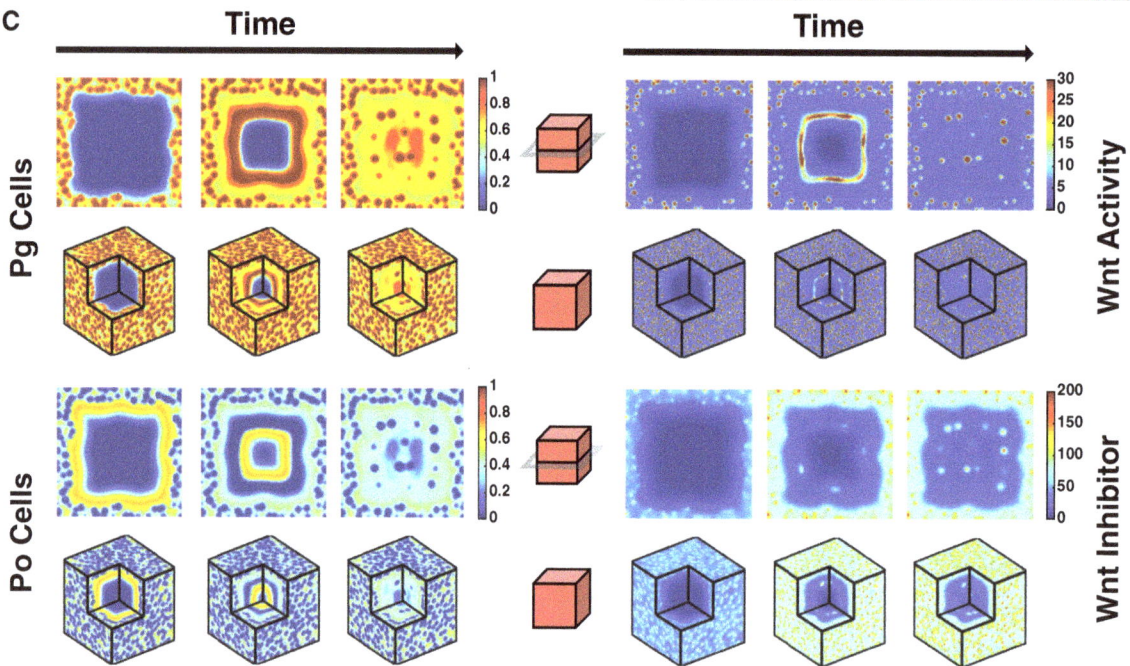

D

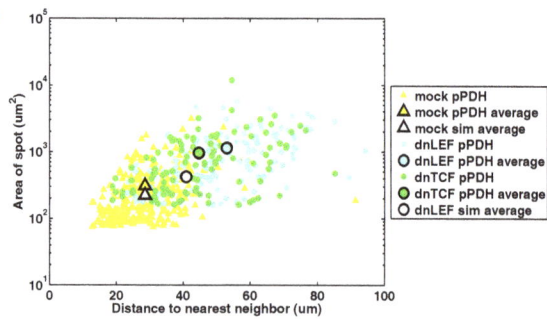

E

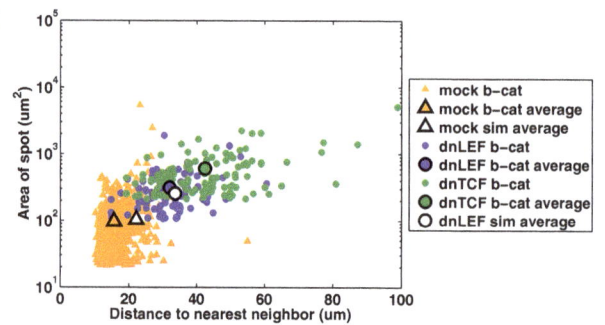

Figure 3.

◀ **Figure 3. Decreasing Wnt signaling leads to changes in metabolic patterning in xenograft tumors.**

A, B   SW480 cells were lentivirally transduced to express dominant negative LEF-1 (dnLEF-1), and transduced cells were injected subcutaneously into immunocompromised mice. Tumor sections were stained for phosphorylated PDH (A) and β-catenin (B) and counterstained with hematoxylin. Compared to mock tumors, the spots are larger and more heterogeneous and the background staining is lighter, which reflects an overall decrease in Wnt signaling. Scale bars are 100 μm. The red curves denote spot contours and the blue curves denote convex hulls, which group spots that are sufficiently close to one another (see Appendix A1).

C   Numerical simulations that lower Wnt signaling activity in the model show an overall decrease in glycolysis and a change in the spotted pattern that closely mimics that observed in the dnLEF tumors. Color bars refer to unitless concentrations.

D   Image analysis of spot size versus distance of spot to nearest neighbor, using analyzed images. Averages for mock and dnLEF spot simulations are denoted in white outlined symbols (pPDH: spot sizes/inter-spot distances: mock simulation average: 225 ± 278 μm^2/29 ± 12 μm; mock xenograft tumor average: 309 ± 367 μm^2/29 ± 10 μm; dnLEF-1 simulation average: 423 ± 327 μm^2/41 ± 14 μm; dnLEF-1 xenograft tumor average: 1,139 ± 1,042 μm^2/53 ± 15 μm). Results are also shown for dominant negative transcription factor 1 (dnTCF-1) tumors (see Appendix A1.10–A1.12 and Appendix Figs S8–S10). The metabolic pattern in dnTCF-1 tumors is consistent with that in dnLEF-1 tumors. The analysis and model predict that the changes in the metabolic spotted pattern (larger spots, greater distance between spots) are due to an increase in the diffusion of Wnt and the Wnt inhibitor.

E   Comparison of mock β-catenin spots to dnLEF-1 and dnTCF-1 β-catenin spots from image analysis. Averages for mock and dnLEF-1 spot simulations are denoted in white outlined symbols (β-catenin spot sizes/inter-spot distances: mock simulation average: 139 ± 145 μm^2/26 ± 11 μm; mock xenograft tumor average: 97 ± 209 μm^2/16 ± 4 μm; dnLEF-1 simulation average: 342 ± 221 μm^2/39 ± 14 μm; dnLEF-1 xenograft tumor average: 312 ± 277 μm^2/32 ± 9 μm; dnTCF-1 xenograft tumor average: 603 ± 578 μm^2/42 ± 14 μm). The analysis and model predict that Wnt signaling diffuses further with dominant negative LEF-1 expression.

Interfering with Wnt signaling alters colon cancer metabolic patterns *in vivo*

Since our model utilizes Wnt signaling, we tested how interference of this pathway would alter metabolic patterning. To disrupt the pathway, we used lentiviral transduction to express dominant negative LEF-1 (dnLEF-1) or dominant negative TCF-1 (dnTCF-1) transcription factors. Both dominant negative versions are naturally occurring LEF/TCF isoforms that lack the β-catenin binding domain and therefore interfere with the activation/expression of Wnt target genes. Expression of moderate, physiological levels of dnLEF-1 or dnTCF-1 expression, partially, but not completely, disrupts Wnt target gene expression in the xenograft tumors (Van de Wetering *et al*, 2002; Hoverter *et al*, 2012; Pate *et al*, 2014). Partial disruption is necessary because complete inhibition of Wnt activity would block cell cycle progression and the formation of tumors altogether.

SW480 colon cancer cells that had been lentivirally transduced and selected for dnLEF-1 or dnTCF-1 expression were subcutaneously injected into immunocompromised mice for tumor formation. Experiments showed that, as a result of dnLEF-1 expression, PDK1 activity was reduced, Warburg metabolism was diminished, and tumor mass was reduced approximately four- to fivefold (Pate *et al*, 2014). Immunohistochemical staining of the levels of phospho-PDH in these tumors (Fig 3A) revealed a lighter background and lower pPDH level overall. Interestingly, pPDH positivity remained easily visible in clusters of cells, but there were striking changes in the spotted pattern. Each pPDH-positive cluster was comprised of a larger number of cells (mock average: ~seven cells/spot and dnLEF average: ~17 cells/spot; see Appendix A1.15 and Appendix Fig S11), and there was a greater distance between each spot (compared to parental, mock-transduced cells; Fig 3D). We also utilized immunohistochemical staining for the Wnt-mediating factor β-catenin in the dnLEF tumors (Fig 3E) (dnLEF-expressing tumors cannot be stained for LEF-1). These stains revealed a spotted pattern, with clusters of cells having higher levels of β-catenin in the nucleus than neighboring cells, although because of the very high levels of β-catenin in SW480 cells, there was an overall strong intensity of the IHC stain. Image analysis showed that while the β-catenin^{HI} spots are, on average, smaller than the pPDH-positive spots, they too had increased in size and distance relative to the pattern of β-catenin^{HI} cell clusters in the mock/parental tumors.

Additional image analyses of staining for pPDH and β-catenin in dnLEF-1- and dnTCF-1-expressing tumors are provided in Appendix A1 (Appendix A1.8–A1.12), together with a quantification of these staining patterns (Appendix A1 and Appendix A1.13–A.1.15). In summary, there were significant changes in both the intensity and distribution of the spotted patterns for pPDH and β-catenin when Wnt signaling was reduced by dnLEF-1/dnTCF-1 expression.

Reaction–diffusion modeling of metabolic patterns under partial disruption of Wnt signaling predicts expression of factors that increase the range of Wnt signaling

To understand the phenotypic changes in the spotted patterns when Wnt signaling was partially disrupted, we used our model to identify changes in parameters that could recapitulate the experimental observations. The simplest change was to reduce S_W, which mimics dnLEF-1 and dnTCF-1 expression in lowering intrinsic Wnt activity throughout the domain, a manipulation that represents the cell-autonomous effect of expressing Wnt-interfering, dominant negative LEF/TCF factors in the nucleus of every cell. However, as described earlier in our parameter study, decreasing S_w lowers the overall background levels of P_g cells, but does not affect the spotted pattern (unless it is taken to be too small in which case the pattern disappears). Thus, solely lowering overall Wnt signaling (S_W) in the model produces outcomes in pattern that are inconsistent with the experimental data.

Clearly, the effects of dnLEF-1/dnTCF-1 expression are more complex than the cell-autonomous manipulation of only decreasing Wnt pathway activity in the nucleus. We considered the possibility that dnLEF-1/dnTCF-1 might also be triggering a cell-extrinsic response that connects collections of cells in the microenvironment. Specifically, our parameter study suggested that the increase in the sizes of pPDH-positive cell clusters might be due to extracellular soluble factors that increase the range of the activator (Wnt ligands) and that the decrease in the number of pPDH-positive cell clusters could be due to factors that increase the range of inhibition. Therefore, we included two additional parameter modifications: increasing D_W, which increases the range of Wnt ligands and makes the spots larger, and concomitantly increasing D_{WI}, which increases the range of Wnt ligand inhibitors and reduces the number of spots.

Changing these two parameters and decreasing S_w simultaneously resulted in a striking recapitulation of the changes in the spotted pattern observed in the dnLEF-1/dnTCF-1-expressing tumors: lower background levels of P_g cells and larger, fewer spots of P_g-glycolysis (Fig 3C). The average sizes and centroid distances of the P_g spots in the simulated tumors correlated very well with the experimental observations (Fig 3D). Further, the simulation showed a decrease in nutrient concentrations throughout the tumor (Appendix A4), a result that is consistent with our previous experimental data as we observe significantly fewer blood vessels in the dnLEF-1 and dnTCF-1 tumors (Pate *et al*, 2014). This is because the nutrient concentration N is linked to the proportion of P_g cells, which are decreasing.

Since in the experiments, we used IHC staining of β-catenin as a direct assessment of patterns in Wnt signaling, in the simulations, we analogously examined patterns of Wnt activity in the model. The results show very good agreement between the simulations and the experiments: The spots of Wnt activity are smaller than the P_g spots but the Wnt-activity spots were increased in size and distance relative to the pattern of Wnt activity in the simulations of the mock tumors (Fig 3E). In summary, our results suggest that stressing the colon cancer cells by interfering with Wnt signaling triggers changes in the expression of factors that increase the diffusion range, or "spread", of Wnt ligands and extend the range of Wnt inhibition.

In vivo validation of model predictions

Only a few studies have directly examined the diffusion range of Wnt ligands in any tissue, a range which is extremely limited, in part because the ligands are post-translationally modified by palmitoylation and are highly lipophilic for membranes and extracellular matrix proteins (Willert *et al*, 2003; Farin *et al*, 2016). There is a growing awareness of proteins that modify the range of ligand diffusion, although their actions and impact are not very well characterized (Fig 4A). Perhaps the best-characterized factors that influence Wnt ligand diffusion are the SFRP protein family, secreted inhibitors that bind directly to Wnt ligands and interfere with receptor binding. Importantly, several studies have shown that even though SFRPs can interfere with Frizzled receptor binding, they are bimodal in their actions, repressing Wnts at high concentrations of ligand but also promoting Wnt actions by increasing their range of diffusion and, in essence, delivering the ligands to cells that are further away (Mii & Taira, 2009, 2011). Given that our mathematical model predicts the diffusion of Wnt ligands and their inhibitors have increased in the dnLEF-1 and dnTCF-1 xenograft tumors, we tested the prediction that one or more candidate regulators of Wnt diffusion were elevated in their expression. Using RNA-seq data as a guide for identifying candidates expressed in SW480 cells, we designed human-specific primers for both diffusion regulators and inhibitors that were detectably expressed in this cell line. Expression analysis of mRNA purified from 2D cultures and 3D xenograft tumors revealed that the Wnt diffusers SPOCK2, GPC4, and SFRP5 are upregulated specifically in dnLEF-1 and dnTCF-1 xenograft tumors but not 2D culture (Fig 4B and C). Since the primers are human specific, the expression changes derive specifically from the human cancer cells and not mouse-derived cell types in the tumor microenvironment.

While small-molecule Wnt inhibitors that mimic the effects of dnLEF-1 and dnTCF-1 are working their way through pre-clinical testing and early-phase clinical trials, there are not yet any available data from patient studies that profile gene expression changes in primary colorectal cancers treated with Wnt inhibitors. However, there are limited data available from patients treated with radio- and chemotherapy regimens, treatments that induce stress and loss of nutrient delivery to the tumor. We analyzed one dataset [(Snipstad *et al*, 2010) NCBI GEO GDS3756], which provided gene expression profiles of a group of colorectal cancer patients before and after radio- and chemotherapy treatment. Figure 4D and E shows that, while the treatment had no significant effect on expression of Wnt ligand regulators in normal rectal tissue, the expression of GPC1 and three SFRP family members (SFRP1, SFRP2, and SFRP4) was strongly and specifically increased in the tumor following treatment (Fig 4E). We checked for changes in expression of Wnt ligands, and although there was a trend toward significantly increased expression of Wnt2, Wnt5b, Wnt8b, and Wnt10b specifically in the tumor and not the neighboring normal tissue, the changes did not quite reach statistical significance (Fig EV2A and B). Interestingly, one glycolytic gene (ENO2) was significantly increased in radiochemotherapy-treated tumor tissue (Fig EV2D), and the glycolytic regulator HIF1A was increased but not to the same level of significance. This suggests that radiochemotherapy may trigger increased expression of proteins that increase the range of Wnt diffusion, a response that we predict might serve to maintain a critical level of glycolytic cells in the tumor.

Modeling a therapeutic treatment for cancer: metabolic targeting

To test whether glycolytic cells are the important subpopulation of cells to target in the tumors, we compared the effectiveness of a hypothetical therapy program that selectively targeted each population by independently varying the death rates of P_o and P_g cells (we introduced additional death terms $\mu_{P_o} P_o$ and $\mu_{P_g} P_g$ in the P_o and P_g equations in Fig 2A). The simulation applied the targeted therapy to a fully developed tumor at steady state for different lengths of time (days), followed by removal of the therapy and a recovery time for tumor development (Fig 5). In this figure, the tumor size (integral of $P_g + P_o$ over the entire domain) is shown relative to that of the untreated tumor (see Fig EV3 for the dynamics of the individual cell populations). The treatment dose refers to cell death rates μ_{P_o} or μ_{P_g}, and targeting means that the death rate is nonzero only for the target cell population. These simulations revealed that regardless of the targeted population, modest rates of cell death suppressed tumor development transiently, followed by full recovery of the system once therapy was removed, a pattern more evident and more robust when cell killing was directed toward the P_o population. At sufficiently large death rates, complete loss of the tumor could be achieved. However, targeting the P_g population led to a complete loss of the simulated tumor at shorter treatment times and smaller death rates than when P_o cells were targeted. Thus, the simulation predicts that P_g cells are the more sensitive population and that targeting these cells could more effectively lead to a full regression of the tumor.

Since selective targeting resulted in full recovery of the simulated tumors unless death rates were sufficiently high and treatment was sufficiently long, we considered dual targeting of two features of cancer cell metabolism as a mechanism for more effective killing.

A

Fly Gene	Human Gene	Reference	Function
	CDC42	Stanganello, et al. *Nature Communications* 2015	CDC42/N-Wasp regulates formation of Wnt-positive filopodia
Swim	LCN7/TINAGL1	Mulligan et al. *PNAS* 2012	Binds to Wg in a lipid-dependent manner to increase range. At high concentrations, can inhibit Wls activity.
Cow	Testican-2/ SPOCK2	Chang, et al. *PloS One* 2014	HSPG that binds to Wg, increasing Wg mobility. Can act as an inhibitor in high Wg concentration.
Dally/Dlp	GPC1-6	Lin, et al. *Nature* 1999	Co-receptor with Fzd2 to modulate range of Wg
	SFRP1-5	Mii & Taira. *Development.* 2009; Mii & Taira. *Dev. Growth & Diff.* 2011	Bimodal actions: repress Wnt by inhibiting Wnt ligand:Fzd receptor binding; promote Wnt by increasing range of ligand diffusion

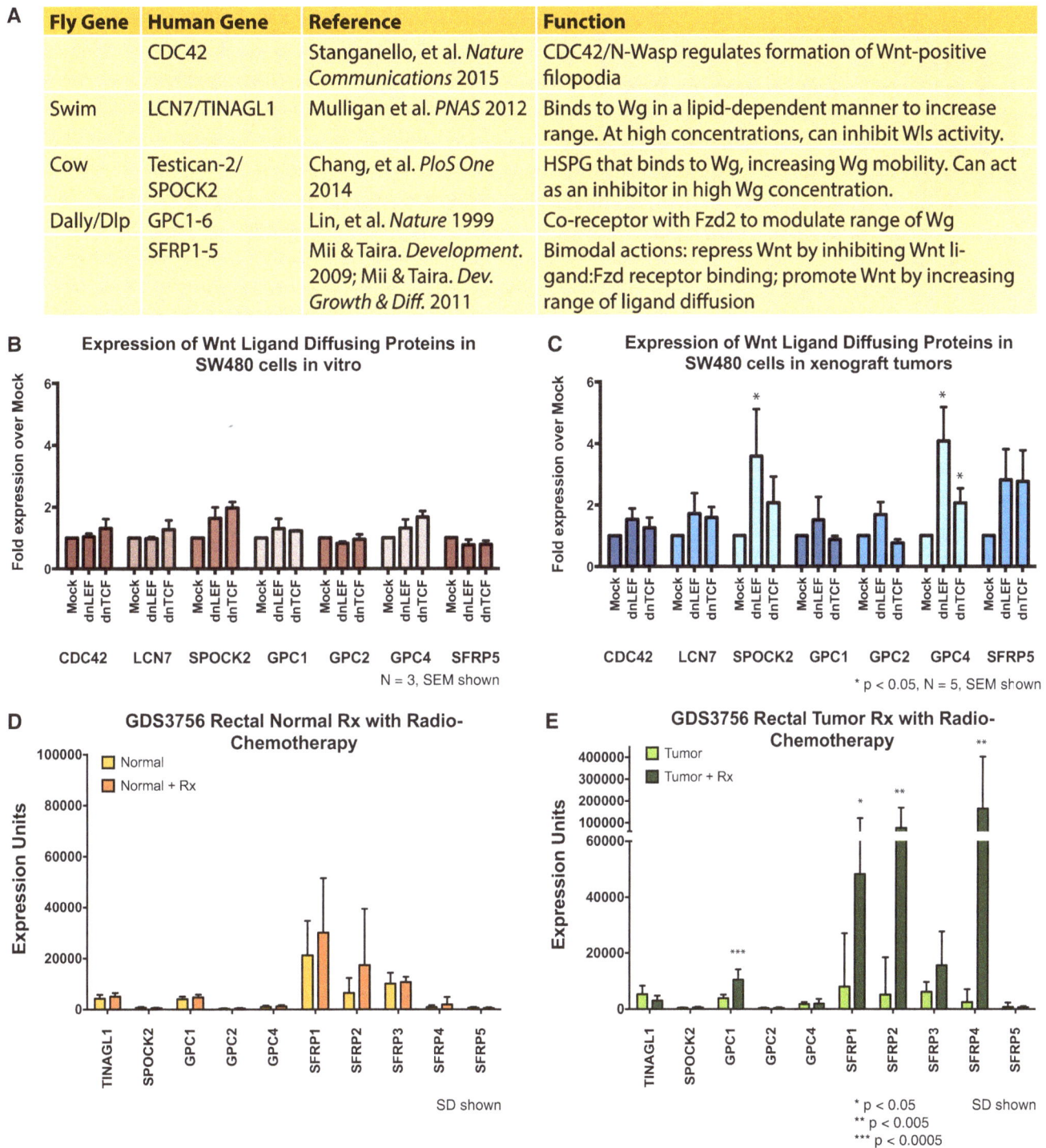

B Expression of Wnt Ligand Diffusing Proteins in SW480 cells in vitro

N = 3, SEM shown

C Expression of Wnt Ligand Diffusing Proteins in SW480 cells in xenograft tumors

* p < 0.05, N = 5, SEM shown

D GDS3756 Rectal Normal Rx with Radio-Chemotherapy

SD shown

E GDS3756 Rectal Tumor Rx with Radio-Chemotherapy

* p < 0.05
** p < 0.005
*** p < 0.0005

SD shown

Figure 4. Model predictions revealed in xenograft tumors and human colorectal cancer.
The model predicted that lowering Wnt signaling results in an increase in the expression of factors that increase the range of diffusion of Wnt and Wnt inhibitors.

A   Known regulators of Wnt ligand diffusion.

B, C   Quantitative PCR of diffusion regulators in SW480 mock, dnLEF-1, and dnTCF-1 (B) transduced cells, and (C) xenograft tumors show human SPOCK2, GPC4, and SFRP5 mRNA are notably upregulated in xenograft tumors but not 2D *in vitro* culture. *In vitro* data represent an average of three sample sets (± SEM), and xenograft tumor data represent the average of five independent tumor sets (± SEM); * denotes $P < 0.05$. Statistical significance was determined using Student's two-tailed *t*-test.

D, E   Gene expression data, from GEO dataset GDS3756, of 21 rectal cancer patient tissue with or without radio-chemotherapy (Snipstad *et al*, 2010). Significant changes in expression levels of GPC1 ($P = 0.00019$), SFRP1 ($P = 0.016$), SFRP2 ($P = 0.0006$), and SFRP4 ($P = 0.0006$) are observed in post-therapy tumor cells compared to before treatment. SD shown. Statistical significance was determined using the Mann–Whitney *U*-test with Benjamini–Hochberg correction for multiple hypotheses (RStudio).

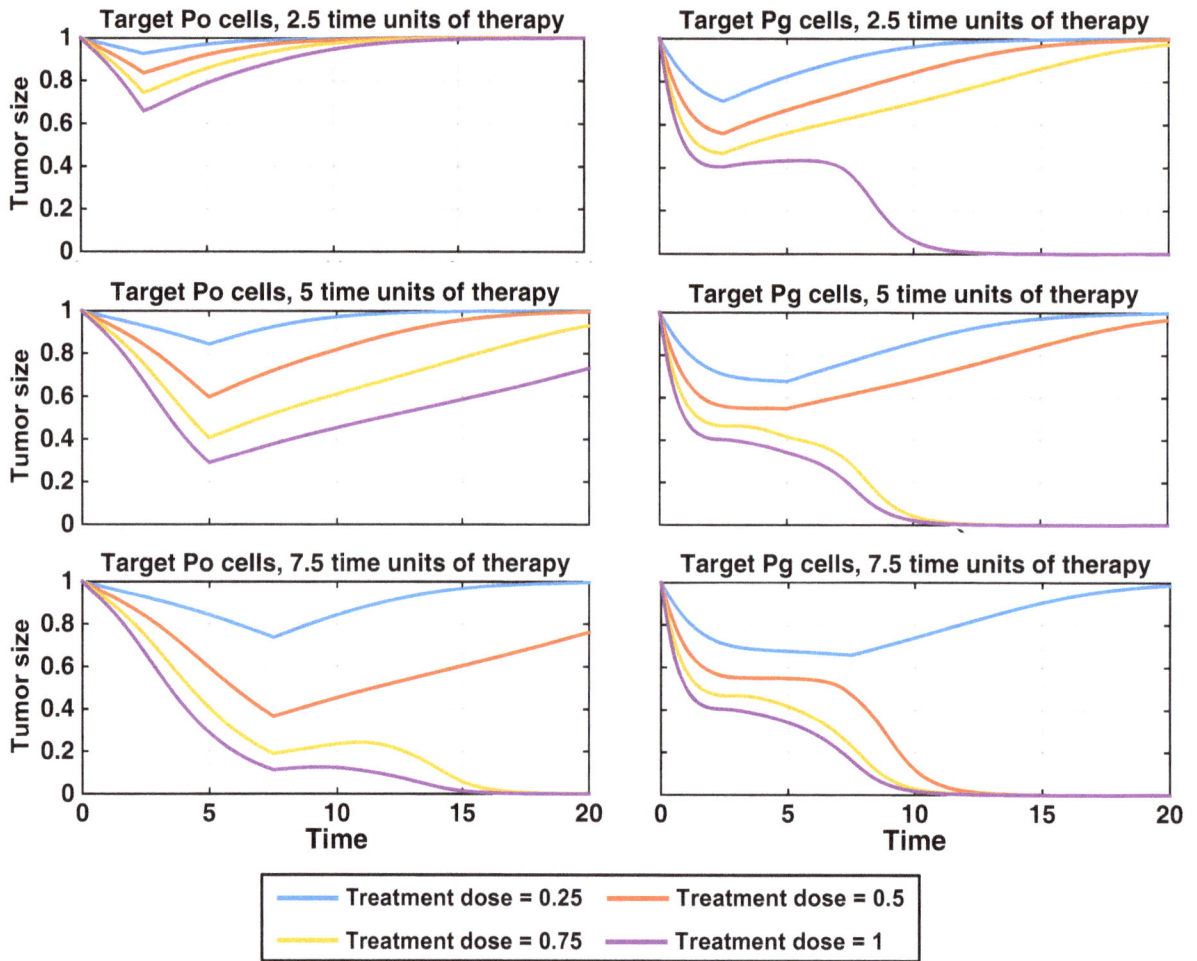

Figure 5.   Simulations identify the glycolytic cell population as a sensitive drug target.
We target either cells with more oxidative phosphorylation (P_o; left) or cells with more glycolysis (P_g; right) selectively, starting from a metabolically patterned state, for 2.5, 5, or 7.5 (arbitrary) time units, with a treatment dose between 0.25 and 1. After therapy is halted, the cells are allowed to evolve according to the original model (Fig 2). The total cell populations, relative to the initial, starting cell population are shown. Corresponding populations of P_o and P_g cells can be found in Fig EV3.

Specifically, we targeted canonical Wnt signaling and PDK enzyme activity, both of which act as regulators of glycolysis (Fig 6A). Dichloroacetate (DCA) inhibits PDK activity and therefore targets cell metabolism directly by releasing inhibition of PDC, which increases OXPHOS capacity. Tankyrase inhibitors such as XAV939 reduce β-catenin levels and hence reduce canonical Wnt signaling. Compounds that target Wnt and PDK are currently being tested in preclinical studies as individual agents in clinical trials, but they have not been tested in combination (Fig 6B). We asked whether targeting glycolytic cells using anti-Wnt and anti-PDK therapies in combination is more effective than single-agent therapy.

Because the inhibition of β-catenin by XAV939 is similar to the effects of dnLEF-1 and dnTCF-1, we modeled treatment by XAV939 using an analogous approach. In particular, we assumed that XAV939 decreases the general Wnt signaling term S_W and increases the ranges of Wnt and its inhibitor (due to upregulation of Wnt and Wnt inhibitor diffusers), which we modeled by increasing D_W and D_{WI} proportionally. To model the effects of DCA, we increased the rate at which cells switched from a glycolytic metabolic phenotype

to an OXPHOS phenotype (e.g., $1/\tau_{go}$ is increased) to reflect the tendency of cells to perform OXPHOS when PDK is inhibited.

In Fig 6C, we simulated a combination therapy applied to a fully developed tumor at steady state. At a fixed dose of XAV939, coupled to increasing doses of DCA, the simulation predicts that the population of oxidative cells will increase initially as cells switch from a glycolytic state (P_g) to an OXPHOS state (P_o) until there is an insufficient level of glycolytic cells to sustain the tumor and all the tumor cells die. Furthermore, the treatment simulations indicate that a combination of the two therapies will be more effective than single therapies as long as one or the other has adequately been applied. For example, a value of $1/\tau_{go} = 12$ is effective in eradicating the cells as long as Wnt signaling has been reduced by more than about 27% (e.g., $\bar{S}_W = 0.73$). In other words, if β-catenin expression has not been sufficiently suppressed by XAV939, then PDK inhibition by DCA must be adequately increased, and vice versa. Similar results are obtained when an *in vitro* version of the mathematical model is used to simulate the growth of colonies in fibrin gels (Fig 7C; see also Appendix A7 and Appendix Fig S17). In the *in vitro* case,

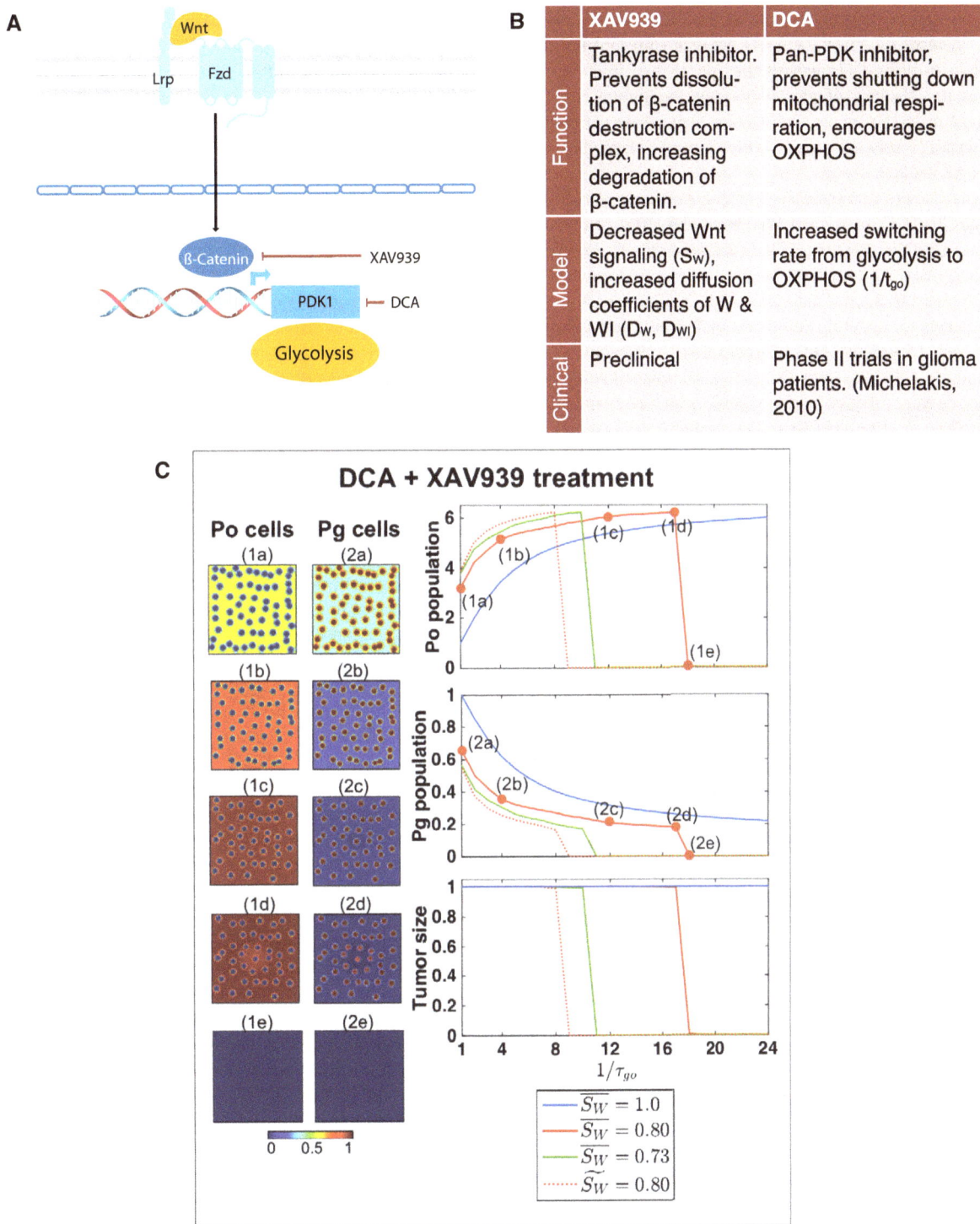

	XAV939	DCA
Function	Tankyrase inhibitor. Prevents dissolution of β-catenin destruction complex, increasing degradation of β-catenin.	Pan-PDK inhibitor, prevents shutting down mitochondrial respiration, encourages OXPHOS
Model	Decreased Wnt signaling (S_W), increased diffusion coefficients of W & WI (D_W, D_{WI})	Increased switching rate from glycolysis to OXPHOS ($1/t_{go}$)
Clinical	Preclinical	Phase II trials in glioma patients. (Michelakis, 2010)

Figure 6. Therapies targeting metabolism and Wnt synergize for tumor death in mathematical simulations.

A, B   Modeled therapies, their targets, and the model parameters influenced by therapy.

C   Starting with a metabolically patterned state, treatment of tumors with dichloroacetic acid (DCA) and XAV939 combined leads to an effective crash in the system, as shown by the complete loss of cells (1e and 2e). The panels on the left show the cell arrangements for the oxidative (P_o) and glycolytic (P_g) populations (metabolic patterning), and the three graphs on the right show the fractions of P_o or P_g cells relative to their initial values, after applying the therapy for 50 (arbitrary) time units. The effects of therapy on the total cell population, relative to the initial cell population, for the same D_W and $1/\tau_{go}$ values, are shown in the third graph. XAV939 treatment is modeled by decreasing S_W and increasing D_W and D_{WI} linearly with respect to the decrease in S_W; legend values are listed relative to mock S_W. The dashed curves labeled $\tilde{S}_W = 0.80$ correspond to the case in which $\tilde{S}_W = 0.80$, but D_W and D_{WI} are unchanged and take the values used in the mock tumor simulations. The panels on the left correspond to the red curve in the graphs and show the effect on patterning for $1/\tau_{go}$ values 1, 4, 12, 17, and 18, respectively [denoted by labels (1a) through (2e)]. Color bar refers to unitless concentrations.

however, the colony does not die out as the DCA concentration is increased. Instead, the effect saturates because sufficient nutrients are available to diffuse throughout the spheroid to maintain cell viability in the absence of glycolytic cells. Additionally, in the *in vitro* version of the model, there is no angiogenesis, but there is cross-feeding between the OXPHOS and glycolytic cells (Appendix A7.1–A7.3).

Given that factors that increase Wnt diffusion were upregulated in the dnLEF/dnTCF tumors, we also tested the effect of increases in Wnt diffusion. We determined that increased expression of Wnt diffusers decreases the sensitivity of the tumors to treatment. For example, decreasing S_W but maintaining D_W and D_{WI} at their pretreatment values to model the inhibition of Wnt diffusers results in more efficient treatment—tumor eradication occurs at smaller concentrations of XAV939 and DCA (e.g., compare the red solid and dashed curves in Fig 6C (right panels), which correspond to a 20% reduction in Wnt signaling; the dashed curve shows the result when the production of Wnt diffusers is inhibited and is labeled as $\tilde{S}_W$).

We performed a preliminary experimental test of model predictions using 3D colony growth of colon cancer cells. A total of 200 single cells were seeded in fibrin gels and cultured under drug treatment for 14 days (Fig 7A). Over this time, the single cells proliferated to give rise to tumor spheroids, and we used image analysis to quantify the increase in colony size as a proxy for proliferation (Fig 7B and C). Treatment with low doses of DCA (0.5 mM, 2 mM) or XAV939 (5 µM) as single agents had no effect on the development and growth of colonies over the 2-week treatment period—in fact, DCA treatment appeared to increase colony size. By contrast, combination therapy had a significant inhibitory effect on colony growth, indicating a strong, negative, and synergistic effect on proliferation.

Synergy can be quantified using the Bliss Independence Model Combination Index (Foucquier & Guedj, 2015), which assumes that XAV939 and DCA treatments act independently (e.g., XAV939 targets Wnt signaling, and DCA targets PDK activity). In particular, if the Combination Index is less than one, this indicates synergy (see Appendix A8 for the definition of the Combination Index and further details). In the *in vivo* simulation and the *in vitro* experiments, the Combination Index is zero because neither XAV939 nor DCA treatment separately affects tumor sizes (provided the concentrations of XAV939 and DCA are not too large; Appendix A8). In the *in vitro* model, the Bliss Combination Index is 0.3462 (using $\tilde{S}_W = 0.80$, $1/\tau_{go} = ¼$), although the Combination Index does depend on the drug concentrations and increases toward one as the DCA concentration increases because the responsiveness to DCA treatment saturates (Appendix A8). Since the Combination Indices in all cases are less than one, this indicates synergy of the XAV939 and DCA combinatorial treatments. As predicted by both the *in vivo* and *in vitro* models, these results suggest that combining Wnt inhibitors and metabolic targeting agents is a promising strategy for treating colon tumors.

Discussion

In this study, we generated colon cancer xenograft tumors and examined changes in cellular metabolism by immunohistochemical staining for phospho-PDH (a marker of PDK activity), and markers of Wnt signaling (LEF-1, β-catenin). We observed that the tumors exhibit a pronounced spotted pattern of metabolic states where the spots indicate clusters of cells in which their mitochondria are inhibited (by PDK action) and thus where glycolysis was likely to be highly active. This is in contrast to the cells in the regions surrounding the spots. In these regions, mitochondria are more active (not inhibited by PDK) and therefore utilize more oxidative phosphorylation. Although we cannot rule out that the spotted pattern is due to the emergence of genetically distinct, clonal populations, the short timescale of the xenografting (14–21 days) and the reproducibility of the pattern in another cell line (SW620) as well as site of injection (subcutaneous and orthotopic within the colon cecum) suggest that what we have observed is a fundamental pattern of tumor heterogeneity that is not genetic in nature, but nongenetic and dynamic (Appendix A9 and Appendix Fig S20).

Metabolic patterning had been previously proposed as a mechanism to facilitate transport of glucose into hypoxic regions of tumors (Sonveaux *et al*, 2008). In particular, cells performing OXPHOS would be located near blood vessels, and, rather than fueling respiration using glucose, these OXPHOS cells would instead use lactate produced by the hypoxic (glycolytic) cells as an alternative nutrient source. Utilization of lactate frees the glucose to travel farther into the hypoxic regions of the tumor where it would be used during glycolysis. Recent studies propose that cancer cell subpopulations segregate and reorganize to survive the sudden loss of nutrients due to antiangiogenic therapy and that this leads to the development of resistance (Allen *et al*, 2016; Jiménez-Valerio *et al*, 2016; Pisarsky *et al*, 2016). However, in our xenograft images, we did not observe this type of spatial relationship between metabolism and vasculature. In fact, we observed that the spotted pattern was denser at the tumor margin where the vascular density was highest.

To investigate the mechanisms underlying the patterning we observed, we proposed a mathematical model based on Turing–Gierer–Meinhardt activator–inhibitor equations that simulated a symbiotic spatiotemporal relationship between these two cell populations (an oxidative population and a glycolytic population). Our model incorporated terms for Wnt as an activator with a short range of diffusion and Wnt inhibitors (e.g., SFRP, DKK) with longer ranges of diffusion. The Wnt and Wnt inhibitor equations describe a feedback relationship, and this lies at the crux of a spotted pattern that emerges in our simulations. Our equations also describe activities of metabolic reprogramming through changes in Wnt levels and availability of nutrients and cell substrates. The model describes a mutualistic interaction between the glycolytic and OXPHOS cells because the glycolytic cells induce the delivery of nutrients from blood vessels to mimic the effects of lactate-induced angiogenesis and these nutrients benefit both cell types. More generally, we can also interpret these nutrients as mutually beneficial cell substrates produced by the glycolytic cells. When we considered the effects of symbiosis by explicitly incorporating cross-feeding between glycolytic and OXPHOS cells in a more general model (Appendix A3) and an *in vitro* model (Appendix A7), we found that Wnt signaling dominates the behavior and the patterning is robust to this form of symbiosis. Although mathematical models have been developed previously to investigate metabolic symbiosis, including spatially homogeneous (Mendoza-Juez *et al*, 2012) and heterogeneous models (McGillen *et al*, 2014; Phipps *et al*, 2015), to our knowledge,

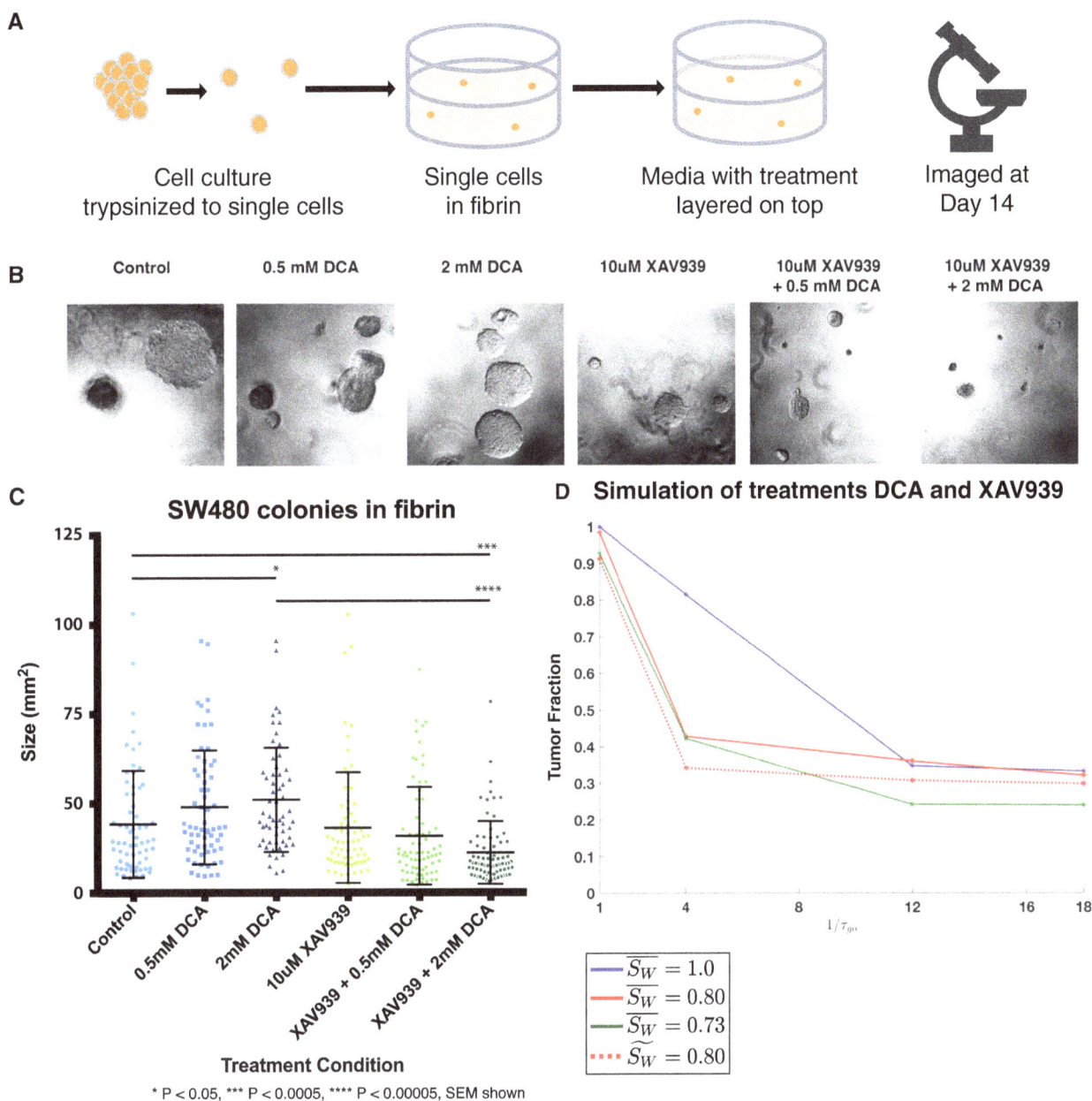

Figure 7. Targeted therapy significantly decreases SW480 tumor spheroid size *in vitro*.

Combined Wnt signaling and glycolysis targeting therapies significantly decrease SW480 spheroid size *in vitro*.

A SW480 cells were embedded in a fibrin gel using the method shown. Media containing a mock treatment, 0.5 mM DCA, 2 mM DCA, 10 μM of XAV939, or a combination of DCA and XAV939.

B Representative 4× images of spheroids each condition, imaged 14 days after treatment, are shown.

C Analysis of 75 spheroids per condition shows 2 mM DCA significantly increases SW480 spheroid size, while combined 2 mM DCA treatment with 10 μM XAV939 significantly decreases their size. Statistical significance was determined using Student's two-tailed *t*-test.

D The effects of therapy on the total cell population, relative to the initial cell population, of combined XAV939 and DCA treatment were simulated using an *in vitro* version of the model (Appendix A7). As in Fig 6, DCA treatment was modeled by increasing the rate at which cells switched from a glycolytic metabolic phenotype to an OXPHOS phenotype (e.g., $1/\tau_{go}$ is increased) to reflect the tendency of cells to perform OXPHOS when PDK is inhibited. XAV939 treatment was modeled by decreasing the general Wnt signaling term S_W and increasing the range of Wnt and its inhibitor (due to upregulation of Wnt and Wnt inhibitor diffusers), which we modeled by increasing D_W and D_{Wi} proportionally. The dashed curves labeled $\tilde{S}_W = 0.80$ correspond to the case in which $\bar{S}_W = 0.80$ but D_W and D_{Wi} are unchanged.

the model presented here is the first to describe a pattern for heterogeneity in tumors that derives from an intricate spatial relationship between metabolic types and Wnt signaling. Though the role of Wnt signaling in cancer growth and development has been studied for many years, only recently has its regulation of Warburg/glycolysis metabolism been described. A better understanding of the link between Wnt and metabolism is crucial for defining how this overactive pathway drives tumorigenesis and progression, and for developing novel cancer treatments that target its key oncogenic actions.

Our mathematical model demonstrated strong qualitative and quantitative (spatial) agreement with the spotted patterns of activity detected in the tumors. The model also predicted that interference with Wnt signaling is not solely the result of a decrease in overall Wnt activity. Making the simple parameter change of decreasing Wnt signaling throughput leads to overall less glycolytic activity (lower background of P_g cells), a prediction that was validated in our xenograft experiments in which Wnt transcription was partially blocked by the overexpression of dominant negative LEF-1. However, the phospho-PDH stains also revealed fewer, but larger regions of PDK activity (i.e., larger, fewer cell clusters). To simulate these observations, coefficients of diffusion for Wnt and its secreted inhibitors were increased—parameter changes that resulted in a "spreading out" of Wnt and Wnt inhibitor activity. This model prediction prompted us to investigate whether the expression of proteins known to increase the range of Wnt ligand and Wnt inhibitor diffusion was increased when dnLEF- or dnTCF-expressing colon cancer cells were developed into xenograft tumors. Consistent with the mathematical model, we observed that the diffusers SPOCK2 and GPC4 were overexpressed in our xenografts but interestingly, not in our 2D *in vitro* culture conditions. The expression of SFRP5, which acts simultaneously as a Wnt inhibitor and diffuser by preventing binding to Frizzled receptors, also shows somewhat higher expression in our Wnt-low xenografts (dnLEF-1). It is important to emphasize that it is human-specific oligonucleotide primers that detect these changes in expression of Wnt ligand modifiers. Thus, the implanted human colon cancer cells appear to adapt to interference with Wnt signaling by directly increasing expression of Wnt ligand regulators. Our analysis of primary human colorectal tumors stressed by radiochemotherapy treatment shows that one consequence of therapy could be a similar increase in the distribution of Wnt ligands through upregulated expression of glypicans (e.g., GPC1) and SFRP proteins (Fig 4E). These observations suggest that there are likely to be significant changes in Wnt signaling dynamics and metabolic programming of treated tumors, perhaps as a means of coping and surviving the loss of nutrient and cellular damage.

The model also suggested that the tumor is most reliant on the glycolytic cells, and we found that inhibitors that target glycolysis and Wnt signaling in combination are more effective than treatments that only target one of these features. We simulated the actions of XAV939, which lowers β-catenin levels, and DCA, which inhibits PDK activity, thereby decreasing glycolysis. Since standard-of-care treatments for colon cancer have not changed significantly for decades, the novel combination of an inhibitor of glycolysis (e.g., DCA) with a Wnt pathway inhibitor (e.g., XAV939) might be an effective treatment to consider. We validated this prediction using 3D colony growth of SW480 colon cancer cells, which have high, intrinsic Wnt signaling and high levels of glycolytic activity. However, given that primary human colon cancers are more complex with respect to intrinsic Wnt pathway activity and cross-talk with the microenvironment, determining which tumor subtypes will be the most sensitive to this drug combination and how *in vivo* tumors respond to treatment will be important issues to resolve in developing effective drug combinations.

Our mathematical model is an abstract idealization and simplification of tumor proliferation and metabolism, and real tumors are much more complex than modeled here. For example, it is likely that Wnt and a Wnt inhibitor are not the only factors contributing to the pattern, although it is clear that Wnt has a strong influence, since the spots change significantly when Wnt signaling is interfered with. In the Appendix, we developed a more complete model that simulated PDK activity, hypoxia-inducible transcription factor concentrations (HIF1α), and lactate concentration (see Appendix A3), and linked these factors to cross-feeding, Wnt signaling, and metabolism. In this more detailed model, the switch between metabolic phenotypes depends on PDK activity, rather than on Wnt directly. However, since Wnt and HIF1α promote PDK expression and activity (Kim *et al*, 2006; Pate *et al*, 2014; Prigione *et al*, 2014), we found that the spatiotemporal distributions of PDK, Wnt, and lactate track closely together and so the results from the more detailed model are qualitatively similar to those presented in the main text for the simplified model.

While we have used a Turing-type model to simulate the spotted pattern, a decision well supported by data, we acknowledge that this type of model is only one possible mechanism to explain the patterning in the tumors. Other alternatives include differential adhesion or cell sorting—a process where cells of different adhesion potential sort away from each other (Amack & Manning, 2012; Foty & Steinberg, 2013), and bet-hedging—a process where different cells are differentially sensitive to external stresses, so that no matter the condition, at least some cells thrive (de Jong *et al*, 2011; Starrfelt & Kokko, 2012; Vogt, 2015). For bet-hedging to generate patterns, cell state changes must be reversible on a time scale slower than cell division, so that cells of like state end up clustered by default. Further experiments are needed to definitively distinguish among these processes. For example, these three types of processes (Turing, differential adhesion, bet-hedging) would be expected to be driven by different types of molecular signals and so these signals would need to be identified and tested.

Finally, although we focused on xenograft tumors, artificial constructs that only partially model tumorigenesis, we observed heterogeneity in metabolism and Wnt signaling in other settings as well. For example, we generated orthotopic tumors by implanting colon cancer cells (mock-parental SW480 and SW620, as well as dnLEF- or dnTCF-expressing variants) in the wall of the mouse cecum and observed patterning (Appendix A9 and Appendix Fig S19). In normal colon epithelia, pPDH stains show a gradient with high PDK activity in the base of the crypt where there is strong Wnt signaling and less PDK activity at the top of the crypt near the mucosal surface where Wnt signaling is not active (Pate *et al*, 2014; Fig 1D). In primary human patient colon tumors, our analysis revealed clear and striking heterogeneity in PDK activity and LEF-1 expression in the epithelial portion of the tumor (Fig 1D). While this heterogeneity was not apparent as a regular array of cell clusters like the xenograft patterns, groups of cells with markedly different activities are clearly evident. Since the tumor microenvironment is more complex in human tumors than xenograft tumors, an inherent pattern of metabolic and Wnt activity may be modified by additional structural and cellular components as well as nutritional stresses and a changing microenvironment. Understanding how these additional components influence metabolic heterogeneity and the symbiosis between neighboring cells and how they might create more complex pattern-on-pattern activities is a challenge going forward.

Materials and Methods

Numerical simulations

The nondimensionalized equations for the rate of change in the population of oxidative (P_o) cells and glycolytic (P_g) cells, respectively, are shown in Fig 2A. See Appendix A2 for details on the nondimensionalization. Figure 2B shows the rate of change in the concentration of Wnt and Wnt inhibitor (W and W_I) activity, respectively.

The first term on the right side of the equality in the P_o and P_g equations refers to diffusion, or random motion of the cells. The next terms are standard logistic proliferation terms, with proliferation dependent on nutrient level N, cell type, and the current total population of cells in the domain. The model sets the proliferation of glycolytic cells to be dependent on Wnt activity according to Michaelis–Menten dynamics, given by the term $W/(\alpha_W+W)$, which saturates at high levels of W. The parameters $1/\tau_o$ and $1/\tau_g$ are proliferation rates and $1/\tau_{go}$ and $1/\tau_{og}$ are switching rates. The last terms in these equations are cell death terms; death is modeled such that it occurs if the nutrient supply N drops below a threshold N_d. The death rates are given by μ_o and μ_g.

The model is designed such that glycolytic and oxidative cells can emphasize, or "switch", metabolism programs depending on W, Wnt activity, which is reflected in the third and fourth terms of each equation, where χ_W and χ_W^* are switch functions. Each switch function is defined by a modified hyperbolic tangent function, such that if Wnt activity falls below a parameter W*, then the cells utilize a more dominant OXPHOS program, and if Wnt activity is above W*, then cells are more likely to utilize a greater level of glycolysis.

We assume that oxidative cells can switch to utilizing more glycolysis only if sufficient nutrient is present, given by parameter N_g^*. Cells will die if nutrient is below the parameter N_d. The steepness of the functions can be adjusted so that they are more step-like and hence more sensitive to W and N. Since we use large values for the steepness of the functions, we could alternatively have used piecewise functions for χ_N and χ_N^*.

The dynamics of dead cells (not shown) is described by a similar reaction–diffusion equation. This is the population of cells that have died from lack of nutrient. These cells can also diffuse and decay. The equations in Fig 2 describe W (Wnt) and W_I (Wnt inhibitor, e.g., DKK or SFRP) activity. D_W and D_{WI} are constant diffusion coefficients. It has been shown in epidermal cells that Wnt target genes produce Wnt signals as well as long-range secreted Wnt inhibitors (Lim et al, 2013), so the inhibitor is assumed to diffuse much longer range than Wnt; that is, D_W must be significantly smaller than D_{WI}. Wnt signaling activity is assumed to be nonlinear with respect to Wnt and is inhibited by the Wnt inhibitor through the term $1/(a+b W_I)$. We assume the Wnt inhibitor is being produced by Wnt activity through both cell types. The terms μ_W and μ_{WI} are decay rates. The term $S_W(P_o+P_g)$ in the Wnt equation refers to constitutive Wnt signaling through the cells.

The equation for nutrient (eq. 1h in Appendix A2) describes the diffusion and uptake, decay, and source of nutrient. The nutrient term has Dirichlet (fixed) boundary conditions and diffuses in from the boundary of the spatial domain, so that the boundary can be considered as regions where vasculature is high. The second and third terms refer to uptake of nutrient by the two different cell types. The term $\mu_N N$ is a natural decay term. The last term, N_S, refers to the nutrient source, which is a small source term applied to the entire domain. This source term is based linearly on the glycolytic activity of the cells and is given by $N_S = N_S(\int P_g) = \gamma_N[(1 - \alpha_N) \int \frac{P_g}{S_x S_y} + \alpha_N]$, where α_N and γ_N are parameters, $\int P_g$ is the integral of P_g cells, and S_x and S_y are the lengths of the sides of the spatial domain. This function was chosen so that $N_s(0) = \gamma_N^{\alpha N}$ and $N_s(S_x S_y) = \gamma_N$ ($S_x S_y$ is the maximum that $\int P_g$ can reach). We chose to have the nutrient source N_s depend on P_g cells because glycolysis induces angiogenesis (Dhup et al, 2012; Porporato et al, 2012; Ruan & Kazlauskas, 2013), allowing more nutrients and growth factors to be delivered to the tumor (Pate et al, 2014). We use a linear function in the model as the simplest form for the dependency between N and P_g, which is consistent with experimental observations.

To summarize, in addition to consideration of Wnt signaling dynamics, biological assumptions for the model include terms for random motion in space (diffusion), terms for each cell type (oxidative and glycolytic, or P_o and P_g, respectively), and their propensity to proliferate, die, and switch to the other cell type. Equations were included to account for dead cells, which consists of P_o and P_g cells that have died from lack of nutrient, and which can diffuse and decay. Terms for Wnt (W) and Wnt inhibitor (W_I) activity were made nonlinear with respect to Wnt, meaning that their rates are proportional to Wnt activity. Nonlinear Wnt activity is dependent on P_g levels, while nonlinear Wnt inhibitor activity is proportional to both P_g and P_o levels. A term for constitutive Wnt signaling was included for both cell types as well as decay terms for W and W_I. The general nutrient term N can diffuse, decay, and be taken up by the different cell populations. A bulk source was included for the nutrient as well as a Dirichlet boundary condition, both of which are dependent on the average level of glycolytic cells in the domain, a simplified way to incorporate increased angiogenesis driven by glycolysis (Pate et al, 2014). This relationship was included to take into account our observation that there is considerably less vasculature in tumors in which Wnt signaling has been blocked by dominant interfering forms of the Wnt transcription factors lymphoid enhancer factor-1 (dnLEF-1) or T cell factor 1 (dnTCF-1) (Pate et al, 2014). Finally, there is a baseline assumption that there is sufficient oxygen available throughout the domain for OXPHOS to operate, even at a minimal level.

In the numerical results presented here, no-flux boundary conditions were used for all terms except N, which is governed by Dirichlet boundary conditions (N at the boundary is equal to the value $\frac{1}{\gamma_N} N_s$ where N_s is described above). Initial conditions were set for a random distribution of P_g cells located near the boundary, and small random values of W and W_I in the same areas where initial P_g cells are located. A constant high level of nutrient throughout the domain was provided (results did not change qualitatively if N was solved as a quasi-steady-state equation), and the initial condition contained no P_o or P_d cells. All parameters are given in Table 1, and a sensitivity analysis is provided in Appendix A5.

Numerical simulations were performed in MATLAB, using a forward difference method for each time derivative. P_o, P_g, W, and W_I equations were solved implicitly in centered diffusion terms. The nutrient equation was solved implicitly in uptake, decay, and centered diffusion terms.

Animal protocols for xenograft and orthotopic tumors

SW480 stable transductants for xenograft or orthotopic injection were prepared through lentiviral infection with pCDH vector from System Biosciences: empty vector (mock), or vector expressing dnLEF-1 or dnTCF-1, followed by selection with 500 µg/ml G418. Transduced cells were collected as a pool for confirmation of expression, and Wnt signaling activity was measured by a SuperTOPFlash luciferase reporter (Pate *et al*, 2014). A total of 2.5×10^6 cells were injected into immunodeficient NSG mice [2-month-old NSG male and female mice were used for the subcutaneous xenograft tumors (JAX™ Mice from Jackson Labs); male and female NSG mice, approximately 3 months old, were used for orthotopic tumors (injection of $5–10 \times 10^3$ cells into the cecum wall)]. Tumors were removed (subcutaneous after 3 weeks, orthotopic after 4 weeks), fixed in paraformaldehyde overnight, and paraffin-embedded 4 weeks after injection. All experiments involving animals were approved by the UCI IACUC (Protocol 2002-2357-4 to R. Edwards).

Immunohistochemistry

Deparaffinized 5- to 6-µm sections of formalin-fixed paraffin-embedded (FFPE) mouse xenograft tumor and human colorectal carcinoma tissues followed by pressure cooker antigen retrieval in citrate buffer were blocked in 3% H_2O_2 and goat or horse serum plus MOM block reagent (if mouse primary antibody was used on mouse tissue), avidin, and biotin blocking reagents (Vector Labs). Sections were incubated in primary antibodies: antiphospho-PDHpSer293 (Calbiochem; 1:50–1:100), anti-β-catenin (BD; 1:500), anti-LEF-1 (Cell Signaling; 1:100), anti-HIF1α (Thermo Scientific; 1:1,000) followed by biotinylated secondary antibodies and visualization using a peroxidase-conjugated avidin-based Vectastain protocol. Slides were then counterstained with hematoxylin and mounted using Permount mounting medium (Fisher). Images were captured using an Olympus FSX100 system and processed in Adobe Photoshop.

Quantitative PCR

RNA was extracted from xenograft tumors and cells using TRIzol (Invitrogen) following the manufacturer's instructions. cDNA was synthesized with 1 µg of total RNA with the High Capacity cDNA Reverse Transcription Kit (Invitrogen), as per the manufacturer's instructions. qPCR was performed in triplicate for each experimental condition using Maxima SYBR Green qPCR Master Mix (Invitrogen), according to the manufacturer's instructions. To normalize mRNA levels, GAPDH probes were used. Primer pairs are as follows: GAPDH forward: TCGACAGTCAGCCGCATCTTCTT, reverse: GCG CCCAATACGACCAAATCC; TINAGL1 forward: ACCAGGTCACTC CTGTCTACC, reverse: GATGCCTCCCTTGTATAGGAAG; CDC42 forward: CCATCGGAATATGTACCGAC, reverse: CTCAGCGGTCG TAATCTGTC; SPOCK2 forward: CCCGGCAATTTCATGGAGG, reverse: GCGGTTCCAGTGCTTGATC; GPC1 forward: GGCTGGTGGCT GCTATGT, reverse: CAGGTTCTCCTCCATCTCGC; GPC2 forward: CACCTGCTGTTCCAGTGAGA, reverse: AGAGAGTGCTGGGCTACT GA; GPC4 forward: GTGGGAAATGTGAACCTGGAA, reverse: CGAG GGACATCTCCGAAGG; DKK4 forward: GGGACACTCTGTGTGAA

CGA, reverse: TGGTTTTCCTGGACTGGGTG; SFRP5 forward: CTGT ACGCGTCATCCTAGCC, reverse: CGGACCAGAAGGGGGTCTAT.

Fibrin gel assay

A total of 200 trypsinized SW480 cells were mixed with 100 µl of 2.5 mg/ml bovine fibrinogen (MP Biomedicals) in DMEM plus 10% FBS and 1% penicillin–streptomycin–glutamine and 1 µl of thrombin (Sigma). The fibrin gels were seeded in 96-well, flat-bottom plates. After the gels solidified, 100 µl of DMEM media containing the desired drug treatment was layered on top (DCA was obtained from Sigma, XAV939 from Stemgent). Wells were imaged after 14 days of incubation. Size measurements were taken using Adobe Photoshop. Data were analyzed using Prism (GraphPad).

Image processing of spots

Image processing (overlay of spot contours and convex hulls) was done using built-in functions in MATLAB's Image Processing Toolbox. Briefly, a color channel of an image was converted to a binary image based on a manually chosen threshold dependent on staining intensity. A noise filter was applied to reduce background staining. Thresholds were then chosen to define cutoff values of spot boundaries. Parameters for built-in tools were chosen manually to give the best fit for pattern contours. Details for this method are provided in Appendix A1 and A1.1–A1.3, and Appendix Figs S1–S10.

Acknowledgements

The authors would like to thank the Waterman, Donovan, and Lowengrub laboratories for their feedback and discussion. We would also like to thank Eric Stanbridge, Peter Donovan, Arthur Lander, Jun Allard, Olivier Cinquin, Jenny Wu, Harry Mangalam, Babak Shababa, and Charless Fowlkes for providing advice and critiques. The work of G.T.C. and M.L.W. was supported by NIH Grants CA096878, CA108697, CA200298, a California CRCC award CRR-17-429379, and a P30CA062203 to the Chao Family Comprehensive Cancer Center. The work of M.L., E.J.P., and J.S.L. was supported in part by NSF Grant DMS-1263796, NIH Grant P50-GM76516 for a Center of Excellence in Systems Biology at the University of California, Irvine, as well as the NIH Grant P30-CA062203 grant. In addition, the authors also received partial support from the University of California, Irvine, through a seed grant. This work was made possible, in part, through access to the Tissue Specimen Shared Resource and Genomics and High Throughput Facility of the Cancer Center Support Grant (CA-62203) at the University of California, Irvine, and NIH shared instrumentation grants 1S10RR025496-01 and 1S10OD010794-01. K.W. and R.A.E. were supported by P30CA062203.

Author contributions

ML and JL conceived the mathematical model; ML and EP coded and ran all simulations; ML, GTC, JL, and MLW wrote the manuscript; MLW and GTC designed the biological experiments; GTC, KW, and RAE performed biological experiments.

References

Allen E, Miéville P, Warren CM, Saghafinia S, Li L, Peng M-W, Hanahan D (2016) Metabolic symbiosis enables adaptive resistance to anti-angiogenic therapy that is dependent on mTOR signaling. *Cell Rep* 15: 1144–1160

Amack JD, Manning ML (2012) Knowing the boundaries: extending the differential adhesion hypothesis in embryonic cell sorting. *Science* 338: 212–215

Cochran WG (1954) Some methods for strengthening the common χ^2 tests. *Biometrics* 10: 417–451

De Saedeleer CJ, Copetti T, Porporato PE, Verrax J, Feron O, Sonveaux P (2012) Lactate activates HIF-1 in oxidative but not in Warburg-phenotype human tumor cells. *PLoS One* 7: e46571

Dhup S, Dadhich RK, Porporato PE, Sonveaux P (2012) Multiple biological activities of lactic acid in cancer: influences on tumor growth, angiogenesis and metastasis. *Curr Pharm Des* 18: 1319–1330

Epstein T, Xu L, Gillies RJ, Gatenby RA (2014) Separation of metabolic supply and demand: aerobic glycolysis as a normal physiological response to fluctuating energetic demands in the membrane. *Cancer Metab* 2: 7

Esen E, Chen J, Karner CM, Okunade AL, Patterson BW, Long F (2013) WNT-LRP5 signaling induces Warburg effect through mTORC2 activation during osteoblast differentiation. *Cell Metab* 17: 745–755

Farin HF, Jordens I, Mosa MH, Basak O, Korving J, Tauriello DVF, de Punder K, Angers S, Peters PJ, Maurice MM, Clevers H (2016) Visualization of a short-range Wnt gradient in the intestinal stem-cell niche. *Nature* 530: 340–343

Foty RA, Steinberg MS (2013) Differential adhesion in model systems. *Wiley Interdiscip Rev Dev Biol* 2: 631–645

Foucquier J, Guedj M (2015) Analysis of drug combinations: current methodological landscape. *Pharmacol Res Perspect* 3: e00149

Gatenby RA, Gillies RJ (2004) Why do cancers have high aerobic glycolysis? *Nat Rev Cancer* 4: 891–899

Giannakis M, Hodis E, Mu XJ, Yamauchi M, Rosenbluh J, Cibulskis K, Saksena G, Lawrence MS, Qian ZR, Nishihara R, Van Allen EM, Hahn WC, Gabriel SB, Lander ES, Getz G, Ogino S, Fuchs CS, Garraway LA (2014) RNF43 is frequently mutated in colorectal and endometrial cancers. *Nat Genet* 46: 1264–1266

Gierer A, Meinhardt H (1972) A theory of biological pattern formation. *Biol Cybern* 12: 30–39

Holcombe R, Marsh J, Waterman M (2002) Expression of Wnt ligands and Frizzled receptors in colonic mucosa and in colon carcinoma. *Mol Pathol* 55: 220–227

Hovanes K, Li TWH, Munguia JE, Truong T, Milovanovic T, Marsh JL, Holcombe RF, Waterman ML (2001) β-catenin–sensitive isoforms of lymphoid enhancer factor-1 are selectively expressed in colon cancer. *Nat Genet* 28: 53–57

Hoverter NP, Ting J-H, Sundaresh S, Baldi P, Waterman ML (2012) A WNT/p21 circuit directed by the C-clamp, a sequence-specific DNA binding domain in TCFs. *Mol Cell Biol* 32: 3648–3662

Jiang Y, Pjesivac-Grbovic J, Cantrell C, Freyer JP (2005) A multiscale model for avascular tumor growth. *Biophys J* 89: 3884–3894

Jiménez-Valerio G, Martínez-Lozano M, Bassani N, Vidal A, Ochoa-de-Olza M, Suárez C, García-Del-Muro X, Carles J, Viñals F, Graupera M, Indraccolo S, Casanovas O (2016) Resistance to antiangiogenic therapies by metabolic symbiosis in renal cell carcinoma PDX models and patients. *Cell Rep* 15: 1134–1143

de Jong IG, Haccou P, Kuipers OP (2011) Bet hedging or not? A guide to proper classification of microbial survival strategies. *Bioessays* 33: 215–223

Kim JW, Tchernyshyov I, Semenza GL, Dang CV (2006) HIF-1-mediated expression of pyruvate dehydrogenase kinase: a metabolic switch required for cellular adaptation to hypoxia. *Cell Metab* 3: 177–185

Kondo S, Miura T (2010) Reaction-diffusion model as a framework for understanding biological pattern formation. *Science* 329: 1616–1620

Lescher B, Haenig B, Kispert A (1998) sFRP-2 is a target of the Wnt-4 signaling pathway in the developing metanephric kidney. *Dev Dyn* 213: 440–451

Lim X, Tan SH, Koh WL, Chau RM, Yan KS, Kuo CJ, van Amerongen R, Klein AM, Nusse R (2013) Interfollicular epidermal stem cells self-renew via autocrine Wnt signaling. *Science* 342: 1226–1230

Mantel N, Haenszel W (1959) Statistical aspects of the analysis of data from retrospective studies of disease. *J Natl Cancer Inst* 22: 719–748

Marcon L, Diego X, Sharpe J, Müller P (2016) High-throughput mathematical analysis identifies turing networks for patterning with equally diffusing signals. *Elife* 5: e14022

Martinez-Outschoorn U, Sotgia F, Lisanti MP (2014) Tumor microenvironment and metabolic synergy in breast cancers: critical importance of mitochondrial fuels and function. *Semin Oncol* 41: 195–216

McGillen JB, Kelly CJ, Martinez-Gonzalez A, Martin NK, Gaffney EA, Maini PK, Perez-Garcia VM (2014) Glucose–lactate metabolic cooperation in cancer: insights from a spatial mathematical model and implications for targeted therapy. *J Theor Biol* 361: 190–203

Mendoza-Juez B, Martínez-González A, Calvo GF, Pérez-García VM (2012) A mathematical model for the glucose-lactate metabolism of in vitro cancer cells. *Bull Math Biol* 74: 1125–1142

Mii Y, Taira M (2009) Secreted Frizzled-related proteins enhance the diffusion of Wnt ligands and expand their signalling range. *Development* 136: 4083–4088

Mii Y, Taira M (2011) Secreted Wnt 'inhibitors' are not just inhibitors: regulation of extracellular Wnt by secreted Frizzled-related proteins. *Dev Growth Differ* 53: 911–923

Murray B, Wilson DJ (2001) A study of metabolites as intermediate effectors in angiogenesis. *Angiogenesis* 4: 71–77

Murray JD (2003) *Mathematical biology II: spatial models and biochemical applications*, Volume II. New York, NY: Springer-Verlag

Obre E, Rossignol R (2015) Emerging concepts in bioenergetics and cancer research: metabolic flexibility, coupling, symbiosis, switch, oxidative tumors, metabolic remodeling, signaling and bioenergetic therapy. *Int J Biochem Cell Biol* 59: 167–181

Pate KT, Stringari C, Sprowl-tanio S, Wang K, Teslaa T, Hoverter NP, Mcquade MM, Garner C, Digman MA, Teitell MA, Edwards RA, Gratton E, Waterman ML (2014) Wnt signaling directs a metabolic program of glycolysis and angiogenesis in colon cancer. *EMBO J* 33: 1454–1473

Pavlides S, Whitaker-Menezes D, Castello-Cros R, Flomenberg N, Witkiewicz AK, Frank PG, Casimiro MC, Wang C, Fortina P, Addya S, Pestell RG, Martinez-Outschoorn UE, Sotgia F, Lisanti MP (2009) The reverse Warburg effect: aerobic glycolysis in cancer associated fibroblasts and the tumor stroma. *Cell Cycle* 8: 3984–4001

Pavlova NN, Thompson CB (2016) The emerging hallmarks of cancer metabolism. *Cell Metab* 23: 27–47

Phipps C, Molavian H, Kohandel M (2015) A microscale mathematical model for metabolic symbiosis: investigating the effects of metabolic inhibition on ATP turnover in tumors. *J Theor Biol* 366: 103–114

Pisarsky L, Bill R, Fagiani E, Dimeloe S, Goosen RW, Hagmann J, Hess C, Christofori G (2016) Targeting metabolic symbiosis to overcome resistance to anti-angiogenic therapy. *Cell Rep* 15: 1161–1174

Porporato PE, Payen VL, De Saedeleer CJ, Préat V, Thissen JP, Feron O, Sonveaux P (2012) Lactate stimulates angiogenesis and accelerates the healing of superficial and ischemic wounds in mice. *Angiogenesis* 15: 581–592

Prigione A, Rohwer N, Hoffmann S, Mlody B, Drews K, Bukowiecki R, Blümlein K, Wanker EE, Ralser M, Cramer T, Adjaye J (2014) HIF1α modulates cell fate reprogramming through early glycolytic shift and upregulation of PDK1-3 and PKM2. *Stem Cells* 32: 364–376

Raspopovic J, Marcon L, Russo L, Sharpe J (2014) Digit patterning is controlled by a BMP-Sox9-Wnt Turing network modulated by morphogen gradients. *Science* 345: 566–570

Roche TE, Baker JC, Yan X, Hiromasa Y, Gong X, Peng T, Dong J, Turkan A, Kasten SA (2001) Distinct regulatory properties of pyruvate dehydrogenase kinase and phosphatase isoforms. *Prog Nucleic Acid Res Mol Biol* 70: 33–75

Rockne R, Rockhill JK, Mrugala M, Spence AM, Kalet I, Hendrickson K, Lai A, Cloughesy T, Alvord EC Jr, Swanson KR (2010) Predicting the efficacy of radiotherapy in individual glioblastoma patients in vivo: a mathematical modeling approach. *Phys Med Biol* 55: 3271

Ruan GX, Kazlauskas A (2013) Lactate engages receptor tyrosine kinases Axl, Tie2, and vascular endothelial growth factor receptor 2 to activate phosphoinositide 3-kinase/AKT and promote angiogenesis. *J Biol Chem* 288: 21161–21172

Seshagiri S, Stawiski EW, Durinck S, Modrusan Z, Storm EE, Conboy CB, Chaudhuri S, Guan Y, Janakiraman V, Jaiswal BS, Guillory J, Ha C, Dijkgraaf GJ, Stinson J, Gnad F, Huntley MA, Degenhardt JD, Haverty PM, Bourgon R, Wang W *et al* (2012) Recurrent R-spondin fusions in colon cancer. *Nature* 488: 660–664

Sherwood V (2015) WNT signaling: an emerging mediator of cancer cell metabolism? *Mol Cell Biol* 35: 2–10

Sick S, Reinker S, Timmer J, Schlake T (2006) WNT and DKK determine hair follicle spacing through a reaction-diffusion mechanism. *Science* 314: 1447–1450

Snipstad K, Fenton CG, Kjæve J, Cui G, Anderssen E, Paulssen RH (2010) New specific molecular targets for radio-chemotherapy of rectal cancer. *Mol Oncol* 4: 52–64

Sonveaux P, Végran F, Schroeder T, Wergin MC, Verrax J, Rabbani ZN, De Saedeleer CJ, Kennedy KM, Diepart C, Jordan BF, Kelley MJ, Gallez B, Wahl ML, Feron O, Dewhirst MW (2008) Targeting lactate-fueled respiration selectively kills hypoxic tumor cells in mice. *J Clin Invest* 118: 3930–3942

Sonveaux P, Copetti T, De Saedeleer CJ, Végran F, Verrax J, Kennedy KM, Moon EJ, Dhup S, Danhier P, Frérart F, Gallez B, Ribeiro A, Michiels C, Dewhirst MW, Feron O (2012) Targeting the lactate transporter MCT1 in endothelial cells inhibits lactate-induced HIF-1 activation and tumor angiogenesis. *PLoS One* 7: e33418

Sprowl-Tanio S, Habowski AN, Pate KT, McQuade MM, Wang K, Edwards RE, Grun F, Lyou Y, Waterman ML (2016) Lactate/pyruvate transporter MCT-1 is a direct Wnt target that confers sensitivity to 3-bromopyruvate in colon cancer. *Cancer Metab* 4: 20

Starrfelt J, Kokko H (2012) Bet-hedging—a triple trade-off between means, variances and correlations. *Biol Rev* 87: 742–755

Turing AM (1952) The chemical basis of morphogenesis. *Philos Trans R Soc Lond B Biol Sci* 237: 37–72

Uhlén M, Fagerberg L, Hallström BM, Lindskog C, Oksvold P, Mardinoglu A, Sivertsson Å, Kampf C, Sjöstedt E, Asplund A, Olsson I, Edlund K, Lundberg E, Navani S, Szigyarto CA, Odeberg J, Djureinovic D, Takanen JO, Hober S, Alm T *et al* (2015) Tissue-based map of the human proteome. *Science* 347: 1260419

Van de Wetering M, Sancho E, Verweij C, De Lau W, Oving I, Hurlstone A, Van der Horn K, Batlle E, Coudreuse D, Haramis AP, Tjon-Pon-Fong M, Moerer P, Van den Born M, Soete G, Pals S, Eilers M, Medema R, Clevers H (2002) The β-catenin/TCF-4 complex imposes a crypt progenitor phenotype on colorectal cancer cells. *Cell* 111: 241–250

Vander Heiden M, Cantley L, Thompson C (2009) Understanding the Warburg effect: the metabolic requirements of cell proliferation. *Science* 324: 1029–1033

Végran F, Boidot R, Michiels C, Sonveaux P, Feron O (2011) Lactate influx through the endothelial cell monocarboxylate transporter MCT1 supports an NF-kB/IL-8 pathway that drives tumor angiogenesis. *Cancer Res* 71: 2550–2560

Vogt G (2015) Stochastic developmental variation, an epigenetic source of phenotypic diversity with far-reaching biological consequences. *J Biosci* 40: 159–204

Voloshanenko O, Erdmann G, Dubash TD, Augustin I, Metzig M, Moffa G, Hundsrucker C, Kerr G, Sandmann T, Anchang B, Demir K, Boehm C, Leible S, Ball CR, Glimm H, Spang R, Boutros M (2013) Wnt secretion is required to maintain high levels of Wnt activity in colon cancer cells. *Nat Commun* 4: 2610

Walton KD, Whidden M, Kolterud A, Shoffner S, Czerwinski MJ, Kushwaha J, Parmar N, Chandhrasekhar D, Freddo AM, Schnell S, Gumucio DL (2015) Villification in the mouse: BMP signals control intestinal villus patterning. *Development* 143: 734–764

Warburg O (1956) On the origin of cancer cells. *Science* 123: 309–314

Willert K, Brown JD, Danenberg E, Duncan AW, Weissman IL, Reya T, Yates JR, Nusse R (2003) Wnt proteins are lipid-modified and can act as stem cell growth factors. *Nature* 423: 448–452

Yan H, Romero-Lopez M, Hughes CCW, Lowengrub JS (2016) Multiscale modeling of glioma suggests that the partial disruption of vessel/cancer stem cell crosstalk can promote tumor regression without invasiveness. *IEEE Trans Biomed Eng* doi: 10.1109/TBME.2016.2615566

Youssefpour H, Li X, Lander AD, Lowengrub JS (2012) Multispecies model of cell lineages and feedback control in solid tumors. *J Theor Biol* 304: 39–59

Zhang L, Lander AD, Nie Q (2012) A reaction–diffusion mechanism influences cell lineage progression as a basis for formation, regeneration, and stability of intestinal crypts. *BMC Syst Biol* 6: 93

5

Alternative TSSs are co-regulated in single cells in the mouse brain

Kasper Karlsson[1] (iD), Peter Lönnerberg[2] & Sten Linnarsson[2,*] (iD)

Abstract

Alternative transcription start sites (TSSs) have been extensively studied genome-wide for many cell types and have been shown to be important during development and to regulate transcript abundance between cell types. Likewise, single-cell gene expression has been extensively studied for many cell types. However, how single cells use TSSs has not yet been examined. In particular, it is unknown whether alternative TSSs are independently expressed, or whether they are co-activated or even mutually exclusive in single cells. Here, we use a previously published single-cell RNA-seq dataset, comprising thousands of cells, to study alternative TSS usage. We find that alternative TSS usage is a regulated process, and the correlation between two TSSs expressed in single cells of the same cell type is surprisingly high. Our findings indicate that TSSs generally are regulated by common factors rather than being independently regulated or stochastically expressed.

Keywords alternative TSS usage; neurons; single-cell RNA sequencing; transcription; UMI

Subject Categories Genome-Scale & Integrative Biology; Transcription

Introduction

Our understanding of TSS usage has increased dramatically over the last decade since the introduction of deep sequencing technologies. It is now clear that most genes are transcribed from multiple distinct TSSs. For example, the FANTOM consortium recently found an average of four robust TSSs per gene, across more than 800 tissues and cell lines (Forrest *et al*, 2014); however, the number of TSS reported was highly dependent on the filtering method used and was complicated by CAGE peaks in enhancer regions, coding regions, and promoter-associated short RNA (de Klerk & t Hoen, 2015). Transcriptome-wide studies using DeepCAGE have found that the hippocampus has a larger number of active TSSs compared to other cell types tested and that generally TSS use is highly tissue specific (de Klerk & t Hoen, 2015). For example, IGF1 and IGF2 are

known to be regulated by multiple TSSs and expressed in various embryonic and adult tissues (Leroith & Roberts, 1993). In *Drosophila*, alternative TSS generally implements distinct regulatory programs during development (Batut *et al*, 2013). Alternative TSS usage can affect protein diversity by incorporating extended or alternative N-terminal polypeptides. One example is NADH-cytochrome b5 reductase, where usage of alternative TSSs results in two protein forms, one membrane-bound and one soluble form (Ayoubi, 2005). However, the majority of alternative TSS do not change the mRNA coding potential, and thus, for the majority the biological effect must come from isoform-specific regulation like mRNA abundance, stability, and localization (Rojas-Duran & Gilbert, 2012). Alternative TSS can also cause differences in translation efficiencies up to a 100-fold when examined in yeast (Rojas-Duran & Gilbert, 2012).

Transcription start sites are activated by a complex chain of events initiated by the binding of transcription factors to proximal sites or distal enhancers. However, the rules by which local or distal transcription factor (TF) binding causes the activation of specific local TSSs are not well understood. It is often assumed that TF binding upstream of a TSS will specifically activate that TSS, but it is also possible that multiple nearby TSSs could be simultaneously activated, or indeed that local and distal enhancers could exhibit preferences for specific TSSs.

Surprisingly, in general, it is not known to what extent alternative TSSs show cell type-specific expression. In the extreme case, it is possible that all TSSs respond equally well to regulatory input, that is, alternative TSSs are regulated by common factors. Alternatively, there may be hitherto unexplored rules that determine a strict preference for each regulatory sequence to a specific TSS, that is, TSSs are regulated by specific factors. In the former case, alternative TSSs would be expected to be always active in the same tissues and cell types, perhaps in some fixed proportion. In the latter case, alternative TSSs would be expected to show largely independent expression.

Due to the stochastic nature of gene expression, alternative TSS usage may also have functional consequences in single cells, even if they are not differentially regulated at the population level. Consider a gene with two TSSs, major and minor, which are both active. If TSSs compete for binding to regulatory elements, and activation events lead to bursts of transcription, this would lead to a stochastic, mutually exclusive, anti-correlated expression pattern, similar to

1   Stanford Cancer Institute, Stanford University School of Medicine, Stanford, CA, USA
2   Laboratory for Molecular Neurobiology, Department of Medical Biochemistry and Biophysics, Karolinska Institutet, Stockholm, Sweden
    *Corresponding author. E-mail: sten.linnarsson@ki.se

that observed for random monoallelic expression (Deng *et al*, 2014). Alternatively, if TSSs were activated independently, this would lead to uncorrelated expression in single cells. Yet another possibility is that both TSSs are simultaneously expressed, or that stochastic TSS activation occurs on a timescale much shorter than mRNA degradation, so that even in single cells these effects would be washed out, and mRNA from alternative TSSs would be correlated and detected at fixed proportions.

Heterogeneity of gene expression has been extensively studied; however, the study of single-cell isoform variation has only just started. For example, Velten *et al* (2015) recently studied single-cell polyadenylation site usage and found that even in homogenous cell populations, individual cells differ in their preferences for 3′ RNA isoform choice. Another study examined single-cell splice isoforms and found that genes having multiple splice isoforms at the population level tended to have only one expressed isoform at the single-cell level (Shalek *et al*, 2013).

Here, we address these questions using single-cell RNA-seq. We take advantage of our recently published analysis of mouse cortex and hippocampus. This extensively validated dataset comprises over 3,000 single-cell transcriptomes, classified into nine major cell types and 47 subtypes. We used STRT, a single-cell RNA-seq method suitable to study TSS usage since it captures and sequences the 5′ end of polyadenylated mRNA transcripts (Islam *et al*, 2011, 2012). The inclusion of unique molecular identifiers (UMI) ensured an increased quantitative accuracy by eliminating most PCR bias and allowed the absolute counting of mRNA molecules (Kivioja *et al*, 2012).

Results

Measuring TSS activity in single cells

We selected a set of 2,816 single-cell transcriptomes representing seven cell types: interneurons, somatosensory cortex pyramidal neurons, hippocampal pyramidal neurons, oligodendrocytes, astrocytes/ependymal cells, microglia, and vascular cells (endothelial cells, pericytes, and smooth muscle cells). Raw reads were mapped to the genome, assigned to FANTOM annotated TSSs, and converted to absolute number of molecules using UMIs (Islam *et al*, 2014). Henceforth, a set of reads mapped to a single genomic position, and with the same UMI, will be referred to as a "molecule" of mRNA. It should be noted that detected molecules likely represent only about 20% of all expressed molecules (Zeisel *et al*, 2015).

In order to accurately measure TSS-specific gene expression in single cells, we first needed to ensure that our protocol indeed preferentially captured 5′ ends of transcripts. We took advantage of ERCC spike-in RNAs present in each single-cell experiment. ERCC transcripts are 250–2,000 bp synthetic polyadenylated RNAs, at known concentrations. We found that 75% of all molecules mapped exactly at the 5′ end, whereas the rest were scattered across the rest of each RNA (Fig 1A).

For endogenous genes, in contrast, most molecules did not map to the annotated 5′ end. In agreement with previous findings (Islam *et al*, 2011), we found a single peak at the 5′ end and a broad 3′-biased distribution with a preference for the 3′ end (Fig 1B). As a consequence, out of the average of 26,500 detected molecules per cell, only 3,800 (14%), could be allocated to an annotated FANTOM TSS (for details, see Table EV1). Of the rest, about half mapped outside of known protein-coding genes (e.g., to expressed transposons), and the rest mapped to genes but outside the annotated TSSs (Fig 1C). The TSSs clearly showed an elevated signal, as expected. However, many molecules mapped all over the gene, including at low levels in intron regions, and some genes showed extensive 3′ UTR expression (Appendix Fig S1). Nevertheless, these findings suggest that there was enough signal specifically at the TSSs for an analysis to be possible.

To determine whether molecules assigned to TSS regions were specific, we examined a region of ± 100 bp around TSSs (here defined as the center of the FANTOM5 TSS interval). Reassuringly, we found a distinct, sharp peak at the ± 0 position, and 97% of all molecules mapped within 50 bp of the putative TSS (Fig 1D). It should be noted that FANTOM TSS intervals are not guaranteed to be centered on the true TSS, so some of the imprecision can be attributed to imperfect annotation.

Molecules mapping outside annotated TSSs could represent degradation products, as (in contrast to CAGE) our methods are not selective for capped 5′ ends. We therefore searched for signs of translation-associated mRNA decay (Pelechano *et al*, 2015), which should lead to a 3-bp repeated pattern aligned with the reading frame. We found no evidence of such degradation in this data, as can be seen in Appendix Fig S2A and B. However, a clear pattern emerged around the start and stop codons. Many molecules (i.e., 5′ ends of transcripts) mapped upstream of the start codon, but not downstream, probably reflecting the fact that most molecules in this region represent bona fide 5′ ends of transcripts (by definition, the true TSS cannot be placed after the start codon). In agreement with this hypothesis, the increase in reads upstream of the TSS closely reflected the prevalence of annotated transcription start sites. A similar, but weaker, pattern was observed around the stop codon. Interestingly, there was an enrichment of mapped reads starting just upstream of the stop codon, which may reflect pausing of the translational machinery at this point. Translation-associated mRNA decay would catch up with the stalled ribosome, leading to a relative depletion of upstream fragments and enrichment of fragments downstream of the stop codon.

To ensure that true TSS and not degradation products were assessed, molecules were counted both upstream and downstream of the TSS within a region of similar size, hereafter referred to as a "reference region". If the number of molecules in the reference region on either side of the TSS constituted more than 20% of the combined TSS and reference region reads, then the TSS was removed from the analysis. This removed around 20% of the TSSs.

Finally, to assess the reproducibility of the data, we compared correlations of gene expression profiles calculated from whole-gene bodies, as in Zeisel *et al* (2015) or only from the major TSS, both for combined and single-cell data (Fig 1F–H). As expected, data from independent pools of randomly selected cells were highly correlated, and the correlation dropped slightly for single cells. Surprisingly, for some genes, expression from the major TSS was more highly correlated compared to expression from whole-gene bodies, despite the fact that major TSSs only contained a fraction of the molecules. As shown in Appendix Figs S3A–C, the effect was rather common and depended in a large part on the expression level of the cells. For two cells with combined high average TSS expression (> 1,000

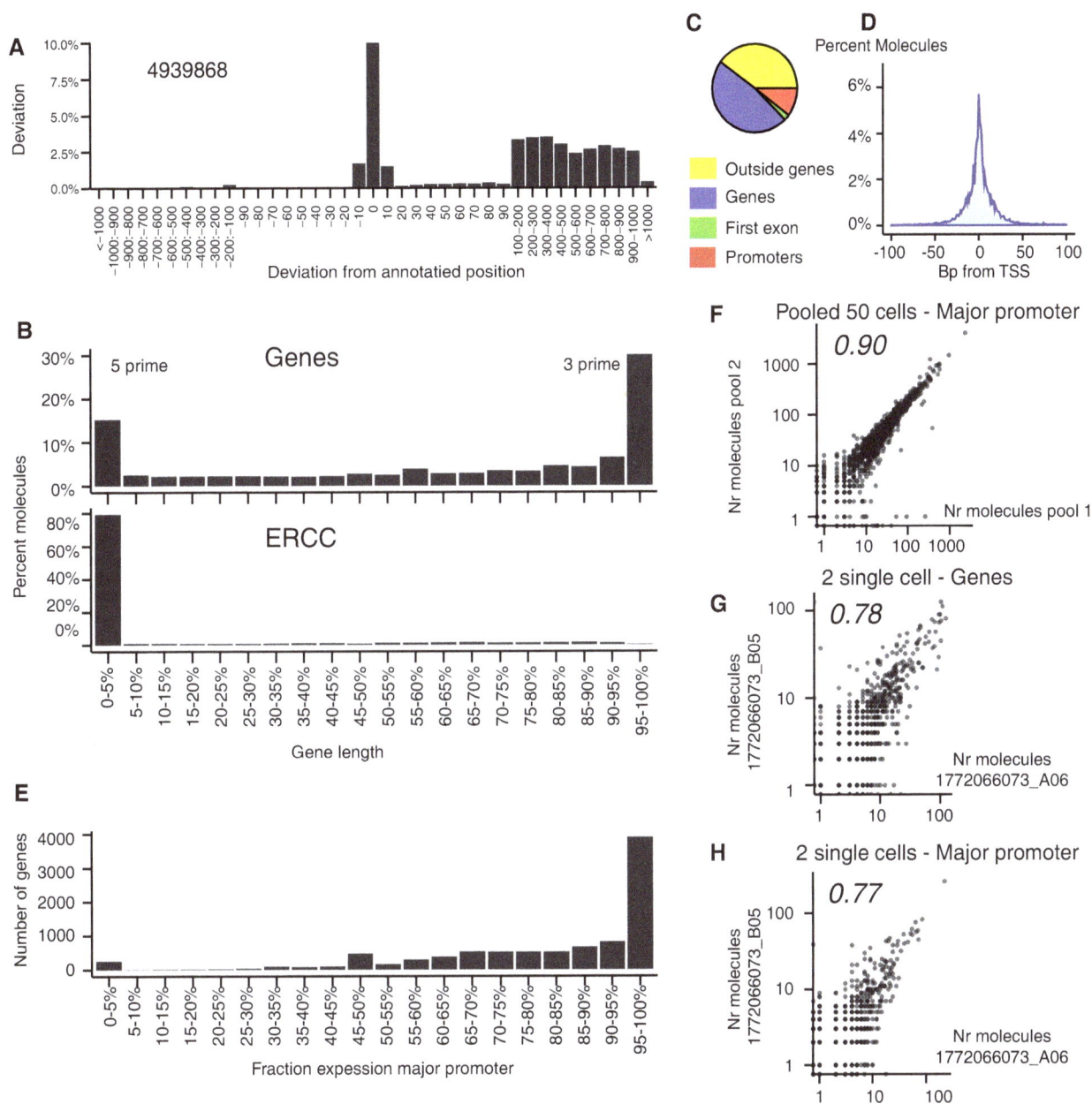

Figure 1. Measuring TSS activity in single cells.

A Number of reads deviating from starting position in ERCC control molecules shown as percent of reads mapping exactly to the starting position. The number in red shows total number of ERCC molecules participating in libraries for CA1 neurons. Note that the scale is broken.

B Read distribution across genes, shown as percent mapped reads (after UMI correction) in 20 length-bins for CA1 neurons (top) and ERCC control RNA (bottom).

C Read distribution for CA1 neurons. A read is only assigned once to one of the four groups.

D CA1 neuron reads mapped to the region ± 100 bp from the TSS, defined as the center of CAGE-annotated TSSs.

E Major TSS preference in pooled single cells from CA1 neurons.

F Correlation of major TSS expression in pooled single CA1 neurons.

G Correlation of whole-gene expression in individual single CA1 neurons.

H Correlation of major TSS expression in individual single CA1 neurons.

Data information: (F–H) Pearson correlation values in red.

molecules), using major TSS counts instead of total gene counts tended to show higher correlation. This was also true when comparing two genes as shown in Appendix Figs S3D–F. The phenomenon can be explained if reads mapping to the major TSSs only reflect new transcription, while reads mapping to the total gene body reflect a number of processes that are not always correlated, such as mRNA degradation, PCR strand invasion (Tang *et al*, 2013), alternative TSS expression, intronic reads, and cryptic 3′ UTR expression.

However, we also noted a few cases where the major promoter was located outside the Refseq gene annotation leading to artefactual low correlation when using gene counts. An example of two genes that show medium correlation ($r = 0.49$) when counting major TSS reads and low correlation ($r = 0.12$) when counting total gene reads is shown in Appendix Fig S4.

Alternative TSS usage in single cells

Having established the quality of the data, we first asked how often genes carried more than one active TSS. As shown in Fig 1E for CA1 pyramidal neurons (hereafter called CA1 neurons), around 40% of all genes showed only or almost only expression from the major TSS, counting all genes with at least one molecule mapped to a TSS. For remaining genes, there was a varying degree of co-expression between major and minor TSS. The same trend held true also for a more conservative subset of genes with expression higher than 1 molecule per cell (Appendix Fig S5). Thus, expression from two or more TSSs is common, but many genes show a strong preference for the major TSS.

Given that two TSSs are active in a cell population, it is natural to ask whether this could be explained by subpopulations of cells, or whether both TSSs are typically simultaneously active in single cells (Fig 2A). To address this, we examined the correlation of expression from the major and minor TSS between individual cells, across genes. If alternative TSSs were expressed in distinct subsets of cells (whether stochastically or by some regulated mechanism), we would expect them to be anti-correlated. In contrast, if they were simultaneously expressed in individual cells, they would be uncorrelated or positively correlated, depending on the rate of transcript degradation.

We found strong positive correlation within all studied cell types for highly expressed genes (Pearson correlation > 0.5 for most genes expressing > 4 molecules per TSS per cell in average) and a weak positive correlation for lowly expressed genes (Pearson correlation 0.1–0.5 for most genes expressing < 4 molecules per TSS per cell in average). In fact, the correlation between the TSSs increased almost linearly with expression (Fig 2E and Appendix Fig S6), indicating that at low levels of expression, noise takes over and reduces the correlation.

Snap25, which encodes a synaptic vesicle membrane fusion protein, was highly expressed in most cells and showed a high major/minor TSS correlation ($r = 0.80$), but also lowly expressed genes like *Dcn* were sometimes highly TSS correlated ($r = 0.85$ Fig 2B and Appendix Fig S7). Genes with very weakly correlated major/minor TSSs ($r \sim 0.1$) were the exception, for example, *Syt1*, and they did not show anti-correlation. These exceptions were probably the result of low expression. For genes with high ratio of major to minor TSS, the major TSS was consistently higher expressed in almost every cell (e.g., *Dcn*, *Snap25*, and *Son*, Appendix Fig S8).

One possible explanation of the finding that alternative TSSs are positively correlated in single cells is that the correlation would depend on degradation of the major TSS, resulting in artefactual reads on the minor TSS. In this case, a higher expressed major TSS would create more artefactual reads on the minor TSS and hence create a false correlation. An argument supporting this reasoning is that most annotated TSSs are less than 100 bp apart. Indeed, the majority of the TSSs that are expressed at the population level

represented such spatially connected TSSs. 56% of the annotated TSS in the FANTOM dataset are located in such composite transcription initiation regions (Forrest *et al*, 2014). We argue that the positive correlation is not due to degradation based on the following observations: First, TSSs with reads mapping to the reference region, constituting putative degradation events have been removed. Reads on valid TSSs were highly specific. Second, if correlation depended on degradation of reads expressed from the major TSS, then there should be an increase of genes where the major TSS was upstream of the minor TSS. However, this was not the case, and there was an almost even distribution between the major TSS being upstream or downstream of the minor (major TSS upstream of minor $n = 88$, downstream of minor $n = 109$, see Appendix Fig S9A). Third, if the correlation depended mostly on degradation of the major TSS, then the case where the major TSS is upstream of the minor should have a higher correlation. The difference was not significant ($P = 0.11$, Student's *t*-test, major TSS upstream average $r = 0.46$, SD = 0.18, major TSS downstream average $r = 0.42$, SD = 0.18, see Appendix Fig S9B). Fourth, TSSs were chosen based on CAGE peaks, an orthogonal method to single-cell RNA-seq. It is unlikely that degradation peaks by chance arises at annotated TSS regions. Fifth, genes with TSSs located far apart showed similar correlation pattern as proximal TSSs (Fig 2E) and were only slightly less correlated (Fig 2D).

Another possible explanation to the positive correlation is that the correlation is an artifact of using read counts, and would depend on sequencing depth. To verify that this is not the case, the same analysis was carried out as in Fig 2E using reads per 10 k (rpk, same as reads per million but multiplied with 10,000 instead of a million for readability) instead of counts. The number of molecules per gene per cell is strongly influenced by the total number of molecules per cell (Appendix Fig S10), which strongly influence noise (example for *Snap25*, Appendix Fig S11A–D). Therefore, only highly expressed cells (> 2,000 molecules) were included in the analysis. For most genes expressing two TSSs, the major and minor TSSs were still positively correlated after normalization (Appendix Fig S11E and F). Correlation was slightly lower after normalization, indicating that some of the correlation indeed could depend on the choice of using read counts instead of rpk, but we believe the difference in expression level between cells to be more influenced by cell size than read depth and therefore prefer to use read counts.

We next sought to quantify for each gene what percentage of cells agreed with our hypothesis that the major and minor TSSs show a fixed ratio. To this end, we used the two-sided binomial test to search for cells whose proportion of major and minor expression significantly deviated from the average proportion among all cells. For most genes, only few CA1 cells deviated from the expected ratio as can be seen in Fig 2F, and this was true for other cell types as well (Appendix Fig S12A). On average 1–3% (depending on cell type) of all cells expressing at least one molecule in any TSS per gene deviated significantly from the expected TSS ratio. However, for the gene cystatin C (*Cst3*), 44% of cells deviated from the expected ratio (Fig 2F), indicating an unusual bimodal expression pattern, and this was true for other cell types as well, albeit at varying degree (Appendix Fig S12B). Notably in vascular endothelial cells, this pattern could not be discerned, and in oligodendrocytes, proteolipid protein 1 (Plp1) showed even more cells deviating from

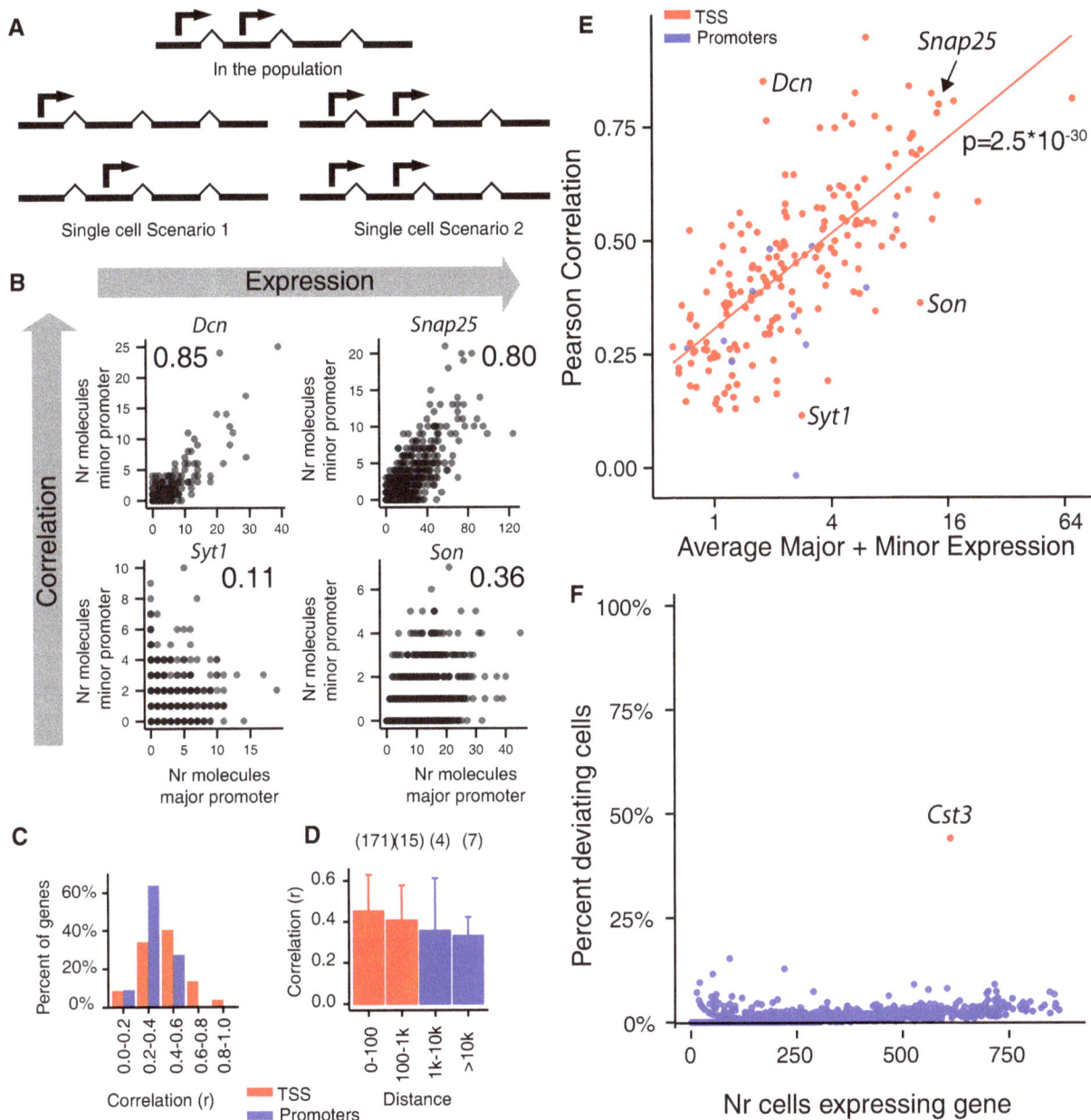

Figure 2. Correlated expression of major and minor TSS in single cells.

A   Illustration of two models for TSS expression in single cells versus their common effect on the population expression.

B   Examples of TSS expression in single CA1 neuron cells. Plots show the number of mRNA molecules detected from the major and minor TSSs in single cells. Each dot is a single cell. Pearson correlation values are indicated in red. Four examples are shown, representing high and low expression, and high and low correlation.

C   Distribution of correlation values showing effect of TSS distance. Histogram based on 197 genes with an expression of at least 0.3 molecules per cell per TSS for CA1 neurons. Genes were divided based on major/minor TSS distance ("TSS", 186 genes with < 1 kb inter-TSS distance; "Promoters", 11 genes with > 1 kb TSS distance).

D   Average correlation between major/minor TSS as a function of TSS distance. Colors represent TSSs and promoters as in (C). The error bars show standard deviation and the numbers in brackets show participating genes.

E   Scatterplot showing total expression (horizontal axis) and major/minor TSS correlation coefficient (vertical). Each dot is a gene. Colors represent genes with "TSS" and "Promoter" TSSs as in (D). Four example genes from (B) are marked in the plot. Linear regression lines are shown with *P*-values of the fit.

F   Percentage of cells that significantly deviates from expected major to minor TSS ratio as a function of cell gene expression. Percentage of deviating cells were calculated using the binomial test. Each dot represents a gene.

the expected ratio than Cst3. The bimodal expression pattern of Cst3 will be discussed in more detail below.

One of the more prominent findings in this data set was the almost complete absence of genes with expression of alternative promoters. Promoter length varies from promoter to promoter; however, 1 kb upstream and a few 100 bp downstream from the TSS is a commonly used measure, for example, in Akan and Deloukas (2008). We therefore defined two TSSs located more than

1 kb apart as putative alternative promoters and examined the correlation of alternative TSS expression as a function of the distance between TSSs. We found that promoters were significantly less correlated than TSSs located in close proximity with a small margin ($P = 0.041$, Student's t-test; promoters $n = 11$, average $r = 0.33$, SD = 0.16; TSS $n = 186$, average $r = 0.44$ and SD = 0.18; Fig 2D and E), but the absolute difference in correlation was small. 7,369 TSSs had an expression of at least 100 molecules across all 2,816 cells. Of these, 5,872 TSSs passed the criteria that the peak at the TSS should be specific. 922 genes had expression of at least two such TSSs and are listed in Table EV2 for CA1 cells. Of these, a modest number (197) of TSS pairs expressed more than an average 0.5 molecules per cell per TSS, and only six were potential alternative promoters (*Cox16*, *Nrxn1*, *Meg3*, *Fis1*, *Grm5*, and *2610017I09Rik*). Alternative TSSs for *Cox16* and *2610017I09Rik* overlapped other genes and can therefore not with certainty be said to constitute true alternative TSSs.

All genes with potential alternative promoter usage were expressed from different exons. This was also true for two genes with alternative TSSs located within 1 kb (*Snca* and *Caly*), perhaps indicating alternative promoters. Of the coding genes, the majority (*Nrxn1*, *Grm5*, *Snca*, and *Caly*) are involved in neuronal signaling and the two other (*Cox16* and *Fis1*) are active in the mitochondria. Only *Nrxn1*, *Cox16*, and *Fis1* codes for different protein products. Four genes with TSS expressed from different exons are shown in Fig 3A–C.

The FANTOM data contain another class of CAGE peaks, located in genes and often in the 3′ UTR which are not annotated as TSSs (see Appendix Fig S1). Interestingly, we found that the correlations between alternative TSSs where one TSS was located in the 3′ UTR was much lower than when TSSs were located in the 5′ end. In these cases, correlation also did not increase with gene expression (Appendix Fig S13). This indicates that a different mechanism of regulation controls the appearance of cryptic 3′ UTR CAGE peaks. For this reason, like FANTOM, we have not included these CAGE peaks as true TSSs. Thus, our data recapitulate the surprising finding done by CAGE sequencing that low aggregates of molecules map to internal exons and sometimes rather large aggregates of molecules map to the 3′ UTR (Carninci *et al*, 2006).

An interesting observation is that genes commonly were expressed in distinct peaks from multiple genomic nucleotide positions within an annotated TSS in single cells, which was the case with, for example, *Snap25*, *Stmn3*, and *Calm1* (Appendix Figs S1 and S14). However, more often reads were scattered across the TSS region and rarely a single peak from a single nucleotide position was seen.

To verify our finding that multiple TSSs are expressed in single cells, we used a previously published dataset where six single oligodendrocyte cells were sequenced for full-length mRNA using PacBio long read technology. Due to the low sequencing depth of PacBio sequencing, not all oligodendrocyte alternative TSS could be verified. A handful of genes with alternative TSS are shown in Appendix Fig S15A–C, and for clarity, genes with long distance between the TSSs were chosen.

Considering the bursty nature of gene expression, the rather low efficiency of single-cell RNA-seq, and the low TSS mapping, we assumed that either there would be no correlation between the TSSs or that they would be anti-correlated due to the bursts. In contrast,

we found a high correlation at the single-cell level even at rather low levels of average TSS expression. In conclusion, minor TSSs were generally expressed at a fixed fraction of the major TSS, regardless of the inter-TSS distance, suggesting that they respond to common distal regulatory signals. Few genes exhibited a different expression pattern, and the most prominent of those were *Cst3*.

Bimodal expression pattern of cystatin C (Cst3)

Cst3 encodes cystatin C (Cst C), which is a member of the cystatin superfamily and its most abundant and potent inhibitor of the cysteine cathepsins. Since Cst C can be secreted, it confers cysteine protease regulation both intra- and extracellularly. Cst C has been implicated in a number of conditions including apoptosis, antigen presentation, atherosclerosis and pathogen invasion, and the level of *Cst3* mRNA can be influenced by different stimuli like inflammatory cytokines, pathogens, growth factors, hormones and oxidative stress (Xu *et al*, 2015).

For CA1 neurons, cells with more than five expressed *Cst3* TSS molecules (in total 251 cells) could be clearly divided into two groups: One group had a high major TSS expression (> 2:1 ratio major/minor, here referred to as *Cst3* major high) from the group with a low major TSS expression (≤ 2:1 ratio major/minor, here referred to as *Cst3* major low), and this separation was consistent in other cell types as well (Appendix Fig S16A). Interestingly, by this separation, the vast majority of astrocytes were labeled as *Cst3* major high, and the vast majority of interneurons were labeled *Cst3* major low. Transcription factors with binding sites close to the *Cst3* gene include Myog, Spi1, Ebf1, Foxo1, Stat5a:Stat5b, AR, and AP1. Of those transcription factors, only androgen receptor (AR) and activator protein 1 (AP1), which is a transcription factor that is composed of several proteins from the Jun, Fos, ATF, and JDP families, were expressed at a level higher than 0.5 molecules per cell in cells expressing *Cst3*, and of those, only Fos was significantly differentially expressed between *Cst3* major high and *Cst3* major low cells with a fold change > 2 (3.8 molecules in average for *Cst3* major high compared to 1.9 for *Cst3* major low, $P = 0.02$ Welch's t-test), which indicates that AP1, which is a transcription factor that regulates gene expression in response to stress, growth factors and cytokines, may be responsible for the high expression of *Cst3* major TSS in some cells. However, the significance of the differential expression was removed after Bonferroni correction for multiple testing.

For CA1 neurons, 7,907 genes were expressed at more than 0.5 molecules per cell among *Cst3* major high and major low cells. Of those, 493 genes were significantly differentially expressed in *Cst3* major high ($P < 0.05$, Welch's t-test with Bonferroni correction) and 127 higher expressed in *Cst3* major low. Gene Ontology terms associated with regulation of cell proliferation and developmental processes, receptor, and transmembrane receptor activities and with extracellular space and plasma region were significantly enriched (Table EV3) among these genes. Single-cell expression of *Cst3* for CA1 neurons is shown in Appendix Fig S16B.

Discussion

We have examined the use of alternative TSSs in single cells, using a large dataset comprising thousands of cells and found that

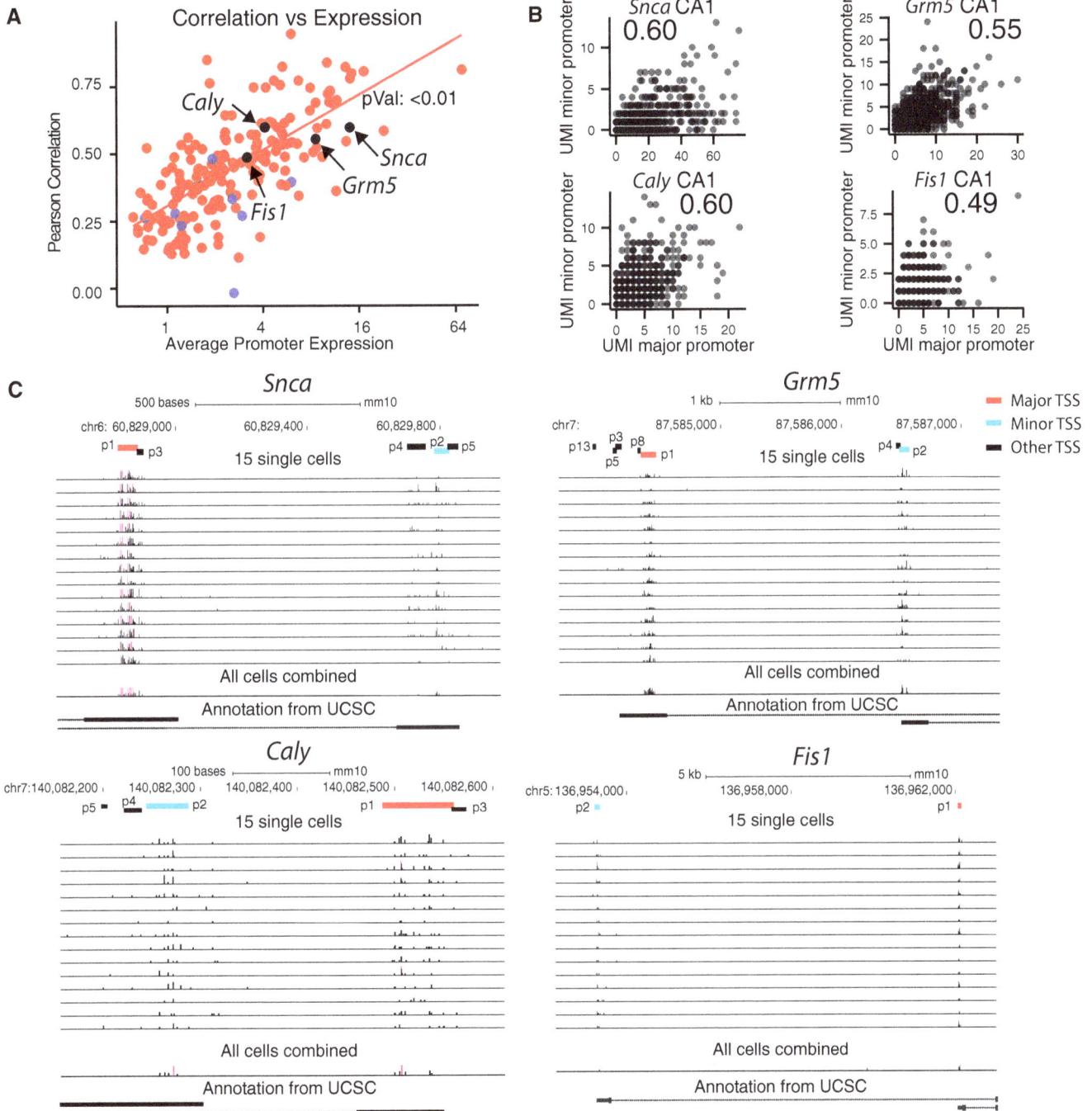

Figure 3. TSS pairs expressed from different exons.

A Scatterplot showing total expression (horizontal axis) and major/minor TSS correlation coefficient (vertical). Each dot is a gene. Colors represent genes with "TSS" and "Promoter" TSSs.

B Examples of TSS expression in single CA1 neuron cells. Plots show the number of mRNA molecules detected from the major and minor TSSs in single cells. Each dot is a single cell. Pearson correlation values are indicated in red.

C Promoter expression of genes with TSS pairs expressed from different exons using the UCSC genome browser. Expression is shown as bars where the y-axis for single cells has a limit of five molecules and for all cells combined has a limit of 500 molecules. Major and minor TSSs are marked in red, while other CAGE peaks associated with a gene are marked in black.

alternative TSSs were almost always co-expressed in single cells. Furthermore, in highly expressed genes, alternative TSSs were expressed in a correlative manner, and the level of correlation was highly dependent on expression level indicating that the correlation in lowly expressed genes was reduced due to noise and would potentially increase with higher mRNA capture efficiency. mRNA degradation is likely contributing, but not substantially, to the high correlation in expression between alternative TSS.

These findings would seem to contradict previous reports on the highly stochastic and bursty nature of gene expression in mammalian cells (Raj *et al*, 2006; Raj & van Oudenaarden, 2008). However, first of all, these previous studies did not measure TSS-specific gene expression, and indeed, we found that often whole-gene expression was noisier than TSS-specific expression, especially for highly expressed genes (Fig 1G and H, and Appendix Fig S3). Second, we have examined the entire transcriptome, whereas imaging-based studies have been limited to studying specific selected genes, which may or may not have been typical of the average gene. Third, it should be noted that we have here measured steady-state levels, not new transcription, and thus, the rate of mRNA degradation will influence our measurements. If the degradation rate is low, any fluctuations in transcriptional rate will tend to be erased by time averaging. Finally, there are many other possible sources of variation in observed total gene expression, such as the total number of mRNA molecules per cell (which varies substantially), and technical differences such as differences in sequencing depth or mRNA recovery. Indeed, the expression level of individual genes is dependent on total mRNA expression level in a cell (Appendix Fig S10). These other factors would not be expected to differ for alternative TSSs of the same gene.

There are at least two possible explanations for the high correlation between alternative TSSs: Either gene expression is not as bursty as previously believed (relative to the degradation rate), or both TSSs participate in each burst of transcription. Our finding that TSSs located far apart were less correlated lends some support to the latter explanation, although the effect was small.

Very few genes were exceptions to the rule that there is a set ratio between the expression of major and minor TSS. The most prominent of these genes was *Cst3* which codes for the gene cystatin C, an inhibitor of cysteine proteases. mRNA of *Cst3* showed a bimodal expression pattern where the major and minor TSS were expressed to a similar degree in some cells, while in other cells, the major TSS was highly selectively expressed. Genes that were highly expressed in the latter cells were associated with GO terms for receptor activity, extracellular space, and the plasma membrane, possibly indicating response to a stress signal. The immediate-early transcription factor Fos was associated with selective major TSS expression of *Cst3*, is a well-known stress response factor, and is regulated by neuronal activity in the brain.

For highly expressed genes, the correlation between two different genes was higher when using only reads mapping to the major TSS as compared to using reads mapping to the full gene body (Appendix Fig S3). This is probably due to the fact that reads mapping to the major TSS only reflect gene expression, while reads mapping to the gene reflect many processes that may not always be correlated, including mRNA degradation, alternative TSS expression, PCR strand invasion, intronic reads, and cryptic 3′ expression. This may have implications for clustering of single cells into cell types since many clustering methods rely on correlation between genes.

Surprisingly, few genes expressed alternative promoters (defined as TSS located more than 1 kb apart), and only three coded for different protein products. One explanation for this may be that there are few occasions where it would be beneficial for a gene to express transcripts from multiple promoters within a cell type. The presence

and levels of TFs vary across tissues and developmental time, and since it is known that specific TFs associate more strongly with certain promoters it is reasonable to believe that this can influence the promoter preference for specific genes (Rach *et al*, 2009). Similarly alternative promoters are known to be expressed across tissues and developmental stages and can, for example, ensure that housekeeping genes keep a similar expression level given a different regulatory landscape, or they can tune the level of expression between different cell types (Ayoubi, 2005). However, the need for a single cell type to express multiple TSS isoforms of a gene may be limited.

In summary, we found a surprising degree of co-expression of alternative TSSs in single cells. These findings provide strong constraints on models of transcriptional regulation.

Materials and Methods

Data collection

This study uses previously published data (Zeisel *et al*, 2015, Gene Expression Omnibus www.ncbi.nlm.nih.gov/geo under accession code GSE60361) from 2,816 single cells from the mouse somatosensory cortex and hippocampal CA1 region of genetically outbred (CD-1) mice. Raw reads were remapped to mm10 using Bowtie I, allowing for three mismatches, and annotated to TSSs as explained below. Reads were converted into mRNA molecule counts using UMIs, as previously explained (Kivioja *et al*, 2012). The UMI sequence is 6 bp long, and reads with the same UMI sequence were collapsed and UMIs with only one read were removed. Number of molecules per cell can be found in Table EV4.

Full-length single oligodendrocyte mRNA sequencing data were taken from previously published data (Karlsson & Linnarsson, 2017, Gene Expression Omnibus www.ncbi.nlm.nih.gov/geo under accession code GSE76026).

This link: http://genome-euro.ucsc.edu/cgi-bin/hgTracks?hgS_do OtherUser = submit&hgS_otherUserName = Kasper&hgS_otherUser SessionName = mm10_public_promoters_CA1neurons provides access to a UCSC track showing promoter expression of 20 single cells as well as the combined expression for all CA1 neurons.

This link: http://genome-euro.ucsc.edu/cgi-bin/hgTracks?hgS_do OtherUser = submit&hgS_otherUserName = Kasper&hgS_otherUser SessionName = mm10_public_promoters_CA1_Oligos provides access to a UCSC track showing promoter expression of 10 single CA1 neuron cells, 10 single oligodendrocyte cells as well as the combined expression for all CA1 neurons and all oligodendrocytes.

Analysis

Input data

Transcription start site regions were defined based on an early access program from the FANTOM 5 project and may therefore differ slightly from the published FANTOM 5 mouse TSS database (Forrest *et al*, 2014). All CAGE tags used are shown in Table EV5. CAGE tags were curated, and tags without association to a gene were discarded. CAGE tags were also moved from mm9 to mm10 using liftOver from UCSC. In the cases where two TSS regions overlapped, the longer TSS region was shortened so there would be no overlap and the minimum distance between two TSSs was 1 bp.

The distance between TSSs was calculated as the distance between the edges of the TSSs, not the distance between the centers of the TSSs. Molecules were allocated to TSSs using a custom program. For each single cell, cDNA molecules with their 5′ end within the curated CAGE-defined FANTOM 5 TSSs were counted.

For each gene, the major TSS was defined as the TSS that had the greatest number of mapped mRNA molecules (UMIs) across the entire dataset (i.e., counting all cell types). The minor TSS was defined as the TSS that had the second-largest number of molecules. Because of this definition, the major TSS need not be the highest expressed TSS in individual cells or cell types.

The major and minor TSSs of each gene were considered to be proximal TSS if the distance between the TSSs was < 1 kb; all other TSS pairs were considered as promoter pairs.

To limit the interference of degradation on the analysis, mapped reads were calculated in a region of equal size as the TSS next to the annotated TSS both upstream and downstream (here called a reference region). If another TSS was located in the area where the reference region should be, then the reference region was moved outside of that TSS. If a reference region (either upstream or downstream) contained more than 20% of the combined reads from the TSS and reference region, then that particular TSS was removed from the analysis.

Normalization to rpk was done for each cell as follows: molecules per TSS/total expressed TSS molecules × 10,000.

RefSeq genes used for annotation in Fig 1B and G were downloaded from the UCSC table browser (21-09-2015), and only the first isoform by the order of the file of each gene was kept.

We used a slightly modified database to map ERCC reads, and the modifications are shown in Table EV6. The reason for this is that we previously have observed that for many ERCC reads, the 5′ end is slightly upstream of the ERCC reference sequences. In Fig 1A, for each ERCC, the median of all mapped reads was used as the starting position and the percentage of reads deviating from the starting position is shown.

Transcription factor binding sites were extracted from the UCSC genome browser based on the ORegAnno database, as well as from a previously published paper (Huh et al, 1995).

Statistical methods

To calculate the difference in correlation between genes with alternative TSS and genes with alternative promoters, the non-parametrical a two-sided Mann–Whitney U-test was used, implemented by the R function wilcox.test since the number of genes with alternative promoters was too few to ensure normal distribution ($n = 11$), but had a similar variance. To calculate the difference in correlation between the upstream and downstream location of the major TSSs (compare with Appendix Fig S7), a two-sided Student's t-test was used, implemented with the R function t.test, since the data were normally distributed. The line and associated P-value for Fig 2E and Appendix Figs S6, S7, and S9 was calculated with the lm method in R. Correlations were calculated using Pearson correlation and implemented in python using the stats.linregress function in scipy or the cor function in R.

To calculate how many cells that significantly deviated from expected major to minor TSS ratio, the binomial test was performed for each gene across all cells and implemented with the python function scipy.stats.binom_test, using number of reads of the major TSS, the total number of TSS reads, and the percentage of major TSS reads across all cells in a cell type as the probability of success and without correction for multiple testing.

To calculate significant alternative TSS usage between cell types, the minor fraction (molecules on minor TSS divided by the sum of molecules on both TSSs) for each gene and cell was first calculated. Cells with no expression in either major or minor TSS were removed and genes expressed in fewer than 20% of cells in any of the two cell types were removed to keep only genes with at least modest expression in the two cell types. Since the data were not normally distributed, a two-sided non-parametrical Mann–Whitney test was applied, implemented in python using the function stats.mannwhitneyu in scipy, with Bonferroni correction on the minor fraction for each gene on the two cell types. The data could not assume equal variance of cell expression between cell types for each gene; however, the combined sample sizes were large ($n \geq 75$) for all genes with a statistical significant differential usage of minor fraction. The number of cells was more than 800 for some genes and cell types, so small differences in minor fraction could be determined as statistically significant. Therefore, we additionally required a difference of at least 0.3 in minor fraction to find genes with potential biological significance.

All examples from a single cell type were taken from CA1 neurons since they had a high number of molecules per cell in average and constituted a large fraction of the total dataset with 876 single cells. Comparisons between cell types were performed between CA1 neurons and oligodendrocytes, unless otherwise stated, because there were many oligodendrocyte cells, and they are a clearly distinct cell type from neurons.

Acknowledgements

We acknowledge funding from the Swedish Research Council (STARGET) and the European Research Council (ERC grant 261063, BRAINCELL).

Author contributions

SL conceived the study, PL mapped sequencing reads to the genome, and KK performed all additional analysis. KK and SL drafted the manuscript with input from PL.

References

Akan P, Deloukas P (2008) DNA sequence and structural properties as predictors of human and mouse promoters. Gene 410: 165–176

Ayoubi T (2005) Alternative Promoter Usage. In eLS, Chichester: John Wiley & Sons Ltd. http://www.els.net

Batut P, Dobin A, Plessy C, Carninci P, Gingeras TR (2013) High-fidelity promoter profiling reveals widespread alternative promoter usage and transposon-driven developmental gene expression. Genome Res 23: 169–180

Carninci P, Sandelin A, Lenhard B, Katayama S, Shimokawa K, Ponjavic J, Semple CA, Taylor MS, Engstrom PG, Frith MC, Forrest AR, Alkema WB, Tan SL, Plessy C, Kodzius R, Ravasi T, Kasukawa T, Fukuda S, Kanamori-Katayama M, Kitazume Y et al (2006) Genome-wide analysis of mammalian promoter architecture and evolution. Nat Genet 38: 626–635

Deng Q, Ramskold D, Reinius B, Sandberg R (2014) Single-cell RNA-seq reveals dynamic, random monoallelic gene expression in mammalian cells. *Science* 343: 193–196

Forrest AR, Kawaji H, Rehli M, Baillie JK, de Hoon MJ, Haberle V, Lassmann T, Kulakovskiy IV, Lizio M, Itoh M, Andersson R, Mungall CJ, Meehan TF, Schmeier S, Bertin N, Jorgensen M, Dimont E, Arner E, Schmidl C, Schaefer U *et al* (2014) A promoter-level mammalian expression atlas. *Nature* 507: 462–470

Huh C, Nagle JW, Kozak CA, Abrahamson M, Karlsson S (1995) Structural organization, expression and chromosomal mapping of the mouse cystatin-C-encoding gene (Cst3). *Gene* 152: 221–226

Islam S, Kjallquist U, Moliner A, Zajac P, Fan JB, Lonnerberg P, Linnarsson S (2011) Characterization of the single-cell transcriptional landscape by highly multiplex RNA-seq. *Genome Res* 21: 1160–1167

Islam S, Kjallquist U, Moliner A, Zajac P, Fan JB, Lonnerberg P, Linnarsson S (2012) Highly multiplexed and strand-specific single-cell RNA 5' end sequencing. *Nat Protoc* 7: 813–828

Islam S, Zeisel A, Joost S, La Manno G, Zajac P, Kasper M, Lonnerberg P, Linnarsson S (2014) Quantitative single-cell RNA-seq with unique molecular identifiers. *Nat Methods* 11: 163–166

Karlsson K, Linnarsson S (2017) Single-cell mRNA isoform diversity in the mouse brain. *BMC Genom* 18: 126

Kivioja T, Vaharautio A, Karlsson K, Bonke M, Enge M, Linnarsson S, Taipale J (2012) Counting absolute numbers of molecules using unique molecular identifiers. *Nat Methods* 9: 72–74

de Klerk E, t Hoen PA (2015) Alternative mRNA transcription, processing, and translation: insights from RNA sequencing. *Trends Genet* 31: 128–139

Leroith D, Roberts CT Jr (1993) Insulin-like growth factors. *Ann N Y Acad Sci* 692: 1–9

Pelechano V, Wei W, Steinmetz LM (2015) Widespread co-translational RNA decay reveals ribosome dynamics. *Cell* 161: 1400–1412

Rach EA, Yuan HY, Majoros WH, Tomancak P, Ohler U (2009) Motif composition, conservation and condition-specificity of single and alternative transcription start sites in the *Drosophila* genome. *Genome Biol* 10: R73

Raj A, Peskin CS, Tranchina D, Vargas DY, Tyagi S (2006) Stochastic mRNA synthesis in mammalian cells. *PLoS Biol* 4: e309

Raj A, van Oudenaarden A (2008) Nature, nurture, or chance: stochastic gene expression and its consequences. *Cell* 135: 216–226

Rojas-Duran MF, Gilbert WV (2012) Alternative transcription start site selection leads to large differences in translation activity in yeast. *RNA* 18: 2299–2305

Shalek AK, Satija R, Adiconis X, Gertner RS, Gaublomme JT, Raychowdhury R, Schwartz S, Yosef N, Malboeuf C, Lu D, Trombetta JJ, Gennert D, Gnirke A, Goren A, Hacohen N, Levin JZ, Park H, Regev A (2013) Single-cell transcriptomics reveals bimodality in expression and splicing in immune cells. *Nature* 498: 236–240

Tang DT, Plessy C, Salimullah M, Suzuki AM, Calligaris R, Gustincich S, Carninci P (2013) Suppression of artifacts and barcode bias in high-throughput transcriptome analyses utilizing template switching. *Nucleic Acids Res* 41: e44

Velten L, Anders S, Pekowska A, Jarvelin AI, Huber W, Pelechano V, Steinmetz LM (2015) Single-cell polyadenylation site mapping reveals 3' isoform choice variability. *Mol Syst Biol* 11: 812

Xu Y, Ding Y, Li X, Wu X (2015) Cystatin C is a disease-associated protein subject to multiple regulation. *Immunol Cell Biol* 93: 442–451

Zeisel A, Munoz-Manchado AB, Codeluppi S, Lonnerberg P, La Manno G, Jureus A, Marques S, Munguba H, He L, Betsholtz C, Rolny C, Castelo-Branco G, Hjerling-Leffler J, Linnarsson S (2015) Brain structure. Cell types in the mouse cortex and hippocampus revealed by single-cell RNA-seq. *Science* 347: 1138–1142

6

CRISPR/Cas9 screening using unique molecular identifiers

Bernhard Schmierer[1,†] (iD), Sandeep K Botla[1,†] (iD), Jilin Zhang[1] (iD), Mikko Turunen[2], Teemu Kivioja[2] (iD) & Jussi Taipale[1,2,*] (iD)

Abstract

Loss-of-function screening by CRISPR/Cas9 gene knockout with pooled, lentiviral guide libraries is a widely applicable method for systematic identification of genes contributing to diverse cellular phenotypes. Here, Random Sequence Labels (RSLs) are incorporated into the guide library, which act as unique molecular identifiers (UMIs) to allow massively parallel lineage tracing and lineage dropout screening. RSLs greatly improve the reproducibility of results by increasing both the precision and the accuracy of screens. They reduce the number of cells needed to reach a set statistical power, or allow a more robust screen using the same number of cells.

Keywords CRISPR/Cas; genetic screening; massively parallel lineage tracing; unique molecular identifiers
Subject Categories Genome-Scale & Integrative Biology; Methods & Resources

Introduction

Pooled CRISPR/Cas9 loss-of-function screening is a powerful approach to identify genes contributing to a wide range of phenotypes (Shalem *et al*, 2015). A library of guide sequences is integrated lentivirally into Cas9-expressing cells, which are then subjected to a selection pressure. Relative guide frequencies in the population before and after selection are quantified by next-generation sequencing (NGS) to determine both depleted and enriched guides.

The approach has been applied successfully (Gilbert *et al*, 2014; Koike-Yusa *et al*, 2014; Shalem *et al*, 2014; Wang *et al*, 2015), but suffers from several shortcomings: First, the presence of a guide does not necessarily cause loss of the corresponding gene, and cells sharing the same guide have distinct genotypes and phenotypes. Second, identification of guides that are under negative selection can be confounded by random drift and undersampling. Third,

growth characteristics of individual cells can vary substantially (Levy *et al*, 2015; Sandler *et al*, 2015) and the site of viral integration can affect the phenotype. For these reasons, each guide needs to be present in a large number of cells. In conventional screens, only the sum of all cells with a specific guide is measured, and no information regarding the distribution of cell behaviors can be obtained. Optimal identification of hit genes would require a method that individually tracks clonal lineages derived from single virus-transduced cells.

Results and Discussion

Here, we address these issues by incorporating an RSL into the guide-library plasmid (Fig 1A) to allow tracing of hundreds of individual virus-transduced cell lineages in a CRISPR screen. In contrast to the use of barcodes in single-cell transcriptome analysis following CRISPR/Cas9 gene editing (Adamson *et al*, 2016; Dixit *et al*, 2016; Datlinger *et al*, 2017), we use unique molecular identifiers (Kivioja *et al*, 2012 and references therein) to either trace single clones (Kalhor *et al*, 2017) of identically edited cells, or very small pools of sublineages composed of cells with different editing outcomes at the same locus (Fig 1B). Such massively parallel lineage tracing enables both lineage dropout analysis (LDA), and the creation and analysis of internal replicates (IRA), while retaining the option of conventional, total read count analysis (TCA, Fig 1C).

To demonstrate the power and flexibility of the approach, we screened the human colorectal carcinoma cell line RKO for essential genes with an RSL-guide library targeting 2,325 genes with 10 guides per gene (Wang *et al*, 2015). Briefly, Cas9-expressing RKO cells were transduced with the lentiviral guide library, and samples were taken at Day 4 and Day 28 after transduction (control and treatment time points, respectively). Guide frequencies in the two time points were then assessed by NGS. The experiments were run in duplicate and at far larger screen size (we define "screen size" as the number of cells per guide sequence) and sequencing depth (reads per guide) than previous screens (Shalem *et al*, 2014; Wang *et al*, 2015). Such redundancy allows subsequent subsampling using the RSL information, and robust testing of different analytical

1 Department of Medical Biochemistry and Biophysics, Karolinska Institutet, Stockholm, Sweden
2 Genome-Scale Biology Research Program, Faculty of Medicine, University of Helsinki, Helsinki, Finland
 *Corresponding author. E-mail: jussi.taipale@ki.se
 †These authors contributed equally to this work

A Library design.

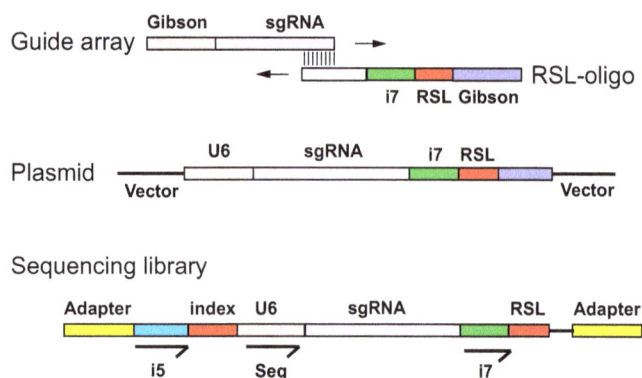

B Lineage drop out and lineage depletion

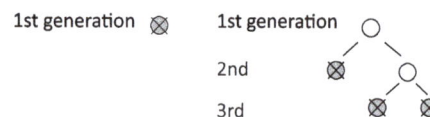

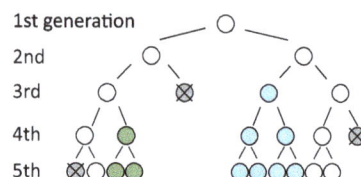

C Levels of analysis

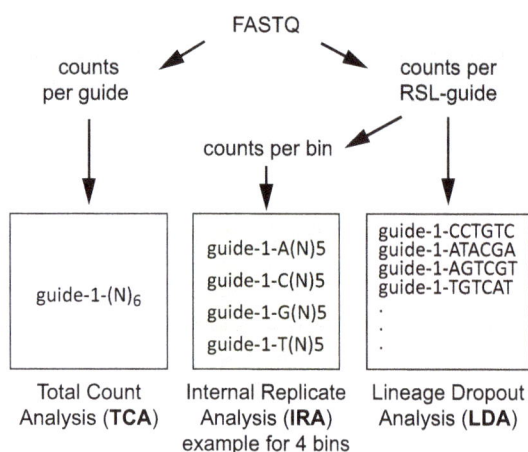

D Screen size and sequencing depth

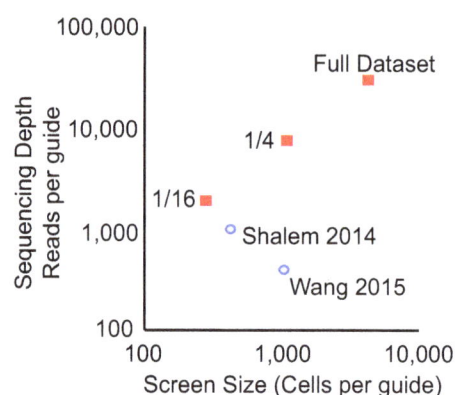

Figure 1. CRISPR/Cas9 screening using unique molecular identifiers.

A   Library design and cloning. Top: The guide library is synthesized as an oligonucleotide array; the RSL-part is synthesized as a single, overlapping oligonucleotide containing a 6-bp random sequence (RSL) and the Illumina index primer (i7) binding sequence. Guide-array and RSL-oligonucleotide are annealed and double-stranded. Homology arms for Gibson assembly are also indicated. Middle: Guide plasmid. The i7 index read primer binding site and the RSL are located downstream of the sgRNA termination signal and are not part of the guide RNA. Bottom: Sequencing library. Sequencing is performed using a custom primer (Seq) placed directly upstream of the guide (gRNA). The sample index and RSL are read as two index reads with Illumina i5 and i7 index primers, respectively (20 + 6 + 6 sequencing cycles).

B   Lineage dropout versus lineage depletion. Depending on the kinetics of editing, single cell lineages harboring a single RSL-guide against an essential gene can either disappear (dropout) or decrease in their abundancy (depletion). Top: Dropout happens if the editing occurs early on, either before the cell can divide, or in several independent events at later time points (gray, dead cell; white, unedited cell). Bottom: In lineage depletion, editing occurs either after several cell divisions and/or with several different outcomes (blue and green edits), some of which will retain gene function of the essential gene. In such cases, the traced lineage is comprised of several sublineages.

C   RSL-guides allow additional methods of analysis. In total count analysis (TCA, left), RSL information is ignored and only the sum of readcounts for all RSL-guides is taken into account. In internal replicate analysis (IRA, middle), readcounts of RSL-guides are binned such that internal replicates are created for each guide. The example shown bins RSL-guides into four internal replicates; however, RSL-guides can be binned in any number of replicates. In lineage dropout analysis (LDA, right), each RSL-guide is monitored separately.

D   Screen size and sequencing depth. The screens were performed at a very large screen size of roughly 4,500 cells per guide and sequenced to a depth of 30,000 reads per guide. Using RSL information, the data from these oversized experiments were then subsampled bioinformatically to approximately one quarter and 1/16, to test different analysis methods at different screen sizes. The corresponding values for two published screens are indicated for comparison (Shalem *et al*, 2014; Wang *et al*, 2015).

methods at varying screen sizes (Fig 1D). Perhaps counterintuitively, analysis of hundreds of RSL-labeled cell lineages per guide neither requires more cells per guide, nor markedly deeper sequencing, because any screen needs to use a relatively large number of cells per guide to achieve statistical power. Tagging each individual lineage incurs no cost. The RSL approach simply splits the total

guide read count obtained to read counts representing individual constituent cell lineages (Fig EV1), thus increasing the amount of information that is obtained, and consequently improving both precision and accuracy of the screen.

The plasmid library input contained 78 million unique RSL-guide combinations, 93% of which were also detected in the virus-transduced

A Internal replicate analysis (IRA) at the guide level

B Internal replicate analysis (IRA) at the gene level

C Lineage dropout analysis (LDA) - Replicate concordance

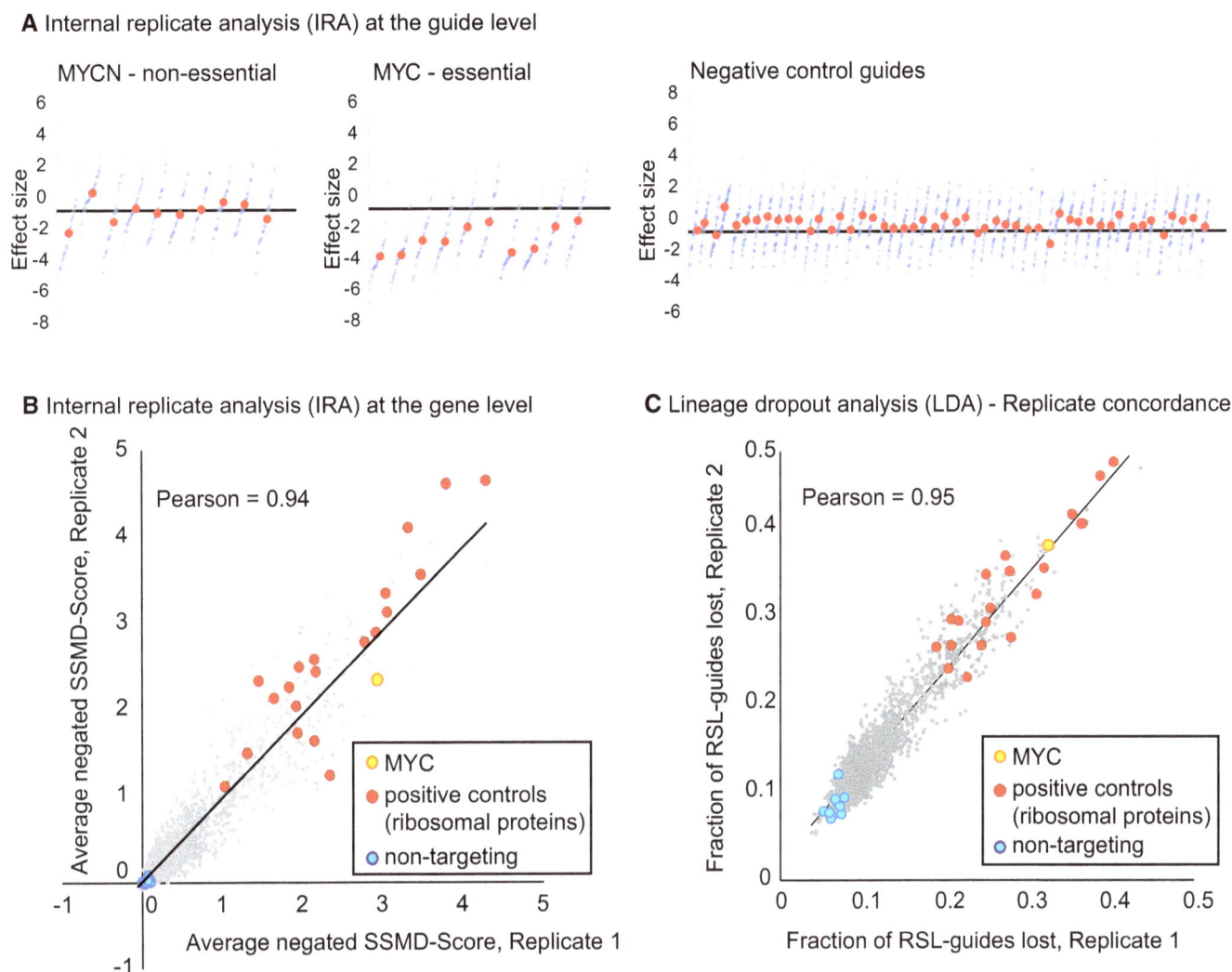

Figure 2. RSLs enable internal replicate and lineage dropout analyses.

A   Internal replicate analysis (IRA) at the guide level. RSL-guides were binned to create 64 internal replicates. Effect sizes (log2 fold change in readcount between Day 4 and Day 28 after virus transduction) for each bin are plotted in ascending order, 10 guides each for MYCN (top left) and MYC (top middle), as well as 50 representative, non-targeting guides (top right, these non-cutters seem to have a small fitness advantage). Red dots, median effect size (MES) of the 64 internal replicates (effect size of each internal replicate is one blue dot); black line, MES of all guides in the library. Hits for this type of data were called by SSMD score, see Materials and Methods for details. More examples are shown in Fig EV3.

B   Internal replicate analysis (IRA) at the gene level. RSL-guides were binned into 64 internal replicates. SSMD scores were calculated for each guide and averaged across all guides targeting the gene to obtain a score for each gene. For plotting, average SSMD scores for each gene were negated for easier comparison with Fig 1C. Red, positive controls (ribosomal proteins); blue non-targeting controls; orange, MYC; black line, linear regression.

C   Lineage dropout analysis (LDA). The fraction of RSL-guides lost from Day 4 to Day 28 in each experimental replicate is plotted for each gene (average over all guides targeting the gene). Red, positive controls (ribosomal proteins); blue, non-targeting controls; orange, MYC; black line, linear regression. The number of virus-transduced cell lineages lost is the most direct readout of the guide effect on cell viability.

cell populations (Fig EV2). Based on the Poisson distribution, this indicates that about half of the RSL-guides were incorporated into one or two cell lineages. Because only a subset of the cells can be harvested at each time point, undersampling is unavoidable, and some cell lineages (and corresponding RSL-guides) were present only in one of the time points (Venn diagram, Fig EV2). Such undersampling and loss of cell lineages occur whether or not RSLs are present, however go undetected in their absence. With RSLs, the effect becomes apparent and can be used in quality control of individual experiments as well as in filtering out inconsistently sampled lineages prior to data analysis.

RSL-labeled, distinguishable guide sequences can be used to split the data into internal replicates, which allow the usage of classical statistical tools to test for significant differences. To demonstrate the approach, RSL-guides were binned into 64 internal replicates per guide. The median effect size (Fig 2A) and a median-based version of strictly standardized mean difference (SSMD; Zhang, 2007) were then used to rank the guides (internal replicate analysis using SSMD, IRA/SSMD, Fig EV3). The average of all guide scores for each gene was used as a gene score (Fig 2B). The relatively high variability within internal replicates (Fig EV4) is consistent with

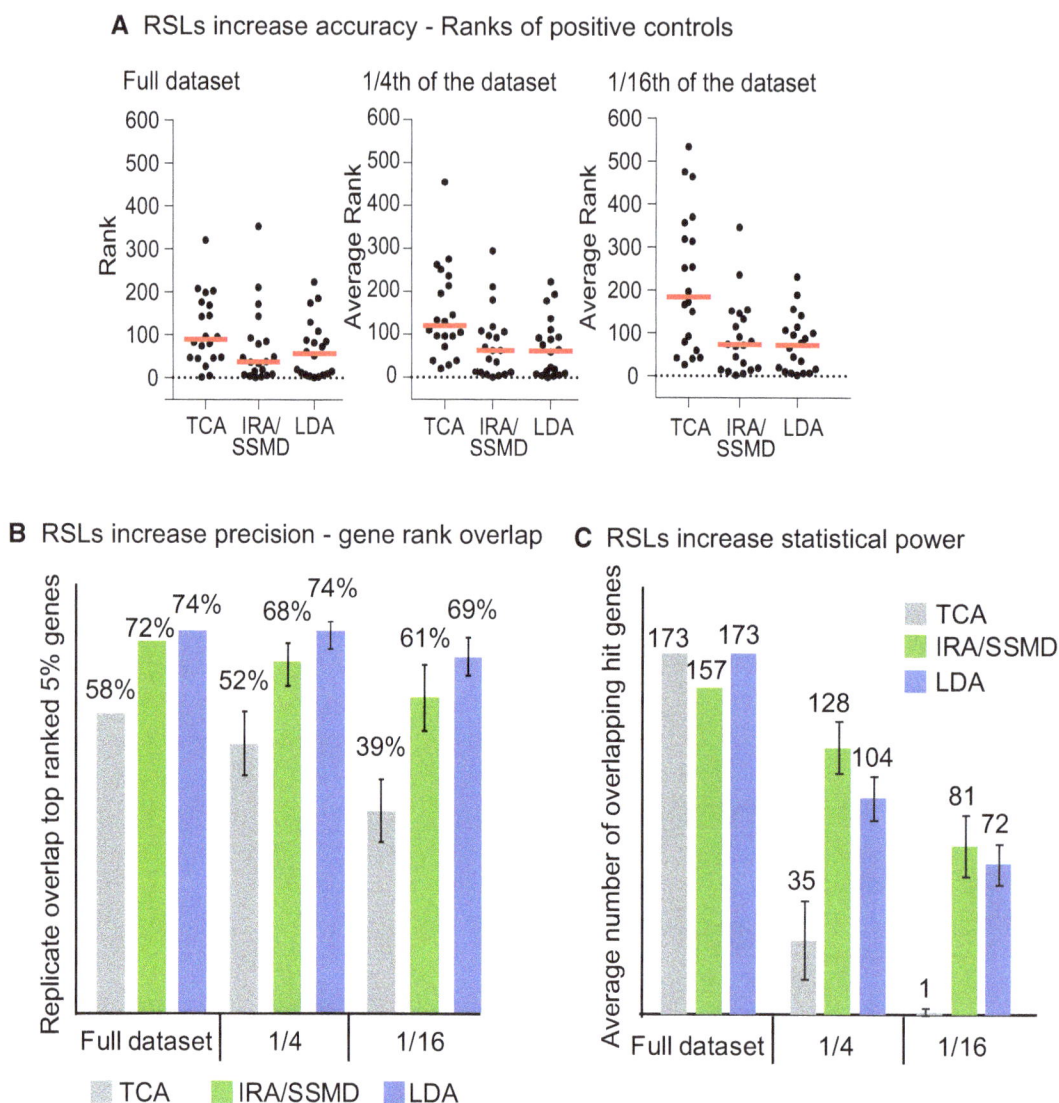

A RSLs increase accuracy - Ranks of positive controls

B RSLs increase precision - gene rank overlap

C RSLs increase statistical power

Figure 3. RSLs improve precision and accuracy of hit calling.

A RSLs increase accuracy of hit calling. Ranks or average ranks of known positive controls (20 ribosomal proteins out of a total of 2,335 interrogated genes) in one experimental replicate for the full screen size (left, rank is plotted), as well as one quarter (middle) and 1/16 (right) of the full screen size (average rank of four subsamples is plotted). Red line, median rank. At all screen sizes, IRA/SSMD analysis and LDA assigned lower ranks to the positive controls than TCA. In TCA, the variance of the ranking increased substantially with decreasing screen size, but not in the two RSL-based methods, which are robust and allow hit calling from fewer cells.

B RSLs increase the precision of gene ranking. Average percent overlap of the top-ranked 5% of genes (116 genes) between two experimental replicates. Error bars, standard deviation of four subsamples. LDA is the most precise method, followed by IRA/SSMD. Again, both RSL-based methods outperform TCA and are more robust at smaller screen sizes.

C RSLs boost statistical power. Hit gene (FDR < 1%) overlap between experimental replicates at full screen size, one quarter, and 1/16 of the full screen size. Error bars, standard deviation for hit gene overlap between four subsamples in experimental replicate 1 and four subsamples in experimental replicate 2 (16 comparisons in total). Only at full screen size, TCA matches the RSL-based analyses. At more practical screen sizes, both RSL-based analyses have much higher statistical power and identify considerably more hit genes.

long culture time and the known variability of cell growth under culture conditions (Levy *et al*, 2015). In addition, variation in Cas9- and guide-RNA expression and distinct repair outcomes are expected to cause initial and long-term variation of growth characteristics of individual lineages. Such variability is present whether or not RSLs are included, however is not readily detected in the conventional total count analysis.

Finally, RSL-labeled guides enable lineage dropout screening, where gene hits are called solely based on the number of lost

RSL-guide lineages (lineage dropout analysis, LDA, Fig 2C). This is the simplest way of analyzing RSL data.

To evaluate IRA/SSMD and LDA, and to compare them with conventional TCA performed with the pipeline MAGeCK (Li *et al*, 2014), we assessed the ranks of a set of known essential genes (accuracy), and the hit gene overlap between experimental replicates (precision). In principle, RSL-based methods should outperform TCA when the number of cells per guide is relatively low, and their benefit should progressively decrease as the number of cells per guide approaches infinity. Thus, the comparisons were performed using the complete dataset, and subsamples of the data that were similar in sample size to published screens (Shalem *et al*, 2014; Wang *et al*, 2015; Fig 3).

Both IRA/SSMD and LDA were more accurate than TCA, as indicated by lower hit ranks of 20 known essential, ribosomal proteins (Figs 3A and EV5). Both IRA/SSMD and LDA were also more precise than TCA, with much improved replicate concordance between the top-ranked 5% of genes (Fig 3B). Consistently with the theoretical considerations, our analysis revealed that the RSL-based methods were far more robust at practically used screen sizes when compared to TCA. The number of highly significant hit genes (as defined by a false discovery rate smaller than 1%) was massively increased in IRA/SSMD and lineage dropout analysis when compared to total read count analysis (Fig 3C). Only at dramatically exaggerated screen size, TCA performed comparably well. Thus, at practicable screen sizes (hundreds of cells per guide), RSL-based methods outperform TCA. The availability of two different RSL-based analysis methods provides increased flexibility; allowing the user to choose the most appropriate method for the specific design of a particular screen.

To summarize, RSLs dramatically improve accuracy, precision, and statistical power in CRISPR/Cas9 screening. The RSL strategy is not limited to CRISPR knockout screening, but can be applied in other screening methods such as CRISPR-dependent inhibition or activation screens (Gilbert *et al*, 2014; Konermann *et al*, 2015). We expect the RSL method to become instrumental in the interrogation of small genomic features, for example, exons, promoters, and even individual transcription factor binding sites. In many of these cases, there is just one possible guide sequence, and in such cases, the inclusion of RSLs is the only way to obtain the replicates that are required for hit calling. In the absence of precise knowledge of both on- and off-target activity, inclusion of multiple guide positions is, however, still important, and rescue experiments and/or analysis of the mutational spectrum of the cutsite are necessary to establish that the mutation induced by the guide results in the observed phenotype. Incorporation of RSLs is technically straightforward and does not require a higher number of cells or sequencing reads compared to conventional approaches. In contrast, RSLs give the same statistical power at a lower number of cells per guide, improving the economy of CRISPR/Cas9 screens. They also improve accuracy and precision at a given number of cells per guide, which is particularly advantageous in cases where cell numbers are limiting, such as in primary cells, or in very large genome-wide screens targeting genes or genomic regulatory regions.

Materials and Methods

Oligo nucleotide synthesis and library cloning

The guide library targets 2,325 genes and contains a total of 23,279 guides (Dataset EV1). The targeted gene set contains all human transcription factors (Vaquerizas *et al*, 2009), other genes of interest as well as ribosomal proteins as positive controls and 101 non-targeting guides as negative controls. All sgRNA sequences used in this library were taken from a previously published, genome-wide library (Wang *et al*, 2014). Oligos were synthesized on an array

(CustomArray). A single overlapping oligo containing random six base pairs as RSLs was annealed to the oligo library, and double-stranded to create the insert for cloning by Gibson assembly into the lentiviral vector pLenti-Puro-AU-flip-3xBsmBI, which was created by modifying lentiGuide-Puro (a gift from Feng Zhang, Addgene #52963) by replacing the sequence

gttttagagctagaaatagcaagttaaaa......TTTTTT with
gtttAagagctagaaatagcaagttTaaa......TTTTTTcgtctct

to create an AU-flip (Chen *et al*, 2013) and an additional BsmBI site downstream of the tracrRNA. The full insert sequence is

**ggctttatatatcttgtgtggaaaggacgaaacaccgnnnnnnnnnnnnnnnnn
nnnnngtttaagagctagaaatagcaagtttaaataa*ggctagtccgttat
caacttgaaaaagtggcaccg*** *agtcggtgctttttttGATCGGAAGAGCAC*
ACGTCTGAACTCCAGTCACnnnnnnnaagcttggcgtaactagatcttgag
acaaa

The fragment from the oligo array is shown in bold; the overlapping fragment containing the RSL and the Illumina i7 index primer (uppercase) was synthesized as a single 119-bp oligo (italics). This oligo was annealed to the oligo library (overlapping region bold italics) and double-stranded using outer primers (underlined).

Gibson assembly, transformation, and amplification of the library

100 ng vector and 12 ng insert were assembled in a total reaction volume of 100 µl (NEBuilder® HiFi DNA Assembly Master Mix, NEB). The reaction was cleaned via a Minelute reaction cleanup column (Qiagen) and transformed into 6 × 50 µl electrocompetent *E. coli* (Endura™ ElectroCompetent Cells, Lucigen) using a 1.0 mm cuvette, 25 µF, 400 Ohms, 1,800 Volts. Bacteria were plated on several 24 × 24 cm agar plates, and colonies were grown overnight at 30°C. Colonies were scraped into LB medium, and the contained plasmids were isolated by Maxiprep.

Library packaging

The library was packaged in HEK 293T cells by cotransfecting the library plasmid and the two packaging plasmids psPAX2 (a gift from Didier Trono, Addgene #12260) and pCMV-VSV-G (a gift from Bob Weinberg, Addgene #8454) in equimolar ratios. After 48 h, the virus-containing supernatant was concentrated 40-fold using Lenti-X concentrator (Clontech), aliquoted for one time use, and stored at −140°C.

Cell lines and cell culture

RKO cells used in this study were purchased from ATCC. Cells were regularly tested for mycoplasma using the Mycoalert detection kit (Lonza; cat# LT07-218).

Creating editing-proficient Cas9 cell lines

To rapidly generate editing-proficient cell lines, we synthesized a lentiviral construct (pLenti-Cas9-sgHPRT1) that encodes a codon optimized WT-SpCas9 that is flanked by two nuclear localization signals (derived from lenti-dCAS-VP64_Blast, a gift from Feng Zhang, Addgene #61425). In addition, the construct codes for blasticidin resistance and carries an sgRNA against HPRT1 (GATGTGATGAAGGAGATGGG). HPRT1 loss confers resistance to the antimetabolite 6-thioguanine (6-TG). Lentivirally transduced cells were selected in 5 μg/ml blasticidin and after one week to 10 days additionally with 5 μg/ml 6-TG until control cells had died. Only cells that both express Cas9 and are editing proficient, as indicated by loss of HPRT1 function, will survive. The method allows rapid establishment of a pool of editing-proficient cells. Compared to single cell clones, this method retains the genetic heterogeneity of the original cell line, avoids potential clonal effects of the particular integration site of Cas9, and greatly accelerates cell line generation. These benefits need to be weighed carefully against possible disadvantages, such as synthetic lethality with HPRT1 loss, or potential effects of the presence of a second guide in the cell.

Library transduction

Per experimental replicate, 100 million RKO Cas9 cells were transduced with the library virus. Two separate replicates were transduced. Cells were then selected for guide integration and expression by 1 μg/ml puromycin selection for 48 h. A proportion of cells will contain more than one guide. Because of the vast number of RSL-guides, any ineffective passenger guides will associate with effective guides randomly and will not be significantly enriched or depleted in the population.

Cell propagation and sample preparation

Cells were kept in culture for a total of 28 days after transduction by sub-culturing them every 3–4 days. 100 million cells were reseeded at each split, and genomic DNA was prepared from at least 50 million cells at Day 4 and Day 28 after transduction. Day 4 after transduction was considered the control time point.

Preparation of the sequencing library from genomic DNA

Genomic DNA was isolated using Blood and Tissue Maxi Kit (Qiagen), and 200 μg, theoretically corresponding to 30 million diploid cells, was used as PCR template in 40 parallel PCR1 reactions (5 μg template DNA each) using KAPA HiFi HotStart polymerase (KAPA Biosystems). After 14 cycles, the reactions were pooled. PCR2 used 5 μl of pooled PCR1 as template and was run for 19 cycles; PCR3 used 2 μl of PCR2 as template and was run for 14 cycles. The resulting product of 288 bp was gel purified and sequenced with a custom primer (CRISPRSeq) and the i5 and i7 index primers by running 20 + 6 + 6 cycles on the Illumina HiSeq4000, where i7 reads the RSL and i5 the illumina sample index.

Primers used for library preparation and sequencing:

PCR1_FW	GGACTATCATATGCTTACCGTAACTTGAAAGTATTTCG
PCR1_REV	CTTTAGTTTGTATGTCTGTTGCTATTATGTCTACTATTCTTTCC
PCR2_FW	TCTTTCCCTACACGACGCTCTTCCGATCTCTTGTGGAAAGGACGAAACAC

PCR2_REV	AGAAGACGGCATACGAGATCTGCCATTTGTCTCAAGATCTAGTTAC
PCR3_FW	AATGATACGGCGACCACCGAGATCTACAC[i5]TCTTTCCCTACACGACGCTCTTCCG
PCR3_REV	CAAGCAGAAGACGGCATACGAGATCTGCCATTTG
CRIPSRSEQ	CGATCTCTTGTGGAAAGGACGAAACACCG

Final amplicon for sequencing (n indicates the guide, bold n represents the sample index, capital N the RSL, sequencing primer is underlined)

aatgatacggcgaccaccgagatctacac**nnnnnn**tctttccctacacga
cgctcttc<u>cgatctcttgtggaaaggacgaaacaccg</u>nnnnnnnnnnnnnn
nnnnnnngtttaagagctagaaatagcaagtttaaataaggctagtccgtt
atcaacttgaaaaagtggcaccgagtcggtgctttttttgatcggaagagca
cacgtctgaactccagtcac*NNNNNN*aagcttggcgtaactagatcttgag
acaaatggcagatctcgtatgccgtcttctgcttg

Scripts used for counting RSL-guides

RSL-guides were counted in the original fastq files with the Perl scripts *Batch-GuideUMI-count-p0.1.pl*, which requires the script *GuideUMI-count-p0.1.pl*.

Binning of RSL-guide counts for creation of internal replicates in IRA/SSMD analysis

Binning was done using the script *Bin-count-TruncatedUMIs.pl*. The script bins according to RSL sequences, taking the first base (4 bins), first two bases (16 bins), etc. into account. Generally, sequences whose sum of readcounts in control and treatment was less than five were filtered out prior to data analysis.

IRA/SSMD analysis of read count data

Data were normalized to total read count: c_{ij} and t_{ij} represent the raw read counts for RSL-guide j in guide-set i for control (Day 4 after lentiviral transduction) and treatment (Day 28 after lentiviral transduction), respectively. The normalized read counts c'_{ij} and t'_{ij} are then

$$c'_{ij} = c_{ij} \frac{\sum_{ij}(c_{ij} + t_{ij})}{2\sum_{ij} c_{ij}}$$

$$t'_{ij} = t_{ij} \frac{\sum_{ij}(c_{ij} + t_{ij})}{2\sum_{ij} t_{ij}}$$

Median effect size and variability of the guide-sets

We defined the effect size ES_{ij} for each RSL-guide or bin j in guide-set i as the log2 of the fold change between treatment count and control count. To handle total loss of an RSL-guide or bin in the treatment sample, we added a pseudo-count of 1 to all counts:

$$ES_{ij} = log_2 \frac{t'_{ij} + 1}{c'_{ij} + 1}$$

Next, we calculated the median effect size for guide-set i, MES_i, and the median of the absolute deviations (MAD) of all RSL-guides or bins j in guide-set i from MES_i

$$MES_i = \underset{j}{median}\, ES_{ij}$$

$$MAD_i = 1.4826\, \underset{j}{median}\, |ES_{ij} - MES_i|$$

The factor 1.4826 was chosen such that the MAD is approximately equal to the standard deviation under the assumption of normal distribution (Zhang, 2011).

Median effect size and variability of the control guide-sets

The RSL library contains 101 non-targeting guide-sets. We calculate a single median effect size and MAD for this control set in the following way:

Median effect size of all non-targeting RSL-guides

$$MES_{CON} = \underset{ij}{median}\, ES_{ij}^{NONT}$$

Median absolute deviation of all non-targeting RSL-guides:

$$MAD_{CON} = 1.4826\, \underset{ij}{median}\, \left|ES_{ij}^{NONT} - MES_{CON}\right|$$

Strictly standardized mean difference

Strictly standardized mean difference is a measure for the significance of the difference in behavior of sample i and the non-targeting controls. It takes into account both the effect size and the variability of the data.

$$SSMD_i = \frac{MES_i - MES_{CON}}{\sqrt{MAD_i^2 + MAD_{CON}^2}}$$

For samples with relatively small effect size, the SSMD can still become large if the spread is small. We thus introduce a score in which the effect size weighs more strongly, and which is used as a ranking parameter:

$$Score_i = MES_i\, |SSMD_i|$$

For hit calling, the average score and standard deviation were calculated for all non-targeting guide-sets.

The script used to do these calculations is *IRA-SSMD.sh*, which calls the script R-script *IRA-SSMD.R*. Guide-sets were then ranked according to their score. A gene hit list was obtained by analyzing the ranked guide list with the "pathway" function of MAGeCK, v0.5.6 (Kolde *et al*, 2012; Li *et al*, 2014b) using Dataset EV2.

Lineage dropout

An RSL-guide was considered a dropout if it had less than two read-counts in the treatment time point. The numbers of RSLs per guide at Day 4 and Day 28 were then used to calculate an effect size (log2 fold change). Guides were ranked according to effect size, and significantly depleted genes were called with the "pathway" function of MAGeCK, v0.5.6 using Dataset EV2.

Subsampling

For subsampling the full data set, RSL-guides were grouped according to their RSL-sequence. For medium screen size, the whole dataset was split into four groups (RSLs starting with A, C, G, and T). For small screen size, the whole dataset was split into 16 groups, the first four of which (AA, AC, AG, AT) were used for analysis. Such subsampling simulates both decreased sequencing depth and a smaller number of cells per guide (smaller screen size). Subsamples were used as replicates in the analyses shown in Fig 3.

Acknowledgements

The authors would like to thank Drs Jenna Persson, Inderpreet Kaur Sur, and Minna Taipale for suggestions on the manuscript. Part of this work was carried out at Karolinska High Throughput Center (KHTC) and the High Throughput Genome Engineering Facility (HTGE) at Science for Life Laboratory, Sweden. This work was supported by the Academy of Finland, grant 250345; the Knut and Alice Wallenberg Foundation, grant KAW 2013.0088; and the Center for Innovative Medicine (CIMED) project grant "Growth Control and Cancer".

Author contributions

BS, SKB, and JT developed the approach; BS, SKB, and MT performed the experiments; BS, SKB, JZ, and TK analyzed the data; BS and JT wrote the manuscript.

References

Adamson B, Norman TM, Jost M, Cho MY, Nunez JK, Chen Y, Villalta JE, Gilbert LA, Horlbeck MA, Hein MY, Pak RA, Gray AN, Gross CA, Dixit A, Parnas O, Regev A, Weissman JS (2016) A multiplexed single-cell CRISPR screening platform enables systematic dissection of the unfolded protein response. *Cell* 167: 1867–1882.e1821

Chen B, Gilbert LA, Cimini BA, Schnitzbauer J, Zhang W, Li GW, Park J, Blackburn EH, Weissman JS, Qi LS, Huang B (2013) Dynamic imaging of genomic loci in living human cells by an optimized CRISPR/Cas system. *Cell* 155: 1479–1491

Datlinger P, Rendeiro AF, Schmidl C, Krausgruber T, Traxler P, Klughammer J, Schuster LC, Kuchler A, Alpar D, Bock C (2017) Pooled CRISPR screening with single-cell transcriptome readout. *Nat Methods* 14: 297–301

Dixit A, Parnas O, Li B, Chen J, Fulco CP, Jerby-Arnon L, Marjanovic ND, Dionne D, Burks T, Raychowdhury R, Adamson B, Norman TM, Lander ES, Weissman JS, Friedman N, Regev A (2016) Perturb-Seq: dissecting molecular circuits with scalable single-cell RNA profiling of pooled genetic screens. *Cell* 167: 1853–1866.e1817

Gilbert LA, Horlbeck MA, Adamson B, Villalta JE, Chen Y, Whitehead EH, Guimaraes C, Panning B, Ploegh HL, Bassik MC, Qi LS, Kampmann M, Weissman JS (2014) Genome-scale CRISPR-mediated control of gene repression and activation. *Cell* 159: 647–661

Kalhor R, Mali P, Church GM (2017) Rapidly evolving homing CRISPR barcodes. *Nat Methods* 14: 195–200

Kivioja T, Vaharautio A, Karlsson K, Bonke M, Enge M, Linnarsson S, Taipale J (2012) Counting absolute numbers of molecules using unique molecular identifiers. *Nat Methods* 9: 72–74

Koike-Yusa H, Li Y, Tan EP, Velasco-Herrera Mdel C, Yusa K (2014) Genome-wide recessive genetic screening in mammalian cells with a lentiviral CRISPR-guide RNA library. *Nat Biotechnol* 32: 267–273

Kolde R, Laur S, Adler P, Vilo J (2012) Robust rank aggregation for gene list integration and meta-analysis. *Bioinformatics* 28: 573–580

Konermann S, Brigham MD, Trevino AE, Joung J, Abudayyeh OO, Barcena C, Hsu PD, Habib N, Gootenberg JS, Nishimasu H, Nureki O, Zhang F (2015) Genome-scale transcriptional activation by an engineered CRISPR-Cas9 complex. *Nature* 517: 583–588

Levy SF, Blundell JR, Venkataram S, Petrov DA, Fisher DS, Sherlock G (2015) Quantitative evolutionary dynamics using high-resolution lineage tracking. *Nature* 519: 181–186

Li W, Xu H, Xiao T, Cong L, Love MI, Zhang F, Irizarry RA, Liu JS, Brown M, Liu XS (2014) MAGeCK enables robust identification of essential genes from genome-scale CRISPR/Cas9 knockout screens. *Genome Biol* 15: 554

Sandler O, Mizrahi SP, Weiss N, Agam O, Simon I, Balaban NQ (2015) Lineage correlations of single cell division time as a probe of cell-cycle dynamics. *Nature* 519: 468–471

Shalem O, Sanjana NE, Hartenian E, Shi X, Scott DA, Mikkelsen TS, Heckl D, Ebert BL, Root DE, Doench JG, Zhang F (2014) Genome-scale CRISPR-Cas9 knockout screening in human cells. *Science* 343: 84–87

Shalem O, Sanjana NE, Zhang F (2015) High-throughput functional genomics using CRISPR-Cas9. *Nat Rev Genet* 16: 299–311

Vaquerizas JM, Kummerfeld SK, Teichmann SA, Luscombe NM (2009) A census of human transcription factors: function, expression and evolution. *Nat Rev Genet* 10: 252–263

Wang T, Wei JJ, Sabatini DM, Lander ES (2014) Genetic screens in human cells using the CRISPR-Cas9 system. *Science* 343: 80–84

Wang T, Birsoy K, Hughes NW, Krupczak KM, Post Y, Wei JJ, Lander ES, Sabatini DM (2015) Identification and characterization of essential genes in the human genome. *Science* 350: 1096–1101

Zhang XD (2007) A pair of new statistical parameters for quality control in RNA interference high-throughput screening assays. *Genomics* 89: 552–561

Zhang XD (2011) Illustration of SSMD, z score, SSMD*, z* score, and t statistic for hit selection in RNAi high-throughput screens. *J Biomol Screen* 16: 775–785

7

Systematic protein–protein interaction mapping for clinically relevant human GPCRs

Kate Sokolina[1,†], Saranya Kittanakom[1,†], Jamie Snider[1,†], Max Kotlyar[2], Pascal Maurice[3,4,5,6] (iD), Jorge Gandía[7,8] (iD), Abla Benleulmi-Chaachoua[3,4,5], Kenjiro Tadagaki[3,4,5], Atsuro Oishi[3,4,5], Victoria Wong[1], Ramy H Malty[9], Viktor Deineko[9], Hiroyuki Aoki[9], Shahreen Amin[9], Zhong Yao[1], Xavier Morató[7,8], David Otasek[2], Hiroyuki Kobayashi[10], Javier Menendez[1], Daniel Auerbach[11], Stephane Angers[12], Natasa Pržulj[13], Michel Bouvier[10], Mohan Babu[9], Francisco Ciruela[7,8], Ralf Jockers[3,4,5], Igor Jurisica[2,14,15] (iD) & Igor Stagljar[1,16,17,*] (iD)

Abstract

G-protein-coupled receptors (GPCRs) are the largest family of integral membrane receptors with key roles in regulating signaling pathways targeted by therapeutics, but are difficult to study using existing proteomics technologies due to their complex biochemical features. To obtain a global view of GPCR-mediated signaling and to identify novel components of their pathways, we used a modified membrane yeast two-hybrid (MYTH) approach and identified interacting partners for 48 selected full-length human ligand-unoccupied GPCRs in their native membrane environment. The resulting GPCR interactome connects 686 proteins by 987 unique interactions, including 299 membrane proteins involved in a diverse range of cellular functions. To demonstrate the biological relevance of the GPCR interactome, we validated novel interactions of the GPR37, serotonin 5-HT4d, and adenosine ADORA2A receptors. Our data represent the first large-scale interactome mapping for human GPCRs and provide a valuable resource for the analysis of signaling pathways involving this druggable family of integral membrane proteins.

Keywords G-protein-coupled receptors; high-throughput screening integrative computational biology; interactome; protein–protein interactions; split-ubiquitin membrane yeast two-hybrid assay
Subject Categories Network Biology; Signal Transduction

Introduction

G-protein-coupled receptors (GPCRs) are seven-transmembrane proteins involved in many signal transduction pathways and in numerous human diseases such as schizophrenia (Moreno *et al*, 2009), Parkinson's disease (Pinna *et al*, 2005; Dusonchet *et al*, 2009; Gandía *et al*, 2013), hypertension (Brinks & Eckhart, 2010), obesity (Insel *et al*, 2007), and multiple cancers (Lappano & Maggiolini, 2011). GPCRs propagate ligand-specific intracellular signaling cascades in response to extracellular stimuli—following ligand activation, GPCRs catalyze the exchange of GDP for GTP on the Gα subunit, leading to a decreased affinity of Gα for Gβγ. The

1 Donnelly Centre, University of Toronto, Toronto, ON, Canada
2 Princess Margaret Cancer Centre, University Health Network, University of Toronto, Toronto, ON, Canada
3 Inserm, U1016, Institut Cochin, Paris, France
4 CNRS UMR 8104, Paris, France
5 Sorbonne Paris Cité, University of Paris Descartes, Paris, France
6 UMR CNRS 7369 Matrice Extracellulaire et Dynamique Cellulaire (MEDyC), Université de Reims Champagne Ardenne (URCA), UFR Sciences Exactes et Naturelles, Reims, France
7 Unitat de Farmacologia, Departament de Patologia i Terapèutica Experimental, Facultat de Medicina, IDIBELL, Universitat de Barcelona, L'Hospitalet de Llobregat, Barcelona, Spain
8 Institut de Neurociències, Universitat de Barcelona, Barcelona, Spain
9 Department of Biochemistry, Research and Innovation Centre, University of Regina, Regina, SK, Canada
10 Department of Biochemistry, Institute for Research in Immunology & Cancer, Université de Montréal, Montréal, QC, Canada
11 Dualsystems Biotech AG, Schlieren, Switzerland
12 Department of Pharmaceutical Sciences, Leslie Dan Faculty of Pharmacy and Department of Biochemistry, Faculty of Medicine, University of Toronto, Toronto, ON, Canada
13 Department of Computing, University College London, London, UK
14 Departments of Medical Biophysics and Computer Science, University of Toronto, Toronto, ON, Canada
15 Institute of Neuroimmunology, Slovak Academy of Sciences, Bratislava, Slovakia
16 Department of Molecular Genetics, University of Toronto, Toronto, ON, Canada
17 Department of Biochemistry, University of Toronto, Toronto, ON, Canada
*Corresponding author. E-mail: igor.stagljar@utoronto.ca
†These authors contributed equally to this work

resulting dissociation of the hetero-trimer allows the GTP-bound Gα and free Gβγ to interact with several downstream effectors, including adenylyl cyclases, phosphodiesterases, phospholipases, tyrosine kinases, and ion channels (Dupré et al, 2009; Ritter & Hall, 2009).

Due to their involvement in signal transmission, GPCRs are highly druggable targets for numerous pharmaceutical compounds used for various clinical indications (Lagerström & Schiöth, 2008). To design successful treatments for these diseases, it is essential to increase the depth and breadth of our understanding of the molecular events occurring during GPCR-mediated signal transduction, and to identify all of the proteins interacting with a particular GPCR relevant for human health.

Over the last decade, numerous biochemical, cell biological, and genetic assays have been used to identify and characterize GPCR-interacting partners (Daulat et al, 2009; Maurice et al, 2011). These studies showed that, in addition to G-proteins, GPCRs also interact with a wide variety of integral membrane proteins (e.g. other GPCRs, ion channels, transporters, and other family receptors) and cytosolic proteins (e.g. arrestins, GPCR kinases, Src homology 2 and 3 (SH2− and SH3−), and PDZ-domain containing proteins; Ritter & Hall, 2009; Marin et al, 2012; Hall & Lefkowitz, 2014). Despite wide usage of biochemical assays such as co-immunoprecipitation (co-IP), pull-down- and affinity purification-linked to mass spectrometry (AP-MS), and protein microarrays to identify GPCR-associated proteins (Daulat et al, 2007, 2011; Maurice et al, 2008; Chung et al, 2013; Benleulmi-Chaachoua et al, 2016), these methods have not been widely applied to assay GPCR-related protein–protein interactions (PPIs) in a systematic manner on a large scale. Furthermore, these methods are technically difficult and time-consuming, involving harsh treatments for cell disruption and membrane protein solubilization, and often require optimization for each target protein complex examined (Chung et al, 2013; Snider et al, 2015).

Technical progress has also been made in developing methods based on fluorescence or bioluminescence resonance energy transfer (FRET or BRET) to study GPCR-interacting partners in live mammalian cells with kinetics that are close to real-time (Lohse et al, 2012; Ayoub & Pin, 2013). Nonetheless, the analysis of GPCR interactors using BRET and FRET is not readily scalable to high-throughput screening (HTS), but is rather more suited to medium-throughput screens involving a limited number of putative hits. Aside from these biochemical and cell biological approaches, genetic methods such as the conventional yeast two-hybrid (YTH) system (Fields & Song, 1989) have been used to identify proteins interacting with the soluble domains of selected GPCRs (Gavarini et al, 2004; Canela et al, 2007; Yao et al, 2015). Unfortunately, while interesting, these studies are restricted to the investigation of only the soluble components of particular human GPCRs for which interacting proteins are selected in the yeast nucleus, which is an unnatural cellular compartment for identifying protein interactors of integral membrane proteins. Thus, our knowledge of the interacting proteins of human GPCRs suffers potentially serious limitations and biases due to the lack of a suitable high-throughput technology to efficiently and comprehensively characterize interacting proteins of integral membrane proteins in their native cellular and membrane environment.

In this study, we used a modified membrane yeast two-hybrid (MYTH) approach (Deribe et al, 2009; Snider et al, 2010; Mak et al, 2012; Usenovic et al, 2012; Huang et al, 2013; Xie et al, 2013), specifically tailored to identify interactors of full-length integral membrane proteins, as well as in-depth bioinformatics analysis to create and annotate an interactome for 48 selected full-length, clinically relevant human GPCRs in their ligand-unoccupied state, localized to their native plasma membrane. Using this rich GPCR-interactome resource, we then prioritized candidates by systematic computational analysis for further biological studies, and carried out functional studies of selected PPIs. The GPCR-interaction network presented here will be a crucial resource for increasing our fundamental understanding of the cellular role and regulation of this important family of integral membrane proteins, and may facilitate development of new disease treatments and a clearer understanding of drug mechanisms of action.

Results

Selection of human GPCRs, generation of GPCR "baits", and their functional validation

We used a modified split-ubiquitin membrane yeast two-hybrid (MYTH; Stagljar et al, 1998; Gisler et al, 2008; Deribe et al, 2009; Snider et al, 2010) assay to define the interactomes of 48 full-length, human, ligand-unoccupied GPCRs localized to the plasma membrane. Specific GPCRs were selected based upon their importance for human health, specifically their direct link to human disease. We screened 44 Class A rhodopsin-like receptors to create a representative interactome of this most abundant family of GPCRs in order to identify physical interaction partners, 2 Class B secretin-like receptors (vasoactive intestinal peptide receptor 2 and retinoic acid-induced gene 2 protein), and 2 Class F receptors (smoothened and Frizzled7; Table EV1 lists the 48 GPCRs and related human diseases). An overview of the complete MYTH workflow is presented in Fig 1.

MYTH GPCR constructs ("baits") were generated from the selected 48 full-length human GPCR ORFs. All baits were N-terminally tagged with the signal sequence of the yeast mating factor α to encourage plasma membrane localization and stable expression in yeast (Deribe et al, 2009), and C-terminally tagged with the C-terminal half of ubiquitin (Cub) fused to an artificial transcription factor (TF) comprised of LexA and VP16 (Fig 2A; Snider et al, 2010). Bait fusion proteins were tested for proper expression at the yeast plasma membrane by immunofluorescence, and for lack of self-activation via the NubG/NubI test using the non-interacting yeast plasma membrane prey protein Fur4p (Snider et al, 2010, 2013; Fig 2B). Functionality of GPCR baits upon addition of the MYTH tag was also demonstrated using two selected GPCR baits by measuring changes in growth rate of bait-expressing yeast in the presence and absence of an agonist (Fig 2C). In summary, all GPCR baits used in this study passed stringent validation tests ensuring they are properly expressed, localized, and are functional prior to their usage in high-throughput MYTH screens to identify protein interaction partners.

Validation of MYTH GPCR baits using known GPCR interactions

To further confirm that the addition of the Cub-TF tag to the C-termini of GPCR proteins does not disrupt their function and that the MYTH system itself represents a suitable tool for use in the

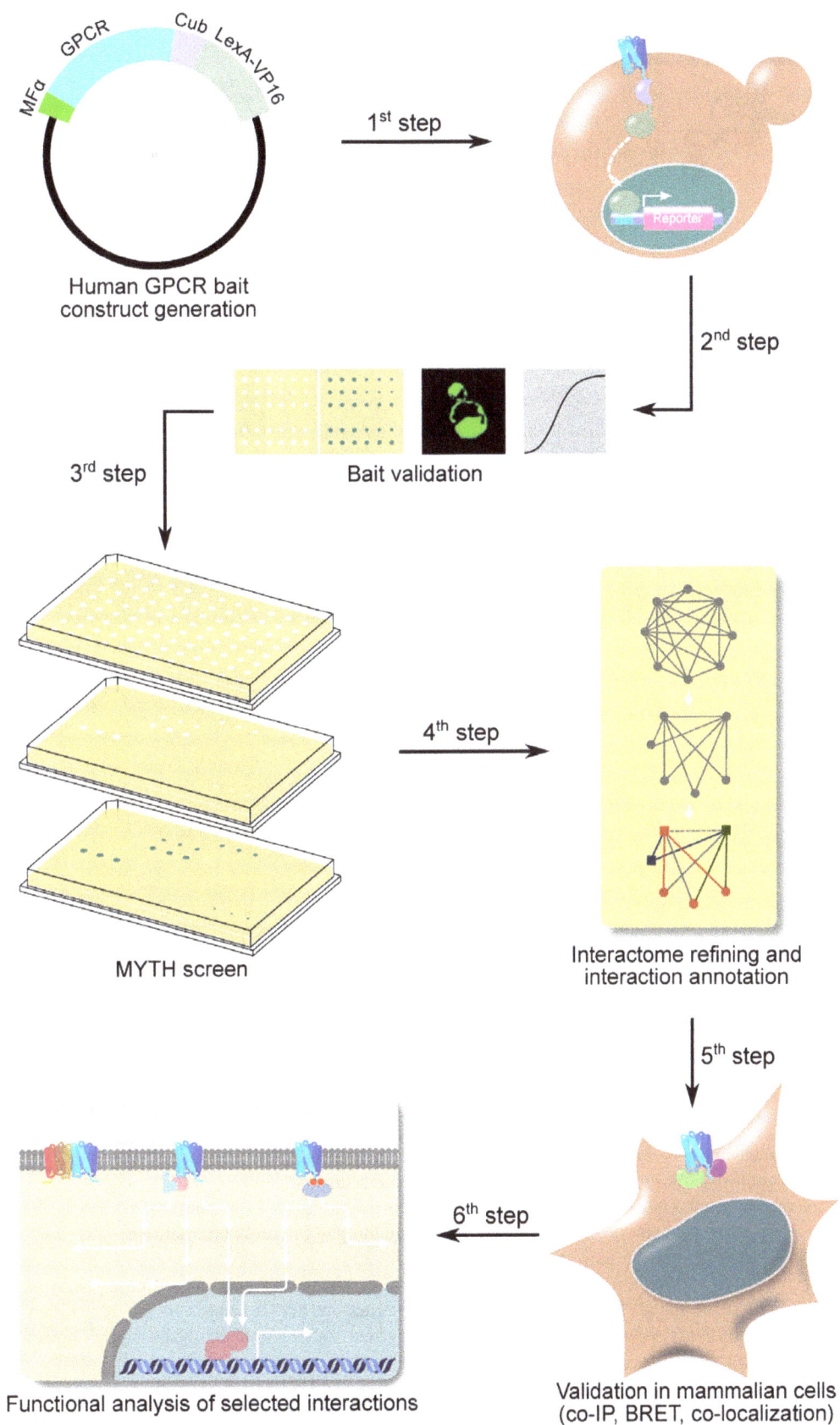

Figure 1. Workflow for generating the human full-length GPCR interactome.

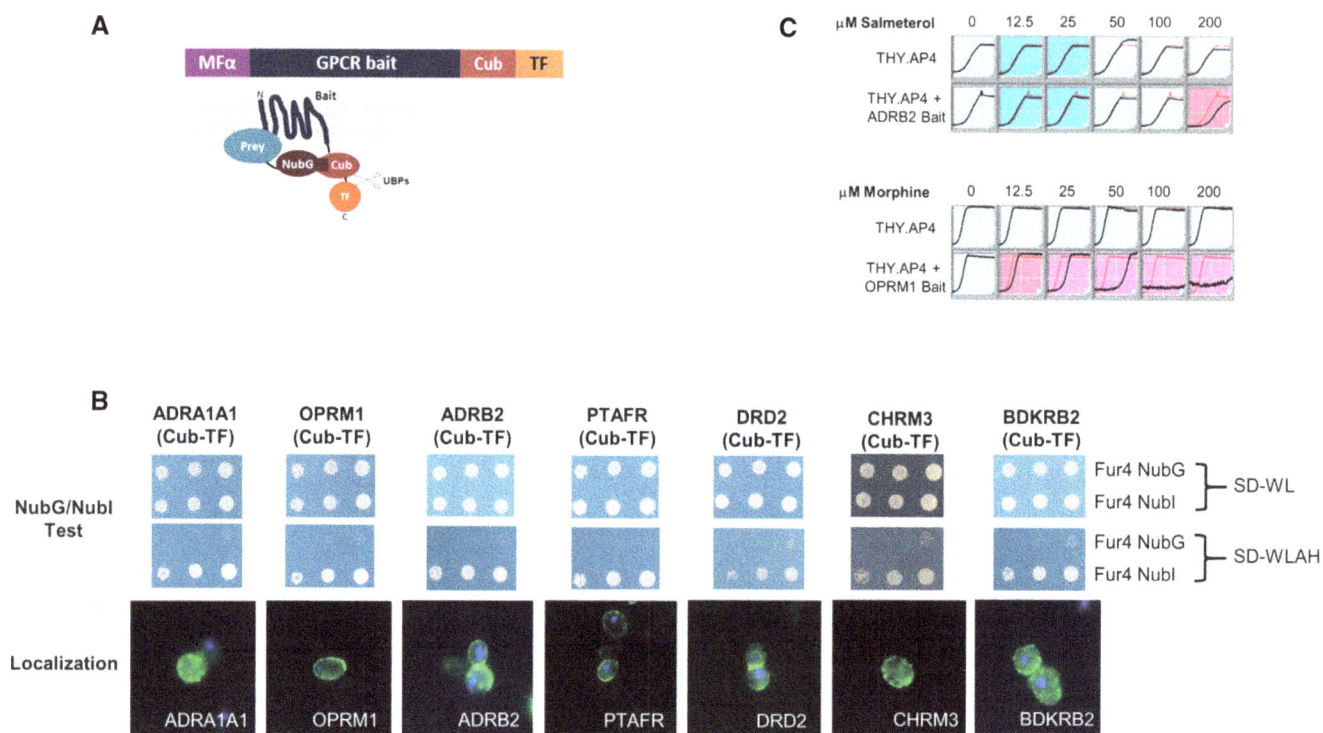

Figure 2. Expression of human MYTH GPCR "baits" in yeast cells.

A The structure of the GPCR bait proteins used in this study is shown. The signal sequence of yeast α-mating pheromone precursor (MFα) was fused to the N-terminus of human GPCR baits, while the C-terminal fragment of ubiquitin (Cub) followed by an artificial transcription factor (TF) was fused to the C-terminus of the baits.

B Representative sample of functional validation/localization tests performed on all GPCR baits used in this study. The top two panels show proper expression and MYTH function of human GPCR-Cub-TF baits demonstrated using the NubG/NubI test. In this test, GPCR-Cub-TF bait and a non-interacting yeast plasma membrane protein (Fur4p), fused to either NubI (Fur4 NubI) or NubG (Fur4 NubG) are co-expressed in yeast MYTH-reporter cells. Growth on minimal SD medium lacking Trp and Leu (SD-WL, top panel) selects only for presence of bait and prey plasmids, while minimal SD medium lacking Trp, Leu, Ade, and His (SD-WLAH, middle panel) selects for interaction between bait and prey. Co-expression of GPCR-Cub-TF bait with Fur4p fused to NubI leads to activation of the reporter system and consequent growth on SD-WLAH medium, since the wild-type NubI leads to reconstitution of ubiquitin independent of a bait–prey interaction, demonstrating that the bait protein is expressed/correctly folded. Co-expression of GPCR-Cub-TF bait and non-interacting Fur4p fused to NubG (which does not spontaneously associate with Cub) does not lead to activation of the reporter system and growth on SD-WLAH medium, demonstrating that the bait is not self-activating. The bottom panel shows localization of human GPCR bait proteins in THY.AP4 yeast reporter strain. Yeast cells expressing given human GPCR baits were fixed by paraformaldehyde and digested by zymolyase. Methanol-acetone-treated yeast spheroplasts were detected using an antibody against the transcription factor (rabbit anti-VP16) and were visualized by Cy3-conjugated secondary antibodies (shown in green). DAPI-stained nuclei can be seen as blue fluorescence. Note that similar NubGI test and localization results were obtained for all GPCR baits used in this study.

C Growth inhibition of the human ADRB2 and OPRM1 baits expressed in yeast THY.AP4 in response to their corresponding agonist. Growth curves were carried out in triplicate, and curves shown are the average of three independent measurements at each individual time point. The red line shows the control yeast growth in the absence of drug, while the black line shows growth in the presence of drug. Inhibited growth in response to drug indicates GPCR activity.

identification of GPCR-interaction partners, we used MYTH to test a subset of 50 previously identified GPCR PPIs (Table EV2). To verify that the absence of interaction is not a false negative due to lack of prey protein expression, we made a side-by-side comparison of the NubG-tagged MYTH prey construct and the prey tagged with the original, spontaneously reconstituting wild-type NubI. Overall, 12 of the 50 (24%) could be confirmed in the MYTH assay (Fig EV1 and Table EV2). Note that not all previously reported interactions can be expected to be validated by our technique, due both to differences in the technical details of the approaches originally used (e.g. working with cell lysates instead of live cells when doing affinity purifications, working with only soluble portions of GPCRs when doing traditional YTH) and assay conditions (e.g. our assay is carried out in the absence of ligand). Our results therefore clearly demonstrate

the robustness and accuracy of the MYTH assay to detect GPCR-interacting partners.

Building of the GPCR interactome

To systematically map interacting partners of human GPCRs, we carried out MYTH screens of the 48 selected human GPCR baits against an N-terminally NubG-tagged human cDNA library, as described previously (Snider et al, 2010). Briefly, yeast cells expressing MYTH baits were transformed with NubG prey pools and plated onto SD-WLAH growth media. Positive colonies were subjected to additional selection steps, and prey DNA was then isolated and sequenced to identify candidate interaction partners. The results of our extensive MYTH screens were assembled into a "preliminary" interactome, which was further refined experimentally using the bait dependency test, which allows us to both retest each interaction (thereby demonstrating reproducibility) and identify/remove spuriously interacting preys which bind to unrelated control bait (Snider *et al*, 2010, 2013; Lam *et al*, 2015). All of the interactions that passed this secondary testing were used in subsequent bioinformatics analysis and filtering (to further identify and remove false positives/spurious interactors, including signal peptide processing and ribosomal proteins which are frequently identified "non-specific" interactors associated with general translation and trafficking processes). All remaining candidates were then assembled into our final GPCR interactome, comprising 987 unique interactions between 686 proteins, including 299 membrane proteins (Fig 3 and Table EV3). Table EV4 lists the false positives/spurious interactors removed from our final interactome.

To further investigate the biological context of the generated interactome, we analyzed its enrichment for pathways, diseases, molecular function, biological process, domains, and drug targets (see Fig 3 and Table EV5). Using pathDIP 2.5 (Rahmati *et al*, 2017), we identified significantly enriched pathways, among baits and preys including transmembrane transport of small molecules (7.0% of baits and preys, adjusted $P = 8.7e-8$), neuroactive ligand–receptor interaction (5.0% of baits and preys, adjusted $P = 2.3e-6$), and calcium regulation in the cardiac cell (7.7% of baits and preys, adjusted $P = 6.9e-6$; Fig EV2A).

We investigated enrichment of diseases, functions, processes, and domains among interacting preys (Table EV5). No diseases were significantly enriched among preys, after adjusting P-values for multiple testing. Diseases with the lowest unadjusted P-values included hereditary spastic paraplegia (1.6% of preys, $P = 4.5e-5$), schizophrenia (13.6% of preys, $P = 1.0e-4$), and neurodegenerative disorders (6.6% of preys, $P = 1.0e-4$; Fig EV2B). Three functions were significantly enriched: calcium ion transmembrane transporter activity (2.3% of preys, adjusted $P = 5.4e-3$), ion channel binding (2.2% of preys, adjusted $P = 1.7e-2$), and cation-transporting ATPase activity (2.5% of preys, adjusted $P = 1.9e-2$). Top enriched processes included transmembrane transport (15.5% of preys, adjusted $P = 1.1e-3$), endoplasmic reticulum calcium ion homeostasis (1.3% of preys, adjusted $P = 1.2e-3$), and ATP hydrolysis coupled proton transport (1.3% of preys, adjusted $P = 2.7e-2$). No domains were enriched after adjusting P-values for multiple testing; top domains based on unadjusted P-values were fatty acid hydroxylase (0.5% of preys, $P = 8.6e-3$), V-ATPase proteolipid subunit C-like domain (0.5% of preys, $P = 1.1e-2$), and TRAM/LAG1/CLN8 homology domain (1.0% of preys, $P = 1.2e-2$). We also investigated whether pairs of protein domains or conserved sites (one on a bait and the other on a prey) were enriched among interacting protein pairs. Top enriched pairs (adjusted $P < 2.7e-12$) included bait domain GPCR, rhodopsin-like, 7TM (IPR017452) paired with prey domains/sites Tetraspanin, conserved site (IPR018503), Tetraspanin, EC2 domain (IPR008952), and Marvel domain (IPR008253).

A significant number of bait GPCRs are already targeted by drugs (28 of 48 proteins, $P = 3.1e-8$ relative to all proteins; Fig 3). These drugs comprise a variety of categories, such as histamine antagonists, antiparkinson agents, and antipruritics agents, and affect diverse organ systems, including nervous, cardiovascular, and respiratory (Fig EV3). In total, 122 baits and preys are targeted by 737

drugs. These proteins and their interactions have substantial medical and economic significance (Fig 3). Drugs that target these proteins include four of the top 100 prescribed drugs and five of the top 100 selling drugs in the United States for 2014, according to data from IMS Health, reported in Medscape (2015). These selected drugs had over 27 million prescriptions and over $14 billion in sales. Using the GPCR interactome, we can gain a more detailed understanding of how these drugs, as well as other compounds, modulate disease-related pathways.

Orthogonal validation of MYTH-identified PPIs in mammalian cells

As a secondary validation of our GPCR interactome, a subset of PPIs selected from our interaction data was tested in mammalian cells using two distinct co-immunoprecipitation (co-IP) approaches. In the first approach, FLAG-tagged GPCR interactors were overexpressed in mammalian cells, pulled-down using anti-FLAG antibody, subjected to SDS–PAGE, transferred to membranes, and probed with commercial antibody raised against their identified endogenous GPCR-interaction partner. We tested a subset of interactions corresponding to 11 different GPCR proteins, using four MYTH-identified interacting preys and two non-interacting negative control preys for each. Of the 11 GPCR baits, five performed well in our analysis, producing no more than background signal in at least one of two negative control samples, from which we were able to confirm a total of 13 (65%) of tested interactions (Figs 4A and EV4, Table EV6). Proper expression of transiently transfected preys in these blots was checked by Western blot (Fig EV5). Of the six remaining blots, two had extremely low levels of bait expression, while four produced signal in both negative controls comparable to that in test samples, under multiple test conditions, preventing meaningful interpretation of results (Fig EV4 and Table EV6). In the second approach, an additional 14 PPIs were selected, and both immunoprecipitation and subsequent Western development were performed using native antibody directed against endogenously expressed bait and prey. Of these 14 PPIs, nine (64.2%) were successfully validated (Fig 4B and Table EV6).

As an additional orthogonal validation, we were also able to use bioluminescence resonance energy transfer (BRET; Hamdan *et al*, 2006) to confirm a small subset of eight interactions, including six not validated using either of our co-IP approaches (Table EV7).

Overall, we were able to validate a substantial number of our tested interactions (28/40, 70%) using either co-IP and/or BRET (Fig 4C), providing strong support for the robustness and quality of our MYTH-generated GPCR interactome.

Functional analysis of novel, MYTH-identified GPCR PPIs

In an attempt to frame our GPCR interactome results in a biological context, as well as demonstrate the utility of the interactome in revealing novel interactions of biological significance, we decided to validate several novel PPIs with potential impact in neurobiology: specifically, the interactions of the hydroxytryptamine (serotonin) 5-HT4d (HTR4) receptor, a promising target for Alzheimer disease (Lezoualc'h, 2007), with both GPRIN2, a G-protein-regulated inducer of neurite outgrowth 2 that interacts with G-proteins (Chen *et al*, 1999), and the Parkinson's disease-associated receptor GPR37

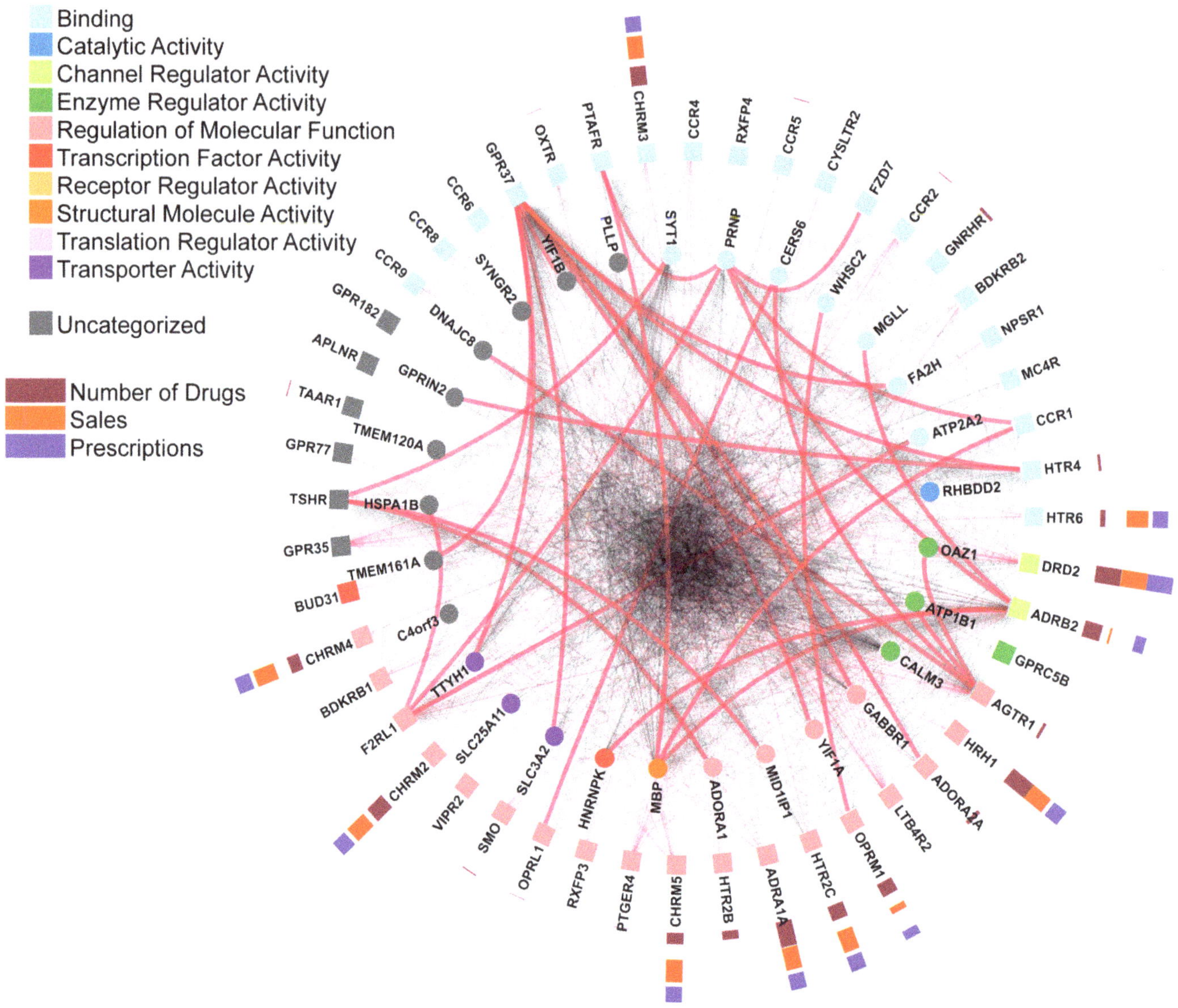

Figure 3. 48 clinically relevant GPCR receptors mapped using MYTH.

GPCR interactome. Validated interactions between baits–preys and preys targeted by drugs are highlighted. Drug targets were downloaded from DrugBank, drugs sales and prescription numbers were obtained from Medscape (2015). Bait–prey interactions are based on the IID database (black edges), MYTH detection (red edges), and validation assays (thick red edges). Nodes are ordered and categorized by NAViGaTOR 3's GO Molecular Function categorizer. Square nodes correspond to GPCR baits, while circular nodes correspond to interacting prey partners.

(Dusonchet *et al*, 2009), as well as the interaction between GPR37 and the adenosine A2A receptor (ADORA2A), also involved in Parkinson's disease (Pinna *et al*, 2005; Gandía *et al*, 2013).

To confirm the interaction of 5-HT4d with GPRIN2 and GPR37 in a mammalian system, we carried out co-IP experiments (Fig 5A) and BRET saturation assays (Fig 5B) in HEK-293 cells. Though the interaction with GPRIN2 is not observed by BRET, it can be detected by co-IP (Fig 5A, lanes 2 and 3), likely because the distance between *R*luc and YFP is greater than the BRET detection threshold of 100 angstroms. The interaction between 5-HT4d and GPR37 was confirmed in both assays. Co-localization of 5-HT4d with GPR37 and GPRIN2 was also observed at the plasma membrane (Fig 5C). Additionally, Erk1/2 phosphorylation and cAMP production, in

response to stimulation of 5-HT4d, were modulated by co-expressed GPR37 and GPRIN2, with ERK1/2 phosphorylation being largely abolished (Fig 5D) and maximal cAMP production potentiated (Fig 5E). This effect occurred without any modification in expression level of 5-HT4d (Fig EV6). Importantly, overexpression of GPR37 and GPRIN2 on their own did not affect cAMP production in response to agonist stimulation (Fig EV7).

Control experiments using overexpressed chemokine CCR5 receptor or a C-terminally truncated form of GPRIN2, which is unable to interact with G-proteins, did not show modulation of 5-HT4d response. GPRIN2 and GPR37 were also unable to modify the ERK and cAMP response elicited by the β2-adrenergic receptor upon isoproterenol stimulation (Fig 5D and E). Collectively, these

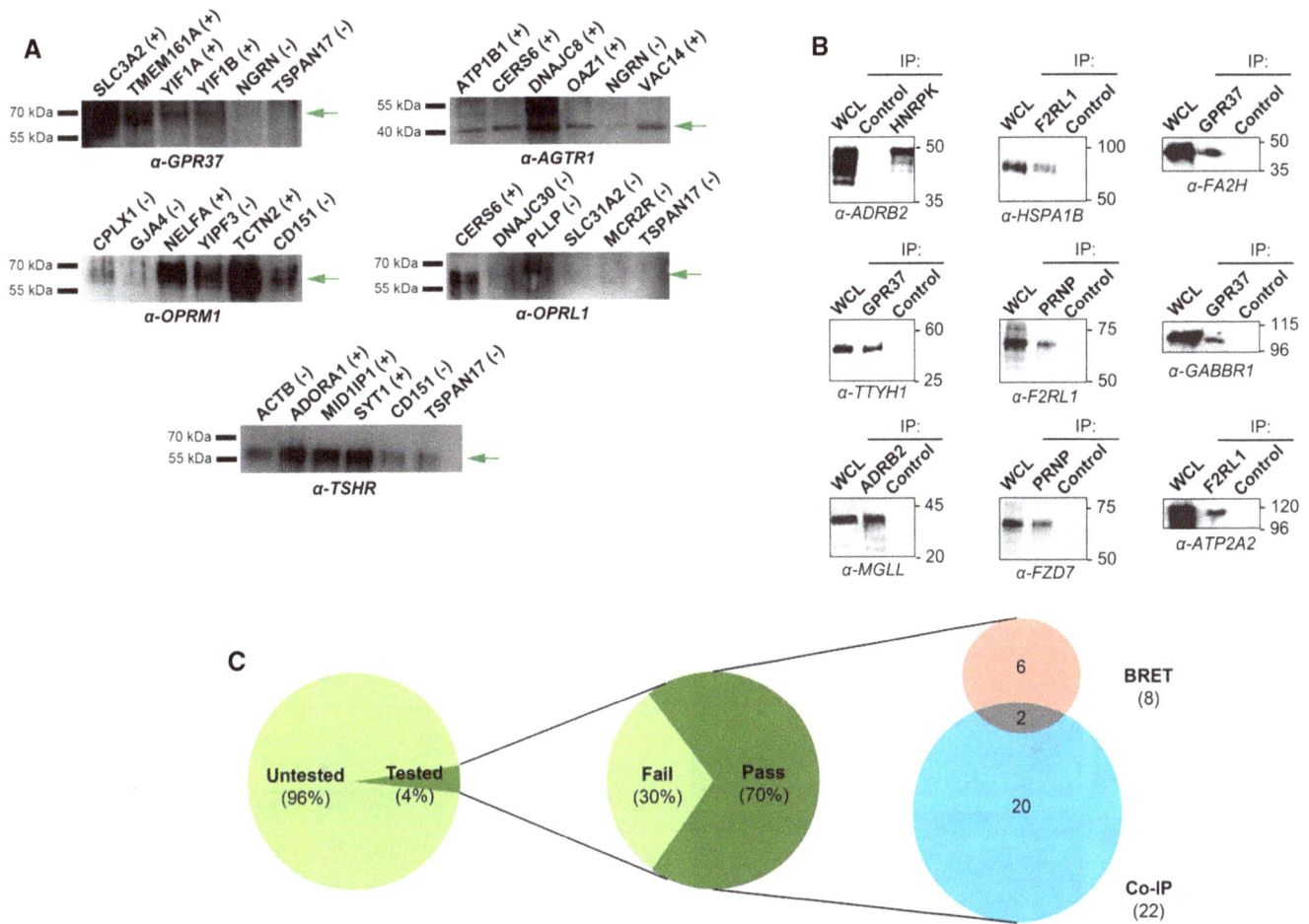

Figure 4. Orthogonal validation of the MYTH-based GPCR interactome.

A Co-immunoprecipitations were performed using α-FLAG antibody directed against overexpressed FLAG-tagged protein corresponding to either MYTH-identified interactor (first four lanes) or negative control (last two lanes), followed by Western blotting using antibody directed against the corresponding putative GPCR protein interaction partner (listed below each blot). All blots shown here produced no more than background signal in at least one negative control sample, making them suitable for use in validation of MYTH-detected interactions. (+) indicates an interaction was detected by co-IP. (−) indicates no interaction was detected by co-IP. Green arrows point to the band corresponding to the indicated GPCR.

B Co-immunoprecipitations were performed using native antibody directed against the interaction partner indicated below each blot, followed by Western blotting using native antibody directed against the other member of the interacting pair. All proteins were endogenously expressed. WCL, whole-cell lysate. Control, pull-down using beads only.

C A total of 40 MYTH-detected interactions were successfully tested by co-immunoprecipitation or BRET and 28 were validated, a success rate of 70%. Of the 40 interactions, 34 were tested by co-immunoprecipitation approaches and 22 of these were validated, a success rate of 64.7%. BRET was used to test eight interactions, including two tested by co-immunoprecipitation, and all were validated.

data demonstrate the specificity of the effect of GPRIN2 and GPR37 on 5-HT4d function.

Another GPCR interactor of GPR37 identified in our MYTH screen was ADORA2A, an adenosine receptor highly expressed in the striatum, a region of the brain involved in Parkinson's disease (Pinna et al, 2005; Gandía et al, 2013). The co-distribution and co-immunoprecipitation of ADORA2A and GPR37 were confirmed in HEK-293 cells (Fig 6A and B). Subsequently, the direct association between ADORA2A and GPR37 was confirmed by BRET saturation experiments (Fig 6C and D). Importantly, we did not observe a positive interaction between GPR37 and ADORA1, a related adenosine receptor (Fig 6C). Furthermore, we explored the impact of the ADORA2A/GPR37 interaction on the cell surface expression of these

receptors (Fig 6E). The levels of GPR37 when expressed alone are particularly low, as previously reported (Gandía et al, 2013). Interestingly, co-expression with ADORA2A markedly enhanced both whole and cell surface expression of GPR37 (Fig 6E), suggesting an ADORA2A chaperone-like function. Importantly, the expression levels of GPR37 were not enhanced by ADORA1 co-expression (Fig 6F), thus providing insight into the specificity of the ADORA2A/GPR37 interaction.

Since the levels of ADORA2A appear to affect GPR37 expression, we next aimed to explore the role of GPR37 in ADORA2A signaling in vivo. To this end, we first validated the ADORA2A/GPR37 interaction in native tissue, namely mouse striatum, by means of co-immunoprecipitation experiments. The immunoprecipitation of striatal

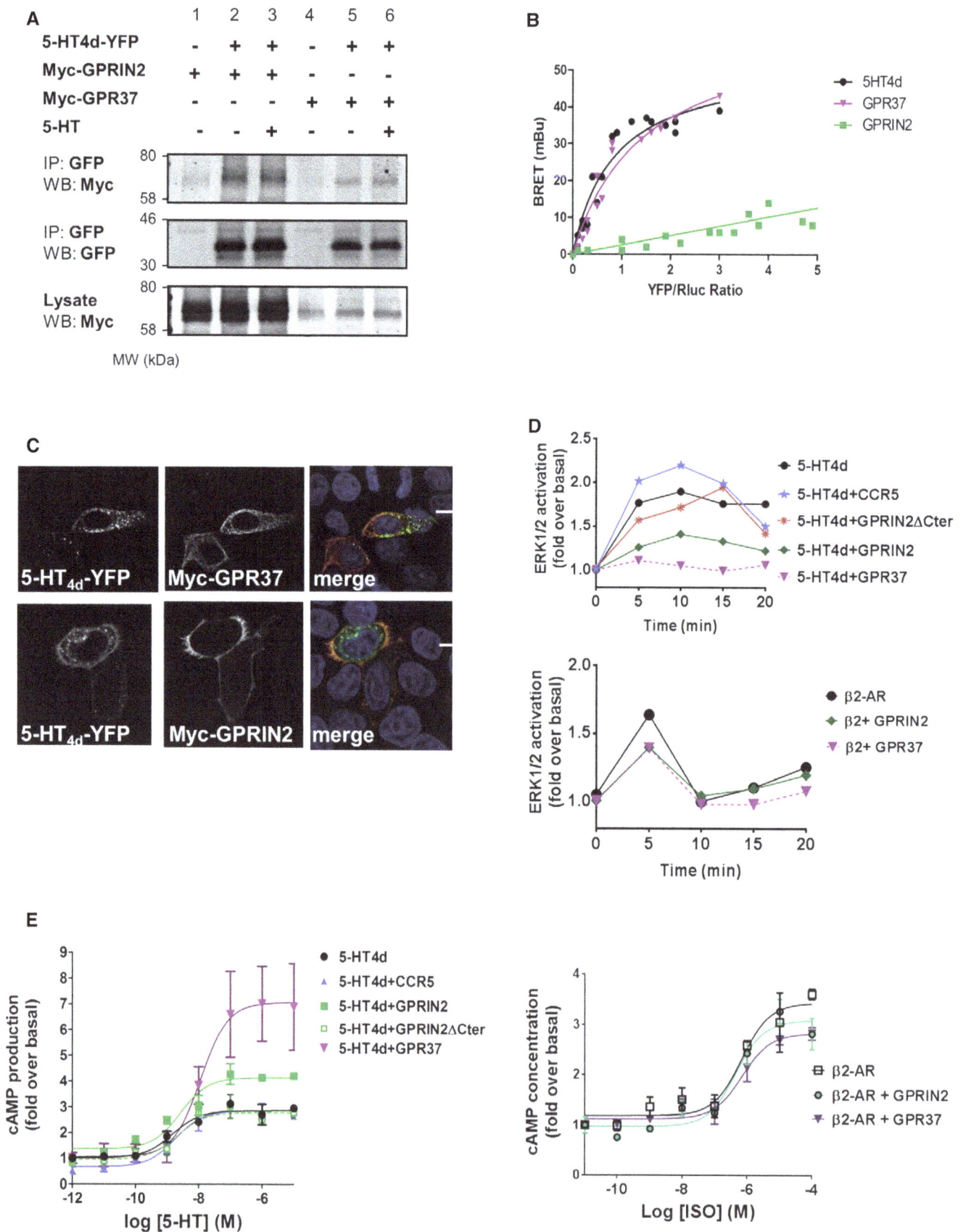

Figure 5.

Figure 5. Functional interactions of GPR37 and GPRIN2 with 5-HT4d in transfected cells.

A Co-immunoprecipitation in the presence and absence of 1 µM 5-HT agonist for 15 min. HEK-293 cells were transiently transfected with 5-HT4d-YFP (lanes 2, 3, 5, 6) and myc-GPRIN2 (lanes 1–3) or GPR37 (lanes 4–6) and processed for immunoprecipitation using an anti-GFP antibody. The crude extracts (lysate) and immunoprecipitates (IP) were analyzed by SDS-PAGE and immunoblotted using a rabbit anti-GFP or anti-Myc antibody. Data are representative of at least two independent experiments.

B BRET donor saturation curves were performed by co-transfecting a fixed amount of 5-HT4d-Rluc and increasing amounts of 5-HT4d-YFP, GPR37-YFP, and GPRIN2-YFP in HEK-293 cells. Data are means of three independent experiments performed in triplicate.

C Co-expression of HeLa cells transfected with 5-HT4d-YFP (green) and myc-GPR37 or myc-GPRIN2 (red) and analyzed by confocal microscopy. Superimposition of images (merge) reveals co-distribution in orange and DAPI-stained nuclei in blue. Scale bar: 15 µm. Data are representative of at least two independent experiments.

D ERK1/2 activation in HEK-293 cells over time in response to 10 µM 5-HT agonist and the presence of overexpressed 5-HT4d and GPRIN2, GPR37, or CCR5. CCR5 is used as a negative control. The bottom panel shows ERK1/2 activation over time, in the presence of overexpressed β_2-adrenergic receptor and GPRIN2 or GPR37. Data are means of three independent experiments performed in triplicate.

E Cyclic AMP levels in HEK-293 cells, in response to increasing concentrations of serotonin agonist and the presence of overexpressed 5-HT4d and GPRIN2, GPR37, or CCR5. CCR5 is used as a negative control. The right panel shows cAMP levels in response to increasing isoproterenol concentrations, in the presence of overexpressed β_2-adrenergic receptor and GPRIN2 or GPR37. Data are means of three independent experiments performed in triplicate. Error bars indicate SEM.

GPR37 yielded a band of ~45 kDa corresponding to the ADORA2A (Fig 6G). Notably, ADORA2A co-immunoprecipitation was not observed when an unrelated antibody was used, or in striatal membranes from GPR37$^{-/-}$ mice, thus validating the specificity of the interaction in native tissue. Next, we assessed the impact of GPR37 expression on ADORA2A functionality *in vivo*. Dopamine (DA) has been implicated in the central processes involved in locomotor activity (LA) regulation and psychomotor behaviors (Beninger, 1983). Interestingly, molecular and functional interactions between Dopamine Receptor 2 (D$_2$R) and ADORA2A in the nucleus accumbens are involved in mediating LA (Ferré & Fuxe, 1992). Since it appears that GPR37 interacts with both ADORA2A (from our interactome) and D$_2$R (Dunham *et al*, 2009), we assessed haloperidol-induced catalepsy in GPR37 knockout mice (GPR37$^{-/-}$) to ascertain the role of this receptor in dopamine-/adenosine-mediated psychomotor behavior. Interestingly, our results showed that in the GPR37$^{-/-}$ mice the catalepsy scores were significantly lower ($P < 0.01$) than in the GPR37$^{+/+}$ mice (Fig 6H). This result suggested a possible role of GPR37 in modulating D$_2$R-mediated neurotransmission. Next, to test the efficacy of ADORA2A in modulating haloperidol-induced catalepsy we treated animals with SCH58261, a selective A$_{2A}$R antagonist (Wardas *et al*, 2003). The administration of SCH58261 (1 mg/kg, i.p.) significantly ($P < 0.01$) reduced the catalepsy score of GPR37$^{+/+}$ animals (Fig 6H), as previously reported (Wardas *et al*, 2003). Importantly, in the GPR37$^{-/-}$ animals, SCH58261 completely abolished the haloperidol-induced catalepsia (Fig 6H). These results suggest that GPR37 might modulate D$_2$R-mediated psychomotor behavior through a putative ADORA2A/GPR37 oligomer *in vivo*.

Taken together, we were able to confirm and functionally characterize two MYTH interactions, thus further demonstrating the utility of our MYTH-based GPCR interactome as a useful resource for disease-related biological research. Annotated interactions from this study are made publicly available in the IID database (Kotlyar *et al*, 2016), with accession number #IID-003170131 (http://ophid.utoron to.ca/iid/SearchPPIs/dataset/IID-003170131).

Discussion

Although GPCRs represent one of the most important protein classes involved in cell signaling, comprehensive studies of their interactors have been lacking because traditional high-throughput interactive proteomics assays do not make use of full-length GPCRs in a natural cellular context. In this study, we report the first systematic interactome analysis of 48 clinically important human GPCRs in their ligand-unoccupied state. We have thus created a foundational GPCR interactome, which is necessary for assessing and understanding complex signaling pathways and for elucidating mechanisms of drug action. Overall, our bioinformatics analysis of the human GPCR interactome, focusing on human diseases, provides critical and focused research directions for GPCR signaling and function.

In establishing the utility of the MYTH system to identify human GPCR interactions, we tested known GPCR-interacting proteins in MYTH, confirming 24% of tested interactions. Though not all tested interactions could be validated using MYTH, this is not unexpected due to differences in the approaches used. For instance, many of the interactions used in our test subset were previously identified using affinity purification (which makes use of cell lysates instead of live cells) or traditional YTH-based approaches (which can typically only be performed using soluble portions of membrane proteins), while MYTH allows for the study of full-length membrane proteins, directly in the membrane environment of a live cell. As such, we expect MYTH to more accurately reflect the natural cellular conditions of membrane proteins, and therefore potentially better identify membrane protein interactions, and detect fewer false positives, than traditional methods. We were still able to recapitulate a substantial percentage of previously identified interactions, however, demonstrating the effectiveness of the MYTH assay for use in the detection of GPCR interactions.

Using our MYTH screening approach, combined with comprehensive bioinformatics analysis, we were able to generate a richly annotated interactome comprised of 987 unique interactions across a total of 686 proteins. Of these, 299 were membrane proteins, demonstrating the effectiveness of MYTH in identifying membrane protein interactions. To further validate our interactome, we successfully carried our orthogonal analysis using co-IP and BRET approaches on a subset of 40 interactions spanning 10 different GPCRs, and were able to confirm a total of 28 of 40 interactions (70%). Failure to validate tested interactions, or identify conditions under which certain interactions could be properly assessed by our orthogonal methods, could be reflective of poor endogenous expression of tested GPCRs and/or aberrant interaction behavior in the unnatural and stringent environment produced upon cellular lysis. Overall, however, the strong confirmation rate obtained using our orthogonal test approaches extensively supports the quality of our MYTH GPCR interactome dataset.

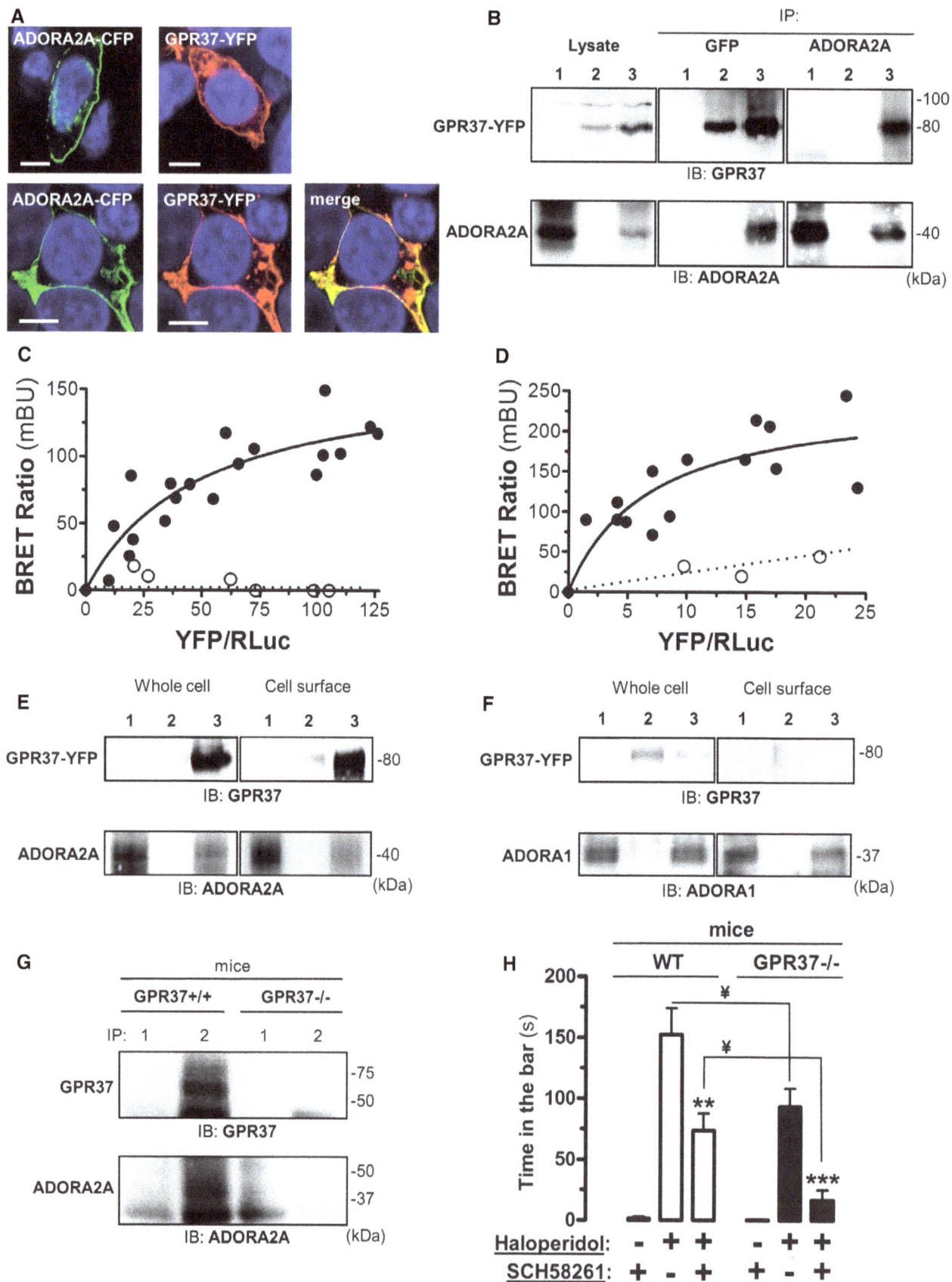

Figure 6.

Figure 6.   Validation of ADORA2A and GPR37 interaction in HEK-293 cells and native tissue.

A   Co-localization of ADORA2A and GPR37 in HEK-293 cells transiently transfected with ADORA2A-CFP, GPR37-YFP, or ADORA2A-CFP plus GPR37-YFP. Transfected cells were analyzed by confocal microscopy. Merged images reveal co-distribution of ADORA2A-CFP and GPR37-YFP (yellow) and DAPI-stained nuclei (blue). Scale bar: 10 μm.

B   Co-immunoprecipitation of ADORA2A and GPR37 from HEK-293 transiently transfected with ADORA2A (lane 1), GPR37-YFP (lane 2) or ADORA2A plus GPR37-YFP (lane 3) using a mouse anti-GFP antibody (2 μg/ml) or a mouse anti-A_{2A}R antibody (1 μg/ml). The crude extracts (Lysate) and immunoprecipitates (IP) were analyzed by SDS–PAGE and immunoblotted (IB) using a rabbit anti-GPR37 (1/2,000) or rabbit anti-A_{2A}R antibody (1/2,000).

C, D   BRET saturation experiments between GPR37-Rluc and ADORA2A-YFP (black circle) or ADORA1-YFP (white circle; C), or ADORA2A-Rluc and GPR37-YFP (black circle) or CD4R-YFP control (white circle; D) in transiently transfected HEK-293. Plotted on the x-axis is the fluorescence value obtained from the YFP, normalized with the luminescence value of the Rluc constructs 10 min after h-coelenterazine (5 μM) incubation, and on the y-axis the corresponding BRET ratio (×1,000). mBU, mBRET units. Data shown are from three independent experiments.

E   Cell surface expression of HEK-293 cells transiently transfected with cDNA encoding ADORA2A (lane 1), GPR37-YFP (lane 2) or ADORA2A plus GPR37-YFP (lane 3). Cell surface proteins were biotinylated and crude extracts (whole cell) and biotinylated proteins were subsequently analyzed by SDS–PAGE and immunoblotted (IB) using a rabbit anti-GPR37 antibody (1/2,000) or a rabbit anti-A_{2A}R antibody (1/2,000).

F   Cell surface expression of HEK-293 cells transiently transfected with cDNA encoding ADORA1 (lane 1), GPR37-YFP (lane 2), or ADORA1 plus GPR37-YFP (lane 3). Cell surface proteins were biotinylated and crude extracts (whole cell) and biotinylated proteins were subsequently analyzed by SDS–PAGE and immunoblotted (IB) using a rabbit anti-GPR37 antibody (1/2,000) or a rabbit anti-A_1R antibody (1/2,000).

G   Co-immunoprecipitation of ADORA2A and GPR37 from C57BL/6J wild-type (GPR37$^{+/+}$) and mutant (GPR37$^{-/-}$) mice striatum using a rabbit anti-FLAG antibody (4 μg/ml; lane 1) or a rabbit anti-GPR37 antibody (4 μg/ml; lane 2). The immunoprecipitates (IP) were analyzed by SDS–PAGE and immunoblotted (IB) using a rabbit anti-GPR37 (1/2,000) or mouse anti-A_{2A}R antibody (1/2,000).

H   Involvement of GPR37 in haloperidol-induced catalepsy. The influence of systemic injection of ADORA2A antagonist SCH 58261 (1 mg/kg, i.p.) on the catalepsy induced by haloperidol (1.5 mg/kg i.p.) was assessed in both WT (GPR37$^{+/+}$) and mutant (GPR37$^{-/-}$) mice as described in Materials and Methods. The data indicate the mean ± SEM (n = 6 per group). Asterisks denote data significantly different from the haloperidol-treated mice: **$P < 0.01$ and ***$P < 0.001$ by one-way ANOVA with Bonferroni multiple comparison post hoc test. In the GPR37$^{-/-}$ mice, the haloperidol plus SCH 58261 group were not significantly different ($P > 0.05$) from the control (i.e. SCH 58261 alone). #$P < 0.01$ by two-way ANOVA with Bonferroni multiple comparison post hoc test for genotype and treatment comparisons.

We also carried out additional, in-depth functional validation on selected GPCR PPIs identified in our interactome using biochemical and cell-based assays as well as knockout and knock-in animals. First, we found that GPRIN2 and GPR37 physically and functionally interact with the 5-HT4d receptor, a promising target for Alzheimer's disease. Activation of 5-HT4d has been shown to modulate α-secretase activity, thus promoting the generation of the amyloid precursor protein (APP)α at the expense of the Alzheimer disease-associated APPβ (Thathiah & De Strooper, 2011). This effect involves the G_s/cAMP signaling pathway (Maillet et al, 2003). Based on our results, the suspected beneficial effect of 5-HT4d on Alzheimer disease development is expected to be amplified in cells co-expressing either GPRIN2 or GPR37.

Another functionally important interactor of GPR37 was ADORA2A, whose co-expression is observed to markedly enhance whole and cell surface expression of GPR37, and whose interaction with GPR37 we validated in native tissue. This interaction is particularly notable in light of a reported interaction between GPR37 and D_2R (Dunham et al, 2009). Both ADORA2A and D_2R are known to co-express (Fuxe et al, 2007) and interact (Hillion et al, 2002) in regions of the brain also expressing GPR37 (i.e. striatum), and are involved in mediating locomotor activity (Ferré & Fuxe, 1992; Lein et al, 2007). Taking our above data, together with our observations pertaining to the effects of GPR37 deletion in mice on haloperidol-induced catalepsy and previous findings that GPR37 affects ligand binding affinities of D_2R (Dunham et al, 2009), we hypothesize that the interaction between GPR37 and ADORA2A (and possibly with D_2R) may play a critical role in D_2R/ADORA2A-mediated psychomotor behavior, and thus may function as a homeostatic regulator of dopaminergic/adenosinergic transmission in vivo.

GPCR–GPCR heterodimerization has been widely reported (Prinster et al, 2005), and the resultant cross-talk and mutual regulation have been important for understanding the functionality of receptors (Fuxe et al, 2014), such as ADORA2A and D_2R in the brain

(Fuxe et al, 2007; Ciruela et al, 2011). Our interactome data, in addition to functionally elucidated receptor interactions described above, report other novel GPCR–GPCR interactions for further investigation by the scientific community, highlighting the importance of large-scale GPCR screens, such as those performed here using MYTH, in identifying new PPIs of potential clinical relevance.

Interestingly, interacting partners were observed to have different effects on GPCR function; for example, GPRIN2 and GPR37 modulate 5-HT4d signaling capacity directly, most likely through an allosteric mechanism, whereas ADORA2A promotes GPR37 expression with important consequences on the well-established and relevant ADORA2A-mediated antagonism of D_2R function in vivo. These focused analyses of novel GPCR interactions further demonstrate the utility of our MYTH-based GPCR interactome as a powerful resource for biological research in this area.

In summary, we report here the largest, most comprehensive interactome study of full-length, human GPCRs carried out directly in the context of living cells. All of the data generated in this work is freely available for use by the scientific community [see the Expanded View and online in the IID database (Kotlyar et al, 2016)]. Additionally, we have performed preliminary functional validation of a selection of PPIs, which should serve as a starting point for further work. Our GPCR-interactome data, particularly when combined with other collaborative projects, such as the GPCR Network (Stevens et al, 2012) and the mapping of GPCR interaction networks performed using other recently developed technologies, such as CHIP-MYTH (Kittanakom et al, 2014) and the mammalian membrane two-hybrid (MaMTH; Petschnigg et al, 2014; Yao et al, 2017), will contribute significantly to our understanding of the chemistry and biology of these clinically relevant proteins, serving as an important tool to further our knowledge of cell signaling processes and helping identify novel biologically important interactions for use in the development and improvement of therapeutic strategies.

Materials and Methods

Full-length human bait generation

Each human GPCR was amplified by PCR and inserted by homologous recombination (Chen *et al*, 1992) in yeast into either of the two bait vectors pCCW-STE or pTMBV (Dualsystems Biotech). The primers used for the pCCW vector are 5'-CCTTTAATTAAGGCCGCC TCGGCCATCTGCAGG-3' (forward) and 5'-CGACATGGTCGACGGT ATCGATAAGCTTGATATCAGCAGTGAGTCATTTGTACTAC-3' (reverse). The primers used for the pTMBV4 vector are 5'-CCAGTGGC TGCAGGGCCGCCTCGGCCAAAGGCCTCCATGG-3' (forward) and 5'-ATGTCGGGGGGGATCCCTCCAGATCAACAAAGATTG-3' (reverse). In MYTH bait vectors, the GPCRs were fused N-terminally to the yeast mating factor alpha signal sequence to target full-length non-yeast membrane proteins to the membrane (King *et al*, 1990). At the C-terminus, the GPCR was fused in-frame with the MYTH tag consisting of a C-terminal ubiquitin (Cub) moiety and LexA-VP16 transcription factor (TF; Fields & Song, 1989; Fashena *et al*, 2000).

Bait validation

The resulting MYTH bait constructs were tested as previously described (Snider *et al*, 2010, 2013). Briefly, the baits were transformed (Gietz & Woods, 2006) into either of the yeast reporter strains THY.AP4 or NMY51. The correct localization of modified baits to the membrane was confirmed by immunofluorescence using (rabbit) anti-VP16 (Sigma Cat# V4388; 1/200); secondary (goat) anti-(rabbit) Cy3 (Cedarlane Cat#111-165-003; 1/500)). Test MYTH was carried out with control interacting (NubI) preys to confirm functionality in MYTH, and with non-interacting (NubG) preys to verify that baits do not self-activate in the absence of interacting prey (Snider *et al*, 2010).

Functionality of select GPCR-Cub-TF baits (Pausch, 1997) was confirmed (Dowell & Brown, 2009) in either wild-type THY.AP4 or the same strain expressing a given GPCR-Cub-TF fusion. Cells were diluted from an overnight culture to an OD_{600} of 0.0625 in minimum SD or SD-Leu media, respectively. The various concentrations of drugs, salmeterol (agonist for ADRB2) or morphine (agonist for OPRM1), were added to a final concentration of 200 μM. The growth rate was monitored by measuring the OD_{600} every 15 min for 24 h by TECAN Sunrise plate reader.

Confirmation of known GPCR interactions by MYTH

Known GPCR-interacting partners were identified from the Integrated Interactions Database (IID) (Kotlyar *et al*, 2016). Gateway compatible ORFs were obtained from the Human ORFeome Collection version 8.1 (Yang *et al*, 2011) and used, via the Gateway system (Life Technologies), to generate either N-terminally tagged preys in pGPR3N (Dualsystems Biotech) or C-terminally tagged preys in pGLigand (created in-house, Stagljar lab) depending on which end is available for tagging. All bait prey interaction tests were carried out using MYTH as previously described (Snider *et al*, 2010) in the NMY51 yeast reporter strain. Note that prior to use in interaction tests with GPCR baits all preys were tested for promiscuity by use of an artificial bait construct that consists of the single-pass transmembrane domain of human T-cell surface glycoprotein

CD4 and the Cub-TF tag (Snider *et al*, 2010) and by use of the yeast protein RGT2.

Membrane yeast two-hybrid (MYTH) screens

Bait containing yeast were transformed in duplicate with the human fetal brain DUALmembrane cDNA library in the NubG-x orientation (DualSystems Biotech) as previously described (Snider *et al*, 2010) and plated onto synthetic dropout minus tryptophan, leucine, adenine, and histidine (SD-Trp-Leu-Ade-His) plates with various amounts of 3-amino-1,2,4-triazole (3-AT) as assessed by the NubG/I control test for each individual bait. Transformants were picked and spotted onto SD-Trp-Leu-Ade-His plates containing 3-AT and X-Gal dissolved in *N,N*-dimethyl formamide. Blue colonies, expressing putative interacting preys, were used to inoculate overnight liquid cultures (SD-Trp) and plasmid DNA extracted. Plasmid DNA was used to transform *E. coli*, DH5alpha strain for amplification. Plasmid DNA was extracted once more and sent for sequencing as well as used in the bait dependency test to rule out spurious interactors, as described previously (Snider *et al*, 2010).

Filtering interactions

To reduce the number of false positives, we eliminated detected interactions involving preys that carry out signal peptide processing (GO:0006465) and ribosomal contaminants (Glatter *et al*, 2009). We identified these preys using Gene Ontology (GO; Ashburner *et al*, 2000) annotations from the UniProt-GO Annotation database (Matthews *et al*, 2009; Dimmer *et al*, 2012), downloaded through the EMBL-EBI QuickGO browser (Binns *et al*, 2009; http://www. ebi.ac.uk/QuickGO/GTerm?id = GO:0006465#term = annotation), on September 10, 2016.

Identifying previously known interactions

Overlap between detected interactions and interactions already reported in previous studies was identified using the IID database (Kotlyar *et al*, 2016) ver. 2016-03 (http://ophid.utoronto.ca/iid).

Annotating interacting proteins: membrane localization

Baits and preys localized to the plasma membrane were identified using GO annotations from the UniProt-GO Annotation database (Dimmer *et al*, 2012), obtained through the EMBL-EBI QuickGO browser (Binns *et al*, 2009; http://www.ebi.ac.uk/QuickGO/GTe rm?id = GO:0006465#term = annotation) on August 31, 2016.

Process annotations and enrichment analysis

Baits and preys were annotated with GO Slim process terms from the *goslim_generic* set (http://www.ebi.ac.uk/QuickGO/GMultiTe rm#tab = choose-terms; Table EV3), downloaded on August 31, 2016.

Pathway annotations

Pathway annotations for baits and preys, as well as pathway enrichment analysis, were performed using the pathDIP database

(Rahmati et al, 2017) ver. 2.5 (http://ophid.utoronto.ca/pathDIP), using the setting "Extended pathway associations" with default parameters. P-values were FDR-corrected using the Benjamini–Hochberg method.

Disease annotations and enrichment analysis

Disease annotations for baits and preys were downloaded from the DisGeNET database (Piñero et al, 2015) v4.0, on Aug. 31, 2016. Disease enrichment of preys was assessed by calculating hypergeometric P-values (using the human genome as the background population), and correcting for multiple testing using the Benjamini–Hochberg method.

Molecular function and biological process annotations and enrichment analysis

Molecular function and biological process Gene Ontology annotations were downloaded from Gene Ontology Consortium (Gene Ontology Consortium, 2015) on November 30, 2016. Enrichment of preys for molecular functions was calculated using the topGO library version 2.24.0 in R version 3.3.1 (Alexa & Rahnenfuhrer, 2016). A topGOdata object was created with nodeSize = 10, and the runTest function was used with the default algorithm (weight01) and statistic = fisher. P-values were adjusted for multiple testing using the Benjamini–Hochberg method. Enrichment of preys for biological processes was calculated the same way.

Domain annotation and enrichment analysis

InterPro domain annotations were obtained from UniProt release 2016_11 (Mitchell et al, 2015; UniProt Consortium, 2015). Domain enrichment of preys was assessed by calculating hypergeometric P-values (using the human proteome as the background population), and correcting for multiple testing using the Benjamini–Hochberg method.

Domain pairs enriched among interacting bait–prey pairs were identified in two steps. First, sets of co-occurring domains were identified for baits; each set comprised domains that always occurred together on baits. Similarly, sets of co-occurring domains were identified on preys. Domains that did not always co-occur with others were considered domain sets of length 1. Enrichment was subsequently calculated for pairs of domain sets—one set on baits and the other on preys. Domain sets were identified for three reasons: (i) to avoid redundant results from different domains representing the same proteins, (ii) to avoid excessive multiple testing penalties from non-independent tests, and (iii) for easier interpretation of results, since a domain set clarifies that enrichment analysis cannot distinguish between domains within the set. After domain sets were identified, P-values were calculated for domain set pairs using hypergeometric probability with the following parameters: N = the number of possible interactions involving baits (number of baits × size of human proteome), M = the number of detected interactions, n = the number of possible pairings between the bait domain set and the prey domain set (number of baits with domain set × number of human proteins with prey domain set), and m = number of interacting bait–prey pairs with corresponding domain sets. Adjusted P-values were calculated using the Benjamini–Hochberg method.

Drug target enrichment and drug category enrichment

Drug targets and drug therapeutic categories were downloaded from DrugBank version 5 (Wishart et al, 2006). Enrichment of drug targets among GPCR baits was calculated as a hypergeometric P-value, using the following parameters: the number of human protein-coding genes in the HGNC database (Gray et al, 2015) (19,008), the number drug targets in DrugBank (4,333), the number of baits (48), and the number of baits that are drug targets (28).

Enrichment of therapeutic categories among baits and preys was calculated as hypergeometric P-values using the following parameters: the number of human protein-coding genes in the HGNC database (Gray et al, 2015) (19,008), the number of targets in a therapeutic category, the number of baits and preys (686), and the number of baits and preys that are targets in the category. We calculated Q-values (P-values adjusted for multiple testing) using the Benjamini–Hochberg method.

Drugs sales and prescription numbers were obtained from Medscape (2015).

PPI predictions

Predictions were obtained using the FpClass algorithm (Kotlyar et al, 2015): a probabilistic method that integrates diverse PPI evidence including compatibility of protein domains, gene co-expression, and functional similarity, as well as other methods integrated in IID (version 2016-03, http://ophid.utoronto.ca/iid; Kotlyar et al, 2016). Resulting networks were visualized in NAVi-GaTOR 3.0 (http://ophid.utoronto.ca/navigator; Brown et al, 2009).

Confirmation of interactions by co-immunoprecipitation

Approach 1—Endogenous baits and transiently transfected FLAG-tagged preys

293T cells were maintained in Dulbecco's modified Eagle's medium (DMEM) containing 10% FBS, 100 U penicillin, and 100 μg/ml streptomycin (Fisher Scientific, cat# SV30010) and split at 80% confluence. To co-immunoprecipitate GPCRs with their preys, plasmids encoding FLAG-tagged preys were transiently transfected in 293T cells and their interaction with GPCR was detected using Western blotting with anti-GPCR antibodies.

Briefly, 293T cells were plated at 40% confluence overnight. On the following morning, cells were transfected using calcium phosphate [$Ca_3(PO_4)_2$] kit ProFection from Promega (cat# E1200) following manufacturer's instructions. 70 μg of plasmid DNA was added to $CaCl_2$ and water, and the mixture was added to HEPES-buffered saline while vortexing. The mixture was incubated at room temperature for 30 min. Prior to adding to cells, the mixture was vortexed again. After 24 h post-transfection, 2 × 150 mm dishes of 293T cells/plasmid were harvested and the cells were washed with ice-cold PBS. After that, cells were cross-linked with 0.5 mM DSP at room temperature for 30 mins followed by quenching excessive DSP with a buffer containing 0.1 M Tris–HCl, pH 7.5, and 2 mM EDTA. Detached cells were centrifuged at 400 g for 10 min at 4°C. The cell pellet was lysed in RIPA buffer containing 1× protease inhibitor cocktail (Sigma Aldrich, cat# P2714) on ice for 30 min with occasional agitation. To aid lysis, cells were passed through a 21G needle 10×. Lysate was cleared by centrifugation at 16,000 g for

15 min at 4°C. A volume of cell lysate containing 10 mg protein was adjusted to 1 ml with RIPA containing 1× protease inhibitor cocktail and 3 μg of each anti-GPCR receptor antibody were added. The tube rotated for 1 h at 4°C followed by addition of 100 μl of μMACS protein-G magnetic microbeads (Miltenyi, cat# 130-071-101) with continued rotation for additional 4 h at 4°C. μMACS columns (Miltenyi, cat# 130-092-444) were equilibrated with RIPA 1× protease inhibitor cocktail. The microbeads suspension was passed through the columns, and the retained microbeads were washed 3× with 800 μl of RIPA 0.1% of detergents and 1× protease inhibitor cocktail followed by another 2× washes with 500 μl detergent-free RIPA containing 1× protease inhibitor cocktail only. Proteins bound to the microbeads were released by addition of 25 μl Laemmli loading buffer at 95°C 2×. Eluates were analyzed using SDS-PAGE and visualized using SuperSignal West Femto Maximum Sensitivity Substrate (Thermo Fisher, cat# 34094).

Approach 2—Endogenous baits and preys

Ten 150-mm dishes of HEK-293 cells were harvested and centrifuged at 400 g for 10 min. The cell pellet was resuspended in 15 ml phosphate-buffered saline (PBS) and mixed with an equal volume of cross-linking reagent (1 mM dithiobis-succinimidyl propionate prepared in PBS). After 30-min incubation, the cross-linked cells pelleted by centrifugation at 400 g were lysed in IPLB (immunoprecipitation lysis buffer containing 1% digitonin and 1× protease inhibitor cocktail) for 30 min. The lysates were then centrifuged at 16,000 g for 15 min at 4°C. The cell lysate containing ~10 mg of protein was adjusted to 1 ml with IPLB (containing 1% digitonin and 1× protease inhibitor cocktail) and 3 μg of antibody specific to the target protein was added to the mixture. The samples were incubated with 100 μl of μMACS protein-G magnetic beads followed by 5-h gentle rotation at 4°C. The bead suspension was passed through the μMACS columns (equilibrated with IPLB containing 1% digitonin and 1× protease inhibitor), and the retained beads were washed three times with 800 μl of IPLB (0.1% digitonin and 1× protease inhibitor) followed by another two washes with 500 μl IPLB (1× protease inhibitor only). Co-purifying protein that bound to the beads was eluted by the addition of 25 μl Laemmli loading buffer at 95°C, and analyzed by SDS-PAGE and immunoblotting using protein-specific antibody.

Antibodies used in co-immunoprecipitation experiments

Santa Cruz: OPRL1 (sc-15309), TSHR (sc-13936), OPRM1 (sc-15310), AGTR1 (sc-1173-G), PTAFR (sc-20732), C5L2 (sc-368573), HRH (sc-20633), CHRM5 (sc-9110), OXTR (sc-33209). Abcam: ADRB2 (ab36956), HNRPK (ab52600), F2RL (ab124227), TTYH1 (ab57582), PRNP (ab52604), MGLL (ab24701), ATP2A2 (ab2861), FA2H (ab54615), HSPA1B (ab79852). Cell Signaling: GABBR1 (3835). ProteinTech: GPR37 (14820-1-AP), FZD7 (16974-1-AP).

Confirmation of interactions by BRET

To confirm select interactions using BRET as an orthogonal validation assay, GPCR interactors identified in MYTH assays were fused to GFP2, a blue-shifted variant of GFP, to act as BRET acceptor, and GPCR receptors to RLucII, a brighter Renilla luciferase mutant, to act as donor, then plotted as increasing BRET levels compared to

GFP/Rluc, as previously described (Mercier et al, 2002; Loening et al, 2006; Breton et al, 2010).

5-HT4d experiments

Materials

The cDNAs encoding human GPR37 and GPRIN2 were purchased from UMR cDNA Resource Center. The 5-HT4d-Rluc, 5-HT4d-YFP, and HA-CCR5 constructs have been described elsewhere (Berthouze et al, 2005; Tadagaki et al, 2012). An N-terminally 6xMyc tagged version of GPRIN2 and GPR37 and C-terminally YFP tagged GPR37-YFP and GPRIN2-YFP fusion proteins were obtained by PCR using the Phusion High-Fidelity DNA Polymerase (Finnzymes). All constructs were inserted in the pcDNA3.1 expression vector and verified by sequencing. The C-terminally deleted GPRIN2ΔCter construct was obtained by mutagenesis by introducing a stop codon resulting in a truncated protein of 149 amino acids.

Co-immunoprecipitation

HEK-293 cells transiently transfected with 5-HT4d-YFP and myc-GPRIN2 or GPR37 were analyzed in the presence and absence of 1 μM 5-HT for 15 min and processed for immunoprecipitation using a monoclonal anti-GFP antibody. Crude extracts and immunoprecipitates were analyzed by SDS–PAGE and immunoblotted using rabbit anti-GFP or anti-myc antibodies.

BRET

BRET donor saturation curves were performed in HEK-293 cells by co-transfecting a fixed amount of 5-HT4d-Rluc and increasing amounts of 5-HT4d-YFP, GPR37-YFP, and GPRIN2-YFP as described previously (Maurice et al, 2010).

Fluorescence microscopy

HeLa cells expressing 5-HT4d-YFP and Myc-GPR37 or Myc-GPRIN2 were fixed, permeabilized with 0.2% Triton X-100, nuclei stained with DAPI (blue) and incubated with monoclonal anti-Myc antibody (Sigma, St Louis, MO; 2 mg/ml) and subsequently with a Cy3-coupled secondary antibody. GFP, Cy3, and DAPI labeling was observed by confocal microscopy.

Signaling assays

ERK1/2 activation and cyclic AMP levels were determined in HEK-293 cells as described previously (Guillaume et al, 2008).

ADORA2A experiments

Materials

The cDNA encoding the human GPR37 (Unigene ID: Hs.725956; Source BioScience, Nottingham, UK) was amplified and subcloned into the HindIII/EcoRI restriction sites of the pEYFP vector (Invitrogen, Carlsbad, CA, USA) using the iProof High-Fidelity DNA polymerase (Bio-Rad, Hercules, CA, USA) and the following primers: FGPR37 (5′-CGCAAGCTTATGCGAGCCCCGG-3′) and RGPRYFP (5′-CGCGAATTCCGCAATGAGTTCCG-3′). GPR37 was also subcloned in the HindIII/KpnI restriction sites of the pRluc-N1 vector (Perkin-Elmer, Waltham, MA, USA) using the following primers: FGPR37 and RGPRLuc (5′-CGCGGTACCGCGCAATGAGTTCCG-3′).

The constructs for the human adenosine A2A receptor (namely, ADORA2A-YFP and ADORA2A-Rluc) were obtained as previously described (Gandia et al, 2008) and ADORA2A-CFP was obtained by subcloning the adenosine receptor from ADORA2A-YFP into the pECFP-N1 plasmid.

A homemade rabbit anti-GPR37 polyclonal antibody (Lopes et al, 2015) was used. Other antibodies used were rabbit anti-$A_{2A}R$ (Ciruela et al, 2004), mouse anti-$A_{2A}R$ (05-717, Millipore, Temecula, CA, USA), rabbit anti-FLAG (F7425, Sigma) and rabbit anti-A_1R (PA1-041A, Affinity BioReagents, Golden, CO, USA).

C57BL/6J wild-type and GPR37$^{-/-}$ mice with a C57BL/6J genetic background (Strain Name: B6.129P2-Gpr37tm1Dgen/J; The Jackson Laboratory, Bar Harbor, ME, U.S.A.) were used. Mice were housed in standard cages with ad libitum access to food and water, and maintained under controlled standard conditions (12-h dark/light cycle starting at 7:30 AM, 22°C temperature and 66% humidity). The University of Barcelona Committee on Animal Use and Care approved the protocol, and the animals were housed and tested in compliance with the guidelines described in the Guide for the Care and Use of Laboratory Animals (Clark et al, 1997) and following the European Community, law 86/609/CCE.

Immunocytochemistry

HEK-293 cells were transiently transfected with ADORA2A-CFP, GPR37-YFP, or ADORA2A-CFP plus GPR37-YFP using the Trans-Fectin Lipid Reagent (Bio-Rad) and following the instructions provided by the manufacturer. The cells were analyzed by confocal microscopy 48 h after transfection. Superimposition of images (merge) reveals co-distribution of ADORA2A-CFP and GPR37-YFP in yellow and DAPI-stained nuclei in blue.

Co-immunoprecipitation

Membrane extracts from HEK-293 cells and C57BL/6J mouse striatum were obtained as described previously (Burgueño et al, 2003). Membranes were solubilized in ice-cold radioimmunoassay (RIPA) buffer (150 mM NaCl, 1% NP-40, 50 mm Tris, 0.5% sodium deoxycholate, and 0.1% SDS, pH 8.0) for 30 min on ice in the presence of protease inhibitor (Protease Inhibitor Cocktail Set III, Millipore, Temecula, CA, USA). The solubilized membrane extract was then centrifuged at 13,000 ×g for 30 min, and the supernatant was incubated overnight with constant rotation at 4°C with the indicated antibody. Then, 50 μl of a suspension of Protein A–agarose (Sigma) or TrueBlot anti-rabbit Ig IP beads (eBioscience, San Diego, CA) was added and incubated for another 2 h. The beads were washed with ice-cold RIPA buffer and immune complexes were dissociated, transferred to polyvinylidene difluoride membranes and probed with the indicated primary antibodies followed by horseradish peroxidase (HRP)-conjugated secondary antibodies. The immunoreactive bands were detected using Pierce ECL Western Blotting Substrate (Thermo Fisher Scientific) and visualized in a LAS-3000 (FujiFilm Life Science).

BRET

For BRET saturation experiments, HEK-293 cells transiently transfected with a constant amount of cDNA encoding the Rluc constructs and increasing amounts of YFP tagged proteins were rapidly washed twice in PBS, detached and resuspended in Hank's balanced salt solution (HBSS) buffer (137 mM NaCl, 5.4 mM KCl, 0.25 mM Na_2HPO_4, 0.44 mM KH_2PO_4, 1.3 mM $CaCl_2$, 1.0 mM $MgSO_4$, 4.2 mM $NaHCO_3$, pH 7.4), containing 10 mM glucose and processed for BRET determinations using a POLARstar Optima plate-reader (BMG Labtech, Durham, NC, USA; Ciruela et al, 2015) or Mithras plate reader (Berthold Technologies; Cecon et al, 2015).

Cell surface expression

HEK-293 cells were transiently transfected with the cDNA encoding ADORA2A, ADORA1, GPR37-YFP, ADORA2A plus GPR37-YFP or ADORA1 plus GPR37-YFP. Cell surface labeling was performed by biotinylation experiments (Burgueño et al, 2003). Crude extracts and biotinylated proteins were subsequently analyzed by SDS–PAGE and immunoblotted using a rabbit anti-GPR37 antibody (1/2,000), a rabbit anti-$A_{2A}R$ antibody (1/2,000), or a rabbit anti-A_1R antibody (1/2,000). The primary bound antibody was detected as described before.

Catalepsy score

Catalepsy behavior was induced by the D_2R antagonist haloperidol (1.5 mg/kg, i.p.), as previously described (Chen et al, 2001). Mice used in the catalepsy test were 2-month-old males. The animals were randomly distributed among the experimental groups. Fifteen min before animals were administered either saline or SCH58261 (1 mg/kg, i.p.), an $A_{2A}R$ antagonist. The cataleptic response was measured as the duration of an abnormal upright posture in which the forepaws of the mouse were placed on a horizontal wooden bar (0.6 cm of diameter) at 4.5 cm high from the floor. The latency to move at least one of the two forepaws was recorded 2 h after haloperidol administration. The test was carried out by an experimenter who was blind to the identity of treatments and the cataleptic time latency was automatically recorded and counted by an independent researcher. A cutoff time of 180 s was imposed. Catalepsy testing was performed under dim (16 lux) light conditions. The sample size was initially set as five determinations per experimental condition. Subsequently, the statistical power was calculated using the IBM SPSS Statistics (version 24) software. Accordingly, the sample size was then designed to achieve a minimum of 80% statistical power.

Acknowledgements

We thank K. Seuwen (Novartis) for discussions during initiation phase of the project and for providing bait cDNA. The work in the Stagljar laboratory was supported by grants from the Canadian Institutes of Health Research (CIHR, #MOP-106527), Canadian Foundation for Innovation, Natural Sciences and Engineering Research Council of Canada, Ontario Genomics Institute, Canadian Cystic Fibrosis Foundation, Canadian Cancer Society and Ontario Research Fund (University Health Network). The work in the Jockers laboratory was performed within the Département Hospitalo-Universitaire (DHU) AUToimmune and HORmonal diseaseS and supported by grants from the Institut National de la Santé et de la Recherche Médicale (INSERM), the Fondation Recherche Médicale (Equipe FRM DEQ20130326503 to R.J.), the Association pour la Recherche sur le Cancer (ARC, SFI20121205906, to R.J.), a doctoral fellowship from the CODDIM 2009 (Région Ile-de-France to A.B.C.) and a research fellowship of the Université Paris Descartes (to K.T.) and the "Who am I?" laboratory of excellence No. ANR-11-LABX-0071 funded by the French Government through its "Investments for the Future" program operated by The French National Research Agency (ANR) under grant ANR-11-IDEX-0005-01. The Jurisica laboratory was supported by Ontario Research Fund (GL2-01-030), Canada Foundation for Innovation (CFI #12301, #203373, #29272,

#225404), Canada Research Chair Program (CRC #203373 and #225404), Natural Sciences Research Council (NSERC #203475) and IBM. The work from Babu's laboratory was supported by the grants from CIHR (MOP# 132191) and Saskatchewan Health Research Foundation (SHRF #2895). The Ciruela laboratory was supported by MINECO/ISCIII (SAF2014-55700-P, PCIN-2013-019-C03-03, and PIE14/00034), the Catalan Government (2014 SGR 1054), ICREA (ICREA Academia-2010), Fundació la Marató de TV3 (Grant 20152031), and FWO (SBO-140028). Also, J.G., X.M., and F.C. belong to the "Neuropharmacology and Pain" accredited research group (Generalitat de Catalunya, 2014 SGR 1251). The Przulj laboratory was supported by the European Research Council (ERC) Starting Independent Researcher Grant 278212, the National Science Foundation (NSF) Cyber-Enabled Discovery and Innovation (CDI) grant OIA-1028394, the Serbian Ministry of Education and Science Project III44006, and ARRS Project J1-5454.

Author contributions

IS designed the project and was involved in the writing of the manuscript, and IJ managed the bioinformatics analysis of the interactome. KS and JS compiled and managed data, were actively involved in the analysis, and wrote the bulk of the manuscript. SK[#] created baits, carried out screening, co-immunoprecipitation, growth curve, and co-localization experiments. VW, DA, and JM carried out bait generation, bait validation, and screening. VW was also involved in bait localization and data compilation, and, with ZY, known PPI confirmations. MK, DO, and IJ performed bioinformatic analysis and generated the interactomes. NP analyzed the structural complexity of the interactome. RHM VD, HA, and SA from the Babu laboratory carried out the co-immunoprecipitation experiments to confirm interactions, and the experiments were overseen by MBo, RJ oversaw the serotonin experiments and critically reviewed the manuscript. PM, AB-C, and AO carried out the serotonin experiments, and KT performed the serotonin BRET experiments. FC oversaw the adenosine experiments carried out by JG and XM, and critically reviewed the manuscript. MBa, SA, and HK were involved in the preparation of the interactome.

References

Alexa A, Rahnenfuhrer J (2016) topGO: enrichment analysis for gene ontology. R Package. Version 2.24.0

Ashburner M, Ball CA, Blake JA, Botstein D, Butler H, Cherry JM, Davis AP, Dolinski K, Dwight SS, Eppig JT, Harris MA, Hill DP, Issel-Tarver L, Kasarskis A, Lewis S, Matese JC, Richardson JE, Rubin GM, Sherlock G (2000) Gene ontology: tool for the unification of biology. *Nat Genet* 25: 25 – 29

Ayoub MA, Pin J-P (2013) Interaction of protease-activated receptor 2 with G proteins and β-arrestin 1 studied by bioluminescence resonance energy transfer. *Front Endocrinol* 4: 196

Beninger RJ (1983) The role of dopamine in locomotor activity and learning. *Brain Res* 287: 173 – 196

Benleulmi-Chaachoua A, Chen L, Sokolina K, Wong V, Jurisica I, Emerit MB, Darmon M, Espin A, Stagljar I, Tafelmeyer P, Zamponi GW, Delagrange P, Maurice P, Jockers R (2016) Protein interactome mining defines melatonin MT1 receptors as integral component of presynaptic protein complexes of neurons. *J Pineal Res* 60: 95 – 108

Berthouze M, Ayoub M, Russo O, Rivail L, Sicsic S, Fischmeister R, Berque-Bestel I, Jockers R, Lezoualc'h F (2005) Constitutive dimerization of human serotonin 5-HT4 receptors in living cells. *FEBS Lett* 579: 2973 – 2980

Binns D, Dimmer E, Huntley R, Barrell D, O'Donovan CAR (2009) QuickGO: a web-based tool for gene ontology searching. *Bioinformatics* 25: 3045 – 3046

Breton B, Sauvageau É, Zhou J, Bonin H, Le Gouill C, Bouvier M (2010) Multiplexing of multicolor bioluminescence resonance energy transfer. *Biophys J* 99: 4037 – 4046

Brinks HL, Eckhart AD (2010) Regulation of GPCR signaling in hypertension. *Biochim Biophys Acta* 1802: 1268 – 1275

Brown KR, Otasek D, Ali M, McGuffin MJ, Xie W, Devani B, Toch IL, van Jurisica I (2009) NAViGaTOR: network analysis, visualization and graphing Toronto. *Bioinformatics* 25: 3327 – 3329

Burgueño J, Blake DJ, Benson MA, Tinsley CL, Esapa CT, Canela EI, Penela P, Mallol J, Mayor F, Lluis C, Franco R, Ciruela F (2003) The adenosine A2A receptor interacts with the actin-binding protein alpha-actinin. *J Biol Chem* 278: 37545 – 37552

Canela L, Luján R, Lluís C, Burgueño J, Mallol J, Canela EI, Franco R, Ciruela F (2007) The neuronal Ca(2+) -binding protein 2 (NECAB2) interacts with the adenosine A(2A) receptor and modulates the cell surface expression and function of the receptor. *Mol Cell Neurosci* 36: 1 – 12

Cecon E, Chen M, Marcola M, Fernandes PAC, Jockers R, Markus RP (2015) Amyloid peptide directly impairs pineal gland melatonin synthesis and melatonin receptor signaling through the ERK pathway. *FASEB J* 29: 2566 – 2582

Chen D, Yang B, Kuo T (1992) One-step transformation of yeast in stationary phase. *Curr Genet* 21: 83 – 84

Chen LT, Gilman AG, Kozasa T (1999) A candidate target for G protein action in brain. *J Biol Chem* 274: 26931 – 26938

Chung KY, Day PW, Vélez-Ruiz G, Sunahara RK, Kobilka BK (2013) Identification of GPCR-interacting cytosolic proteins using HDL particles and mass spectrometry-based proteomic approach. *PLoS ONE* 8: e54942

Ciruela F, Gómez-Soler M, Guidolin D, Borroto-Escuela DO, Agnati LF, Fuxe K, Fernández-Dueñas V (2011) Adenosine receptor containing oligomers: their role in the control of dopamine and glutamate neurotransmission in the brain. *Biochim Biophys Acta* 1808: 1245 – 1255

Daulat AM, Maurice P, Froment C, Guillaume J-L, Broussard C, Monsarrat B, Delagrange P, Jockers R (2007) Purification and identification of G protein-coupled receptor protein complexes under native conditions. *Mol Cell Proteomics* 6: 835 – 844

Daulat AM, Maurice P, Jockers R (2009) Recent methodological advances in the discovery of GPCR-associated protein complexes. *Trends Pharmacol Sci* 30: 72 – 78

Daulat AM, Maurice P, Jockers R (2011) Tandem affinity purification and identification of GPCR-associated protein complexes. *Methods Mol Biol* 746: 399 – 409

Deribe YL, Wild P, Chandrashaker A, Curak J, Schmidt MHH, Kalaidzidis Y, Milutinovic N, Kratchmarova I, Buerkle L, Fetchko MJ, Schmidt P, Kittanakom S, Brown KR, Jurisica I, Blagoev B, Zerial M, Stagljar I, Dikic I (2009) Regulation of epidermal growth factor receptor trafficking by lysine deacetylase HDAC6. *Sci Signal* 2: ra84

Dimmer EC, Huntley RP, Alam-Faruque Y, Sawford T, O'Donovan C, Martin MJ, Bely B, Browne P, Mun Chan W, Eberhardt R, Gardner M, Laiho K, Legge D, Magrane M, Pichler K, Poggioli D, Sehra H, Auchincloss A, Axelsen K, Blatter M-C et al (2012) The UniProt-GO annotation database in 2011. *Nucleic Acids Res* 40: D565 – D570

Dowell SJ, Brown AJ (2009) Yeast assays for G protein-coupled receptors. In *G protein-coupled receptors in drug discovery*, Leifert WR (ed.), pp 213 – 229. Totowa, NJ: Humana Press

Dunham JH, Meyer RC, Garcia EL, Hall RA (2009) GPR37 surface expression enhancement via N-terminal truncation or protein-protein interactions. *Biochemistry* 48: 10286–10297

Dupré DJ, Robitaille M, Rebois RV, Hébert TE (2009) The role of Gbetagamma subunits in the organization, assembly, and function of GPCR signaling complexes. *Annu Rev Pharmacol Toxicol* 49: 31–56

Dusonchet J, Bensadoun J-C, Schneider BL, Aebischer P (2009) Targeted overexpression of the parkin substrate Pael-R in the nigrostriatal system of adult rats to model Parkinson's disease. *Neurobiol Dis* 35: 32–41

Fashena SJ, Serebriiskii IG, Golemis EA (2000) LexA-based two-hybrid systems. *Methods Enzymol* 328: 14–26

Ferré S, Fuxe K (1992) Dopamine denervation leads to an increase in the intramembrane interaction between adenosine A2 and dopamine D2 receptors in the neostriatum. *Brain Res* 594: 124–130

Fields S, Song O (1989) A novel genetic system to detect protein-protein interactions. *Nature* 340: 245–246

Fuxe K, Ferré S, Genedani S, Franco R, Agnati LF (2007) Adenosine receptor-dopamine receptor interactions in the basal ganglia and their relevance for brain function. *Physiol Behav* 92: 210–217

Fuxe K, Borroto-Escuela DO, Romero-Fernandez W, Palkovits M, Tarakanov AO, Ciruela F, Agnati LF (2014) Moonlighting proteins and protein-protein interactions as neurotherapeutic targets in the G protein-coupled receptor field. *Neuropsychopharmacology* 39: 131–155

Gandía J, Fernández-Dueñas V, Morató X, Caltabiano G, González-Muñiz R, Pardo L, Stagljar I, Ciruela F (2013) The Parkinson's disease-associated Gpr37 receptor-mediated cytotoxicity is controlled by its intracellular cysteine-rich domain. *J Neurochem* 125: 362–372

Gandia J, Galino J, Amaral OB, Soriano A, Lluís C, Franco R, Ciruela F (2008) Detection of higher-order G protein-coupled receptor oligomers by a combined BRET-BiFC technique. *FEBS Lett* 582: 2979–2984

Gavarini S, Bécamel C, Chanrion B, Bockaert J, Marin P (2004) Molecular and functional characterization of proteins interacting with the C-terminal domains of 5-HT2 receptors: emergence of 5-HT2 "receptosomes". *Biol Cell* 96: 373–381

Gene Ontology Consortium (2015) Gene Ontology Consortium: going forward. *Nucleic Acids Res* 43: D1049–D1056

Gietz RD, Woods RA (2006) Yeast transformation by the LiAc/SS Carrier DNA/PEG method. *Methods Mol Biol* 313: 107–120

Gisler SM, Kittanakom S, Fuster D, Wong V, Bertic M, Radanovic T, Hall RA, Murer H, Biber J, Markovich D, Moe OW, Stagljar I (2008) Monitoring protein-protein interactions between the mammalian integral membrane transporters and PDZ-interacting partners using a modified split-ubiquitin membrane yeast two-hybrid system. *Mol Cell Proteomics* 7: 1362–1377

Glatter T, Wepf A, Aebersold R, Gstaiger M (2009) An integrated workflow for charting the human interaction proteome: insights into the PP2A system. *Mol Syst Biol* 5: 237

Gray KA, Yates B, Seal RL, Wright MW, Bruford EA (2015) Genenames.org: the HGNC resources in 2015. *Nucleic Acids Res* 43: D1079–D1085

Guillaume J-L, Daulat AM, Maurice P, Levoye A, Migaud M, Brydon L, Malpaux B, Borg-Capra C, Jockers R (2008) The PDZ protein mupp1 promotes Gi coupling and signaling of the Mt1 melatonin receptor. *J Biol Chem* 283: 16762–16771

Hall RA, Lefkowitz RJ (2014) Regulation of G protein-coupled receptor signaling by scaffold proteins. *Circ Res* 91: 672–680

Hamdan FF, Percherancier Y, Breton B, Bouvier M (2006) Monitoring protein-protein interactions in living cells by bioluminescence resonance energy transfer (BRET). *Curr Protoc Neurosci* 34: 5.23.1–5.23.20

Hillion J, Canals M, Torvinen M, Casado V, Scott R, Terasmaa A, Hansson A, Watson S, Olah ME, Mallol J, Canela EI, Zoli M, Agnati LF, Ibanez CF, Lluis C, Franco R, Ferre S, Fuxe K (2002) Coaggregation, cointernalization, and codesensitization of adenosine A2A receptors and dopamine D2 receptors. *J Biol Chem* 277: 18091–18097

Huang X, Dai FF, Gaisano G, Giglou K, Han J, Zhang M, Kittanakom S, Wong V, Wei L, Showalter AD, Sloop KW, Stagljar I, Wheeler MB (2013) The identification of novel proteins that interact with the GLP-1 receptor and restrain its activity. *Mol Endocrinol* 27: 1550–1563

Insel PA, Tang C-M, Hahntow I, Michel MC (2007) Impact of GPCRs in clinical medicine: monogenic diseases, genetic variants and drug targets. *Biochim Biophys Acta* 1768: 994–1005

King K, Dohlman HG, Thorner J, Caron MG, Lefkowitz RJ (1990) Control of yeast mating signal transduction by a mammalian beta 2-adrenergic receptor and Gs alpha subunit. *Science* 250: 121–123

Kittanakom S, Barrios-Rodiles M, Petschnigg J, Arnoldo A, Wong V, Kotlyar M, Heisler LE, Jurisica I, Wrana JL, Nislow C, Stagljar I (2014) CHIP-MYTH: a novel interactive proteomics method for the assessment of agonist-dependent interactions of the human β_2-adrenergic receptor. *Biochem Biophys Res Commun* 445: 746–756

Kotlyar M, Pastrello C, Pivetta F, Lo Sardo A, Cumbaa C, Li H, Naranian T, Niu Y, Ding Z, Vafaee F, Broackes-Carter F, Petschnigg J, Mills GB, Jurisicova A, Stagljar I, Maestro R, Jurisica I (2015) In silico prediction of physical protein interactions and characterization of interactome orphans. *Nat Methods* 12: 79–84

Kotlyar M, Pastrello C, Sheahan N, Jurisica I (2016) Integrated interactions database: tissue-specific view of the human and model organism interactomes. *Nucleic Acids Res* 44: D536–D541

Lagerström MC, Schiöth HB (2008) Structural diversity of G protein-coupled receptors and significance for drug discovery. *Nat Rev Drug Discov* 7: 339–357

Lam MHY, Snider J, Rehal M, Wong V, Aboualizadeh F, Drecun L, Wong O, Jubran B, Li M, Ali M, Jessulat M, Deineko V, Miller R, Lee ME, Park H-O, Davidson A, Babu M, Stagljar I (2015) A comprehensive membrane interactome mapping of sho1p reveals Fps1p as a novel key player in the regulation of the HOG pathway in *S. cerevisiae. J Mol Biol* 427: 2088–2103

Lappano R, Maggiolini M (2011) G protein-coupled receptors: novel targets for drug discovery in cancer. *Nat Rev Drug Discov* 10: 47–60

Lein ES, Hawrylycz MJ, Ao N, Ayres M, Bensinger A, Bernard A, Boe AF, Boguski MS, Brockway KS, Byrnes EJ, Chen L, Chen L, Chen T-M, Chin MC, Chong J, Crook BE, Czaplinska A, Dang CN, Datta S, Dee NR *et al* (2007) Genome-wide atlas of gene expression in the adult mouse brain. *Nature* 445: 168–176

Lezoualc'h F (2007) 5-HT4 receptor and Alzheimer's disease: the amyloid connection. *Exp Neurol* 205: 325–329

Loening AM, Fenn TD, Wu AM, Gambhir SS (2006) Consensus guided mutagenesis of *Renilla luciferase* yields enhanced stability and light output. *Protein Eng Des Sel* 19: 391–400

Lohse MJ, Nuber S, Hoffmann C (2012) Fluorescence/bioluminescence resonance energy transfer techniques to study G-protein-coupled. *Pharmacol Rev* 64: 299–336

Lopes JP, Morató X, Souza C, Pinhal C, Machado NJ, Canas PM, Silva HB, Stagljar I, Gandía J, Fernández-Dueñas V, Luján R, Cunha RA, Ciruela F (2015) The role of parkinson's disease-associated receptor GPR37 in the hippocampus: functional interplay with the adenosinergic system. *J Neurochem* 134: 135–146

Maillet M, Robert SJ, Cacquevel M, Gastineau M, Vivien D, Bertoglio J, Zugaza JL, Fischmeister R, Lezoualc'h F (2003) Crosstalk between

Rap1 and Rac regulates secretion of sAPPalpha. *Nat Cell Biol* 5: 633–639

Mak AB, Nixon AML, Kittanakom S, Stewart JM, Chen GI, Curak J, Gingras A-C, Mazitschek R, Neel BG, Stagljar I, Moffat J (2012) Regulation of CD133 by HDAC6 promotes β-catenin signaling to suppress cancer cell differentiation. *Cell Rep* 2: 951–963

Marin P, Becamel C, Dumuis A, Bockaert J (2012) 5-HT receptor-associated protein networks: new targets for drug discovery in psychiatric disorders? *Curr Drug Targets* 13: 28–52

Matthews L, Gopinath G, Gillespie M, Caudy M, Croft D, de Bono B, Garapati P, Hemish J, Hermjakob H, Jassal B, Kanapin A, Lewis S, Mahajan S, May B, Schmidt E, Vastrik I, Wu G, Birney E, Stein L, D'Eustachio P (2009) Reactome knowledgebase of human biological pathways and processes. *Nucleic Acids Res* 37: D619–D622

Maurice P, Daulat AM, Broussard C, Mozo J, Clary G, Hotellier F, Chafey P, Guillaume J-L, Ferry G, Boutin JA, Delagrange P, Camoin L, Jockers R (2008) A generic approach for the purification of signaling complexes that specifically interact with the carboxyl-terminal domain of G protein-coupled receptors. *Mol Cell Proteomics* 7: 1556–1569

Maurice P, Daulat AM, Turecek R, Ivankova-Susankova K, Zamponi F, Kamal M, Clement N, Guillaume J-L, Bettler B, Galès C, Delagrange P, Jockers R (2010) Molecular organization and dynamics of the melatonin MT$_1$ receptor/RGS20/G(i) protein complex reveal asymmetry of receptor dimers for RGS and G(i) coupling. *EMBO J* 29: 3646–3659

Maurice P, Guillaume J-L, Benleulmi-Chaachoua A, Daulat AM, Kamal M, Jockers R (2011) GPCR-interacting proteins, major players of GPCR function. *Adv Pharmacol* 62: 349–380

Medscape (2015) 100 best-selling, most prescribed branded drugs through March. http://www.medscape.com/

Mercier J-F, Salahpour A, Angers S, Breit A, Bouvier M (2002) Quantitative assessment of beta 1- and beta 2-adrenergic receptor homo- and heterodimerization by bioluminescence resonance energy transfer. *J Biol Chem* 277: 44925–44931

Mitchell A, Chang H-Y, Daugherty L, Fraser M, Hunter S, Lopez R, McAnulla C, McMenamin C, Nuka G, Pesseat S, Sangrador-Vegas A, Scheremetjew M, Rato C, Yong S-Y, Bateman A, Punta M, Attwood TK, Sigrist CJA, Redaschi N, Rivoire C et al (2015) The InterPro protein families database: the classification resource after 15 years. *Nucleic Acids Res* 43: D213–D221

Moreno JL, Sealfon SC, González-Maeso J (2009) Group II metabotropic glutamate receptors and schizophrenia. *Cell Mol Life Sci* 66: 3777–3785

Pausch MH (1997) G-protein-coupled receptors in high-throughput screening assays for drug discovery. *Trends Biotechnol* 15: 487–494

Petschnigg J, Groisman B, Kotlyar M, Taipale M, Zheng Y, Kurat CF, Sayad A, Sierra JR, Mattiazzi Usaj M, Snider J, Nachman A, Krykbaeva I, Tsao M-S, Moffat J, Pawson T, Lindquist S, Jurisica I, Stagljar I (2014) The mammalian-membrane two-hybrid assay (MaMTH) for probing membrane-protein interactions in human cells. *Nat Methods* 11: 585–592

Piñero J, Queralt-Rosinach N, Bravo À, Deu-Pons J, Bauer-Mehren A, Baron M, Sanz F, Furlong LI (2015) DisGeNET: a discovery platform for the dynamical exploration of human diseases and their genes. *Database* 2015: bav028

Pinna A, Wardas J, Simola N, Morelli M (2005) New therapies for the treatment of Parkinson's disease: adenosine A2A receptor antagonists. *Life Sci* 77: 3259–3267

Prinster SC, Hague C, Hall RA (2005) Heterodimerization of G protein-coupled receptors: specificity and functional significance. *Pharmacol Rev* 57: 289–298

Rahmati S, Abovsky M, Pastrello C, Jurisica I (2017) PathDIP: an annotated resource for known and predicted human gene-pathway associations and pathway enrichment analysis. *Nucl Acids Res* 45: D419–D426

Ritter SL, Hall RA (2009) Fine-tuning of GPCR activity by receptor-interacting proteins. *Nat Rev Mol Cell Biol* 10: 819–830

Snider J, Kittanakom S, Damjanovic D, Curak J, Wong V, Stagljar I (2010) Detecting interactions with membrane proteins using a membrane two-hybrid assay in yeast. *Nat Protoc* 5: 1281–1293

Snider J, Hanif A, Lee ME, Jin K, Yu AR, Graham C, Chuk M, Damjanovic D, Wierzbicka M, Tang P, Balderes D, Wong V, Jessulat M, Darowski KD, San Luis B-J, Shevelev I, Sturley SL, Boone C, Greenblatt JF, Zhang Z et al (2013) Mapping the functional yeast ABC transporter interactome. *Nat Chem Biol* 9: 565–572

Snider J, Kotlyar M, Saraon P, Yao Z, Jurisica I, Stagljar I (2015) Fundamentals of protein interaction network mapping. *Mol Syst Biol* 11: 848

Stagljar I, Korostensky C, Johnsson N, te Heesen S (1998) A genetic system based on split-ubiquitin for the analysis of interactions between membrane proteins *in vivo*. *Proc Natl Acad Sci USA* 95: 5187–5192

Stevens RC, Cherezov V, Katritch V, Abagyan R, Kuhn P, Rosen H, Wüthrich K (2012) The GPCR Network: a large-scale collaboration to determine human GPCR structure and function. *Nat Rev Drug Discov* 12: 25–34

Tadagaki K, Tudor D, Gbahou F, Tschische P, Waldhoer M, Bomsel M, Jockers R, Kamal M (2012) Human cytomegalovirus-encoded UL33 and UL78 heteromerize with host CCR5 and CXCR4 impairing their HIV coreceptor activity. *Blood* 119: 4908–4918

Thathiah A, De Strooper B (2011) The role of G protein-coupled receptors in the pathology of Alzheimer's disease. *Nat Rev Neurosci* 12: 73–87

UniProt Consortium (2015) UniProt: a hub for protein information. *Nucleic Acids Res* 43: D204–D212

Usenovic M, Knight AL, Ray A, Wong V, Brown KR, Caldwell GA, Caldwell KA, Stagljar I, Krainc D (2012) Identification of novel ATP13A2 interactors and their role in α-synuclein misfolding and toxicity. *Hum Mol Genet* 21: 3785–3794

Wardas J, Pietraszek M, Dziedzicka-Wasylewska M (2003) SCH 58261, a selective adenosine A2A receptor antagonist, decreases the haloperidol-enhanced proenkephalin mRNA expression in the rat striatum. *Brain Res* 977: 270–277

Wishart DS, Knox C, Guo AC, Shrivastava S, Hassanali M, Stothard P, Chang Z, Woolsey J (2006) DrugBank: a comprehensive resource for *in silico* drug discovery and exploration. *Nucleic Acids Res* 34: D668–D672

Xie L, Gao S, Alcaire SM, Aoyagi K, Wang Y, Griffin JK, Stagljar I, Nagamatsu S, Zhen M (2013) NLF-1 delivers a sodium leak channel to regulate neuronal excitability and modulate rhythmic locomotion. *Neuron* 77: 1069–1082

Yang X, Boehm JS, Yang X, Salehi-Ashtiani K, Hao T, Shen Y, Lubonja R, Thomas SR, Alkan O, Bhimdi T, Green TM, Johannessen CM, Silver SJ, Nguyen C, Murray RR, Hieronymus H, Balcha D, Fan C, Lin C, Ghamsari L et al (2011) A public genome-scale lentiviral expression library of human ORFs. *Nat Methods* 8: 659–661

Yao Z, Petschnigg J, Ketteler R, Stagljar I (2015) Application guide for omics approaches to cell signaling. *Nat Chem Biol* 11: 387–397

Yao Z, Darowski K, St-Denis N, Wong V, Offensperger F, Villedieu A, Amin S, Malty R, Aoki H, Guo H, Xu Y, Iorio C, Kotlyar M, Emili A, Jurisica I, Neel

8

Drug detoxification dynamics explain the postantibiotic effect

Jaydeep K Srimani[1], Shuqiang Huang[2], Allison J Lopatkin[1] & Lingchong You[1,3,4,*] (iD)

Abstract

The postantibiotic effect (PAE) refers to the temporary suppression of bacterial growth following transient antibiotic treatment. This effect has been observed for decades for a wide variety of antibiotics and microbial species. However, despite empirical observations, a mechanistic understanding of this phenomenon is lacking. Using a combination of modeling and quantitative experiments, we show that the PAE can be explained by the temporal dynamics of drug detoxification in individual cells after an antibiotic is removed from the extracellular environment. These dynamics are dictated by both the export of the antibiotic and the intracellular titration of the antibiotic by its target. This mechanism is generally applicable for antibiotics with different modes of action. We further show that efflux inhibition is effective against certain antibiotic motifs, which may help explain mixed cotreatment success.

Keywords antibiotic tolerance; postantibiotic effect; systems biology
Subject Categories Microbiology, Virology & Host Pathogen Interaction; Quantitative Biology & Dynamical Systems

Introduction

The postantibiotic effect (PAE) refers to the temporary suppression of bacterial growth following transient exposure to antibiotics. This transient inhibition has been observed since the first studies of penicillin against *Pneumococcus* and *Streptococcus* in the 1940s. Even after the antibiotic had been degraded by a penicillinase, the target populations exhibited a significant lag before resuming growth (Bigger, 1944; Parker & Marsh, 1946; Eagle, 1949; Eagle & Fleischman, 1950). Subsequent studies have observed PAE following treatment with a variety of antibiotics, including aminoglycosides (Zhanel & Craig, 1994), β-lactams (Hanberger *et al*, 1990; Odenholt-Tornqvist & Löwdin, 1991), fluoroquinolones (Athamna, 2004; Mizunaga, 2005), and others (Zhanel & Hoban, 1991;

Odenholt-Tornqvist, 1993), and against both Gram-positive and Gram-negative bacterial species (Eagle & Musselman, 1949; Eagle *et al*, 1950; Bundtzen *et al*, 1981). PAE has also been observed in animal models (Craig, 1993; Gudmundsson & Einarsson, 1993), where, in addition to suppressing growth, transient antibiotic treatment can render the surviving population more susceptible to innate immune responses and result in decreased virulence expression (Eagle, 1949). Moreover, the extent of PAE is of vital importance in the design and optimization of periodic and multi-dose antibiotic regimens (Eagle *et al*, 1950; AliAbadi & Lees, 2000). For example, antibiotics that induce a long PAE can be dosed less frequently. However, the high concentrations required to reduce dosing frequency may result in adverse consequences, for example, toxicity (Avent *et al*, 2011). Accordingly, PAE is a standard metric used to evaluate novel antibiotics (Beam *et al*, 1992).

Despite widespread observations of PAE, its underlying mechanisms are not well established. Previous studies have speculated on a number of possible explanations, including nonspecific binding and nonlethal damage induced by antibiotic treatment (Craig & Vogelman, 1987; Li *et al*, 1997), antibiotic persistence within the periplasmic space, or the resynthesis of essential enzymes (MacKenzie & Gould, 1993). Moreover, these studies have not ruled out the possibility of multiple concurrent mechanisms, or different mechanisms being applicable for different antibiotics. More recently, zur Wiesch and colleagues proposed that antibiotic–target binding kinetics are sufficient to explain PAE (zur Wiesch *et al*, 2015), and fit their model to bacterial responses to tetracycline treatment.

Given that many, if not all, antibiotics lead to PAE, we asked whether there exists a common core mechanism that dictates the generation of PAE. In this study, we propose that a minimal unifying titration-based interaction, emphasizing target titration and antibiotic efflux, is sufficient to account for PAE observed in response to a wide variety of antibiotics. Our results indicate that efflux inhibition, an established antibiotic adjuvant strategy, may be effective only in conjunction with certain antibiotics. Moreover, understanding transient dynamics in antibiotic response is essential to designing effective combinations of drug and efflux inhibitor.

1 Department of Biomedical Engineering, Duke University, Durham, NC, USA
2 Center for Synthetic Biology Engineering Research, Shenzhen Institutes of Advanced Technology, Chinese Academy of Sciences, Shenzhen, China
3 Center for Genomic and Computational Biology, Duke University, Durham, NC, USA
4 Department of Molecular Genetics and Microbiology, Duke University School of Medicine, Durham, NC, USA
*Corresponding author. E-mail: you@duke.edu

Results

To quantify population recovery dynamics, we first defined population recovery time (RT_{pop}) as the doubling time relative to the end of antibiotic treatment (Fig 1A, left panel). PAE, then, corresponds to the prolonged recovery time in response to antibiotic treatment, as compared to the control (Fig 1A, right panel). In the absence of PAE, populations would resume normal growth immediately after the antibiotic was removed. The corresponding recovery time would be independent of the antibiotic treatment. This metric is preferable to typical definitions of PAE, which measure the time required for a population to increase 10-fold after treatment via counting colony-forming units (CFUs) (Eagle & Musselman, 1949);

by measuring twofold increases in cell density, our metric more precisely captures the effect of transient population dynamics immediately following antibiotic treatment.

In addition to potentially masking recovery dynamics on short timescales, previous studies of PAE [including those quantifying recovery by measuring rates of ATP synthesis (Hanberger et al, 1990) and DNA and protein synthesis (Stubbings, 2006)] suffer from a common disadvantage: Due to sparse time series data, they do not yield temporally precise estimates of recovery time. Moreover, they rely on relatively large populations of cells, which could potentially lead to inoculum effects and skew recovery dynamics (Tan et al, 2012). To overcome these technical limitations, we used a custom-made microfluidic device to quantify the response of bacterial

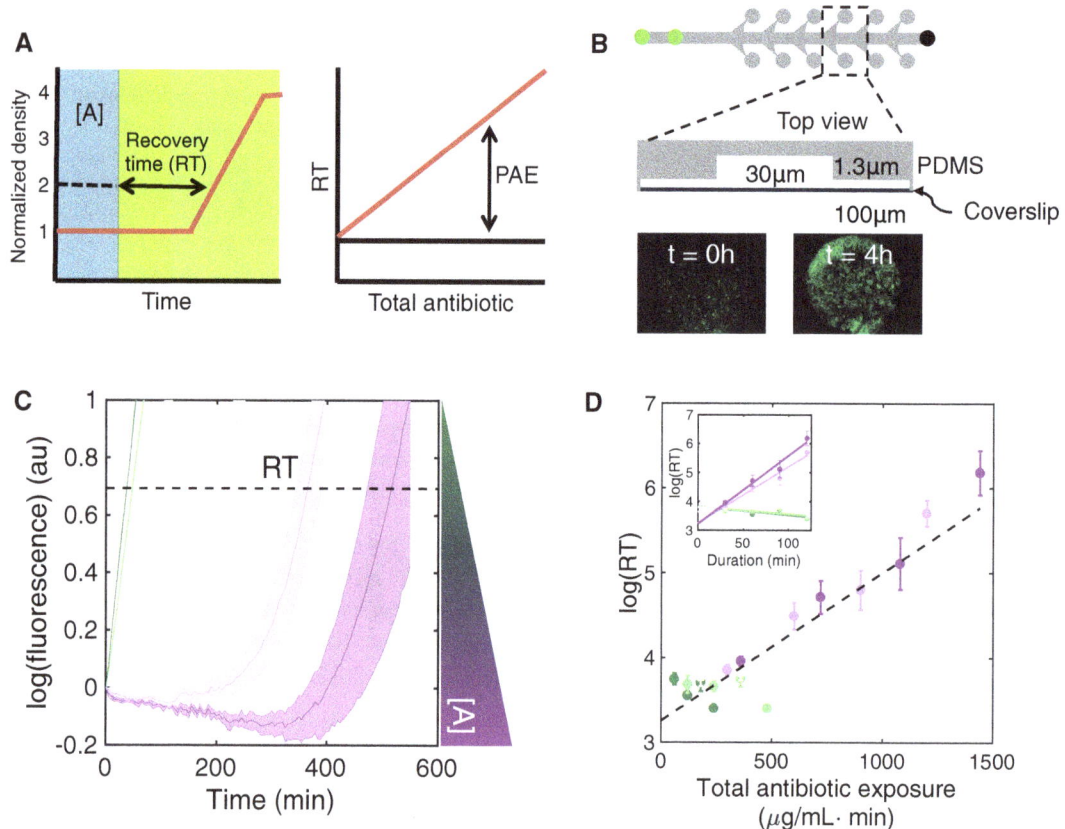

Figure 1. Recovery time increases exponentially as a function of total antibiotic exposure.

A   (Left panel) We define the recovery time (RT) as the time required for a population (red line) to double in response to a transient antibiotic treatment (blue shading). (Right panel) The postantibiotic effect (PAE) induced by an antibiotic treatment refers to the additional time required for a population to recover (red line) in comparison with the untreated control (black line).

B   A microfluidic device for quantitative recovery time measurements. Each PDMS-fabricated chip consists of six independent channels, one of which is shown, from top view (top row). Green circles indicate media inflow; black circle indicates outflow. Bacteria are manually loaded in inflow ports and are trapped in individual culturing chambers (gray circles), and the height of which ensures that bacteria are imaged in a monolayer (bottom row). Growth conditions (e.g., antibiotic dose profile) are controlled via programmable syringe pumps. Fluorescent images (bottom row) show representative growth of a monolayer of *Escherichia coli* BW25113 cells constitutively expressing GFP over four hours.

C   Representative time series fluorescence data showing dose-dependent population recovery in response to transient streptomycin treatment. Here, time zero corresponds to the end of treatment (120 min). Trajectories show mean and standard deviation for five replicates; colors indicate increasing antibiotic concentration (2, 4, 6, 8, 10, 12 µg/ml). Fluorescence values are normalized to those at time zero. Dotted line indicates a twofold increase, corresponding to the recovery time for each population.

D   Recovery time increases with total antibiotic exposure. Inset shows recovery time as a function of dose duration for increasing streptomycin concentrations (as in panel C). When plotted against total antibiotic exposure (calculated as $\int_0^D A(t)dt$), all recovery time values collapse onto a single exponential function (black line, $R^2 = 0.83$). Recovery time was measured in response to streptomycin concentrations of 0, 2, 4, 6, 8, 10, and 12 µg/ml, and treatment durations 30, 60, 90, and 120 min. Error bars indicate standard deviation, calculated from five replicates; y-axis uses the natural logarithm.

populations to a range of antibiotic concentrations and treatment durations (Lopatkin *et al*, 2016). As shown in Fig 1B, the microfluidic chip consists of six independent channels (top row, top view), each with two media inputs (green circles) and one output (black circle). Bacteria are manually loaded into each channel and trapped in individual cylindrical culturing chambers (middle row). The height of the chambers (~1.0 μm) constrains bacterial growth to a single monolayer, facilitating precise quantification (bottom row) using time-lapse fluorescence microscopy. Growth media and antibiotic dosing protocols are controlled by programmable syringe pumps.

Figure 1C shows typical time courses of *Escherichia coli* strain BW25113 (Grenier *et al*, 2014) constitutively expressing GFP, exposed to 120-min streptomycin treatment at increasing concentrations (where $t = 0$ corresponds to the end of treatment, and the observed IC_{50} for streptomycin was 1.99 μg/ml). We found that the fluorescence signal serves as a reliable surrogate measure of the population density over the range of dosing conditions used (Appendix Fig S1). It is particularly suited for the imaging-based characterization of the antibiotic response in the microfluidic device. The recovery time drastically increased with increasing antibiotic concentrations, demonstrating the generation of PAE, where the dashed line indicates a twofold increase relative to $t = 0$. We note that for high streptomycin concentrations, the population fluorescence decreases before recovery; this is likely due to continued inhibition after the removal of antibiotic. At a fixed concentration, the recovery time drastically increased with treatment duration (Fig 1D, inset). Remarkably, by combining the dose and duration for each treatment, we found that the recovery time was determined by the total antibiotic exposure for each treatment, regardless of the dose profile (Fig 1D, main). That is, every treatment (i.e., combination of concentration and duration) that delivered a given total antibiotic resulted in a comparable recovery time, which increased approximately exponentially with the total antibiotic exposure.

A potential caveat of using fluorescence reporters is the effect of incomplete population death during antibiotic treatment. Indeed, under some treatment conditions, we observed cells that expressed GFP at comparable levels to other population members but did not divide for the duration of the recovery phase (~18 h). Therefore, although these cells contributed to total fluorescence, they did not contribute to population recovery. To account for this, we measured population viability as a function of streptomycin concentration and duration in the absence of growth (Appendix Fig S2A). We adjusted the twofold cutoff to account for this death and ensure that doubling times reflect the signal from only viable cells; PAE remained a function of total antibiotic exposure (Appendix Fig S2B).

To gain insight into generation of PAE, we adopted a kinetic model of antibiotic-mediated inhibition of ribosomes (e.g., streptomycin; Tan *et al*, 2012) (Fig 2A). The model consists of six ordinary differential equations (ODEs) that account for the transport of the antibiotic across the cell membrane (A_{in} or A_{out}), synthesis of the ribosome (C) through a positive feedback loop, and antibiotic-mediated ribosome inhibition and potential degradation (equations 4–9 and Appendix Table S1). The ribosome concentration C sets the population growth rate μ (Neidhardt, 1996), connecting the intracellular drug–cell interaction with overall population recovery. Here, we assume homogenous populations that interact (i.e., drug influx and efflux) with a common extracellular environment.

Because the population growth rate is dependent on the ribosome concentration, we can investigate recovery on two scales: on the population level (RT_{pop}, analogous to Fig 1D) as the time required for the cell number to double, and on the individual level, as the time required for the ribosome concentration to achieve its half-maximal synthesis rate (RT_{cell}). In terms of the core structure, our model is similar to one developed by zur Wiesch *et al* (2015), but provides a more detailed description of the underlying kinetics.

Figure 2B shows the typical dynamics of various components, in three phases, in response to the addition and removal of an extracellular antibiotic. During antibiotic treatment (Fig 2B, left schematic and blue shading), $A_{out} \gg A_{in}$ and the dominant dynamic is the accumulation of intracellular antibiotic due to influx. Neglecting the efflux dynamics and titration due to binding of antibiotic to the ribosome, and denoting the dose duration as D, we have: $A_{in}(D) = \frac{k_{in}}{k_{out}} A_{out}(1 - e^{-k_{out}D}) \approx k_{in}A_{out}D$ (for sufficiently small D). That is, at the end of an antibiotic pulse, *the intracellular antibiotic concentration is proportional to total amount of the antibiotic used* (Fig 2C, panel i), consistent with previous studies (Hancock, 1962; Hurwitz & Rosano, 1962; Bryan & Van Den Elzen, 1976; Damper & Epstein, 1981; Muir *et al*, 1984).

Upon removal of the antibiotic, $A_{in} \gg A_{out} = 0$ and the transport dynamics of the antibiotic are dominated by efflux (Fig 2B, middle schematic and green shading). In addition to efflux, binding kinetics between ribosome and antibiotic serve as an intracellular reservoir for A_{in} and can further delay recovery. If the cell survives treatment, it can revert to a basal growth rate when it is sufficiently detoxified (i.e., when A_{in} is sufficiently small to not inhibit the ribosome effectively) (Fig 2C, panel ii). The balance between these three processes (efflux, binding, and degradation) sets the timescale of detoxification and subsequent growth, leading to generation of PAE. Furthermore, because $C \ll A_{in}$, efflux dominates this interplay. Consistent with this notion, for different dosing protocols, the correlations between C and A_{in} approximately collapse to a single line (Fig 2D). This mechanism recaptures the dependence between population recovery time and the total antibiotic exposure (Fig 2E). Moreover, modeling indicates that this dependence is independent of initial cell density (Appendix Fig S3). We note that this model assumes that antibiotic is freely transported across the cell membrane and binds to targets either in the cytoplasm or periplasm, as opposed to those that damage the cell membrane. Although inoculum effects could potentially be induced by these antibiotics, the range of treatment conditions used ensure that all populations undergo first inhibition followed by recovery.

These results provide a simple explanation for the emergence of PAE: Fundamentally, PAE arises as the time required for individual cells to recover by exporting antibiotic, such that ribosome-mediated positive feedback can be activated. This "detoxification" process occurs on the individual cell level, and intracellular recovery dynamics correlate strongly with population recovery (Fig 2F, $R^2 = 0.86$). Moreover, this correlation is robust to changes in the antibiotic-mediated ribosome degradation rate (Appendix Fig S4). Here, we do not consider antibiotic dilution by cell growth, which would result in lower recovery times without changing our qualitative conclusions.

A mechanistic understanding of PAE is critical, as recovery time to a single dose of antibiotic represents a critical metric that can predict the efficacy of long-term multi-dose treatments using the

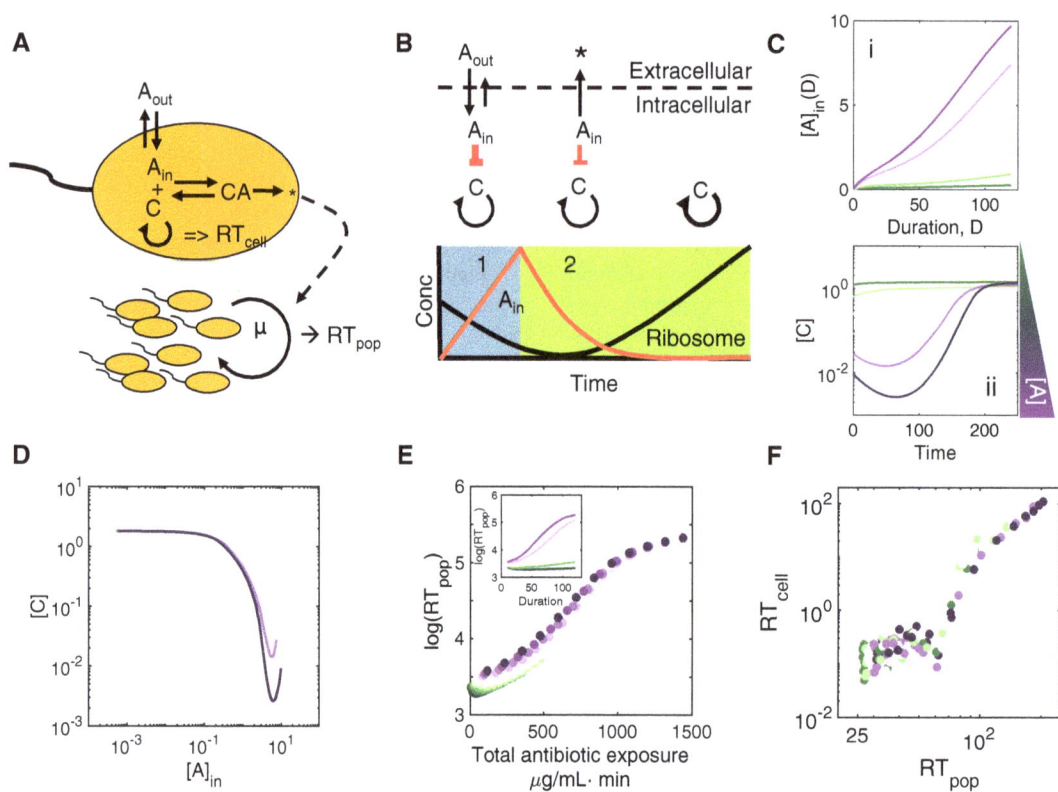

Figure 2. The timescale of overall intracellular antibiotic dictates individual and population recovery.

A   A minimal model of antibiotic action. Antibiotic is transported between the intracellular and extracellular environments (A_{in} and A_{out}) by influx and efflux rates (k_{in} and k_{out}). A_{in} reversibly binds to target ribosomes C to form the complex CA, with binding and dissociation rates k_f and k_b, respectively. This complex can then be degraded through the intermediate CA'. These intracellular dynamics influence the overall population recovery rate; the maximum growth rate μ is dependent on the ribosome concentration C. Thus, recovery can be quantified both on the individual level, as a function of C, or on the population level, in terms of cell density.

B   Concurrent antibiotic transport and ribosome inhibition dictate recovery dynamics. During treatment, $A_{out} \gg A_{in}$, and the intracellular antibiotic concentration can be linearly approximated by $A_{in}(D) \approx k_{in}A_{out}D$, where A_{out} is assumed to be constant and D represents the treatment duration (leftmost schematic and panel 1). Antibiotic influx is greater than efflux; A_{in} binds to target ribosomes and strongly inhibits the upregulation of ribosome synthesis. When extracellular antibiotic is removed, A_{in} decreases (middle and rightmost schematics, green shading and panel 2), and ribosome synthesis resumes when A_{in} is sufficiently small.

C A_{in} accumulates linearly with treatment duration for sufficiently short treatment durations (panel i). After the removal of A_{out}, efflux and inhibition dynamics combine to delay the synthesis of ribosomes in a concentration-dependent manner (panel ii). Colors indicate increasing antibiotic concentration, as shown in panel ii.

D   Antibiotic turnover timescale sets intracellular recovery RT_{cell}. Regardless of the antibiotic treatment history, the relationship between intracellular antibiotic A_{in} and ribosome concentration C approaches the same asymptote, indicating that the timescale of individual detoxification sets the timescale of ribosome synthesis. Colors indicate increasing antibiotic concentration, as in Fig 2C; for these representative trajectories, treatment duration was set to 120 min.

E   Population recovery time is dictated by total antibiotic exposure. Inset shows that RT_{pop} is an increasing function of dose duration; these data collapse onto a single relationship as a function of total antibiotic exposure (main figure).

F   Individual (RT_{cell}) and population recovery time (RT_{pop}) are strongly correlated, suggesting that intracellular dynamics lead to population level recovery ($R^2 = 0.86$).

same antibiotic (Meredith *et al*, 2015). Here, periodic treatments are specified by concentration A, the dose duration T_1, the interval between doses T_2, and the total number of doses N_{dose} (Appendix Fig S5A). RT_n is defined as the recovery time in response to a given N_{dose} treatment; larger values of RT_n correspond to treatment success (Appendix Fig S5B). Appendix Fig S5C shows a transition in treatment outcome at $\frac{T_2}{RT_1} \approx 1$. Specifically, any treatment that delivers successive doses more frequently than the corresponding single dose recovery time (RT_1) will induce a significant RT_n. However, regardless of the frequency of antibiotic dosing, population recovery time remains a function of total antibiotic exposure (Appendix Fig S5D).

The model also predicts the consequences of modulating key processes (Fig 3). In particular, decreasing the rate of antibiotic-mediated ribosome degradation, increasing the efflux

rate, or increasing ribosome synthesis rate all lead to more rapid recovery for the same total antibiotic exposure; microfluidic experiments confirmed these model predictions. To investigate the effect of ribosome degradation, we compared the recovery time induced by streptomycin, which has been shown to induce the heat-shock response (HSR) and lead to rapid ribosome degradation, to chloramphenicol (IC$_{50}$ 1.2 μg/ml), which does not induce HSR (Tan *et al*, 2012), and therefore induces slow ribosome degradation. Consistent with modeling predictions, chloramphenicol treatment resulted in significantly shorter recovery time than streptomycin treatment (Fig 3A and D shows modeling and experimental results, respectively).

To inhibit efflux pump activity, we used carbonyl cyanide 3-chlorophenylhydrazone (CCCP), a phosphorylation inhibitor that inhibits a variety of efflux pumps that rely on the proton motive force (PMF), including resistance nodulation division pumps and others (Kinoshita *et al*, 1984; Cohen *et al*, 1988; Singh *et al*, 2011). We confirmed the concentration-dependent efflux inhibition (Appendix Fig S6). Modeling predicted that inhibiting efflux pump activity would result in longer recovery times (Fig 3B); therefore, we tested the effect of a subinhibitory CCCP concentration with chloramphenicol treatment, which alone induced a minimal recovery delay (Fig 3D). The addition of CCCP indeed caused significantly longer recovery times (Fig 3E); however, CCCP alone did not increase population recovery time.

Finally, we modulated the ribosome synthesis rate by modulating the richness of the growth media, specifically by adjusting the concentration of casamino acids: A higher concentration of the casamino acids leads to faster bacterial growth, all else being equal (Lazzarini & Dahlberg, 1971). Indeed, with faster growth, corresponding to rapid ribosome synthesis, populations recovered significantly faster after streptomycin treatment (Fig 3C and F shows modeling and experimental results, respectively).

These results suggest that drug response dynamics can be modulated to induce longer PAEs, which is advantageous from a clinical perspective (Spivey, 1992; Fishman, 2006; Talpaert *et al*, 2011); antibiotic efflux inhibition is an active area of research (Marquez, 2005; Askoura *et al*, 2017). In particular, resistance nodulation division (RND) pump systems are prevalent in Gram-negative species and have been shown to efflux a variety of antibiotics, including penicillins and cephalosporins, macrolides, aminoglycosides, fluoroquinolones, and tetracyclines (Ma *et al*, 1994; Nikaido, 1994, 1998). Moreover, many resistant strains have been shown to overexpress these efflux pumps (Okusu *et al*, 1996; Mcmurry & Oethinger, 1998; Kriengkauykiat *et al*, 2005). In this context, efflux pump inhibitors (EPIs) have emerged as potential adjuvants for antibiotic treatment. Recent studies have proposed a number of candidate EPIs, via both rational design (Amaral *et al*, 2007) and natural isolation (Stavri *et al*, 2007); these are often based on studies of efflux pump structure, competitive binding, or disruption of transmembrane electrical gradients (Poole & Lomovskaya, 2006; Mahamoud *et al*, 2007). Intuitively, we would expect that efflux inhibition would be universally beneficial in lengthening PAEs in response to a given antibiotic treatment. However, these approaches have been met with limited clinical success (Van Bambeke & Lee, 2006; Opperman & Nguyen,

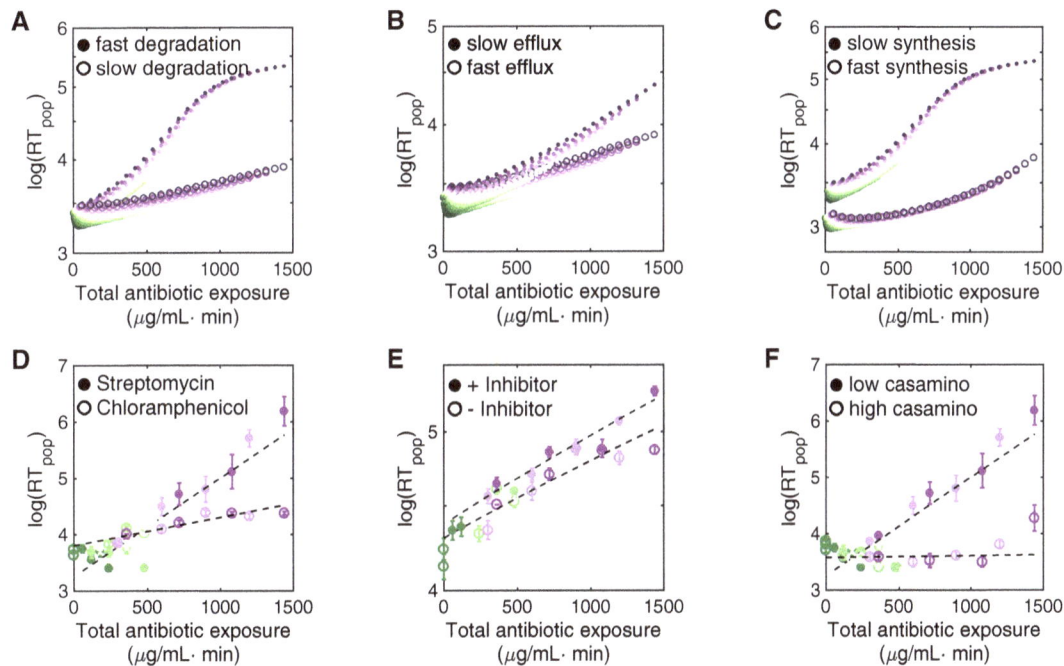

Figure 3. Key parameter perturbations confirm model validity.

A Modeling predicts that decreasing the ribosome degradation rate leads to shorter recovery times in response to equal amounts of total antibiotic exposure. Here, we use $k_d = 0.1$ and $k_d = 0.2$ for high and low degradation, respectively.

B Modeling predicts that decreasing the antibiotic efflux rate leads to longer recovery times in response to equal amounts of total antibiotic exposure. Here, we use $k_{out} = 0.01$ and $k_{out} = 0.001$ for fast and slow efflux, respectively.

C Modeling predicts that increasing the ribosome synthesis rate leads to shorter recovery times in response to equal amounts of total antibiotic exposure. Here, we use $k_1 = 0.2$ and $k_1 = 0.4$ for slow and fast ribosome synthesis, respectively.

D Streptomycin treatment (closed data points, $R^2 = 0.83$) results in significantly longer recovery times than chloramphenicol treatment (open data points). Here, the difference between two responses is statistically significant ($P < 0.05$ by ANOVA). Error bars indicate standard deviation of five replicates.

E The addition of efflux pump inhibitor (CCCP) (closed data points, $R^2 = 0.91$) increases population recovery time in response to chloramphenicol treatment (open data points, $R^2 = 0.80$). Here, CCCP was added at subinhibitory concentrations (3 μg/ml); in the absence of antibiotic treatment, CCCP alone did not inhibit population recovery. The CCCP-mediated increase in recovery time is statistically significant ($P < 0.01$ by ANOVA). Error bars indicate standard deviation of five replicates.

F The ribosome synthesis rate was increased by increasing the concentration of the casamino acids in the media from 0.01% w/v (closed data points) to 0.05% w/v (open data points). Faster synthesis resulted in lower recovery times ($P < 0.001$ by ANOVA) in response to streptomycin treatment. Error bars indicate standard deviation of five replicates. In each of these panels, the color scheme indicates increasing antibiotic concentration, as in Fig 1C.

2015); optimal cotreatment strategies remain an open question (Lomovskaya & Bostian, 2006).

Therefore, we asked the question, under what conditions is efflux inhibition an effective strategy for extending PAE? We used modeling to examine the sensitivity of various parameters (e.g., ribosome degradation and synthesis rates, as well as positive feedback strength), and antibiotic motifs, to changes in efflux. We reasoned that the induction of PAE by antibiotic accumulation and detoxification dynamics (Fig 2A) is applicable to any drug that has an intracellular target (including those in the periplasm). As such, we expect that qualitatively similar dynamics can lead to PAE for these other antibiotics, and may be sensitive to efflux inhibition.

To test this hypothesis, we developed simplified kinetic models of intracellular dynamics to investigate three motifs of antibiotic action that encompass a wide variety of common antibiotics (equations 10–12 and Appendix Table S2). In each case, the antibiotic is transported between the intracellular and extracellular spaces. In the intracellular space, the antibiotic binds to its target reversibly and the concentration of the free target indicates the viability of the cell. However, these different motifs differ in how the target is regulated (corresponding to a change in one relevant parameter in each case). In the first motif, the target synthesis is driven by a positive feedback loop, and binding to the antibiotic leads to enhanced degradation of the target (Fig 4A, left column). This motif accounts for antibiotics that target ribosome synthesis and induce fast ribosome degradation, including aminoglycosides such as streptomycin and kanamycin. The second motif is identical to the first except that the antibiotic does not enhance degradation of the target (Fig 4A, middle column). This motif accounts for the action by chloramphenicol and tetracycline (Tan *et al*, 2012), as well as fluoroquinolones such as ciprofloxacin. The third motif is identical to the second except that the target is synthesized at a constant rate, with no feedback (Fig 4A, right column). This motif can account for inhibition by β-lactams. For each of these simplified motifs, recovery time remains a function of total antibiotic exposure (Fig 4A, second row). However, decreasing efflux rates (Fig 4, open circles) results in a significant increase in recovery time for the aminoglycoside motif only (Fig 4A, left panel). Efflux inhibition was less effective when target degradation was negligible (middle panel) or the target did not undergo positive feedback (right panel). These results suggest that PAE is dictated by total antibiotic exposure and minimal binding/transport rates, independent of an antibiotic's specific mechanism of action. Rather, target binding and efflux are critical processes underlying this relationship.

To quantify the effects of efflux inhibition, we define sensitivity as the total change in recovery time, over a range of antibiotic doses, in response to a change in efflux rate (Appendix Fig S7A). With this definition, we first examined the effect of efflux inhibition in conjunction with the drug-mediated ribosome degradation rate (Appendix Fig S7B). We observed that faster degradation resulted in increased efflux sensitivity. Similarly, we found that increasing the nonlinearity of the target positive feedback loop also increased efflux sensitivity (Appendix Fig S7C). These results suggest that the efficacy of efflux inhibition as an adjuvant treatment depends on the particular antibiotic used.

To test the model predictions, we used an *E. coli* strain constitutively expressing bioluminescence (Andreu *et al*, 2010) (see

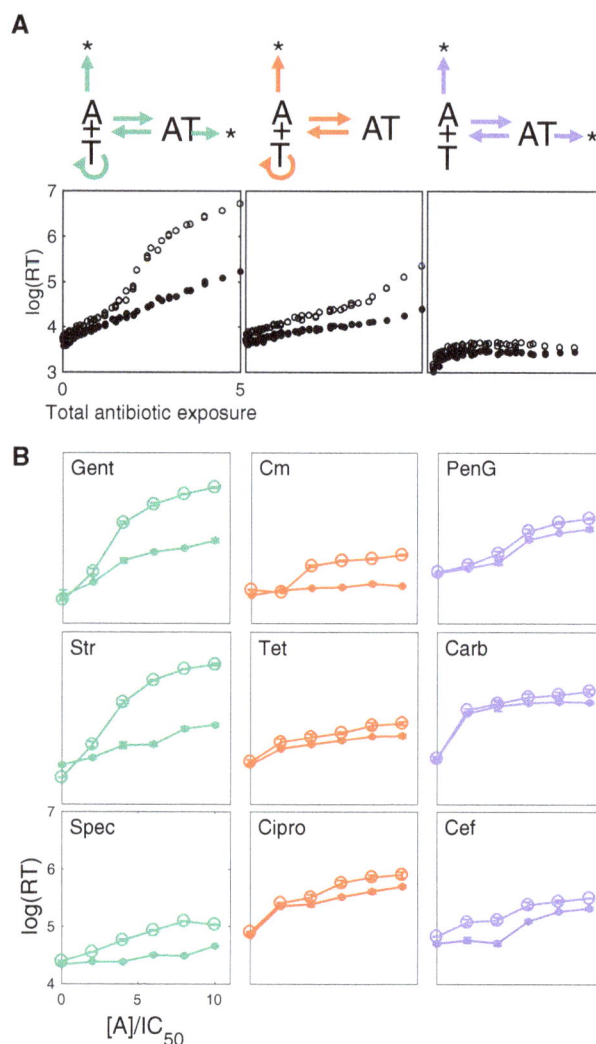

Figure 4. Efflux inhibition is an effective cotreatment strategy for certain antibiotics.

A Three general motifs of intracellular antibiotic action. Left and middle motifs correspond to ribosome-inhibiting antibiotics that induce rapid and minimal ribosome degradation, respectively. In these motifs, the target molecule is subject to nonlinear positive feedback (i.e., transcription and translation, in the case of ribosomes). Right motif corresponds to antibiotics that inhibit other targets that are not subject to positive feedback, for example, β-lactams. In each case, recovery time is a function of total antibiotic exposure (closed circles) and inhibiting efflux results in longer recovery times (open circles).

B Inhibiting antibiotic efflux with CCCP significantly increased recovery time for antibiotics that induced rapid target degradation and involved a positive feedback loop. Antibiotics are as follows: penicillin G (PenG), spectinomycin (Spec), gentamicin (Gent), streptomycin (Str), chloramphenicol (Cm), tetracycline (Tet), and carbenicillin (Carb). Antibiotic concentrations used have been scaled by their respective IC_{50} values (solid points and open circles show response with no efflux inhibition and 2 μg/ml CCCP, respectively). Colors correspond to the motifs of action shown in Fig 4A.

Appendix Supplementary Methods for full details), which is commonly used as a reporter of viable cell density (Prosser *et al*, 1996), accurately reflects cell density over the full range of conditions tested (Appendix Fig S1B), and bypasses the need to control for antibiotic-induced partial population death, as in Fig 1. This reporter allows us to test PAE in a higher throughput compared to the microfluidic device by using a microplate reader. We tested a number of antibiotics (See Appendix Table S3) over the same range of IC_{50} values (Appendix Fig S8). Although the addition of CCCP at a subinhibitory concentration (2 µg/ml) increased the extent of PAE in all cases (Fig 4B), this effect was most significant with the three aminoglycoside antibiotics (streptomycin, gentamicin, and spectinomycin), as compared to drugs that fall in either of the latter two motifs. These results confirm our modeling predictions; the three aforementioned antibiotics induce rapid ribosome degradation. Moreover, we observed minimal inhibition with the β-lactams penicillin G, carbenicillin, and cefotaxime, as suggested by sensitivity analysis.

Discussion

In general, there are two strategies to combat the rapid rise of drug-resistance microbial pathogens: novel antibiotic development or more effective use of existing drugs (Fishman, 2006; Kaki et al, 2011). Given the prohibitive time and financial costs of the former, the latter strategy is becoming increasingly critical. In doing so, it is important to move beyond steady-state measures of efficacy, for example, IC_{50}, and emphasize the importance of temporal dynamics in drug response. PAE is one such response; our results suggest that a minimal but common motif is sufficient to account for this phenomenon in response to a wide variety of antibiotics. These results suggest that efflux-mediated recovery could be a unifying motif to explain PAE in response to a wide variety of antibiotics. This would tie together wide-ranging literature reports of PAE in many antibiotic–bacterium combinations, without requiring distinct drug-dependent explanations. Here, we note that, in addition to antibiotic binding and transfer dynamics, dynamics of PAE in vivo are dependent on pharmacokinetic/pharmacodynamics (PK/PD) factors including availability, toxicity, and plasma vs total drug concentrations (Toutain et al, 2002). Indeed, in vitro PAE measurements often underestimate in vivo observations (Renneberg & Walder, 1989). Optimal antibiotic dosing treatments will need to take these considerations into account to ensure that PAE remains predictive of treatment outcome (Vogelman et al, 1988; Papich, 2014).

Our analysis represents a parsimonious explanation of PAE, based on the combination of the binding kinetics between an antibiotic and its target and the transport of the antibiotic. We do not consider nonspecific binding to secondary antibiotic targets, or drug dilution by cell growth (zur Wiesch et al, 2015). While these factors are not essential to the underlying relationship between antibiotic dosing parameters and recovery, they would affect the extent of PAE: Nonspecific binding would enhance PAE; drug dilution by growth would attenuate PAE.

While these effects can contribute to, and modulate the extent of, PAE, they are not essential to its emergence. Indeed, cell division is extremely slow during detoxification; therefore, in our model, drug dilution exerts minimal effects. On the other hand, nonspecific binding would increase recovery time in response to a given amount of antibiotic; we focus on the minimal network dynamics that explain PAE for the widest possible variety of antibiotics. Furthermore, our model assumes homogeneous populations

and no cell death during dosing. Finally, intra-population variability in cell division and death rates, both during and after antibiotic treatment, can influence PAE (Gottfredsson & Erlendsdóttir, 1998; Wiuff et al, 2005). In this case, antibiotic selection would enrich more tolerant subpopulations and result in lower recovery times. However, accounting for biochemical noise does not qualitatively change our conclusions.

Moreover, we show that efflux inhibition is a more effective strategy to induce PAE for certain antibiotic mechanisms of action, namely those that rapidly degrade a target subject to nonlinear positive feedback, than for others. These observations provide a potential guidance for the use of EPIs as antibiotic adjuvants. Indeed, efflux-mediating adjuvants could themselves be used in combination, provided that concentrations used remain subinhibitory. However, rather than using them in all cases, it is vital to understand the transient population dynamics induced by particular antibiotics to determine whether EPI usage would result in a positive outcome.

Materials and Methods

Strains, media, and growth conditions

Unless otherwise noted, E. coli strain BW25113 (F⁻, DE(araD-araB) 567, lacZ4787(del)::rrnB-3, LAM⁻, rph-1, DE(rhaD-rhaB)568, hsdR514) was used throughout this study. Unless otherwise noted, all experiments were conducted in M9 media supplemented with 0.4% w/v glucose and 0.1% w/v casamino acids, and at 37°C unless otherwise indicated. BW25113 cells were transformed with a constitutively expressed GFP plasmid (kanamycin resistant) for microfluidic and viability experiments; all cultures were supplemented with 50 µg/ml kanamycin for selection purposes. For luminescence assays, BW25113 was transformed with a constitutively expressed luminescence reporter plasmid (kanamycin resistant; Andreu et al, 2010). For all experiments, overnight cultures were grown from single colonies, picked from streaked plates, in 3 ml Luria-Bertani (LB) broth for 16 h at 37°C with 250 rpm shaking. Streaked plates were stored in 4°C when not in use, and remade from glycerol stocks every 2 weeks.

Fluorescence/luminescence reporter calibration

Overnight cultures were diluted 100-fold into fresh M9 media containing selecting antibiotics (kanamycin, 50 µg/ml). Cells were grown in 3 ml aliquots in the presence of increasing streptomycin concentrations (0, 2, 4, 6, 8, 10, 12 µg/ml) for 2 h at 37°C and 250 rpm shaking. Following treatment, cells were serially diluted and plated on selective agar plates to measure CFU counts; GFP fluorescence and luminescence were measured in 96-well black-walled plates (Corning) using a Tecan InfinitePro M200 plate reader (GFP emission/excitation 485/535 nm). Plates were incubated overnight at 37°C and CFUs counted the following morning. All measurements were done in quadruplicate.

Microfluidic platform fabrication and experimental protocol

Transient antibiotic dosing experiments were carried out using a previously published microfluidic platform (Lopatkin et al, 2016).

Briefly, microfluidic chips were fabricated with polydimethylsilox-ane (PDMS) using a silicon mold. Each chip is comprised of six parallel replicate units, each of which consists of a central flow channel and 24 branched culturing chambers. The height of these chambers (~1.3 μm) traps a monolayer of bacterial cells, thereby enabling accurate image processing and population quantification. Each flow channel has two inputs and one output; by controlling media conditions (flow rate, antibiotic concentration) using exter-nally programmed syringe pumps (New Era Pump Systems NE-1600), various antibiotic dose profiles can be implemented.

Overnight cultures were diluted 100-fold into 3 ml M9 and grown for 2 h at 37°C with 250 rpm shaking, reaching an OD_{600} of ~0.2. Cells were then condensed 10-fold. Chips were briefly vacuumed to create negative pressure, which enabled cells to be manually injected into an input using a P2 pipette. Each experiment started with roughly 100 cells per culturing chamber. Media flow in the chip was controlled with syringe pumps. Cells were allowed to grow for 30 min in clean media flow (no antibiotics, 120 μl/h flow rate), followed by antibiotic treatment (500 μl/h with applicable antibiotic concentrations) and a recovery phase (no antibiotics, 120 μl/h flow rate). These flow rates were selected to satisfy two physical limits: Excessively high flow rates resulted in cell washout, and insuffi-ciently low flow rates resulted in minimal antibiotic effect over the range of dose durations used. The entire device was kept at 37°C. Five replicates were used per experimental condition. Culturing chambers were imaged every 5 min using a DeltaVision Elite decon-volution microscope.

Image processing and analysis

All image data from microfluidic experiments were analyzed using custom MATLAB scripts. Following background subtraction, total fluorescence was quantified and normalized to the initial time point following antibiotic treatment. Figures show mean and standard deviation for five replicates.

Quantifying population viability

Overnight cultures were regrown to mid-exponential phase by 100-fold dilution in 3 ml M9 media and incubating for 2 h at 37°C with 250 rpm shaking. Cells were then washed and resuspended in a standard 1× M9 buffer (M9 not containing glucose or casamino acids); 500 μl aliquots in 1.5-ml Eppendorf tubes were incubated with appropriate antibiotic concentrations at 37°C with 250 rpm shaking. Samples were then serially diluted to 10^7-fold and plated on selecting agar plates (50 μg/ml kanamycin). Plates were incu-bated at 37°C overnight and CFUs counted the next day. Cell density was also measured prior to antibiotic treatment (~7E8 CFU/ml). All data points show the mean and standard deviation of six replicates. Viability curves for each streptomycin concentration were normal-ized to initial densities and fit exponential decay curves. The time constants of these fits were used to account for population death.

Accounting for population death in microfluidic experiments

Viability curves were used to adjust the twofold cutoff used to calcu-late recovery time. As an example, suppose a particular antibiotic treatment killed a fraction (X) of cells. Then, the population

fluorescence immediately following antibiotic treatment can be expressed as a combination of dead and viable cells: $XF_0 + (1 - X)F_0$, assuming dead cells contribute equally to the fluo-rescence signal. Therefore, when the viable fraction doubles, the total fluorescence would be $XF_0 + 2(1 - X)F_0 = (2 - X)F_0$.

Determining IC$_{50}$ values for various antibiotics

Single colonies were grown overnight as above and diluted 100-fold into M9 media. Antibiotic concentration gradients were created by serially diluting from 100 μg/ml, and included 0 μg/ml. Growth was measured in 96-well plates (Corning) with 200 μl liquid per well; wells were covered with 50 μl mineral oil (Sigma-Aldrich Chemicals) to prevent evaporation and density (OD_{600}) data were collected every 10 min using a Tecan M200 Infinite Pro plate reader at 37°C. IC_{50} values were determined by first calculating growth rates by iteratively finding the linear region of increase; these growth rates were then fit to a Hill equation, where μ_{max}, n, and A are the maximum growth rates, Hill coefficient, and antibiotic concentration, respectively (equation 1):

$$\mu(A) = \frac{\mu_{max}IC_{50}^n}{IC_{50}^n + A^n} \tag{1}$$

Luminescence reporter and experimental protocol

Overnight cultures were diluted 100-fold into M9 media supple-mented with selection antibiotics; 1-ml aliquots were apportioned in 1.5-ml Eppendorf tubes, and experimental antibiotic concentra-tions were added as appropriate. Cultures were incubated at 30°C with shaking at 250 rpm for 2 h; cells were then resuspended in fresh M9 media containing only selection antibiotics; 200 μl aliquots were transferred to black 96-well plates, and time course luminescence measurements were taken using a Tecan Infinite M200 Pro plate reader every 10 min. Three replicates were used per experimental condition, and only alternate wells were used on 96-well plates to avoid capturing luminescence signal from neighboring wells. Time series data were normalized to initial conditions and population doubling time was determined relative to the initial time point.

Fluorometric assay of efflux activity and modulation with CCCP

Efflux rates were measured by adapting an ethidium bromide (EtBr)-based assay for efflux pump activity (Viveiros et al, 2008; Paixão et al, 2009). At sub-toxic concentrations, ethidium bromide (EtBr) accumulates intracellularly, where it can be detected using plate reader fluorescence measurements (excitation/emission 530 nm/585 nm); previous studies have shown that this signal is significantly stronger intracellularly, as ethidium bromide binds to bacterial DNA. EtBr stock solution was prepared at 1 mg/ml in water and stored at room temperature. To measure efflux in BW25113, overnight cultures were washed and diluted 50-fold into a minimal M9 buffer media (M9 as described above, without glucose or casamino acid); EtBr was added to a final concentration of either 2 μg/ml. Efflux pump activity was inhibited by adding increasing concentrations of carbonyl cyanide m-chlorophenyl hydrazone (CCCP); stock solutions were prepared by dissolving

CCCP in dimethyl sulfoxide (DMSO) to a final concentration of 1 mg/ml and stored at 4°C. Cells were plated and covered with 50 µl mineral oil to prevent evaporation. Intracellular EtBr signal and cell density (OD_{600}) were measured every 10 min at 30°C using a Tecan InifintePro M200 plate reader (EtBr excitation/emission 530/585 nm). Fluorescence time courses were normalized to OD_{600}. Efflux rates were calculated assuming that intracellular EtBr concentration is dictated by the following equation:

$$\frac{dE_{in}}{dt} = k_{in}E_{out} - k_{out}E_{in}, \tag{2}$$

where E_{in} and E_{out} represent the intracellular and extracellular EtBr concentrations, respectively. We assume $E_{total} = E_{in} + E_{out}$ to be constant, in accordance with Paixão et al (2009). Thus, equation (2) can be solved to yield:

$$E_{in}(t) = \frac{k_{in}}{k_{in} + k_{out}}E_{total} + \left(E_{in}(0) - \frac{k_{in}}{k_{in} + k_{out}}E_{total}\right)e^{-(k_{in}+k_{out})t}. \tag{3}$$

Time series fluorescence data were fit to equation (3), and efflux rates were determined, assuming $k_{out} = 0$ at high CCCP concentrations.

Model development and assumptions

We modeled aminoglycoside action similarly to a previously published framework (Tan et al, 2012), using six ordinary differential equations (ODEs):

$$\frac{d[C]}{dt} = \frac{k_1[C]}{V_1 + [C]} - k_f[C][A_{in}] - k_u[C] + k_b[CA], \tag{4}$$

$$\frac{d[A_{in}]}{dt} = k_{in}[A_{out}] - k_{out}[A_{in}] - k_f[C][A_{in}] + k_b[CA] + k_r[CA'], \tag{5}$$

$$[A_{out}] = A_0 \text{ if } t < D, \text{ otherwise } [A_{out}] = 0, \tag{6}$$

$$\frac{d[CA]}{dt} = k_f[C][A_{in}] - k_b[CA] - k_d[CA], \tag{7}$$

$$\frac{d[CA']}{dt} = k_d[CA] - k_r[CA']. \tag{8}$$

$$\frac{dN}{dt} = \mu_0\left(\frac{[C]}{V + [C]}\right)N\left(1 - \frac{N}{N_m}\right) \tag{9}$$

In this formulation, $[C]$ represents the free ribosome concentration, which is synthesized following Michaelis–Menten dynamics. Antibiotic shuttles between the extracellular ($[A_{out}]$) and intracellular ($[A_{in}]$) environment according to the rates of influx and efflux, k_{in} and k_{out}, respectively. Ribosomes reversibly bind to intracellular antibiotic (rates k_f and k_b) yielding the complex $[CA]$. This complex undergoes a two-step degradation process, releasing antibiotic to the intracellular concentration (Kaplan & Apirion, 1975; Edmunds & Goldberg, 1986; Zundel et al, 2009). k_u represents the basal rate of ribosome turnover. Finally, the maximum population growth rate μ_0 is scaled by the ribosome concentration (Neidhardt, 1996). All rate constants used are listed in Appendix Table S2.

For transient dosing simulations, we assumed that the extracellular antibiotic concentration A_0 did not significantly decrease during the dosing interval D. That is, $[A_{out}] = A_0$ for $t < D$, otherwise $[A_{out}] = 0$.

Recovery time was quantified on both the individual and population levels, both relative to the end of the treatment period. In the former case, we define the recovery time RT_{cell} as the time required to achieve $0.5 * \max\left(\frac{d[C]}{dt}\right)$; in the latter, the recovery time RT_{pop} is defined as the time required for the cell density to double.

Periodic dosing simulations

Periodic dosing simulations were specified by the following parameters: the antibiotic concentration A, the duration of one dose T_1, the time between doses T_2, and the total number of doses N_{dose}. For each treatment, recovery time was determined relative to the end of the final dose; we denote the recovery time following N_{dose} as RT_n. Thus, large values of RT_n correspond to successful treatments, whereas minimal value of RT_n indicates treatment failure. Therefore, the results presented in Fig 2 correspond to RT_1; as shown in Appendix Fig S1, we find that RT_1 can predict the efficacy of multi-dose treatments; our results indicate a sharp transition in treatment efficacy at $\frac{T_2}{RT_1} \sim 1$. That is, any treatment where doses are spaced more frequently than the corresponding single dose recovery time (RT_1) will remain effective.

Modeling arbitrary dose profiles

To demonstrate the generality of the dependence between recovery time and total antibiotic, implemented both triangular and exponentially decaying dose profiles. For triangular doses, given the initial antibiotic concentration A_0 and duration D, we assumed a symmetric profile, with a maximum value of A_0 at $t = D/2$ For exponentially decaying doses, we assumed a rate constant such that $A_{out}(D) = 0.01A_0$.

Comparing antibiotic mechanisms of action

We adopted a minimal set of ODEs capable of representing various mechanisms of action (equations 10–12, Appendix Table S2 shows parameter values used). Here, $[T]$ and $[A]$ represent the drug target and intracellular antibiotic concentration, respectively; $[P]$ represents the antibiotic–target binding product. k_0 represents the basal target synthesis rate, k_f and k_b represent the binding and dissociation rates between antibiotic and target, k_p and K_T represent the maximum synthesis and half-maximal values of $[T]$, d_T and d_p represent the degradation rates of target and antibiotic–target complex, respectively. Finally, A_{out} represents the extracellular antibiotic concentration during a dose, which we assume to be constant; k_{in} and k_{out} represent the influx and efflux rates across the cell membrane, respectively. By setting various parameters to zero, these equations can be used to compare diverse modes of action. Setting $k_0 = 0$ results in target synthesis by positive feedback with rapid antibiotic-mediated degradation; setting $d_p = 0$ corresponds to antibiotic-inducing minimal target degradation; and setting $d_p = 0 = k_t$ corresponds to antibiotics whose targets are not subject to positive feedback. As with the full model, intracellular recovery time was calculated as the time to achieve $0.5 * \max\left(\frac{d[T]}{dt}\right)$ during recovery.

$$\frac{\mathrm{d}[T]}{\mathrm{d}t} = k_0 + \frac{k_t[T]^n}{K_t + [T]^n} - k_f[T][A] + k_b[P] - d_T[T] \tag{10}$$

$$\frac{\mathrm{d}[A]}{\mathrm{d}t} = k_{\mathrm{in}}A_{\mathrm{out}} - k_{\mathrm{out}}[A] - k_f[T][A] + k_b[P] + d_p[P] \tag{11}$$

$$\frac{\mathrm{d}[P]}{\mathrm{d}t} = k_f[T][A] - k_b[P] - d_P[P] \tag{12}$$

Acknowledgements

The authors wish to acknowledge assistance from the Duke University Shared Materials Instrumentation Facility (SMIF) in fabricating the microfluidic platform, the Duke University Light Microscopy Core Facility (LMCF, particularly Dr. Sam Johnson) in carrying out experiments, Hannah Meredith for helpful discussions on PAE, and Dr. Kui Zhu for input and advice on efflux inhibition. This study was partially supported by the US Army Research Office (W911NF-14-1-0490, LY), the National Institutes of Health (2R01-GM098642, LY), a David and Lucile Packard Fellowship (LY), a Duke Center for Biotechnology and Tissue Engineering Fellowship (JKS), and the Howard G. Clark fellowship (AJL).

Author contributions

JKS conceived the work, designed and carried out experimental and modeling studies and data analysis, and wrote the manuscript. SH designed and fabricated the microfluidic platform, assisted in experimental studies, and contributed to manuscript revisions. AJL fabricated microfluidic chips, assisted in carrying out experiments and data analysis, and contributed to manuscript revisions. LY conceived the work and assisted in research design, data interpretation, and writing and revising the manuscript.

References

AliAbadi FS, Lees P (2000) Antibiotic treatment for animals: effect on bacterial population and dosage regimen optimisation. Int J Antimicrob Agents 14: 307–313

Amaral L, Engi H, Viveiros M, Molnar J (2007) Comparison of multidrug resistant efflux pumps of cancer and bacterial cells with respect to the same inhibitory agents. In vivo 21: 237–244

Andreu N, Zelmer A, Fletcher T, Elkington PT, Ward TH, Ripoll J, Parish T, Bancroft GJ, Schaible U, Robertson BD, Wiles S (2010) Optimisation of bioluminescent reporters for use with mycobacteria. PLoS One 5: e10777

Askoura M, Mattawa W, Abujamel T, Taher I (2017) Efflux pump inhibitors (EPIs) as new antimicrobial agents against Pseudomonas aeruginosa. Libyan J Med 6: 5870

Athamna A (2004) In vitro post-antibiotic effect of fluoroquinolones, macrolides, -lactams, tetracyclines, vancomycin, clindamycin, linezolid, chloramphenicol, quinupristin/dalfopristin and rifampicin on Bacillus anthracis. J Antimicrob Chemother 53: 609–615

Avent ML, Rogers BA, Cheng AC, Paterson DL (2011) Current use of aminoglycosides: indications, pharmacokinetics and monitoring for toxicity. Intern Med J 41: 441–449

Beam TR, Gilbert DN, Kunin CM (1992) General guidelines for the clinical evaluation of anti-infective drug products. Clin Infect Dis 15: S5–S32

Bigger JW (1944) Treatment of staphylococcal infections with penicillin by intermittent sterilisation. Lancet 244: 497–500

Bryan LE, Van Den Elzen HM (1976) Streptomycin accumulation in susceptible and resistant strains of Escherichia coli and Pseudomonas aeruginosa. Antimicrob Agents Chemother 9: 928–938

Bundtzen RW, Gerber AU, Cohn DL (1981) Postantibiotic suppression of bacterial growth. Rev Infect Dis 3: 28–37

Cohen SP, Hooper DC, Wolfson JS (1988) Endogenous active efflux of norfloxacin in susceptible Escherichia coli. Antimicrob Agents Chemother 32: 1187–1191

Craig WA, Vogelman B (1987) The postantibiotic effect. Ann Intern Med 106: 900–902

Craig WA (1993) Post-antibiotic effects in experimental infection models: relationship to in-vitro phenomena and to treatment of infections in man. J Antimicrob Chemother 31: 149–158

Damper PD, Epstein W (1981) Role of the membrane potential in bacterial resistance to aminoglycoside antibiotics. Antimicrob Agents Chemother 20: 803

Eagle H (1949) The recovery of bacteria from the toxic effects of penicillin. J Clin Invest 28: 832

Eagle H, Musselman AD (1949) The slow recovery of bacteria from the toxic effects of penicillin. J Bacteriol 58: 475

Eagle H, Fleischman R (1950) The bactericidal action of penicillin in vivo: the participation of the host, and the slow recovery of the surviving organisms. Ann Intern Med 33: 544–571

Eagle H, Fleischman R, Musselman AD (1950) Effect of schedule of administration on the therapeutic efficacy of penicillin: importance of the aggregate time penicillin remains at effectively bactericidal levels. Am J Med 9: 280–299

Edmunds T, Goldberg AL (1986) Role of ATP hydrolysis in the degradation of proteins by protease La from Escherichia coli. J Cell Biochem 32: 187–191

Fishman N (2006) Antimicrobial stewardship. Am J Infect Control 34: S55–S63

Gottfredsson M, Erlendsdóttir H (1998) Characteristics and dynamics of bacterial populations during postantibiotic effect determined by flow cytometry. Antimicrob Agents Chemother 42: 1005–1011

Grenier F, Matteau D, Baby V, Rodrigue S (2014) Complete genome sequence of Escherichia coli BW25113. Genome Announc 2: e01038-14

Gudmundsson S, Einarsson S (1993) The post-antibiotic effect of antimicrobial combinations in a neutropenic murine thigh infection model. J Antimicrob Chemother 31: 171–191

Hanberger H, Nilsson LE, Kihlström E (1990) Postantibiotic effect of beta-lactam antibiotics on Escherichia coli evaluated by bioluminescence assay of bacterial ATP. Antimicrob Agents Chemother 34: 102–106

Hancock R (1962) Uptake of 14G-streptomycin by Bacillus megaterium. J Gen Microbiol 28: 503

Hurwitz C, Rosano CL (1962) Accumulation of label from C14-streptomycin by Escherichia coli. J Bacteriol 83: 1193–1201

Kaki R, Elligsen M, Walker S, Simor A, Palmay L, Daneman N (2011) Impact of antimicrobial stewardship in critical care: a systematic review. J Antimicrob Chemother 66: 1223–1230

Kaplan R, Apirion D (1975) The fate of ribosomes in Escherichia coli cells starved for a carbon source. J Biol Chem 250: 1854–1863

Kinoshita N, Unemoto T, Kobayashi H (1984) Proton motive force is not obligatory for growth of Escherichia coli. J Bacteriol 160: 1074–1077

Kriengkauykiat J, Porter E, Lomovskaya O, Wong-Beringer A (2005) Use of an efflux pump inhibitor to determine the prevalence of efflux pump-mediated fluoroquinolone resistance and multidrug resistance in Pseudomonas aeruginosa. Antimicrob Agents Chemother 49: 565–570

Lazzarini RA, Dahlberg AE (1971) The control of ribonucleic acid synthesis during amino acid deprivation in Escherichia coli. J Biol Chem 246: 420–429

Li RC, Lee SW, Kong CH (1997) Correlation between bactericidal activity and postantibiotic effect for five antibiotics with different mechanisms of action. J Antimicrob Chemother 40: 39–45

Lomovskaya O, Bostian KA (2006) Practical applications and feasibility of efflux pump inhibitors in the clinic—a vision for applied use. *Biochem Pharmacol* 71: 910–918

Lopatkin AJ, Huang S, Smith RP, Srimani JK, Sysoeva TA, Bewick S, Karig DK, You L (2016) Antibiotics as a selective driver for conjugation dynamics. *Nat Microbiol* 1: 16044

Ma D, Cook DN, Hearst JE, Nikaido H (1994) Efflux pumps and drug resistance in gram-negative bacteria. *Trends Microbiol* 2: 489–493

MacKenzie FM, Gould IM (1993) The post-antibiotic effect. *J Antimicrob Chemother* 32: 519–537

Mahamoud A, Chevalier J, Alibert-Franco S, Kern WV, Pages JM (2007) Antibiotic efflux pumps in Gram-negative bacteria: the inhibitor response strategy. *J Antimicrob Chemother* 59: 1223–1229

Marquez B (2005) Bacterial efflux systems and efflux pumps inhibitors. *Biochimie* 87: 1137–1147

Mcmurry LM, Oethinger M (1998) Overexpression of marA, soxS, or acrAB produces resistance to triclosan in laboratory and clinical strains of *Escherichia coli*. *FEMS Microbiol Lett* 166: 305–309

Meredith HR, Lopatkin AJ, Anderson DJ, You L (2015) Bacterial temporal dynamics enable optimal design of antibiotic treatment. *PLoS Comput Biol* 11: e1004201

Mizunaga S (2005) Influence of inoculum size of *Staphylococcus aureus* and *Pseudomonas aeruginosa* on *in vitro* activities and *in vivo* efficacy of fluoroquinolones and carbapenems. *J Antimicrob Chemother* 56: 91–96

Muir ME, van Heeswyck RS, Wallace BJ (1984) Effect of growth rate on streptomycin accumulation by *Escherichia coli* and *Bacillus megaterium*. *Microbiology* 130: 2015–2022

Neidhardt FC (1996) *Escherichia coli* and *Salmonella*: cellular and molecular biology. Washington, DC: ASM Press

Nikaido H (1994) Prevention of drug access to bacterial targets: permeability barriers and active efflux. *Science* 264: 382–388

Nikaido H (1998) Antibiotic resistance caused by gram-negative multidrug efflux pumps. *Clin Infect Dis* 27: S32–S41

Odenholt-Tornqvist I, Löwdin E (1991) Pharmacodynamic effects of subinhibitory concentrations of beta-lactam antibiotics *in vitro*. *Antimicrob Agents Chemother* 35: 1834–1839

Odenholt-Tornqvist I (1993) Studies on the postantibiotic effect and the postantibiotic sub-MIC effect of meropenem. *J Antimicrob Chemother* 31: 881–892

Okusu H, Ma D, Nikaido H (1996) AcrAB efflux pump plays a major role in the antibiotic resistance phenotype of *Escherichia coli* multiple-antibiotic-resistance (Mar) mutants. *J Bacteriol* 178: 306–308

Opperman TJ, Nguyen ST (2015) Recent advances toward a molecular mechanism of efflux pump inhibition. *Front Microbiol* 6: 11086

Paixão L, Rodrigues L, Couto I, Martins M, Fernandes P, de Carvalho CC, Monteiro GA, Sansonetty F, Amaral L, Viveiros M (2009) Fluorometric determination of ethidium bromide efflux kinetics in *Escherichia coli*. *J Biol Eng* 3: 18

Papich MG (2014) Pharmacokinetic–pharmacodynamic (PK–PD) modeling and the rational selection of dosage regimes for the prudent use of antimicrobial drugs. *Vet Microbiol* 171: 480–486

Parker RF, Marsh HC (1946) Action of penicillin on *Staphylococcus*. *J Bacteriol* 51: 181

Poole K, Lomovskaya O (2006) Can efflux inhibitors really counter resistance? *Drug Discov Today Ther Strateg* 3: 145–152

Prosser J, Killham K, Glover L, Rattray E (1996) Luminescence-based systems for detection of bacteria in the environment. *Crit Rev Biotechnol* 16: 157–183

Renneberg J, Walder M (1989) Postantibiotic effects of imipenem, norfloxacin, and amikacin *in vitro* and *in vivo*. *Antimicrob Agents Chemother* 33: 1714–1720

Singh M, Jadaun G, Srivastava K (2011) Effect of efflux pump inhibitors on drug susceptibility of ofloxacin resistant *Mycobacterium tuberculosis* isolates. *Indian J Med Res* 133: 535

Spivey J (1992) The postantibiotic effect. *Clin Pharm* 11: 865–875

Stavri M, Piddock LJV, Gibbons S (2007) Bacterial efflux pump inhibitors from natural sources. *J Antimicrob Chemother* 59: 1247–1260

Stubbings W (2006) Mechanisms of the post-antibiotic effects induced by rifampicin and gentamicin in *Escherichia coli*. *J Antimicrob Chemother* 58: 444–448

Talpaert MJ, Gopal Rao G, Cooper BS, Wade P (2011) Impact of guidelines and enhanced antibiotic stewardship on reducing broad-spectrum antibiotic usage and its effect on incidence of Clostridium difficile infection. *J Antimicrob Chemother* 66: 2168–2174

Tan C, Smith RP, Srimani JK, Riccione KA, Prasada S, Kuehn M, You L (2012) The inoculum effect and band-pass bacterial response to periodic antibiotic treatment. *Mol Syst Biol* 8: 617

Toutain PL, del Castillo JRE, Bousquet-Mélou A (2002) The pharmacokinetic–pharmacodynamic approach to a rational dosage regimen for antibiotics. *Res Vet Sci* 73: 105–114

Van Bambeke F, Lee VJ (2006) Inhibitors of bacterial efflux pumps as adjuvants in antibiotic treatments and diagnostic tools for detection of resistance by efflux. *Recent Pat Antiinfect Drug Discov* 1: 157–175

Viveiros M, Martins A, Paixão L, Rodrigues L, Martins M, Couto I, Fähnrich E, Kern WV, Amaral L (2008) Demonstration of intrinsic efflux activity of *Escherichia coli* K-12 AG100 by an automated ethidium bromide method. *Int J Antimicrob Agents* 31: 458–462

Vogelman B, Gudmundsson S, Leggett J, Turnidge J, Ebert S, Craig W (1988) Correlation of antimicrobial pharmacokinetic parameters with therapeutic efficacy in an animal model. *J Infect Dis* 158: 831–847

zur Wiesch PA, Abel S, Gkotzis S (2015) Classic reaction kinetics can explain complex patterns of antibiotic action. *Sci Transl Med* 7: 287ra273

Wiuff C, Zappala RM, Regoes RR, Garner KN, Baquero F, Levin BR (2005) Phenotypic tolerance: antibiotic enrichment of noninherited resistance in bacterial populations. *Antimicrob Agents Chemother* 49: 1483–1494

Zhanel GG, Hoban DJ (1991) The postantibiotic effect: a review of *in vitro* and *in vivo* data. *DICP* 25: 153–163

Zhanel GG, Craig WA (1994) Pharmacokinetic contributions to postantibiotic effects. *Clin Pharmacokinet* 27: 377–392

Zundel MA, Basturea GN, Deutscher MP (2009) Initiation of ribosome degradation during starvation in *Escherichia coli*. *RNA* 15: 977–983

9

Capturing protein communities by structural proteomics in a thermophilic eukaryote

Panagiotis L Kastritis[1,†] (iD), Francis J O'Reilly[1,2,†], Thomas Bock[1,†] (iD), Yuanyue Li[1] (iD), Matt Z Rogon[1], Katarzyna Buczak[1], Natalie Romanov[1] (iD), Matthew J Betts[3], Khanh Huy Bui[1,4] (iD), Wim J Hagen[1], Marco L Hennrich[1] (iD), Marie-Therese Mackmull[1] (iD), Juri Rappsilber[2,5] (iD), Robert B Russell[3], Peer Bork[1], Martin Beck[1,*] (iD) & Anne-Claude Gavin[1,**] (iD)

Abstract

The arrangement of proteins into complexes is a key organizational principle for many cellular functions. Although the topology of many complexes has been systematically analyzed in isolation, their molecular sociology *in situ* remains elusive. Here, we show that crude cellular extracts of a eukaryotic thermophile, *Chaetomium thermophilum*, retain basic principles of cellular organization. Using a structural proteomics approach, we simultaneously characterized the abundance, interactions, and structure of a third of the *C. thermophilum* proteome within these extracts. We identified 27 distinct protein communities that include 108 interconnected complexes, which dynamically associate with each other and functionally benefit from being in close proximity in the cell. Furthermore, we investigated the structure of fatty acid synthase within these extracts by cryoEM and this revealed multiple, flexible states of the enzyme in adaptation to its association with other complexes, thus exemplifying the need for *in situ* studies. As the components of the captured protein communities are known—at both the protein and complex levels—this study constitutes another step forward toward a molecular understanding of subcellular organization.

Keywords computational modeling; cryo-electron microscopy; fatty acid synthase; interaction proteomics; metabolon
Subject Categories Metabolism; Post-translational Modifications, Proteolysis & Proteomics; Structural Biology

Introduction

As the molecular machines of the cell, protein complexes are the cornerstones of most biological processes, and are the smallest, basic functional and structural units of proteome organization (Duve, 1975; Gavin *et al*, 2002; Krogan *et al*, 2006). Many individual studies and extensive proteome-wide screens in a variety of organisms have identified comprehensive repertoires of protein complexes and have provided insights into their molecular composition and anatomy (Gavin *et al*, 2002; Krogan *et al*, 2006; Kuhner *et al*, 2009; Amlacher *et al*, 2011; Havugimana *et al*, 2012; Lapinaite *et al*, 2013; von Appen *et al*, 2015; Hoffmann *et al*, 2015; Wan *et al*, 2015; Yan *et al*, 2015). These studies relied on extensive biochemical purification, often including multiple sequential steps or dimensions, and so inherently selected for the most biophysically stable assemblies. However, protein complexes—as an organizational principle—cannot account alone for the complex integration of the many cellular processes *in situ*. Additional layers of functional organization, beyond free diffusion and random collision of functional biomolecules within organelles, are required to ensure, for example, the efficient transfer of substrates along enzymatic pathways (dubbed metabolons; Srere, 1987; Wu & Minteer, 2015; Wan *et al*, 2015; Wheeldon *et al*, 2016), the effective transduction of signals (Wu, 2013), and the synthesis of proteins according to the local cellular needs (Gupta *et al*, 2016). This requires spatially and temporally synchronized sets of protein complexes—protein communities (Barabasi & Oltvai, 2004; Menche *et al*, 2015)—which we define as higher-order, often dynamically associated, assemblies of multiple macromolecular complexes that benefit from their close proximity to each other in the cell. To date, protein communities have not been properly conceptualized because experimental frameworks to capture this higher-order proteome organization are missing.

1 European Molecular Biology Laboratory, Structural and Computational Biology Unit, Heidelberg, Germany
2 Chair of Bioanalytics, Institute of Biotechnology, Technische Universität Berlin, Berlin, Germany
3 Cell Networks, Bioquant & Biochemie Zentrum Heidelberg, Heidelberg University, Heidelberg, Germany
4 Department of Anatomy and Cell Biology, McGill University, Montreal, QC, Canada
5 Wellcome Trust Centre for Cell Biology, School of Biological Sciences, University of Edinburgh, Edinburgh, UK
    *Corresponding author. E-mail: martin.beck@embl.de
    **Corresponding author. E-mail: gavin@embl.de
    †These authors contributed equally to this work

We used cell fractions from a thermophilic eukaryote, *Chaetomium thermophilum* (Amlacher *et al*, 2011), to delineate and characterize protein communities in crude extracts that retain aspects of cellular complexity. Our experimental design, in particular our choice of a thermophilic organism to minimize the disassembly of protein–protein interactions and the respective fractionation conditions, favors the identification of especially higher molecular weight species. To cope with the complexity of such samples, we combined quantitative mass spectrometry (MS) with electron microscopy (EM) and computational modeling approaches. We computed a network capturing various communities and demonstrate its usefulness for further analysis. We used cross-linking mass spectrometry (XL-MS) and EM to validate our approach, which shows that crude cellular extracts retain the basic principles of proteome organization. They are amenable to high-resolution cryoEM analyses of the sociology of protein complexes within their higher-order assemblies. As the proteins can be readily identified within these extracts, our methodological framework complements the emerging single-cell structural biology approaches that provide high-resolution snapshots of subcellular features (Beck & Baumeister, 2016; Mahamid *et al*, 2016) but are currently unable to pinpoint the underlying biomolecular entities.

Results

Cellular fractions serve as a proxy for the cellular environment and retain basic principles of cellular organization

Many fundamental components of the cell were first structurally investigated from thermophilic archaea because protein interactions in thermophiles have higher stability compared to their mesophilic counterparts. We chose to study the thermophilic eukaryote, *Chaetomium thermophilum*, a promising model organism for structurally investigating eukaryotic cell biology, because protein communities may be more robust than those from other model systems.

Large-scale analyses based on extensive, multi-dimensional fractionation have been applied to characterize protein complexes from various organisms and cell lines. These have all demonstrated that protein complexes—as biochemically highly stable entities—are an ubiquitous organizational principle (Wan *et al*, 2015). Our goal here was to capture more transient, higher-order associations and to characterize the functional organization of a eukaryotic proteome under conditions that mimic the native, cellular state. To achieve this, we obtained simple and crude cellular fractions (simplified cell lysates) from the thermophilic fungus *C. thermophilum* by single-step analytical size exclusion chromatography (SEC; Fig 1). The chromatographic method used here achieves relatively high resolution compared with gel filtration methods commonly used on a preparative scale (Kristensen *et al*, 2012) and the resulting 30 fractions span molecular weights ranging from ~0.2 to ~5 MDa. We first analyzed these fractions in biological triplicate by label-free quantitative liquid chromatography–mass spectrometry (LC-MS/MS) to characterize co-eluting proteins, complexes, and communities. We identified 1,176 proteins across all fractions that were present in at least two of the triplicates (Dataset EV1, Appendix Fig S1A), which account for 27.4% of the expressed proteome of *C. thermophilum* (Bock *et al*, 2014). For comparison, in human HeLa and U2OS cell lines, 19 and 29% of the proteome elutes in these high molecular weight SEC fractions, respectively (Kristensen *et al*, 2012; Kirkwood *et al*, 2013). Of these 1,176 proteins, 97% have a molecular weight < 200 kDa as a monomer but were still reproducibly identified in fractions corresponding to larger molecular masses, suggesting that most are engaged in large macromolecular assemblies.

Next, we determined an experimental elution profile for each protein by quantifying protein abundance based on iBAQ scoring (Schaab *et al*, 2012). The abundance of the detected proteins spans five orders of magnitude (Appendix Fig S1B and C), demonstrating that relatively rare complexes are also captured in this process. The elution profiles correlate well across the biological triplicates (squared Pearson coefficient; $0.82 < r^2 < 0.88$; Appendix Fig S1B and C, and Dataset EV1). Similarly, the protein composition of each SEC fraction was generally highly reproducible (Pearson coefficient; $0.61 < r < 0.98$; Appendix Fig S1D and Dataset EV1). To further assess the quality and effectiveness of the biochemical separation, we determined whether the observed elution profiles matched the composition, molecular weight, and stoichiometries of well-characterized and conserved protein complexes as contained in the Protein Data Bank (PDB; Berman *et al*, 2000). We generated 3D interaction models for 378 out of the identified 1,176 *C. thermophilum* proteins using comparative structural modeling that takes into account species-specific differences (cutoffs: > 30% sequence coverage, > 30% sequence identity; Appendix Figs S2 and S3, Dataset EV2, details in the Materials and Methods). The resulting benchmark set of structurally known protein complexes comprises 34 heteromers (involving 212 proteins) and 166 homomers, the latter mainly consisting of metabolic enzymes (Appendix Fig S2E). As expected, the subunits of the heteromultimeric complexes typically co-eluted in the same biochemical fractions (Fig 2A, Dataset EV2 and Appendix Fig S4), although a considerable number of proteins showed multiple elution peaks indicating that they are engaged in various complexes (Kuhner *et al*, 2009). For 102 protein complexes that eluted in a single peak (Dataset EV2), we also compared their predicted molecular weights to those estimated from their retention time (tR) during SEC elution (Fig 2B). In 52 well-characterized cases —for example, the chaperonin-containing TCP-1 (CCT) complex or the 19S proteasome—we observed a good agreement between the expected and observed tRs, further validating the general efficiency of the cell lysate separation procedure. However, 50 protein complexes eluted at much higher molecular weights than anticipated from their structural models. These shifts are unlikely to be due to non-specific post-lysis protein aggregation as no visible aggregates were formed under our experimental conditions (EM analysis, see below). They are therefore probably functionally relevant as we observed that co-eluting complexes share the same functional ontology (independent two-sample *t*-test *P*-value = 3.88E-50, Appendix Fig S5) or directly interact (cross-linking experiments, see below), suggesting a functional relationship. This is consistent with the view that protein complexes might self-assemble with higher stoichiometries, contain additional components—*that is*, RNA, DNA, metabolites, or proteins—and/or form uncharacterized, protein communities. An interesting example is the glycolytic enzyme enolase (EC 4.2.1.11) that forms a structurally characterized dimer *in vitro* (2 × 47.7 = 95.4 kDa; (Kuhnel & Luisi, 2001); PDB:2AL2) but seems to be part of a ~4-MDa assembly in the cellular fractions of *C. thermophilum* (Fig 2B). This supports previous

Separation of native higher-order assemblies from a eukaryotic thermophile

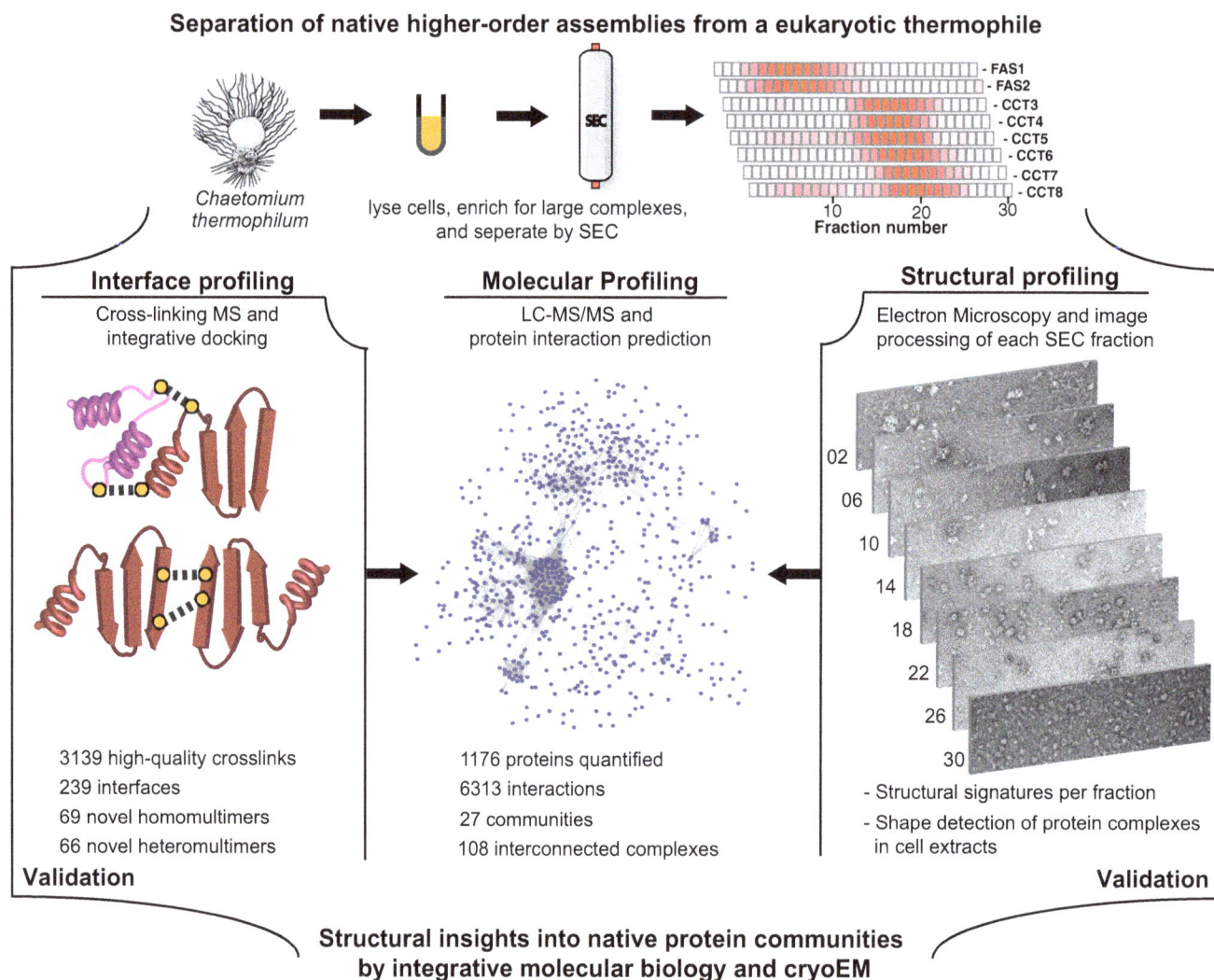

Figure 1. Overview of integrative structural network biology of native cell extracts in a thermophilic eukaryote.

We combined computational modeling approaches adapted from network biology (molecular profiling) with molecular biophysics, electron microscopy (EM; structural profiling), and quantitative and cross-linking mass spectrometry (interface profiling) to systematically chart and characterize the organization of protein complexes into functional, local communities. Large-scale electron microscopy and cross-linking mass spectrometry are used as validation tools.

indications that enolase participates in higher-order multienzyme assemblies, such as the somewhat elusive eukaryotic glycolytic metabolon (Menard *et al*, 2014). Overall, our operational definition of protein communities using a reproducible and sensitive structural proteomics approach captures important snapshots of the functional organization of cellular proteomes.

A compendium of *C. thermophilum* protein complexes within protein communities

We next used the protein elution profiles in conjunction with known functional associations to systematically define protein communities. Correlations between profiles can indicate membership of the same complex (Havugimana *et al*, 2012; Kristensen *et al*, 2012) or of protein communities that perform functions in a spatiotemporal context. For all possible protein pairs in the dataset, we calculated a Pearson correlation coefficient (cross-correlation co-elution (CCC)

score), to measure the similarity of their elution profiles (see Materials and Methods for details). Although distinct complexes can share similar and overlapping elution profiles (Havugimana *et al*, 2012), CCC scores discriminate between random co-eluting and interacting protein pairs (Appendix Fig S6). To improve the assignment of interaction probabilities, we also exploited a set of indirect interactions (e.g. genetic interaction, colocalization) from the STRING database (v.9.1; Franceschini *et al*, 2013). These are based on orthologs from *Saccharomyces cerevisiae* (Dataset EV3) and a set of non-redundant structural interfaces that share homology with *C. thermophilum* predicted interfaces using Mechismo (Betts *et al*, 2015; Materials and Methods; Dataset EV3). We combined these two datasets with the interaction probabilities derived from the elution profiles. We used a random forest classifier trained with randomly sampled sets of true-positive ($N = 5,000$) and true-negative ($N = 5,000$) interactions that we extracted from public sources after manual curation (PDB (Berman *et al*, 2000) and affinity purification–mass

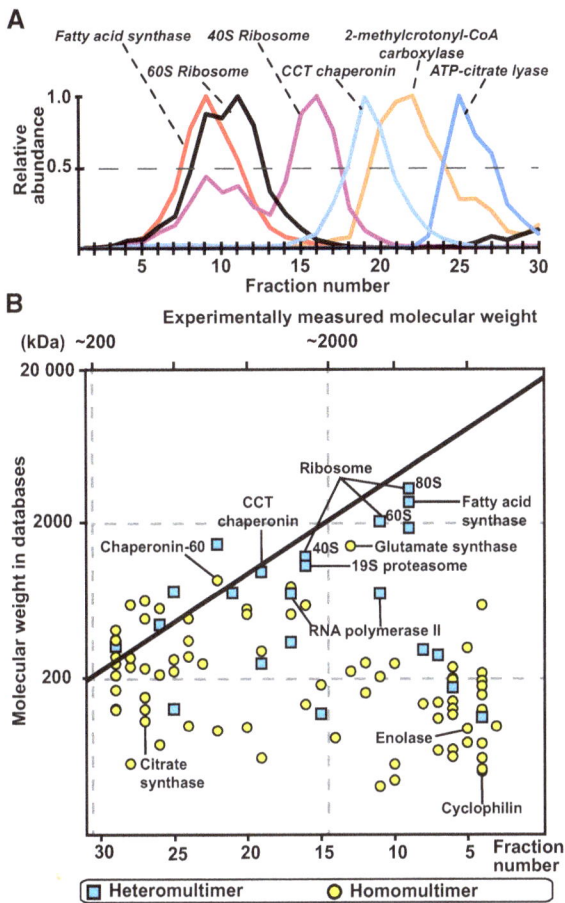

Figure 2. Identification of protein complexes and communities in the cellular extracts.

A Elution of selected protein complexes as a function of their retention times (see Appendix Fig S4 for their corresponding subunit elutions).

B Scatter plot indicating discrepancies in the expected and measured molecular weights of 102 protein complexes that elute as a single peak; 50% of protein complexes are observed to have higher molecular weights than structurally characterized, indicating that they are organized in higher-order assemblies.

account for protein complexes and 27 clusters accounting for protein communities that contain 108 interconnected protein complexes as subsets (Fig 3). Importantly, varying the parameters had only marginal impact on the final protein content (Dataset EV3 and Materials and Methods), highlighting the robustness of the protein communities. Overall, the protein communities include 62% of the set of known protein complexes (the set of known PDB and AP-MS data, Dataset EV2) with 90% average coverage of their components (Fig 3 and Dataset EV4). Of these communities, a well-known example is the ribosome protein community, which comprises not only the stable 60S and 40S ribosomal complexes but also the translation initiation factor eIF2B that is only transiently associated with the ribosome (Fig 3, Appendix Fig S8A). Other examples are novel such as the physical interaction between the Tup1-Cyc8 corepressor and a histone deacetylase complex (community #22), which is consistent with recent functional data demonstrating that these two complexes indeed cooperate to robustly repress transcription in yeast (Fleming *et al*, 2014). The analysis also captured a lipid anabolism metabolon (community #23), which not only includes the homomultimeric complexes of a cytochrome b reductase (Cbr1, which regulates the catalysis of sterol by biosynthetic enzymes) and a choline-phosphate cytidylyltransferase (Pct1, which is a rate-determining enzyme of the CDP-choline pathway for phosphatidylcholine synthesis), but also several enzymes in the sterol synthesis pathway. The transmembrane protein suppressor of choline sensitivity 2 (Scs2) is also observed, which is a known regulator of phospholipid metabolism. Its presence may seem peculiar at first; however, this provides physical evidence for a role for this community in validating the interconnectivity of lipid and sterol metabolism in fungi (Parks & Casey, 1995). Such coordinated regulatory effects may functionally optimize membrane plasticity and specificity (Ramgopal & Bloch, 1983). This community presumably localizes at the endoplasmic reticulum (ER)–plasma membrane (PM) interface as this is thought to be the location of all five predicted transmembrane proteins (Dataset EV4).

The protein communities include associations that have been reported as transient, non-stoichiometric or of low abundance in other organisms. For example, the 19S regulatory particle of the proteasome was found to be associated with two known components, Upb6 and Nas6, and the 20S core particle with two mutually exclusive alternative cap proteins, Blm10 and Cdc48 (Kish-Trier & Hill, 2013; Fig 3, Appendix Fig S8A). The protein communities also capture transient interactions between nuclear transport receptors and transport channel nucleoporins—specifically, the interactions between karyopherins and the Nsp1 complex and the Nup159 complex (Appendix Fig S8B)—that have been elusive in standard biochemical experiments (Patel & Rexach, 2008) and that were recently found to have high off-rates (Milles *et al*, 2015). Elsewhere, RNA polymerase II is found in a community with several splicing complexes, the U2 snRNP, the U4/U6.U5 tri-snRNP, and the smD3 complexes (Appendix Fig S8B, Dataset EV4). These spliceosomal machineries are known to interact with RNA polymerase II via the carboxy-terminal domain of its largest subunit, ensuring the tight coupling of mRNA transcription and splicing (Martins *et al*, 2011). We thus consider that our approach successfully identifies higher-order associations of complex core modules.

This compendium of *C. thermophilum* protein communities (Dataset EV4), which are precisely assigned to specific and highly

spectrometry (AP-MS) data (Benschop *et al*, 2010); Dataset EV3). We took a minimum interaction probability of 0.85 to construct a protein–protein interaction network (Appendix Figs S7–S10) that contains 679 proteins, 427 of which are not known to be members of protein complexes as their orthologs in yeast are not in any complex defined by PDB (Berman *et al*, 2000), AP-MS data (Benschop *et al*, 2010), or the Saccharomyces Genome Database (SGD; www.yeastgenome.org).

From this network, we used a clustering method that efficiently discovers densely connected overlapping regions that represent protein complexes and communities (ClusterONE; Nepusz *et al*, 2012). We systematized the recovery of protein complexes by an exhaustive parameter search and benchmarking (Sardiu *et al*, 2009) with the set of known structures (from the PDB) and yeast complexes (from AP-MS data; Dataset EV2; Materials and Methods). The optimal set of clustering parameters defines 21 clusters that

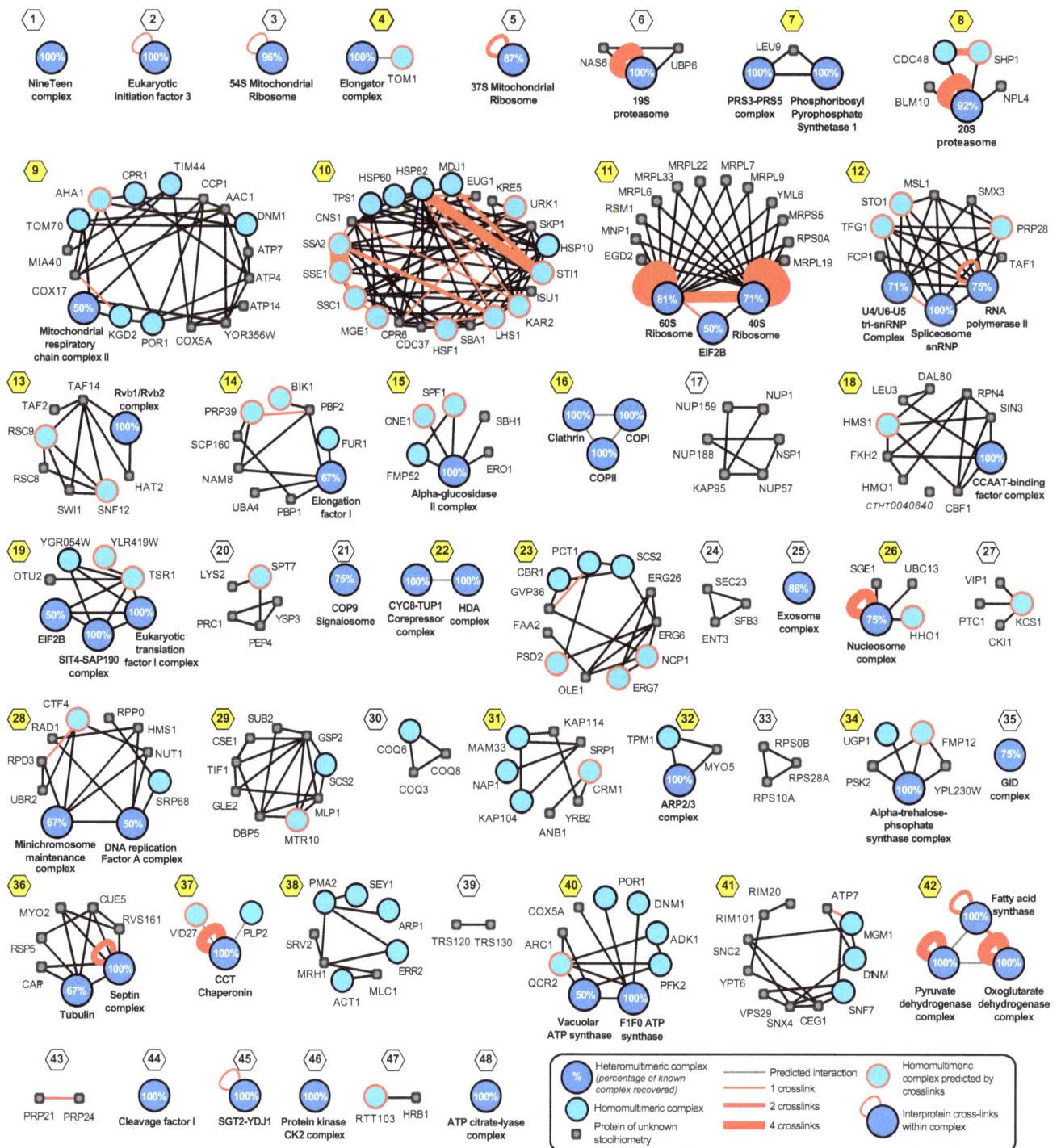

Figure 3. Network derived from large-scale fractionation predicts 48 protein complexes and communities.

Integration of experimental elution data, known functional associations, and predicted interaction interfaces from homologous proteins allow the creation of a high-quality network with interconnected protein complexes (Appendix Figs S8–S10). Here, known protein complexes are shown in blue and other physically associated proteins in gray, predicted interactions of complexes as gray lines and cross-links as red lines, and cross-links between different subunits of a heteromultimeric complex are represented with red loops (see insert). Communities containing multiple complexes are highlighted with yellow; numbering and naming of complexes and communities are described in the legend of Appendix Fig S9.

reproducible cellular fractions, represents an important resource for structural biologists (Appendix Fig S10). It not only captures transient associations but also identifies subunits of known complexes

that have so far remained elusive. Due to the evolutionary distance between *C. thermophilum* and most well-established model organisms, subunits of even highly conserved core complexes are not necessarily identified (or unambiguously identified) by sequence alignments. As exemplified in Appendix Fig S8A, the co-elution data can be used to identify such subunits and to assign orthology (details in Dataset EV4) by narrowing down a set of protein complex member candidates based on their experimental profiles (e.g. Appendix Fig S8A).

Characterization of new interaction interfaces by cross-linking mass spectrometry

Physical interactions inferred from co-occurrences can also be indirect, and so next we characterized the interaction interfaces occurring between members of the predicted protein communities by applying proteome-wide cross-linking MS (XL-MS) to the fractions (see Materials and Methods). To capture a large fraction of the interactome, we integrated three independent XL-MS datasets, which we acquired using different complementary protocols, for example, using different chemical cross-linkers, and both sequence-based and structure-based estimates of the false discovery rate (FDR; see Materials and Methods and Appendix Fig S11). We identified 3,139 high-quality cross-links (177 intermolecular and 2,962 intramolecular; Table 1) with sequence-based and structure-based FDRs of 10.0 and 12.0%, respectively (Dataset EV5). To validate the data, we checked which cross-linked peptide identifications are satisfied at the structural level, *that is,* correspond to distances between C_a atoms of cross-linked lysine residues smaller than 33 Å ($Lys(c_{a\text{-}}c_a) < 33$ Å). A comparison with all structurally known complexes (see above and Dataset EV5) revealed that 73% of intermolecular and 84% of intramolecular cross-links were satisfied. In addition, the measured $Lys(c_{a\text{-}}c_a)$ distances effectively recapitulated the expected log-normal distributions covered by the disuccinimidyl suberate (DSS) and BS3 cross-linkers, which further validates the calibration method we employed (Fig 4A, Appendix Fig S11). A significant fraction of the cross-linked peptides ($N = 2,732$) mapped interactions within single polypeptide chains and therefore probably define intramolecular contacts (Dataset EV5). The remaining 407 cross-linked peptides define 118 heteromultimeric (177 cross-links) and 121 homo-multimeric (230 cross-links) interfaces (Dataset EV5), which is largely consistent with our network analysis of protein communities (Fig 3) and the proteins forming the interconnected complexes (Appendix Fig S8A). Our analysis indicates that 135 (i.e. 56%) of these interfaces were previously unknown, and among the novel ones, 11 are between different complexes within the same community (Dataset EV5, Appendix Figs S11 and S12).

Overall, the cross-linking benchmarking methodology presented here suggests strict, but high-quality, structural validation that may be applicable to any cross-linking study on complex mixtures of proteins or complexes. For example, the XL-MS dataset validates a community of heat-shock complexes that elute with apparent molecular weights in the mega-dalton range (i.e. much higher than known complexes). We mapped nine new interfaces within this community, based on XL-MS data that suggest the existence of a complex interaction network or a chaperone community that comprises chaperones and co-chaperone complexes (Dataset EV5). Our XL-MS analysis further validates the notion of identifiable protein communities and is suggestive of several previously unknown interfaces that might be targeted for high-resolution structural studies.

Characterization of structural signatures of protein communities from cell extracts using fatty acid synthase as an example

To demonstrate that crude cellular fractions are amenable to the structural characterization of protein communities, we examined the different fractions for recurring structural signatures using EM (Fig 4B–D) without adding any cross-linker for further stabilization of interactions. Specifically, we acquired a large set of negatively stained electron micrographs of all fractions, identified single particles, and subjected them to 2D classification. We used cross-correlation to identify structural signatures recurring across neighboring fractions and the number of single particles contained within a class as a proxy for abundance (see Materials and Methods for details). Several structural signatures were observed, some of which were clearly recognizable as corresponding to known protein complexes, for example, the fatty acid synthase (FAS), the proteasome, and the 40S and 60S ribosome (Fig 4C). In these cases, both the quantitative MS and EM data were highly consistent with the molecular weight and size of the given complexes (Fig 4B and C). These results confirm the high quality of our profiling data and illustrate how compositionally complex samples might be rapidly annotated on the structural level in the future. We also observed several potentially novel structural signatures using orthogonal biochemical separation (Fig 4D), demonstrating that a wealth of structural information can be mined with this approach.

We next analyzed one of these structural signatures—fungal FAS—in more detail. In our analysis, FAS is a structurally prominent, 2.6-MDa complex that contains six copies of all eight catalytic centers comprising the complete metabolic pathway for 16- and 18-carbon fatty acid production. It is known to functionally interact with various other enzymes (FAS1 and FAS2 have 16 high-confidence interactors in *S. cerevisiae* according to STRING). Consistent with this notion, additional electron optical densities, probably corresponding to associated protein complexes, are observed that sometimes form linear elongated arrangements (Fig 5A and Appendix Fig S13). The majority locate outside the reaction chambers of the central wheel that is clearly manifested in 2D class

Table 1. Cross-linking statistics at a false discovery rate of 10%.

FDR 10%	Cross-links	Structurally mapped	Total interfaces covered	Novel interfaces
Total cross-links	3,139	931	239	135
Cross-links on monomers	2,732	851	–	–
Cross-links on homomultimers	230	36[a]	121	69
Cross-links on heteromultimers	177	44	118	66

[a]These cross-links show decrease in intra-residue distance when measured on known homomultimers by 26.3 ± 13.4 Å.

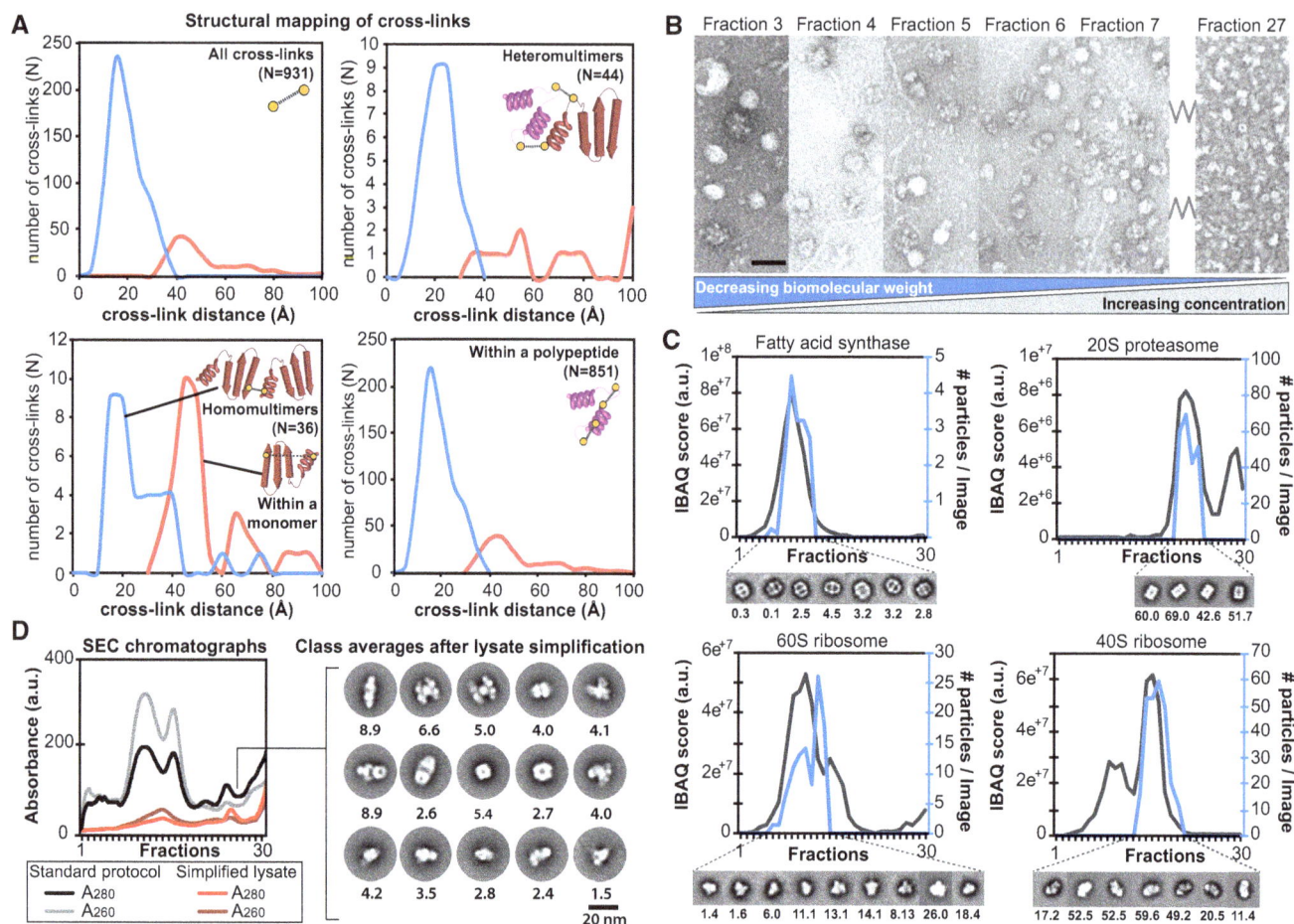

Figure 4. Higher-order assemblies identified by proteome-wide cross-linking mass spectrometry, biomolecular modeling, and negative-stain electron microscopy.

A Distance distributions of identified cross-links on top of the modeled protein complexes and identification of novel interactions. Satisfied distances are shown in blue and over-length cross-links are shown in red.

B Negatively stained electron micrographs of fractions 3–7 and 27 directly derived from size exclusion chromatography showing the structural signatures and their structural integrity within the fractions. Decreasing molecular weight correlates with increased protein concentration as a function of protein complex elution is highlighted. Scale bar: 60 nm.

C Abundance profiles as determined by quantitative mass spectrometry correlate with the number of observed single particles of the corresponding structural signature within the negative-staining electron micrographs; shown for fatty acid synthase, 20S proteasome, 60S and 40S ribosome (the number of particles per image per fraction is indicated below the class averages).

D Simplification of lysate (collecting only the flow-through from anion exchange chromatography) prior to SEC separation allows class averaging of structural signatures from complex fractions that were previously too low abundant. The number of particles per image per fraction is indicated below the class averages.

averages (Fig 5B and Appendix Fig S13). These additional electron optical densities proximal to FAS are seen more frequently than would be expected by random chance (Fig 5C). Their positioning at the entrance/exit tunnel of FAS (the malonyl transacylase domain) suggests the formation of a metabolon with other enzymes that deliver and accept substrates and products (as, for example, observed with acetyl-coA carboxylase (Acc1) in yeast using light microscopy; Suresh *et al*, 2015). To biochemically validate this observation, we utilized the fact that unlike FAS, many enzymes involved in fatty acid metabolism are covalently modified with the co-factor biotin. We therefore affinity-purified biotinylated proteins using avidin beads with subsequent XL-MS. The majority of the proteins in the eluate were known to be natively biotinylated except

for CTHT_0013320 (MCC2; the non-biotin-containing subunit of a carboxylase) and both subunits of FAS (CTHT_0037740, CTHT_0037750). We found the flexible acyl carrier protein (ACP) and malonyl/palmitoyl transferase (MPT) domains (that catalyze the first step in FA synthesis) to be cross-linked with the two sub-units of a carboxylase [CoA carboxylase beta-like; MCC2 and CTHT_0015140 (DUR1,2)] (Fig 5D). This interface characterized by cross-linking matches the one seen on the original cryoEM images and the 2D class averages and further supports the notion that a metabolon comprising other enzymes that deliver and accept substrates and products has been captured. Further corroborating the abovementioned findings, our SEC-MS co-elution data suggest an association of FAS with the same carboxylase (Dataset EV1). The

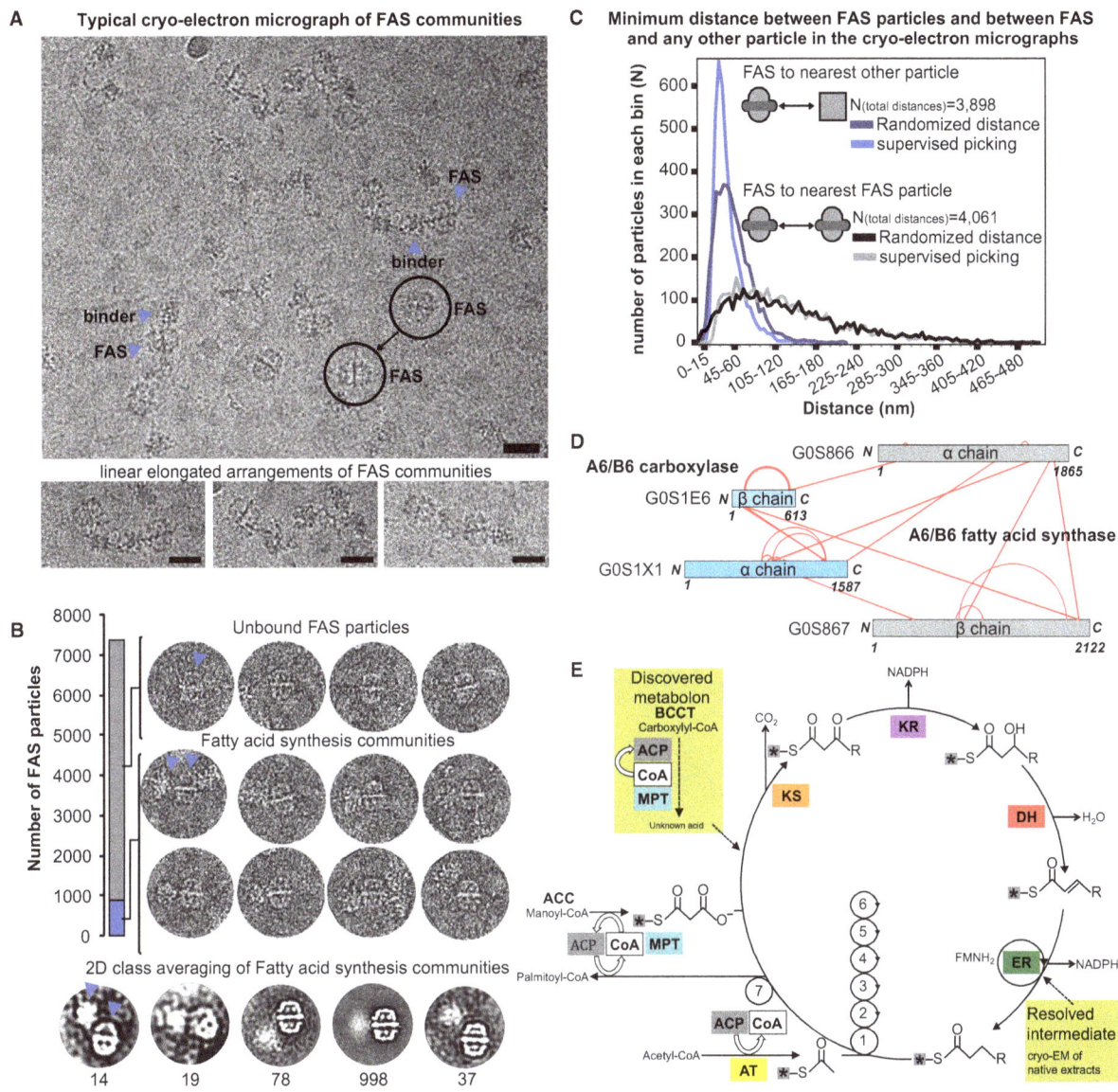

Figure 5. Visualization of transient interactions in fatty acid synthesis.

A   Communities in fatty acid metabolism and the quantification of intra-community distances within cryo-electron micrographs. Fatty acid synthase (FAS) frequently interacts with other sizeable protein complexes in a linear "pearl-string-like" arrangement and usually localizes at the edges of the community. Scale bars correspond to 25 nm. FAS particles (circles) and their nearest neighbors (arrow heads) are indicated.

B   Additional density outside of the ctFAS dome is observed in ~10% of the single particles; 2D class averages shown at the bottom. The arrow heads show typical assemblies within the pool of particles. In 90% of all cases, isolated FAS particles are seen (unbound state). In the remaining 10%, higher-order protein assemblies comprised of FAS particles and high molecular weight binders are seen (bound state).

C   Related to (A). Calculation of minimum distances between pairs of FAS molecules as well as FAS molecules and their closest non-FAS neighbors in comparison with random distributions. Whereas FAS molecules are randomly distributed, their binders are not, confirming physical interactions. Supervised picking means that all single particles were manually picked from the images. Randomized distance means that these manually picked particles were assigned random coordinates in each image (randomization of x, y coordinates considering image borders) and then their distance is calculated.

D   Cross-linking mass spectrometry data show that the binder is a carboxylase that is bound to the malonyl transacylase domain and acyl carrier protein (ACP) is in the vicinity, considering cross-link length and the positions of the lysine on the ctFAS structure. Cross-links come from both affinity-purified and fractionated cell extracts.

E   The molecular mechanisms in fatty acid synthesis (Wakil et al, 1983), and the relevance of the position of the ACP (see Fig 6 for details) and carboxylase to the catalytic cycle is indicated (see text). ACP, acyl carrier protein; CoA, acetyl-coenzyme A; MPT, malonyl/palmitoyl transferase; KS, ketoacyl synthase; KR, ketoacyl reductase; DH, dehydratase; ER, enoyl reductase; AT, acetyltransferase. Asterisks represent the acyl carrier protein (ACP).

organization of these domains in close proximity to each other implies a mechanism of substrate delivery from the carboxylase to FAS (Fig 5E). It is likely that this could be an alternate substrate for either odd-chain fatty acid synthesis (Fulco, 1983) or, less likely, fatty acid branching (Kolattukudy et al, 1987), provided via direct substrate channeling (Fig 5E). FAS and the carboxylase are known

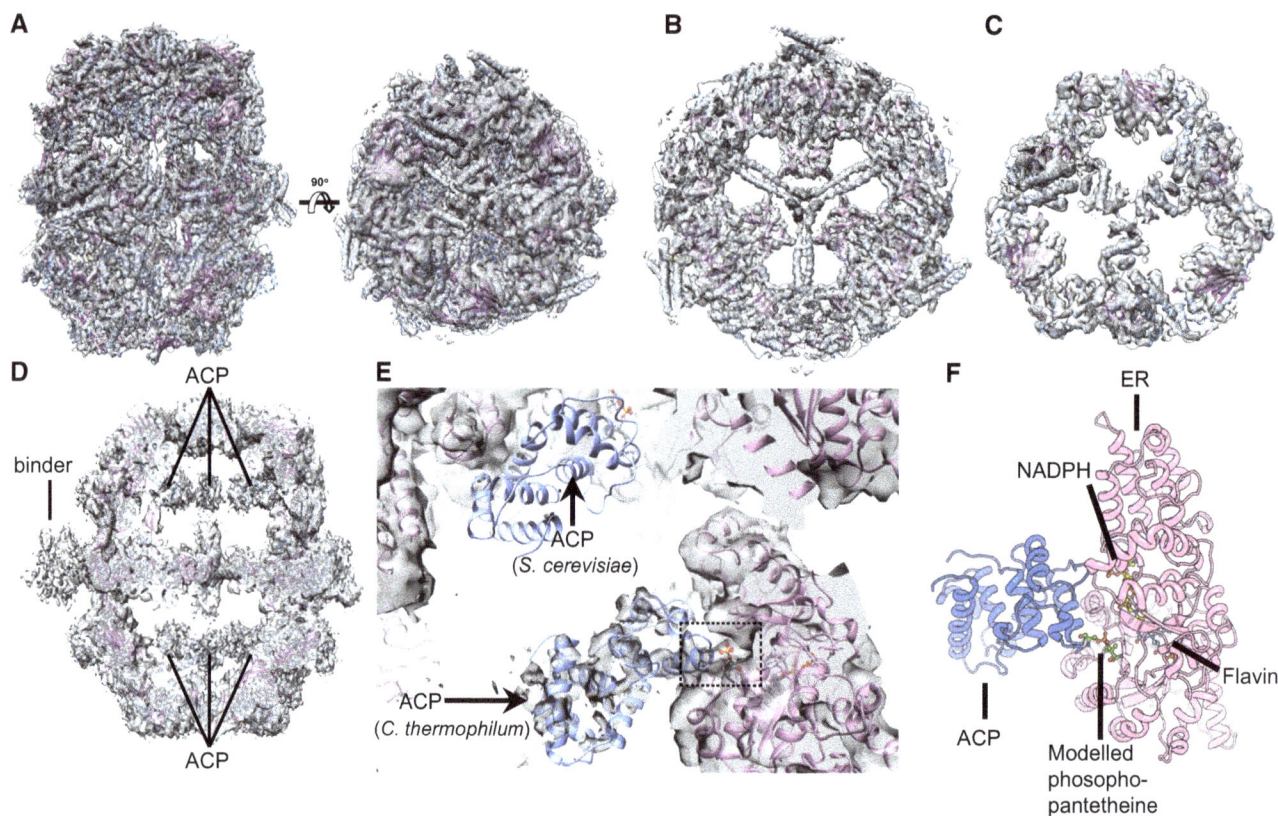

Figure 6. CryoEM structure of fatty acid synthase resolved to 4.7 Å as obtained from cryo-electron micrographs of fraction numbers 7–9.

A   The cryoEM map of *Chaetomium thermophilum* fatty acid synthase (ctFAS) is shown isosurface rendered and superimposed with the fitted X-ray structure of yeast FAS (Jenni *et al*, 2007). The domes and the cap show the unambiguous fit of α-helices and β-strands.

B   A slice through the central wheel of fungal FAS. The pitch of α-helices is resolved.

C   As for (B) but sliced through the dome structure.

D   Location of acyl carrier protein (ACP) within the cryoEM map of ctFAS and the position of additional density outside the dome.

E   Fit of ACP in the cryoEM map of ctFAS and comparison with the crystallographically determined location in yeast FAS; additional density within the active site possibly resembling the acyl chain bound on the ACP is observed.

F   Molecular model of the interaction between the ACP and the ER domains of ctFAS in cartoon representation. The model was derived from a rigid fit from (B) and subsequently flexibly refined for clash removal and interface energetics optimization.

to be two independent complexes and would therefore fit our definition of a community.

We next set out to test whether high-resolution structure determination is possible in these crude extracts. A high-resolution structure of FAS in isolation has been determined by X-ray crystallography (Jenni *et al*, 2007; Leibundgut *et al*, 2007; Lomakin *et al*, 2007). Using cryoEM (Gipson *et al*, 2010; Boehringer *et al*, 2013), certain regions—in particular the lid—remained unresolved, probably due to intrinsic flexibility. We acquired 1,917 cryo-electron micrographs of the relevant biochemical fraction and identified 7,370 single particles displaying the relevant structural signature in 1,597 micrographs (∼83%). Structural analysis and 3D classification resulted in a reconstruction at ∼4.7 Å containing only 3,933 particles (Appendix Fig S14), demonstrating that high-resolution structural analysis by cryoEM is feasible in complex cellular fractions. Overall, the cryoEM map of ctFAS recapitulated the X-ray structure of fungal FAS relatively well (Fig 6A–C), including high-resolution details such as the helical pitch in the central wheel (Fig 6B, Appendix Fig S15). In contrast to previous cryoEM structures of fungal FAS, even

the lid region was clearly resolved (Fig 6C, Appendix Fig S15). Thermophilic proteins are more susceptible to structural analysis by X-ray crystallography and NMR because they contain less flexible loops (Amlacher *et al*, 2011; Lapinaite *et al*, 2013). Our data indicate that this also extends to cryoEM, possibly because of reduced flexibility. Strikingly, the cryoEM structure did exhibit additional low-resolution density outside the reaction chambers that probably corresponds to the community discussed above (Fig 6D). Further, the ACP that iteratively shuttles the substrate within the catalytic chamber of FAS (Jenni *et al*, 2007) was captured at a different active site, albeit at slightly reduced resolution (Fig 6E and Appendix Fig S15). In previous structures, ACP located near the ketoacyl synthase domain involved in the first step in fatty acid synthesis (Jenni *et al*, 2007). Here, ACP is located in the vicinity of the enoyl reductase (ER; Fig 6E and F, and Appendix Fig S15) that reduces the α-β-double-bond of the acyl chain to a single bond. This final catalytic step in acyl chain metabolism is targeted by important antibacterial and antifungal drugs (e.g. Triclosan and Triclocarban, Atromentin and Leucomelone).

Discussion

The hypothesis of an intermediate layer of molecular sociology between supramolecular assemblies and organelles (Srere, 1987; Wu & Minteer, 2015) states that protein complexes spatially and temporally co-exist and directly interact with each other or individual proteins to form higher-order assemblies within specific cellular compartments, referred to here as protein communities (see Box 1). Such communities would be capable of channeling substrates for efficiency, could regulate pathway flux by transient binding kinetics, and would be formed by higher-order interactions (e.g. macromolecular crowding, excluded volume effects, "stickiness" of the cytoplasm, hydrodynamic interactions; Srere, 1987) and are attractive targets for biotechnology to increase reaction efficiencies (Wheeldon *et al*, 2016). Until now, a direct visualization or comprehensive analysis of such complexity was missing.

Although cellular fractions are more similar to the cellular environment than highly purified samples, they are less so than vitreous sections of the true cellular environment that nowadays can be studied using cryo-electron tomography but not by MS methods. We have shown that it is possible to capture at least some aspects of these protein communities in a systematic way using an integrative structural biology approach on cell fractions of the eukaryotic thermophile *C. thermophilum*, a model organism for structural biology as its proteins exhibit superior biochemical stability (Amlacher *et al*, 2011; Lin *et al*, 2016). The fractionation of cell extracts was postulated to retain close to native cellular interactions decades ago (Mowbray & Moses, 1976), but due to the molecular heterogeneity of these extracts, it was long thought to be prohibitive to structural characterization. In this study, we demonstrate that cellular fractions preserve basic principles of proteome organization and enable the identification of protein communities that are directly amenable to high-resolution cryoEM analyses. As a case in point for the latter, we structurally captured a specific catalytic step in fatty acid synthesis as well as some of the interfaces between FAS and other molecules using cryoEM in this setting. A wealth of other recurring structural signatures was identified, some readily recognizable but others novel and requiring further molecular characterization—a promising finding for structural and molecular biologists.

Overall, we designed an integrative approach specifically designed to identify and structurally characterize higher-order biomolecular assemblies. The specific elements implied the use of a chromatographic column to separate high molecular weight cellular assemblies and the choice of a thermophilic organism (to minimize the disassembly of protein–protein interactions upon lysis). The follow-up analyses are also tuned to cope with the large size of the communities (i.e. XL-MS with a cross-linker that identifies interactions up to 3 nm distance, and EM methods which are advantageous for higher-order assemblies of large molecular weight). The method described here is dedicated to the identification of protein communities, although of course other biomolecules such as nucleic acids or lipids might be part of the identified communities and contribute to their association. The combination with other identification strategies such as RNA sequencing and small molecule MS might further enlighten this aspect in the future. The broader applicability of cryoEM to non-purified samples will be limited by the abundance and the stability of the

Box 1. Organization models of a community involving enzymatic pathways.

Definition: Protein communities are higher-order, often dynamically associated, assemblies of multiple macromolecular complexes that benefit from their close proximity to each other in the cell. Protein communities imply spatially and temporally synchronized sets of protein complexes. They ensure, for example, the efficient transfer of substrates along enzymatic pathways (dubbed metabolons and illustrated in the bottom panel), the effective transduction of signals, and the synthesis of proteins according to the local cellular needs. The concept goes beyond the classical linear representations of pathways that imply freely diffusing and randomly colliding biomolecules (bottom right panel). The assembly of protein communities sometimes require molecular scaffolds (e.g. RNA, biological membranes, or structural proteins), and can be regulated by post-translational modifications. Shapes 1–4 in the panel below show different enzymes of a sequential pathway. In contrast to free diffusion, these enzymes might also multimerize within a community to increase reaction efficiencies. *Methods*: The characterization of protein communities implies their retrieval from *in vivo*, cellular and physiological contexts, and the choice of a thermophilic organism is expected to minimize their disassembly. Their biochemical purification can be achieved via affinity purification coupled to mass spectrometry (Gavin *et al*, 2002; Huttlin *et al*, 2017), or, more efficiently, directly in crude cellular fractions that retain aspects of cellular complexity and favor the identification of especially higher molecular weight species (this study). The latter is also amenable to the systematic characterization of protein communities through integrative structural biology approaches, implying for example, quantitative cross-linking mass spectrometry (XL-MS), electron microscopy (EM), and molecular, biophysical modeling.

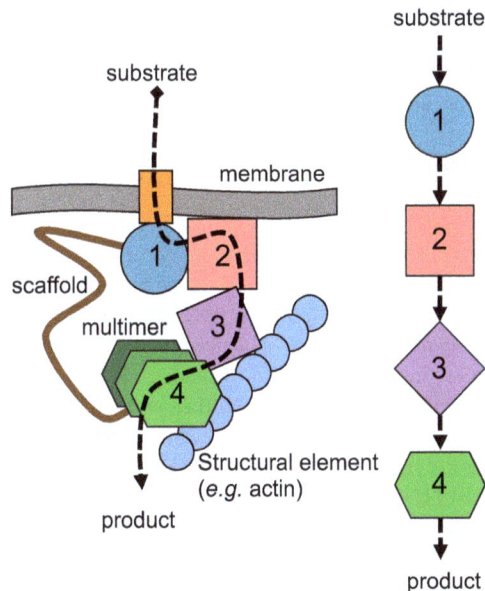

protein communities during the lysis procedure. However, methods to improve the stability of these interactions, potentially with cross-linking prior to fractionation or lysis, would allow discovery of further dynamic interactions and protein communities, and would allow further simplification of the protein mixtures for structural study using this pipeline. Advances in EM acquisition and data analysis methods might further improve the coverage and identification of protein communities in the future.

The emerging *in cellulo* structural biology approaches, based on the electron tomographic analyses of entire cells, have already started to produce the next generation of "big data" (Beck & Baumeister, 2016). These approaches hold great potential to structurally define protein communities in their native environment, the cell. They however fall short in the biochemical and molecular identification of these communities, as single-cell mass spectrometry is likely to remain limited to the few most abundant proteins for the near future. We anticipate that our approach that targets crude cellular extracts of intermediate molecular complexity as a proxy for the cellular milieu will crucially complement *in cellulo* methods because it allows a direct correlation between structural and molecular signatures.

Materials and Methods

Separation of *C. thermophilum* communities

Chaetomium thermophilum communities were enriched from cell lysates by spin filtration and fractionated using a Biosep SEC-S4000 (7.8 × 600) size exclusion chromatography (SEC) column from Phenomenex, Germany (see Appendix Supplementary Methods).

Protein co-elution prediction and mass spectrometry

Protein abundances were recorded from each SEC fraction by liquid chromatography–mass spectrometry (LC-MS). Prediction of protein co-elution was performed by Pearson correlation of protein abundance profiles. LC-MS data were processed using the MaxQuant (Cox & Mann, 2008).

Cross-linking/mass spectrometry

The cross-linking datasets searched with xQuest: Isotope-coded disuccinimidyl suberate (DSS; Creative Molecules) was used to perform cross-linking reactions as described previously (Walzthoeni et al, 2012). Cross-linked peptides were enriched by gel filtration before LC-MS analysis. All LC-MS data were obtained from an Orbitrap Velos Pro instrument (Thermo Scientific). Search and FDR were performed with the xQuest/xProphet (Leitner et al, 2014) software. For the cross-linking dataset searched with Xi, samples were cross-linked using a 1:1 w:w ratio of protein to BS3 (Thermo Scientific). Cross-linked peptides were enriched by gel filtration before LC-MS analysis. All LC-MS data were obtained from an Orbitrap Fusion Lumos Tribrid mass spectrometer (Thermo Scientific). Search and FDR were performed with the Xi (Giese et al, 2016) and XiFDR (Fischer & Rappsilber, 2017) software suites.

Prediction of protein communities

For each protein pair, interactions based on structural homologs were predicted using Mechismo (Betts et al, 2015), *Saccharomyces cerevisiae* orthologs were found using eggNOG (Jensen et al, 2008), and interaction data (excluding physical interactions) were downloaded from String (v.9.1; Franceschini et al, 2013). These data were combined with co-elution data from the SEC analysis using a Random Forest (RF) classifier and a manually curated training set of

reference interactions to filter out spurious connections and infer a network of high-confidence predicted interactions. Protein complexes and communities were inferred using ClusterONE (Nepusz et al, 2012). The cross-linking ld score (Walzthoeni et al, 2012) was calibrated on distance restraints imposed by the cross-linker. Cross-linking distances were calculated by Xwalk (Kahraman et al, 2011) using structural models.

Structure prediction of proteins participating in high molecular weight assemblies

Prediction of the structure of all 1,176 identified proteins was performed with iTASSER v4.2 (Yang et al, 2015) and Modeller 9v2 (Sali & Blundell, 1993). The best predicted model was selected according to its respective c-score (Roy et al, 2010). Details for model quality (for those with > 30% of sequence identity and coverage) are shown in Appendix Fig S2.

Protein complex assignment using Protein Data Bank and calibration of cross-linking quality

Each of the 1,176 proteins found in total in all three biological replicates was submitted to the NCBI BLAST server (http://blast.ncbi.nlm.nih.gov/Blast.cgi) and searched against the Protein Data Bank (PDB; www.pdb.org). A threshold of 30% of sequence identity was assigned. A decision on the assembly was taken after back-BLASTing the rest of the subunits, if any, of the PDB structure to the *C. thermophilum* proteome. All results are included in Dataset EV2.

Modeling of protein interfaces using cross-linking data

HADDOCK was used for modeling protein interfaces (de Vries et al, 2010; van Zundert et al, 2016). Cross-linking data were implemented as interaction restraints, set to have an effective (and maximum) Cα-Cα distance of 35.2 Å, whereas the minimum distance was defined only by energetics.

Negative-stain electron microscopy and 2D class averaging

Samples were directly deposited on glow-discharged (60 s) Quantifoil®, type 300 mesh grids and negative-stained with uranyl acetate 2% (w/w) water. Recording of data was performed with a side-mounted 1K CCD Camera (SIS). After data acquisition (pixel size = 7.1 Å), E2BOXER was used for particle picking (37,424 particles were picked out of 30 fractions). Class averaging was performed using RELION 1.2 (Scheres, 2012a,b). Cross-correlation of final class averages was performed using MATLAB v7.4.

*ct*FAS enzyme preparation and vitrification for cryoEM

ctFAS was ~50% enriched (see Appendix Fig S14) and overall protein concentration was determined to be ~40 ng/μl. Samples were then deposited on glow-discharged (60 sec) carbon-coated holey grids from Quantifoil®, type R2/1. A FEI Vitrobot® was used for plunge-freezing. In short, humidity was set to 70%, blotting and drain time to 3 and 0.5 s, respectively. Sample volume applied was 3 μl and blot offset was set to −3 mm.

CryoEM image acquisition, data processing, and 3D reconstruction

The vitrified samples were recorded on a FEI Titan Krios microscope at 300 kV. Pixel size was set to 2.16 Å and a FEI Falcon 2 camera was used in movie mode. Total dose applied was summed to 48 e$^-$/ Å^2, but the last frame was used only for particle picking. A total number of 13,419 micrographs were acquired in 21 h (1 frame/6 s; 1 movie/42 s). Motion correction was applied to acquired micrographs (Li *et al*, 2013). E2BOXER was used for particle picking. CTFFIND was used for CTF correction (Rohou & Grigorieff, 2015). The RELION 1.2 package (Scheres, 2012a,b) was then used for 2D class averaging, 3D classification, and 3D reconstruction of the density map. Default Gaussian mask from RELION 1.2 gave a calculated resolution (gold standard FSC = 0.143) of 4.7 Å.

Modeling of the ACP–enoyl reductase domain interaction and the FAS–carboxylase metabolon

Models of *C. thermophilum* acyl carrier protein (ACP) and enoyl reductase (ER) domains were generated using Modeller 9v2 and chosen structural homologs were selected from the yeast homolog with resolved densities for both (Leibundgut *et al*, 2007). Additional density of ACP was observed close to the ER domain of fatty acid synthase (FAS); thus, coarse placement of the ACP was performed using CHIMERA (Pettersen *et al*, 2004) and subsequently fitted to the density. Energy calculations were performed as previously described (Kastritis & Bonvin, 2010; Kastritis *et al*, 2014). Correlation of van der Waals energy with experimentally measured equilibrium dissociation constants for known complexes is derived from Kastritis *et al* (2014).

Acknowledgements

PLK and TB acknowledge Marie Curie Actions for the EMBL Interdisciplinary Postdoc (EIPOD) fellowship. The authors acknowledge Vera van Noort (KU Leuven) for kindly providing the orthologous genes in yeast for interaction mapping. The EMBL Electron Microscopy and Proteomics Core Facilities are acknowledged. The authors thank the Gavin and Beck laboratory members for valuable discussions. We thank all groups and group leaders in the EMBL Structural and Computational Biology Unit for inspiring discussions and creating a stimulating and vibrant environment. Part of this work was supported by the Wellcome Trust through a Senior Research Fellowship to JR [grant number 103139]. The Wellcome Centre for Cell Biology is supported by core funding from the Wellcome Trust [grant number 203149]. This work was supported by CellNetworks (Excellence Initiative of the University of Heidelberg) with funds given to RBR, PB, MB and A-CG.

Author contributions

A-CG, MB, PB, RBR, and JR supervised and administered the project and secured funding. A-CG, MB, PB, PLK, RBR, FJO'R, and TB wrote the manuscript. PLK designed and performed the electron microscopy experiments, carried out computation and modeling, analyzed the data, wrote the code, and solved the structure. FJO'R and TB designed and performed the SEC and the quantitative MS/MS and XL-MS experiments. MZR, FJO'R and PLK designed and analyzed the network. KHB and WJH contributed to cryoEM data acquisition and analysis. NR computationally analyzed data. YL, KB, M-TM, and MLH contributed to the SEC and MS experiments. MJB and RBR contributed the homologous interfaces for the network. All authors read and approved the final manuscript.

References

Amlacher S, Sarges P, Flemming D, van Noort V, Kunze R, Devos DP, Arumugam M, Bork P, Hurt E (2011) Insight into structure and assembly of the nuclear pore complex by utilizing the genome of a eukaryotic thermophile. *Cell* 146: 277–289

von Appen A, Kosinski J, Sparks L, Ori A, DiGuilio AL, Vollmer B, Mackmull MT, Banterle N, Parca L, Kastritis P, Buczak K, Mosalaganti S, Hagen W, Andres-Pons A, Lemke EA, Bork P, Antonin W, Glavy JS, Bui KH, Beck M (2015) *In situ* structural analysis of the human nuclear pore complex. *Nature* 526: 140–143

Barabasi AL, Oltvai ZN (2004) Network biology: understanding the cell's functional organization. *Nat Rev Genet* 5: 101–113

Beck M, Baumeister W (2016) Cryo-electron tomography: can it reveal the molecular sociology of cells in atomic detail? *Trends Cell Biol* 26: 825–837

Benschop JJ, Brabers N, van Leenen D, Bakker LV, van Deutekom HW, van Berkum NL, Apweiler E, Lijnzaad P, Holstege FC, Kemmeren P (2010) A consensus of core protein complex compositions for *Saccharomyces cerevisiae*. *Mol Cell* 38: 916–928

Berman HM, Westbrook J, Feng Z, Gilliland G, Bhat TN, Weissig H, Shindyalov IN, Bourne PE (2000) The protein data bank. *Nucleic Acids Res* 28: 235–242

Betts MJ, Lu Q, Jiang Y, Drusko A, Wichmann O, Utz M, Valtierra-Gutierrez IA, Schlesner M, Jaeger N, Jones DT, Pfister S, Lichter P, Eils R, Siebert R, Bork P, Apic G, Gavin AC, Russell RB (2015) Mechismo: predicting the mechanistic impact of mutations and modifications on molecular interactions. *Nucleic Acids Res* 43: e10

Bock T, Chen WH, Ori A, Malik N, Silva-Martin N, Huerta-Cepas J, Powell ST, Kastritis PL, Smyshlyaev G, Vonkova I, Kirkpatrick J, Doerks T, Nesme L, Bassler J, Kos M, Hurt E, Carlomagno T, Gavin AC, Barabas O, Muller CW *et al* (2014) An integrated approach for genome annotation of the eukaryotic thermophile *Chaetomium thermophilum*. *Nucleic Acids Res* 42: 13525–13533

Boehringer D, Ban N, Leibundgut M (2013) 7.5-A cryo-em structure of the mycobacterial fatty acid synthase. *J Mol Biol* 425: 841–849

Cox J, Mann M (2008) MaxQuant enables high peptide identification rates, individualized p.p.b.-range mass accuracies and proteome-wide protein quantification. *Nat Biotechnol* 26: 1367–1372

Duve C (1975) Exploring cells with a centrifuge. *Science* 189: 186–194

Fischer L, Rappsilber J (2017) Quirks of error estimation in cross-linking/mass spectrometry. *Anal Chem* 89: 3829–3833

Fleming AB, Beggs S, Church M, Tsukihashi Y, Pennings S (2014) The yeast Cyc8-Tup1 complex cooperates with Hda1p and Rpd3p histone deacetylases to robustly repress transcription of the subtelomeric FLO1 gene. *Biochem Biophys Acta* 1839: 1242–1255

Franceschini A, Szklarczyk D, Frankild S, Kuhn M, Simonovic M, Roth A, Lin J, Minguez P, Bork P, von Mering C, Jensen LJ (2013) STRING v9.1: protein-protein interaction networks, with increased coverage and integration. *Nucleic Acids Res* 41: D808–D815

Fulco AJ (1983) Fatty acid metabolism in bacteria. *Prog Lipid Res* 22: 133–160

Gavin AC, Bosche M, Krause R, Grandi P, Marzioch M, Bauer A, Schultz J, Rick JM, Michon AM, Cruciat CM, Remor M, Hofert C, Schelder M, Brajenovic M, Ruffner H, Merino A, Klein K, Hudak M, Dickson D, Rudi T *et al* (2002)

Functional organization of the yeast proteome by systematic analysis of protein complexes. *Nature* 415: 141–147

Giese SH, Fischer L, Rappsilber J (2016) A study into the Collision-induced Dissociation (CID) behavior of cross-linked peptides. *Mol Cell Proteomics* 15: 1094–1104

Gipson P, Mills DJ, Wouts R, Grininger M, Vonck J, Kuhlbrandt W (2010) Direct structural insight into the substrate-shuttling mechanism of yeast fatty acid synthase by electron cryomicroscopy. *Proc Natl Acad Sci USA* 107: 9164–9169

Gupta I, Villanyi Z, Kassem S, Hughes C, Panasenko OO, Steinmetz LM, Collart MA (2016) Translational capacity of a cell is determined during transcription elongation via the Ccr4-Not complex. *Cell Rep* 15: 1782–1794

Havugimana PC, Hart GT, Nepusz T, Yang H, Turinsky AL, Li Z, Wang PI, Boutz DR, Fong V, Phanse S, Babu M, Craig SA, Hu P, Wan C, Vlasblom J, Dar VU, Bezginov A, Clark GW, Wu GC, Wodak SJ et al (2012) A census of human soluble protein complexes. *Cell* 150: 1068–1081

Hoffmann NA, Jakobi AJ, Moreno-Morcillo M, Glatt S, Kosinski J, Hagen WJ, Sachse C, Muller CW (2015) Molecular structures of unbound and transcribing RNA polymerase III. *Nature* 528: 231–236

Huttlin EL, Bruckner RJ, Paulo JA, Cannon JR, Ting L, Baltier K, Colby G, Gebreab F, Gygi MP, Parzen H, Szpyt J, Tam S, Zarraga G, Pontano-Vaites L, Swarup S, White AE, Schweppe DK, Rad R, Erickson BK, Obar RA et al (2017) Architecture of the human interactome defines protein communities and disease networks. *Nature* 545: 505–509

Jenni S, Leibundgut M, Boehringer D, Frick C, Mikolasek B, Ban N (2007) Structure of fungal fatty acid synthase and implications for iterative substrate shuttling. *Science* 316: 254–261

Jensen LJ, Julien P, Kuhn M, von Mering C, Muller J, Doerks T, Bork P (2008) eggNOG: automated construction and annotation of orthologous groups of genes. *Nucleic Acids Res* 36: D250–D254

Kahraman A, Malmstrom L, Aebersold R (2011) Xwalk: computing and visualizing distances in cross-linking experiments. *Bioinformatics* 27: 2163–2164

Kastritis PL, Bonvin AM (2010) Are scoring functions in protein-protein docking ready to predict interactomes? Clues from a novel binding affinity benchmark. *J Proteome Res* 9: 2216–2225

Kastritis PL, Rodrigues JP, Folkers GE, Boelens R, Bonvin AM (2014) Proteins feel more than they see: fine-tuning of binding affinity by properties of the non-interacting surface. *J Mol Biol* 426: 2632–2652

Kirkwood KJ, Ahmad Y, Larance M, Lamond AI (2013) Characterization of native protein complexes and protein isoform variation using size-fractionation-based quantitative proteomics. *Mol Cell Proteomics* 12: 3851–3873

Kish-Trier E, Hill CP (2013) Structural biology of the proteasome. *Annu Rev Biophys* 42: 29–49

Kolattukudy PE, Rogers LM, Balapangu A (1987) Synthesis of methyl-branched fatty acids from methylmalonyl-CoA by fatty acid synthase from both the liver and the harderian gland of the guinea pig. *Arch Biochem Biophys* 255: 205–209

Kristensen AR, Gsponer J, Foster LJ (2012) A high-throughput approach for measuring temporal changes in the interactome. *Nat Methods* 9: 907–909

Krogan NJ, Cagney G, Yu H, Zhong G, Guo X, Ignatchenko A, Li J, Pu S, Datta N, Tikuisis AP, Punna T, Peregrin-Alvarez JM, Shales M, Zhang X, Davey M, Robinson MD, Paccanaro A, Bray JE, Sheung A, Beattie B et al (2006) Global landscape of protein complexes in the yeast *Saccharomyces cerevisiae*. *Nature* 440: 637–643

Kuhnel K, Luisi BF (2001) Crystal structure of the *Escherichia coli* RNA degradosome component enolase. *J Mol Biol* 313: 583–592

Kuhner S, van Noort V, Betts MJ, Leo-Macias A, Batisse C, Rode M, Yamada T,

Maier T, Bader S, Beltran-Alvarez P, Castano-Diez D, Chen WH, Devos D, Guell M, Norambuena T, Racke I, Rybin V, Schmidt A, Yus E, Aebersold R et al (2009) Proteome organization in a genome-reduced bacterium. *Science* 326: 1235–1240

Lapinaite A, Simon B, Skjaerven L, Rakwalska-Bange M, Gabel F, Carlomagno T (2013) The structure of the box C/D enzyme reveals regulation of RNA methylation. *Nature* 502: 519–523

Leibundgut M, Jenni S, Frick C, Ban N (2007) Structural basis for substrate delivery by acyl carrier protein in the yeast fatty acid synthase. *Science* 316: 288–290

Leitner A, Walzthoeni T, Aebersold R (2014) Lysine-specific chemical cross-linking of protein complexes and identification of cross-linking sites using LC-MS/MS and the xQuest/xProphet software pipeline. *Nat Protoc* 9: 120–137

Li X, Mooney P, Zheng S, Booth CR, Braunfeld MB, Gubbens S, Agard DA, Cheng Y (2013) Electron counting and beam-induced motion correction enable near-atomic-resolution single-particle cryo-EM. *Nat Methods* 10: 584–590

Lin DH, Stuwe T, Schilbach S, Rundlet EJ, Perriches T, Mobbs G, Fan Y, Thierbach K, Huber FM, Collins LN, Davenport AM, Jeon YE, Hoelz A (2016) Architecture of the symmetric core of the nuclear pore. *Science* 352: aaf1015

Lomakin IB, Xiong Y, Steitz TA (2007) The crystal structure of yeast fatty acid synthase, a cellular machine with eight active sites working together. *Cell* 129: 319–332

Mahamid J, Pfeffer S, Schaffer M, Villa E, Danev R, Cuellar LK, Forster F, Hyman AA, Plitzko JM, Baumeister W (2016) Visualizing the molecular sociology at the HeLa cell nuclear periphery. *Science* 351: 969–972

Martins SB, Rino J, Carvalho T, Carvalho C, Yoshida M, Klose JM, de Almeida SF, Carmo-Fonseca M (2011) Spliceosome assembly is coupled to RNA polymerase II dynamics at the 3′ end of human genes. *Nat Struct Mol Biol* 18: 1115–1123

Menard L, Maughan D, Vigoreaux J (2014) The structural and functional coordination of glycolytic enzymes in muscle: evidence of a metabolon? *Biology* 3: 623–644

Menche J, Sharma A, Kitsak M, Ghiassian SD, Vidal M, Loscalzo J, Barabasi AL (2015) Disease networks. Uncovering disease-disease relationships through the incomplete interactome. *Science* 347: 1257601

Milles S, Mercadante D, Aramburu IV, Jensen MR, Banterle N, Koehler C, Tyagi S, Clarke J, Shammas SL, Blackledge M, Grater F, Lemke EA (2015) Plasticity of an ultrafast interaction between nucleoporins and nuclear transport receptors. *Cell* 163: 734–745

Mowbray J, Moses V (1976) The tentative identification in *Escherichia coli* of a multienzyme complex with glycolytic activity. *Eur J Biochem* 66: 25–36

Nepusz T, Yu H, Paccanaro A (2012) Detecting overlapping protein complexes in protein-protein interaction networks. *Nat Methods* 9: 471–472

Parks LW, Casey WM (1995) Physiological implications of sterol biosynthesis in yeast. *Annu Rev Microbiol* 49: 95–116

Patel SS, Rexach MF (2008) Discovering novel interactions at the nuclear pore complex using bead halo: a rapid method for detecting molecular interactions of high and low affinity at equilibrium. *Mol Cell Proteomics* 7: 121–131

Pettersen EF, Goddard TD, Huang CC, Couch GS, Greenblatt DM, Meng EC, Ferrin TE (2004) UCSF Chimera—a visualization system for exploratory research and analysis. *J Comput Chem* 25: 1605–1612

Ramgopal M, Bloch K (1983) Sterol synergism in yeast. *Proc Natl Acad Sci USA* 80: 712–715

Rohou A, Grigorieff N (2015) CTFFIND4: fast and accurate defocus estimation from electron micrographs. *J Struct Biol* 192: 216–221

Roy A, Kucukural A, Zhang Y (2010) I-TASSER: a unified platform for automated protein structure and function prediction. *Nat Protoc* 5: 725–738

Sali A, Blundell TL (1993) Comparative protein modelling by satisfaction of spatial restraints. *J Mol Biol* 234: 779–815

Sardiu ME, Florens L, Washburn MP (2009) Evaluation of clustering algorithms for protein complex and protein interaction network assembly. *J Proteome Res* 8: 2944–2952

Schaab C, Geiger T, Stoehr G, Cox J, Mann M (2012) Analysis of high accuracy, quantitative proteomics data in the MaxQB database. *Mol Cell Proteomics* 11: M111.014068

Scheres SH (2012a) A Bayesian view on cryo-EM structure determination. *J Mol Biol* 415: 406–418

Scheres SH (2012b) RELION: implementation of a Bayesian approach to cryo-EM structure determination. *J Struct Biol* 180: 519–530

Srere PA (1987) Complexes of sequential metabolic enzymes. *Annu Rev Biochem* 56: 89–124

Suresh HG, da Silveira Dos Santos AX, Kukulski W, Tyedmers J, Riezman H, Bukau B, Mogk A (2015) Prolonged starvation drives reversible sequestration of lipid biosynthetic enzymes and organelle reorganization in *Saccharomyces cerevisiae*. *Mol Biol Cell* 26: 1601–1615

de Vries SJ, van Dijk M, Bonvin AM (2010) The HADDOCK web server for data-driven biomolecular docking. *Nat Protoc* 5: 883–897

Wakil SJ, Stoops JK, Joshi VC (1983) Fatty acid synthesis and its regulation. *Annu Rev Biochem* 52: 537 579

Walzthoeni T, Claassen M, Leitner A, Herzog F, Bohn S, Forster F, Beck M, Aebersold R (2012) False discovery rate estimation for cross-linked peptides identified by mass spectrometry. *Nat Methods* 9: 901–903

Wan C, Borgeson B, Phanse S, Tu F, Drew K, Clark G, Xiong X, Kagan O, Kwan J, Bezginov A, Chessman K, Pal S, Cromar G, Papoulas O, Ni Z, Boutz DR, Stoilova S, Havugimana PC, Guo X, Malty RH *et al* (2015) Panorama of ancient metazoan macromolecular complexes. *Nature* 525: 339–344

Wheeldon I, Minteer SD, Banta S, Barton SC, Atanassov P, Sigman M (2016) Substrate channelling as an approach to cascade reactions. *Nat Chem* 8: 299–309

Wu H (2013) Higher-order assemblies in a new paradigm of signal transduction. *Cell* 153: 287–292

Wu F, Minteer S (2015) Krebs cycle metabolon: structural evidence of substrate channeling revealed by cross-linking and mass spectrometry. *Angew Chem* 54: 1851–1854

Yan C, Hang J, Wan R, Huang M, Wong CC, Shi Y (2015) Structure of a yeast spliceosome at 3.6-angstrom resolution. *Science* 349: 1182–1191

Yang J, Yan R, Roy A, Xu D, Poisson J, Zhang Y (2015) The I-TASSER Suite: protein structure and function prediction. *Nat Methods* 12: 7–8

van Zundert GC, Rodrigues JP, Trellet M, Schmitz C, Kastritis PL, Karaca E, Melquiond AS, van Dijk M, de Vries SJ, Bonvin AM (2016) The HADDOCK2.2 web server: user-friendly integrative modeling of biomolecular complexes. *J Mol Biol* 428: 720–725

10

Transcription factor family-specific DNA shape readout revealed by quantitative specificity models

Lin Yang[1,†], Yaron Orenstein[2,†,‡], Arttu Jolma[3], Yimeng Yin[3], Jussi Taipale[3], Ron Shamir[2,*] (ID) & Remo Rohs[1,**] (ID)

Abstract

Transcription factors (TFs) achieve DNA-binding specificity through contacts with functional groups of bases (base readout) and readout of structural properties of the double helix (shape readout). Currently, it remains unclear whether DNA shape readout is utilized by only a few selected TF families, or whether this mechanism is used extensively by most TF families. We resequenced data from previously published HT-SELEX experiments, the most extensive mammalian TF–DNA binding data available to date. Using these data, we demonstrated the contributions of DNA shape readout across diverse TF families and its importance in core motif-flanking regions. Statistical machine-learning models combined with feature-selection techniques helped to reveal the nucleotide position-dependent DNA shape readout in TF-binding sites and the TF family-specific position dependence. Based on these results, we proposed novel DNA shape logos to visualize the DNA shape preferences of TFs. Overall, this work suggests a way of obtaining mechanistic insights into TF–DNA binding without relying on experimentally solved all-atom structures.

Keywords binding specificity; DNA shape; feature selection; quantitative modeling; transcription factor
Subject Categories Genome-Scale & Integrative Biology; Structural Biology; Transcription

Introduction

Protein–DNA interactions play a central role in gene regulation. Transcription factors (TFs) are proteins that recognize specific DNA sequences. They bind to regulatory regions in the genome and consequently activate or repress transcription of target genes. TFs can bind various DNA sequences with different DNA-binding affinities or specificities. In the last decade, technologies for measuring protein DNA-binding specificities have advanced tremendously (Slattery *et al*, 2014). Platforms based on microarray technology, such as protein-binding microarray (PBM; Berger *et al*, 2006), and high-throughput sequencing technology, such as high-throughput SELEX (HT-SELEX; Jolma *et al*, 2010) or SELEX-seq (Slattery *et al*, 2011), have enabled measurements of protein binding against thousands or even millions of different DNA sequences. The computational challenges are to develop accurate and quantitative models of protein–DNA binding specificities from these massive datasets and to infer binding mechanisms.

Position weight matrix (PWM) or PWM-like models are widely used to represent DNA-binding preferences of proteins (Stormo, 2000). In these models, a matrix is used to represent the TF-binding site (TFBS), with each element representing the contribution to the overall binding affinity from a nucleotide at the corresponding position. An inherent assumption of traditional PWM models is position independence; that is, the contribution of different nucleotide positions within a TFBS to the overall binding affinity is assumed to be additive. Although this approximation is broadly valid, nevertheless, it does not hold for several proteins (Man & Stormo, 2001; Bulyk *et al*, 2002). To improve quantitative modeling, PWM models have been extended to include additional parameters, such as *k*-mer features, to account for position dependencies within TFBSs (Zhao *et al*, 2012; Mathelier & Wasserman, 2013; Mordelet *et al*, 2013; Weirauch *et al*, 2013; Riley *et al*, 2015). Interdependencies between nucleotide positions have a structural origin. For example, stacking interactions between adjacent base pairs form the local three-dimensional DNA structure. TFs have preferences for sequence-dependent DNA conformation, which we call DNA shape readout (Rohs *et al*, 2009, 2010).

Based on this rationale, an alternative approach to augment traditional PWM models is the inclusion of DNA structural features. Models of TF–DNA binding specificity incorporating these DNA shape features achieved comparable performance levels to models

---

1   Molecular and Computational Biology Program, Departments of Biological Sciences, Chemistry, Physics & Astronomy, and Computer Science, University of Southern California, Los Angeles, CA, USA
2   Blavatnik School of Computer Science, Tel Aviv University, Tel Aviv, Israel
3   Division of Functional Genomics and Systems Biology, Department of Medical Biochemistry and Biophysics, Karolinska Institutet, Stockholm, Sweden
    *Corresponding author. E-mail: rshamir@tau.ac.il
    **Corresponding author. E-mail: rohs@usc.edu
    † These authors contributed equally to this work
    ‡ Present address: Computer Science and Artificial Intelligence Laboratory, Massachusetts Institute of Technology, Cambridge, MA, USA

incorporating higher-order k-mer features, while requiring a much smaller number of parameters (Zhou *et al*, 2015). We previously revealed the importance of DNA shape readout for members of the basic helix-loop-helix (bHLH) and homeodomain TF families (Dror *et al*, 2014; Yang *et al*, 2014; Zhou *et al*, 2015). We were also able, for Hox TFs, to identify which regions in the TFBSs used DNA shape readout, demonstrating the power of the approach to reveal mechanistic insights into TF–DNA recognition (Abe *et al*, 2015). This capability was extensively shown for only two protein families, due to the lack of large-scale high-quality TF–DNA binding data. With the recent abundance of high-throughput measurements of protein–DNA binding, it is now possible to dissect the role of DNA shape readout for many TF families.

In this study, we used the most extensive mammalian TF–DNA binding affinity datasets available to date, derived from HT-SELEX experiments (Jolma *et al*, 2013), to inform DNA shape-based binding models. To improve statistical robustness of the analysis, we augmented each experiment by increasing the sequencing depth of existing HT-SELEX data (Jolma *et al*, 2013). We implemented a pipeline to derive accurate TF-binding intensities for all possible DNA M-words (sequences of length M) from HT-SELEX reads. Using these preprocessed data, we trained machine-learning models of TF–DNA binding specificities. Finally, using feature selection, we pinpointed positions in the TFBSs where DNA shape readout is most likely to occur.

Results

HT-SELEX experimental data provide accurate M-word scores for diverse TF families

We analyzed HT-SELEX data, including 548 experiments covering 410 human and mouse proteins from 40 different TF families, to produce M-word binding scores. Increased sequencing depth allowed us to derive accurate scores for longer M-words. This aspect is particularly important because DNA shape is affected by the flanking regions of TFBSs. Therefore, we augmented the original dataset (Jolma *et al*, 2013) with additional sequencing to increase the read depth of the experiments by almost 10-fold (from an average of ~168,000 reads per sequencing file to ~1,656,000 reads). Experimental data were filtered by rigorous quality control (QC) criteria to identify cases with sufficient library complexity and read counts to allow the building of multiparametric models. A total of 218 TFs from 29 families passed the first filter based on high variability and large sample size of the data, and a total of 215 TFs from 27 different families passed the QC step based on regression performance (Fig 1).

For each TF, we selected a core-binding motif, to enable identification of the most probable binding site within M-words and filter out oligonucleotides that are likely to be unbound. The motifs used were derived from a previous study (Jolma *et al*, 2013). These motifs generally contain long flanks in addition to the core consensus sequence, which would prevent us from getting robust M-word scores due to low read coverage for long sequences. To overcome this difficulty, we used motifs from the catalogue compiled by Weirauch and Hughes (Weirauch & Hughes, 2011) to identify and use only the core positions. We calculated the binding score for each M-word that included the core motif in the center (allowing for a

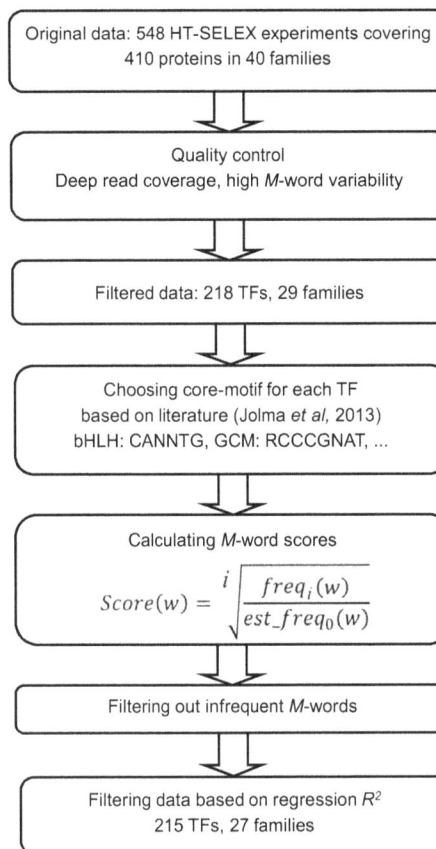

Figure 1. Pipeline used to generate HT-SELEX M-word scores and filter datasets.

M-word scores were derived for cycles $i \geq 3$. For the calculation of the scores, $freq_i(w)$ is the frequency of M-word w in cycle i, and $est\_freq_0(w)$ is its estimated frequency in cycle 0.

few mismatches) and any possible flanking sequences 5′ and 3′ of the motif. We sought to avoid the possibility of cooperative TF–DNA binding, in which multiple copies of the TF occupy different DNA-binding sites (BSs) on the same sequence, as well as to minimize noise caused by inaccurate alignment of M-words based on the core motif. Thus, we excluded HT-SELEX reads that contained multiple instances of the core motifs.

Next, we derived M-word binding scores based on observed experimental enrichment. Each HT-SELEX experiment included several rounds of binding site (BS) selection by the TF, with the binding specificity of selected DNA sequences increasing in each round. We calculated the M-word score as the ratio of the frequency of the M-word in round i over its estimated frequency in the initial round, using a fifth-order Markov model (Slattery *et al*, 2011). The final output of this process was the M-word scores of the core sequence and its flanks for each HT-SELEX experiment (Appendix Fig S1A).

To evaluate the accuracy of our M-word scoring scheme and the value of deeper sequencing, we compared scores derived by HT-SELEX to those measured by genomic-context PBMs (gcPBMs). The gcPBMs use arrays specifically designed with the core sequence in the center, flanked by a genomic context (Gordân *et al*, 2013). These probes are intended to measure the effect of flanking sequences and, therefore, provide an accurate gold standard for long *M*-word ($M \geq 12$) binding scores. The only protein for which both gcPBM and HT-SELEX experimental data exist was the Max homodimer (Zhou *et al*, 2015). Appendix Fig S1B shows the good correlation ($r = 0.64$) of 12-word scores produced by the two technologies, demonstrating the accuracy of our process in producing *M*-word scores from HT-SELEX data. To test how much we gain with respect to gcPBM binding scores by using the new data, we examined three different *M*-word scores: frequency, ratio compared with the initial round, and ratio compared with the estimated initial round. Deeper sequencing improved the correlation of these three scores to gcPBM 12-word scores, and the ratio-to-estimated score achieved the highest correlation (Appendix Fig S1C). Notably, when processing the data previously published in (Jolma *et al*, 2013) with the same pipeline, only 22 proteins passed the quality control, compared with 218 with the higher coverage, showing the advantage of deeper sequencing.

Principal component analysis (PCA) reveals TF family-specific DNA-binding specificities and heterogeneities within TF families

We performed PCA to visualize TF family-specific DNA-binding specificities. The DNA-binding preference of each TF was represented by the DNA *M*-word with the highest binding affinity for this TF. We encoded this *M*-word into numeric feature vectors that included (i) only mononucleotide (i.e., 1-mer) features, and (ii) both 1-mer and DNA shape features. DNA shape features include minor groove width (MGW), Roll, propeller twist (ProT), and helix twist (HelT) and are predicted with our DNAshape approach (Zhou *et al*, 2013). Figure 2A and B shows the first two principal components obtained using each feature vector.

Different TF families tended to form distinct clusters in the PCA scatter plots. To compare the clustering quality in the two plots, we obtained the two-dimensional Euclidean distances between all pairs of TFs from Fig 2A and B. Distances were classified as intra- or inter-family and visualized as boxplots (Fig 2C and D). Inter-family distances were generally larger than intra-family distances. When we used both 1-mer and DNA shape features, the difference between the medians of the inter- and intra-family groups was slightly larger than the difference obtained when using 1-mer features alone (Fig 2C and D). This result was consistent with Fig 2A and B, indicating that more variance could be explained by introducing DNA shape features, in part due to the better separation of the homeodomain family (Fig 2B). To test whether such effects were simply due to the higher dimensionality introduced by the additional DNA

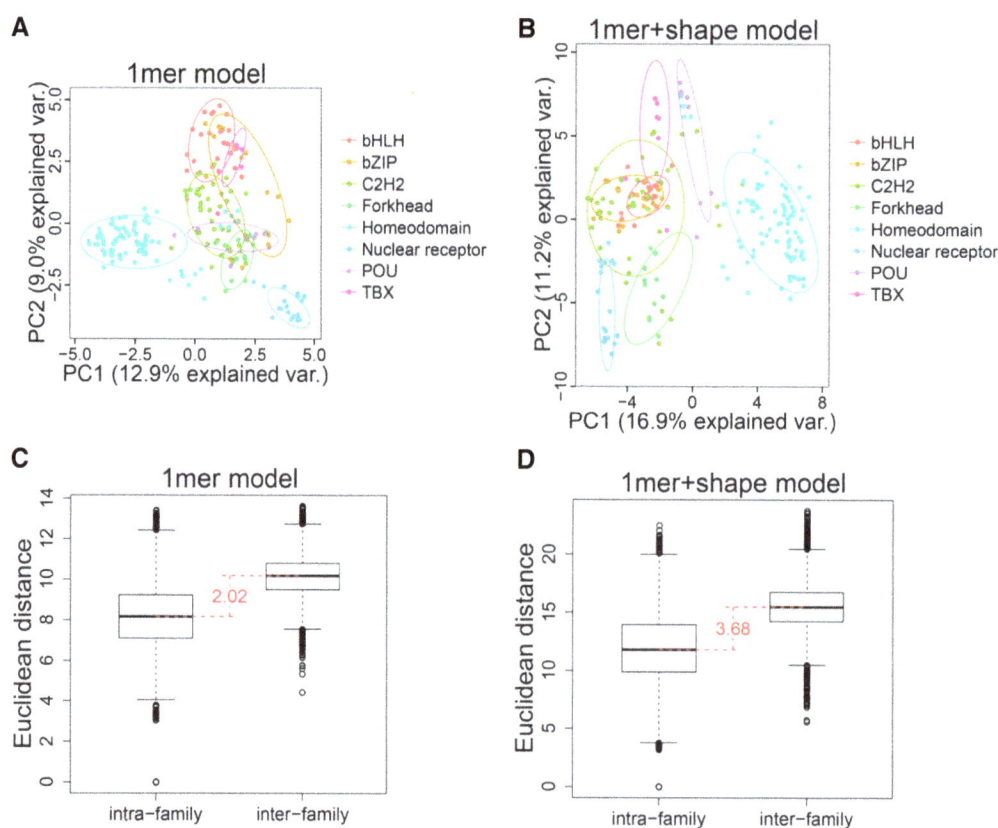

Figure 2. PCA reveals different DNA-binding specificities between TF families.

A   PCA using 1-mer features. Each dot represents a TF. Dots of the same color belong to the same TF family. An ellipse was drawn for each TF family. The ellipse is a contour of a fitted two-variate normal distribution that encloses 0.68 probability (R package default).

B   PCA using 1-mer and shape features, annotated in the same way as described in (A).

C   Boxplots of inter- and intra-family TF distances derived from (A). Difference between medians of inter- and intra-family distances is 2.02 (red).

D   Boxplots of inter- and intra-family TF distances derived from (B). Difference between medians of inter- and intra-family distances is 3.68 (red).

shape features, we added randomly generated shape features based on Gaussian distribution with mean and standard deviation of the original shape features. Both the variance explained and the distance between intra- and inter-family groups were lower in this test (Appendix Fig S2).

DNA shape features improve modeling of DNA-binding specificities across TF families

We tested the importance of the recognition of DNA shape by each TF through quantitative modeling of DNA-binding specificities and comparison of model performance in terms of the R^2 between predicted and experimental M-word scores. Similar to the methodology in Yang et al (2014) and Zhou et al (2015), we built regression models that used only DNA mononucleotide features (i.e., 1mer models) or that combined DNA mononucleotide and shape features (i.e., 1mer+shape models). A result in which the 1mer+shape model outperforms the 1mer model indicates that DNA shape readout might play a role in TF binding.

Based on an analysis of 215 TFs from 27 different families, we found that 1mer+shape models generally outperformed 1mer models (Fig 3A), indicating the prevalence of DNA shape readout across different TF families (for a complete list of datasets used in Fig 3, see Table EV1). With DNA sequence readout playing a dominant role in TF binding, the importance of DNA shape recognition as additional contribution varied both between and within TF families. For example, model performance for homeodomain TFs was generally more substantially improved than for C2H2 TFs. Within the homeodomain TF family, there was a large variance among individual members. Homeodomain and bHLH TFs have been previously observed to be sensitive to DNA shape features (Slattery et al, 2011; Gordân et al, 2013; Yang et al, 2014; Zhou et al, 2015). Here, we confirmed and extended this observation to the bZIP, CENPB, CP2, CUT, ETS, HSF, IRF, MYB, NFAT, nuclear receptor, PAX, POU, PROX, TBX, and TEA TF families. At least half of the members in each of these families, covered by our data, showed greater than 10% performance improvement when DNA shape features were added to the model. However, some families were underrepresented in the data with only one TF present (Table EV1; for full names and detailed information of the TF families, see Table EV2).

To test the robustness of the experimental data and our computational pipeline, we repeated the above analysis on replicate experimental data for three TFs from the bHLH and homeodomain families. Our results consistently showed contributions of DNA shape readout for these two families (Appendix Fig S3A). To test whether the performance gain is simply a result of the increased number of model parameters due to the added DNA shape features, we shuffled the query table for DNA shape features. Shape models based on the shuffled query table generally have poorer performance than those based on the original query table (Fig 3B). We also tested whether the results were robust to the motif seeds used during data preprocessing. We repeated the above analyses using the Weirauch and Hughes seeds (Weirauch & Hughes, 2011) as the final seeds instead of using them for identifying the core positions of the HT-SELEX-based motifs published by Jolma et al (2013). We calculated Pearson's correlation coefficients between the performance of models that were based on the Weirauch and Hughes seeds (Weirauch & Hughes, 2011) and the Jolma et al (2013) seeds.

The high correlation between the two sets of motif seeds indicated that the results were robust to the choice of motif seeds (Appendix Fig S3B). We also tested the robustness of the results under slight changes in the mismatch threshold (see Materials and Methods) and length of the flanking regions. Both tests showed high correlation between different parameter settings, demonstrating sufficient robustness (Appendix Fig S3C and D).

The homeodomain TFs in this study presumably bind DNA as monomers, whereas our previous studies demonstrated the importance of DNA shape for Exd–Hox heterodimers (Slattery et al, 2011). X-ray and nuclear magnetic resonance (NMR) structures of homeodomain DNA-binding domains in complex with DNA repeatedly show that the N-terminal tail of the homeodomain DNA-binding domain interacts with the DNA through minor groove and backbone contacts, which is a signature of DNA shape readout (Joshi et al, 2007).

DNA shape features in flanking regions are important for different TF families

We previously observed that 1mer+2mer+3mer models usually outperform 1mer+shape models (Zhou et al, 2015). Here, we gained additional clues for possible explanations of this observation. As noted previously (Zhou et al, 2015), both 2-mer and 3-mer features are indirect representations of DNA shape characteristics. The 2-mer features describe stacking interactions between adjacent base pairs, whereas 3-mer features describe short structural elements, such as A-tracts that tend to form narrow minor groove regions. Thus, it is not surprising that 1mer+2mer+3mer models can capture TF–DNA binding specificities with high accuracy.

Using our high-quality HT-SELEX data, we observed that, for most TFs, 1mer+2mer+3mer models outperformed 1mer+shape models (Fig 3C). As our prediction of local DNA shape features was based on a sliding window of 5 base pairs (Zhou et al, 2013), we were unable to predict shape features for the two extreme positions at the 5′ and 3′ ends of each DNA sequence. This limitation could give an edge to 1mer+2mer+3mer models. However, we could encode 2-mer and 3-mer features for those terminal positions, which in turn would work as a proxy for DNA shape. To test this hypothesis, we added 3-mer features from only the two end (E2) positions (i.e., 3merE2 features) to the 1mer+shape model. Performance of the resulting 1mer+shape+3merE2 model was indeed comparable to that of the 1mer+2mer+3mer model (Fig 3D). As an additional test, we removed 2-mer and 3-mer features at the end positions from the 1mer+2mer+3mer model, which resulted in the 1mer+2merNoE2+3merNoE2 model that showed similar performance to the 1mer+shape model (Fig 3E).

We also hypothesized that if longer flanking sequences were available for predicting shape features, then 1mer+shape models would perform similar to 1mer+2mer+3mer models without adding 3merE2 features. To verify this possibility, we used an independent dataset generated by the gcPBM platform (Zhou et al, 2015). As expected, 1mer+shape models performed comparable to 1mer+2mer+3mer models for the data without additional 3merE2 features (Appendix Fig S3E). These results imply that DNA shape features in the flanking regions contribute to TF–DNA binding specificities, which was previously known for bHLH TFs (Gordân et al, 2013; Yang et al, 2014; Zhou et al, 2015). Here, we showed for the first

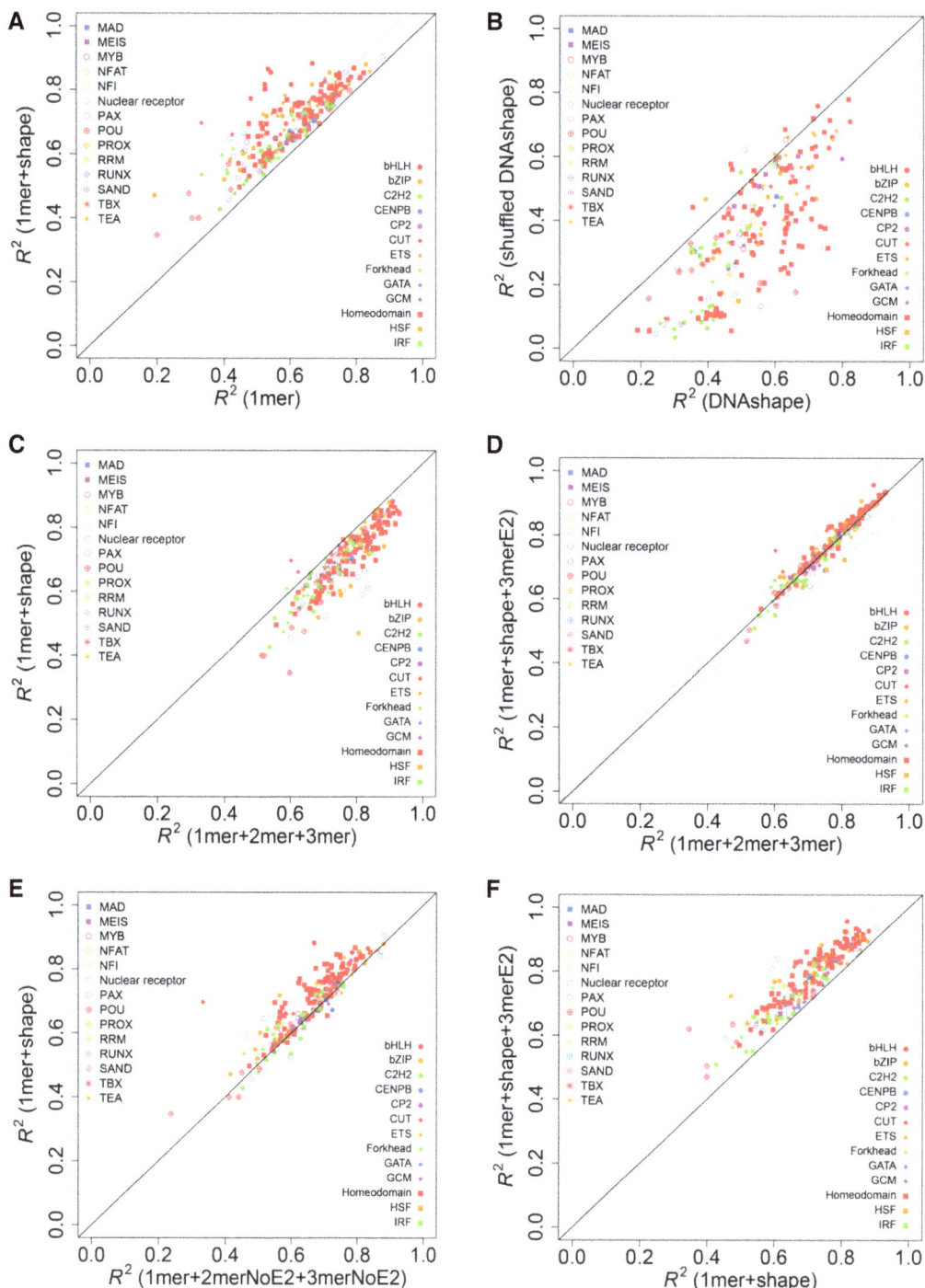

Figure 3. Performance comparisons between models using different features.

A   Comparison between 1mer and 1mer+shape models.

B   Comparison between shape models that are based on the original DNAshape method (Zhou *et al*, 2013) and randomly shuffled pentamer query tables.

C   Comparison between 1mer+2mer+3mer and 1mer+shape models.

D   Comparison between 1mer+2mer+3mer and 1mer+shape+3merE2 models. The label 3merE2 represents 3mer features from the two end positions at the 5' and 3' terminal of each DNA sequence.

E   Comparison between 1mer+2merNoE2+3merNoE2 and 1mer+shape models. The labels 2merNoE2 and 3merNoE3 indicate that 2mer and 3mer features, respectively, were removed from the end positions.

F   Comparison between 1mer+shape and 1mer+shape+3merE2 models.

Data information: Each dot represents one dataset. Coordinates of the dot are determined by the performance, measured in R^2 based on 10-fold cross-validation, of the corresponding models indicated in parentheses. Shape and color of the dots indicate the TF family. Dashed lines in (A and F) have a slope of 1.1, indicating 10% performance increase. Dashed lines in (D) have slopes of 1.1 and 0.9.

time that this phenomenon is of general nature, as adding 3merE2 features as proxy for missing DNA shape features consistently improved the model performance for various TF families (Fig 3F).

Beyond better interpretability of shape-augmented models, an important distinction between the models is the different number of features required to achieve similar performance. The 1mer+shape model requires 12 features (including second-order DNA shape features) per nucleotide position compared with the 84 features required by the 1mer+2mer+3mer model per nucleotide position (Zhou *et al*, 2015). Although we previously included lower-order 1-mers and 2-mers in our 1mer+2mer+3mer models for reasons of interpretability, nevertheless, the 3-mer features actually contain all of the information of the 1-mers and 2-mers. Thus, a 3mer model is equivalent to a 1mer+2mer+3mer model (Materials and Methods and Appendix Fig S3F). This choice, however, would still leave the 3mer model with 64 required features per nucleotide position compared with a maximum of only 12 features in the 1mer+shape model.

Feature selection can provide insights into TF–DNA readout mechanisms

We performed feature selection to identify BS positions where DNA shape features contribute to TF-binding specificities. The method is similar to the one we previously introduced for the analysis of SELEX-seq data for Hox proteins (Abe *et al*, 2015). For each TF, we evaluated the R^2 performance of the baseline 1mer model, denoted R^2_{1mer}. Next, we evaluated models that combined 1-mer features with DNA shape features individually at single nucleotide positions i, denoted 1mer+shape$_i$ models. We denoted the performance as $R^2_{1mer+shape_i}$. We calculated the difference in model performance $\Delta R^2_i = R^2_{1mer+shape_i} - R^2_{1mer}$ for each nucleotide position i (Fig 4A). The $\Delta R^2_i/R^2_{1mer}$ ratio indicates the percentage change in performance due to the availability of DNA shape features at nucleotide position i, with a positive ratio suggesting performance gain. The ratio at position i compared with other positions reflects the relative importance of DNA shape features at different nucleotide positions. We visualized the $\Delta R^2_i/R^2_{1mer}$ ratio as a function of position i for each TF in the form of a heat map (Fig 5A and Appendix Fig S4).

To avoid interference from DNA sequence information, we devised a second feature-selection approach in which we removed DNA shape features at individual positions from a shape-only model. The $\Delta R^2_i/R^2_{shape}$ ratio was then used for generating the heat map (Figs 4B and 5B, and Appendix Fig S4), where $\Delta R^2_i = R^2_{shape} - R^2_{shape_i}$. These two different approaches can sometimes yield conflicting heat maps as discussed below. To address such cases and facilitate the use of these heat maps, we also generated a combined heat map based on the cell-by-cell minimum of the two heat maps (Fig 5C and Appendix Fig S4). Quantitative information about the importance of the position-dependent DNA shape in TF–DNA recognition at single-base pair resolution provides the means to determine the structural protein–DNA readout mechanisms based on sequence data. To achieve this goal, we further expanded our feature-selection method to test each individual DNA shape feature category, which enabled us to gauge the importance of each DNA shape feature, that is, MGW, Roll, ProT, or HelT, at every position (Appendix Fig S5). To date, obtaining such information required experimentally solved structures.

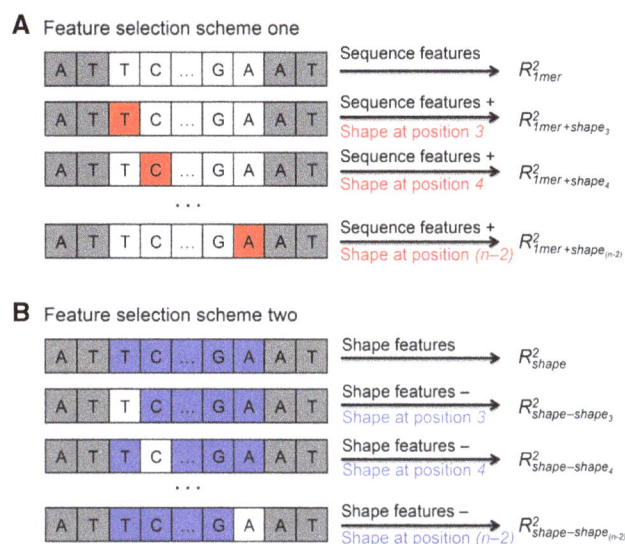

Figure 4. Schematic representation of feature-selection process.

A Feature-selection scheme for adding DNA shape features at one individual position to a sequence-only model.
B Feature-selection scheme for removing DNA shape features from one single position from a shape-only model.

Figure 5 shows the position-dependent DNA shape importance for homeodomain TFs that recognize a TAAT motif. For most of these TFs, DNA shape was more important at the 3′ side of the core motif, as indicated by the darkness of colors (Fig 5). Homeodomain TFs that recognize a different motif, for example, TCRTAAA, were shown to have a different positional DNA shape preference (Appendix Fig S4F). Positional preferences were also protein-family specific. For example, for bHLH TFs DNA shape features in both flanking regions were important, whereas for nuclear receptors that bind to an ACANNNTGT motif the central motif region was generally important (Appendix Fig S4A and H). In comparison, bZIP TFs that bind to a TTRCGC motif and homeodomain TFs were generally sensitive to DNA shape features at only one flanking side of the core motif (Appendix Fig S4B and F).

The exact positions where DNA shape features are important were not unambiguously pinpointed for the bHLH TFs and the nuclear receptors that bind to an ACANNNTGT motif (Appendix Fig S4A and H). Both Appendix Fig S4A and H relate to a scenario where the red heat map shows prominent shape effects in multiple consecutive positions, whereas the blue heat map shows almost no effects. We believe that this is due to false positives in the red heat map, that is, positions that are not important for shape readout but identified as such, and false negatives in the blue heat map, that is, positions that are important for shape readout that were not identified. We conclude in this case that DNA shape is important in some positions in the consecutively red regions, but we failed to locate it, even with the help of the blue heat map.

We illustrated the relevance of feature importance heat maps derived from feature-selection approaches by considering experimental structures of the homeodomain proteins PITX2 (PDB ID 2LKX) and GBX1 (PDB ID 2ME6) in complex with DNA (Fig 6A and B). These structures provide possible explanations for entries representing PITX3 and GBX1 on the heat maps (Fig 5). As no experimental structure for PITX3 is available, we used an NMR structure for PITX2 (Chaney et al, 2005), which shares the same

DNA-binding domain as PITX3. In the heat maps, PITX3 has darker colors at the 3′ side of the TAAT motif, indicating a more important role of DNA shape at these positions. In the PITX2

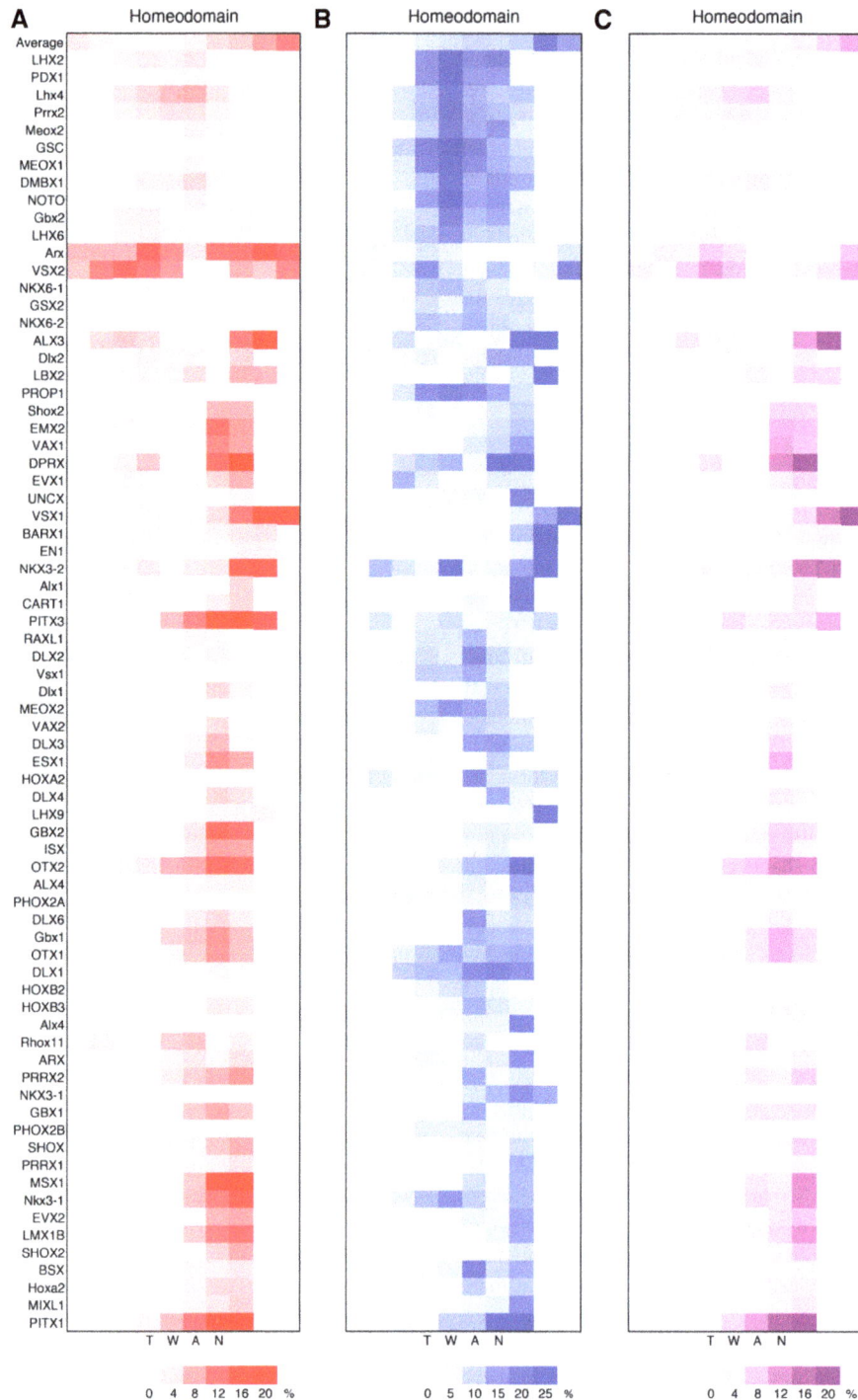

Figure 5. Importance of DNA shape features as a function of nucleotide positions revealed by feature selection with machine learning.

A   Heat map based on adding DNA shape features to a sequence-only model.
B   Heat map based on removing DNA shape features from a shape-only model.
C   Combined heat map that takes cell-by-cell minimum of heat maps in (A and B).

Data information: Case of letters in TF names indicates species, with uppercase being human and lowercase being mouse.

structure, the N-terminal tail of the protein interacts with DNA in the minor groove of the TAAT motif. The structure contains a narrow minor groove region near the second A within the TAAT motif (Fig 6A). In this case, the protein might exploit the DNA structural characteristics at positions highlighted in the heat maps to achieve its binding specificity.

We observed similar concurrence between heat map and structural analyses for the TF GBX1, where the structure has a narrow minor groove region at the 3′ flank (Fig 6B). Although the positions indicated by the heat maps do not match the positions in the structure in an exact way, the heat maps successfully highlighted those nearby positions. Moreover, the heat maps were consistent with our conclusion that DNA shape features in flanking regions are important for TF–DNA binding specificities (Fig 3D–F). In addition to the homeodomain family, we used a structure of the human progesterone receptor (PDB ID 2C7A) from the nuclear receptor family to illustrate how the heat maps can provide hints to the structural mechanisms of protein–DNA

binding. In the structure (Roemer *et al*, 2006), MGW, Roll, and ProT show distinct characteristics in the central region of the DNA-binding site, which potentially explains the central "red" regions in the heat maps (Appendix Fig S6).

DNA shape logos represent structural readout mechanisms

To visualize the detailed DNA shape preferences of individual TFs, we propose a new visualization, *DNA shape logos*, analogous to sequence logos for PWMs. In these logos, we used the letters H, M, P, and R to represent DNA shape features HelT, MGW, ProT, and Roll, respectively. The height of each letter indicates the importance derived from the feature-selection analysis for the corresponding DNA shape feature at a specific position (Fig 6). As an example, we used ΔR^2, that is, the performance gain due to adding an individual DNA shape feature to a 1mer model, to generate shape logos for PITX3 and GBX1 (Fig 6C and D). For PITX3, a prominent M at positions 7, 8, 9, and 10 overlaps with the narrow minor groove region

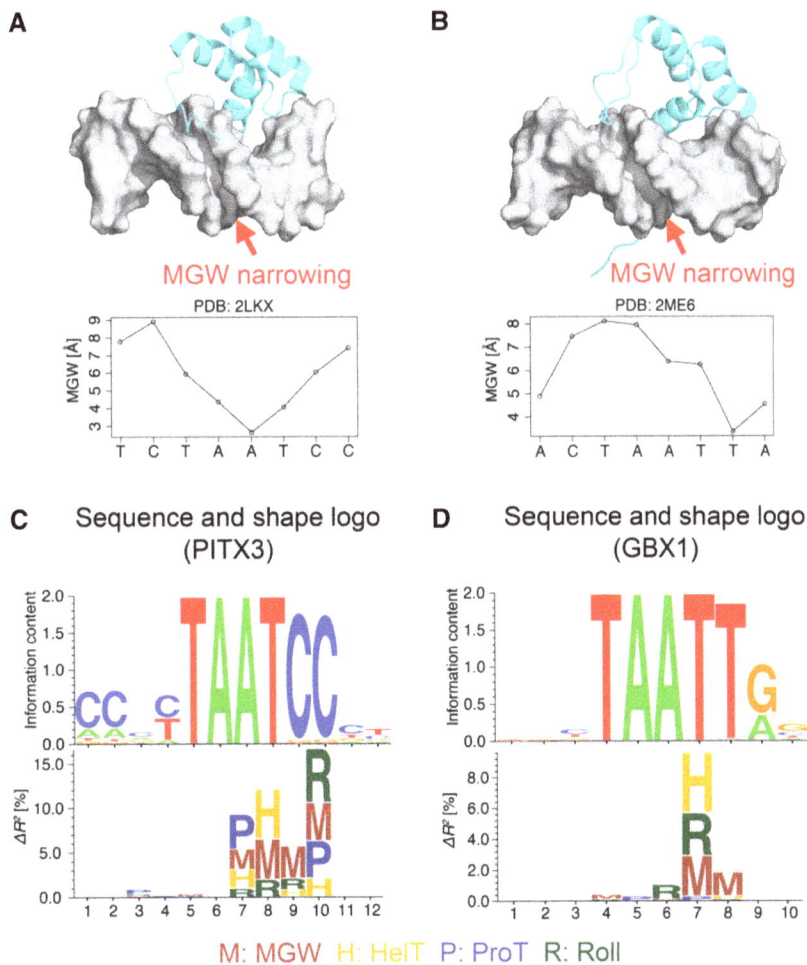

Figure 6. Three-dimensional structure and DNA sequence and shape logos for the homeodomain TFs PITX2/PITX3 and GBX1.

A   NMR structure of PITX2 in complex with DNA (PDB ID 2LKX) and the CURVES (Lavery & Sklenar, 1989) derived plot for the MGW of the bound DNA.
B   NMR structure of GBX1 in complex with DNA (PDB ID 2ME6) and the CURVES (Lavery & Sklenar, 1989) derived plot for the MGW of the bound DNA.
C   DNA sequence and shape logos for PITX3.
D   DNA sequence and shape logos for GBX1.

in the structure. Similarly, for GBX1, a prominent M at positions 7 and 8 overlaps with the narrow minor groove in the structure. DNA shape information was missing for the two nucleotide positions at each end of the TFBS; thus, no letters are shown at these positions in the shape logo. DNA shape logos can facilitate the integration of structural information in motif finding tools. Sequence and shape logos for all the TFs studied in this work are provided as Datasets EV1 and EV2, respectively.

Discussion

Protein–DNA binding models have evolved tremendously in the last decade (Slattery et al, 2014). In the past, binding models were based on a few high-affinity BSs. These models enabled the identification and prediction of the most likely BSs in vivo, but missed many potential low-affinity sites (Stormo, 2000; Tanay, 2006). Weak and suboptimal TFBSs play important roles in transcriptional regulation (Crocker et al, 2015; Farley et al, 2015), emphasizing the necessity of a quantitative understanding of TF–DNA binding specificities. Structures obtained through X-ray crystallography and NMR spectroscopy allow us to determine the detailed mechanisms of protein–DNA binding involving single DNA target sites and have greatly advanced our perception of protein–DNA recognition (Rohs et al, 2010). However, it is inherently difficult to apply these insights at a high-throughput level. Protein crystallization is a time-consuming process, and deriving distance constraints using NMR experiments is costly and likewise time-consuming. As a consequence, structural information is limited to a subset of TFs and individual DNA-binding sites.

In the genomics field, sequencing- and microarray-based high-throughput methods have made it possible to study systematically in vitro TF–DNA binding specificities by simultaneously measuring binding affinities to millions of different DNA sequences. In vitro platforms such as HT-SELEX and PBM provide effective solutions to gain quantitative knowledge of TF–DNA binding (Berger et al, 2006; Zhao et al, 2009; Jolma et al, 2010), as the confounding factors in vivo are not present. With sequencing depth being further improved by an average of 10-fold compared with the original data (Jolma et al, 2013), the HT-SELEX data generated in this study currently represent the most extensive set of TF–DNA binding measurements for mammalian TFs. We constructed an analysis pipeline that derives binding affinities for different DNA M-words from these HT-SELEX data, gaining a much more detailed view of the binding energy landscape than simple PWM models. This approach enabled us to explore, through statistical machine-learning methods, how the mechanisms of DNA shape readout are employed by various TF families. With feature-selection techniques, we revealed TF family-specific positional DNA shape importance at base pair resolution. The results concur with available experimental structures. Overall, this study provides a means to derive binding mechanisms from sequence data without relying on solved structures.

Despite these methodological advances, we see several limitations in our preprocessing of the data. First, while increasing the sequencing depth improved statistical robustness of binding affinities derived for the short M-words used here, the amount of sequencing data may still be insufficient for models using longer M-words. Although the sequencing depth could be increased further

(Slattery et al, 2011), this endeavor would be expensive, considering the large number of TFs that were studied. Second, HT-SELEX technology can be influenced by oligonucleotide synthesis and PCR bias. In addition, TFs may bind in different binding modes, resulting in enrichment of a mixture of oligonucleotides containing one or more binding motifs. To identify features of single binding events, we based our analysis on known core motifs, allowing only one core motif within each oligonucleotide, removing PCR duplicates, and normalizing by the initial round.

Moreover, we note the limitations in the shape readout profiles and their visualization. First, DNA shape alone is obviously insufficient to explain TF binding (Zhou et al, 2015). Second, the shape logos are not equivalent to sequence logos, as they are based on positional scores that do not represent a probability distribution or energy parameters. An alternative way for generating DNA shape logos is to use feature weights derived from models. However, due to the interdependencies between features, such weights are not directly interpretable. In our analysis, we gauged the importance of each individual DNA shape feature by adding it to the 1mer baseline model and observed its effect on the model performance. We believe that DNA shape logos based on such extensive computation are more robust. Although such logos do not yet lead to the prediction of a protein–DNA structure model, they are a step forward and provide a general guide for revealing DNA shape preferences. Third, although the TF–DNA structures supported the heat map results, the correlation is not at all conclusive. Experimentally solved structures in the PDB are not available for most of the studied TFs. Both the "red" and "blue" heat maps aim to summarize the DNA shape importance at individual positions. However, the red heat maps can contain false-positive cells, and the blue heat maps can contain both false-positive and false-negative ones (see definition in Results section). The DNA shape features at a position essentially reflect the pentamer context at that position. Shape features of adjacent positions may contain redundant information. As a result, a position indicated as important in the red heat map may be due to the fact that DNA shape features at the position adjacent to it are important, inducing false positives in the heat map (Appendix Fig S4A and H). On the other hand, for the same reason, a position may not be indicated as important in the blue heat map due to the fact that its directly adjacent position is making up for it, inducing false negatives in the heat map (Appendix Fig S4A and H). Moreover, the DNA shape features used here are derived from sequence, so a position indicated as important in the blue heat map may be due to the loss of sequence information encoded indirectly in the shape features, inducing false positives in the heat map, for example, TBX15 in Appendix Fig S4J. The combined version of the heat maps improves the accuracy to some extent. In addition, the feature-selection analysis that breaks down the DNA shape contribution into individual DNA shape features helps locate the effective shape features. Despite these limitations, we believe that in the future, such heat map analysis, when combined with TF–DNA binding measurements of improved quality, will allow us to gain more clues of TF-binding mechanisms from DNA sequencing data.

Finally, although understanding of in vitro protein–DNA binding mechanisms is a critical step toward understanding in vivo binding, the in vivo scenario consists of multiple layers of complexity, such as the three-dimensional genomic architecture (Rao et al, 2015), DNA accessibility (Neph et al, 2012), nucleosome competition

(Barozzi *et al*, 2014), and TF cooperativity and co-factors (Slattery *et al*, 2011; Crocker *et al*, 2015). Full understanding of gene regulation will require the integration of knowledge obtained in different fields using various technologies.

In conclusion, while the DNA sequence describes opportunities to form hydrogen bonds and other direct contacts between amino acids and bases, DNA shape can provide an important additional contribution to TF binding (Rohs *et al*, 2009, 2010). We systematically explored here, we believe for the first time, the role of DNA shape readout for many TF families, using high-quality HT-SELEX data, and obtained results at base pair resolution. We produced a valuable TF–DNA binding data resource by increasing the sequencing depth of previous HT-SELEX experiments (Jolma *et al*, 2013) and developing tools for deriving TF–DNA binding affinities and mechanisms from DNA sequencing data.

Materials and Methods

HT-SELEX binding data

HT-SELEX experiments were comprised of previously published data (Jolma *et al*, 2013) complemented by new sequencing data. The new data were produced by repooling existing PCR-amplified SELEX ligands into new Illumina sequencing libraries, where samples were multiplexed to a lesser extent (~55× vs. ~800×) than in the previous study. Libraries were sequenced using the Illumina Hiseq2 platform, as in the previous study (Jolma *et al*, 2013). The additional sequencing coverage used in the analysis has been submitted in the European Nucleotide Archive (ENA; http://www.ebi.ac.uk/ena) under study identifier PRJEB14744. The complete dataset comprises 548 experiments covering 410 different TFs, including mouse/human full-length protein–DNA binding domain differences. Forty protein families were represented. Protein family membership can be found in Jolma *et al* (2013). For the three TFs in the validation set, new HT-SELEX experiments were performed essentially as described in (Nitta *et al*, 2015).

The gcPBM data were downloaded from GEO accession number GSE59845 (Zhou *et al*, 2015). Max protein 12-word scores were the average log-normalized fluorescence intensities of probe sequences that included these 12-words.

Choosing core motifs

For each TF, we defined a core-binding sequence to enable identification of the most likely binding site and filter out unbound oligonucleotides. We used the seeds published in Jolma *et al* (2013) as the core motifs, but removed their flanks. To pinpoint the core positions as opposed to the flanks, we used motifs compiled in Weirauch and Hughes (2011), which are consensus sequences for only the core motifs collected for different TF families. Substring positions that have the most agreement to any of the corresponding Weirauch and Hughes motifs (Weirauch & Hughes, 2011) were chosen as the core positions. We used the IUPAC character representation for nucleotide sequence. It was sufficient for positions to agree if they represented the same nucleotide. When using the Weirauch and Hughes (Weirauch & Hughes, 2011) motifs as core seeds, most TF families had only one core motif, which would be the assigned motif for TFs

from the family. For TF families having several motifs, we compared the Weirauch and Hughes motifs (Weirauch & Hughes, 2011) to the published consensus seeds (Jolma *et al*, 2013) and calculated *score1*, the portion of matched nucleotides. The core motif with the highest *score1* was assigned to a TF, respectively. If multiple options remained after this step, then we calculated *score2*, a stricter similarity score such that the IUPAC symbols matched exactly (e.g., R matches R but not A). The core motif with the highest *score2* was then selected. This process ensured that almost all TFs were assigned only one motif. In some rare cases, two motifs survived. For both Jolma *et al* seeds and Weirauch and Hughes motifs, when multiple seeds were selected, a dataset for the TF was derived according to each selected seed, but only the dataset with highest R^2 was included in the analysis in Fig 3. For a complete list of datasets, see Table EV1.

A few TF families were not covered by Weirauch and Hughes (2011). For C2H2 TFs, we used the seeds published in Jolma *et al* (2013) without removing the flanks, as zinc fingers bind different sequences based on the specificity of each finger (noted in Weirauch and Hughes, 2011). For six TF families not covered by (Weirauch & Hughes, 2011), we used other published resources for the seed of each family, as specified here: RRM (Fernandez-Miranda & Mendez, 2012), NFI (Whittle *et al*, 2009), NRF (http://AtlasGeneticsOncology.org), TFAP (http://AtlasGeneticsOncology.org), and znf_BED (http://www.genecards.org). For the complete list of core consensus motifs, see Table EV3.

M-word scores

We derived *M*-word binding scores based on observed experimental enrichment counts. HT-SELEX experiments included several rounds of enrichment of bound DNA sequences by a specific protein. Initially, the experiment began from a pseudo-random DNA oligonucleotide library. The protein was allowed to bind to DNA sequences in the randomized pool. Next, bound ("selected") sequences were isolated and amplified for sequencing and reiteration of the process. The frequency of DNA sequences that have higher binding affinities increased exponentially. It is possible to derive the binding affinity for DNA sequences based on their change in frequencies throughout the rounds (Levine & Nilsen-Hamilton, 2007). In the HT-SELEX experiments, the oligonucleotide length (excluding constant ends) was 14, 20, 30, or 40 base pairs.

M-word scores were produced for each core motif, with the following parameters: number of core-flanking positions to derive, selected round, and number of core mismatches that were allowed. For each HT-SELEX oligonucleotide, at most one BS was accounted for. An *M*-word with a number of matches to the core motif above the threshold was chosen as the BS (if there were several, the oligonucleotide was discarded to avoid multiple modes of binding). Only for occurrences in which the *M*-word had sufficiently long flanks to include, the required side positions were used. The reverse complement strand was also considered and, in cases of hits on both strands, the one with the larger number of matches was used. If no *M*-word matched the core motif given the allowed number of mismatches, the oligonucleotide was discarded.

To produce accurate *M*-word ratio scores, counts were divided by estimated frequencies in the initial pool, as previously described (Slattery *et al*, 2011). Estimated frequencies were generated using a fifth-order Markov model of observed frequencies in the initial pool, following the SELEX-seq protocol (Slattery *et al*, 2011). The score was the ith root of the ratio, where i was the round of selection. This approach was based on the assumption that M-word frequencies increased by the same factor between two consecutive selection cycles (Slattery *et al*, 2011). To compare different alternative scores, we considered the frequency at round i and the ratio of the frequency at round i over the (observed) frequency in the initial round. In all cases, an oligonucleotide was only counted once to avoid PCR duplication bias.

Length of core-motif and flanking regions, number of mismatches allowed, and selected rounds

For each experiment, M-word scores were derived per round for round 3 and later rounds. As the first few rounds did not show a profound enrichment, we did not consider them. Later rounds showed enrichment and varied in quality and read depth. Thus, data were collected per round from round 3 onwards, and selection of the round was deferred to a later stage.

Similarly, we generated datasets for different values of M. There is an inherent tradeoff between increasing M and reducing the accuracy of the scores. While greater M values provide information on binding to longer flanks, counts of M-words decrease as M grows, leading to less accurate binding scores. Keeping this tradeoff in mind, we considered the initial length and the maximum length of flanking regions. The initial length was set to $\lfloor (10 - core\_length)/2 \rfloor$ so that M is at least 10, allowing DNA shape prediction for at least 6 positions (the two positions at each flank are not available due to the pentamer model). For example, for core TAAATTA of length 7, the initial flank length was 1. We called an M-word reliable if its count was > 8. M was set to be the largest value for which the number of reliable M-words was ≥ 1,000, and the maximum M-word count was ≥ 100. When all M-word counts are < 100, the scores may be inaccurate, and samples with less than 1,000 reliable M-words are considered small and excluded from our analysis. For example, for the same core of TAAATTA, if GAGTAAAT-TACTC was the most frequent 13-word and it appeared only 89 times, whereas the 11-word AGTAAATTACT appeared 1,540 times, assuming there are more than 1,000 reliable M-words in both, the maximum length would be 3 (the core is of length 7, leaving 3 flanking positions on each side). Datasets were created for all flanking region lengths, starting at the initial and up to the maximum length.

Another tradeoff exists in the number of mismatches: up to a point, allowed mismatches increase the variability of M-words, and thus add useful information. Too many mismatches would lead to the introduction of M-words that do not represent BSs, resulting in added noise. With this tradeoff in mind, we set the number of mismatches allowed to depend on the length of the core motif. Generally, the number was $\lfloor (core\_length - 4)/2 \rfloor + 1$. In case the core motif contains degenerate characters, that is, those that represent multiple nucleotides, we counted these characters differently in the core length. The weight of a character in this count was $1/nucleotides\_it\_represents$, and the length of a core was the sum of its characters' weights. For example, for ATAAAA, we allowed two mismatches as there are six characters of weight 1. For CANNTG, we allowed only one mismatch (in addition to the two central fully degenerate positions), because its total weight is $4*1+2*0.25 = 4.5$. By applying this threshold, on average, $74 \pm 25\%$ of the oligonucleotides were retained, which suggests that it can detect probable BSs while

removing oligonucleotides that are less likely to be bound. The above threshold was used as a first step in order to exclude unbound oligonucleotides. In the second step, a stricter threshold allowing one less mismatch was used to filter out oligonucleotides that have multiple motif occurrences, in order to exclude cooperative binding events from our analysis. The stricter threshold ensures that not too many oligonucleotides are filtered out in the second step. Finally, the oligonucleotides were aligned according to the core motif.

Dataset filtering

In large-scale experimental data, it is inherently difficult to ensure that every dataset has equivalent diversity and enrichment level. Although PWM models can be constructed from low-quality experimental data, complex models require high levels of enrichment and sequence diversity. To reach reliable conclusions, we used multiple data filtering procedures to discard datasets of insufficient quality. We performed two stages of QC for these datasets. In the first stage, we used four QC criteria to ensure high counts for accurate score estimates, large sample size, and score variability.

1. All M-words with count ≤ 8 were discarded because low counts lead to inaccurate estimates of binding scores.
2. If the number of different M-words after step 1 was < 1,000, then the dataset was filtered out, to ensure that datasets have sufficient numbers of samples for the learning algorithm.
3. Datasets were tested for variable scores. The score of the 90th percentile had to be at least 0.2 greater than the score of the 10th percentile.
4. The maximum M-word had to appear at least 100 times; otherwise, counts would be too small and estimates inaccurate for most of the M-words.

We filtered out datasets based on R^2 performance criteria. We ran L2-regularized multiple linear regression (MLR) on each of the remaining datasets using different combinations of features. Due to their linearity, we would expect that, for MLR models, model A would perform at least as well as model B, given that B uses a subset of features used by A. We defined a dataset as invalid only when the performance of model A was smaller than that of model B by more than 3%, given that B uses a subset of features used by A. This process reduced the number of valid datasets to 533. Datasets for which even the best model had $R^2 < 0.5$ were excluded from the analyses. Finally, 512 datasets covering 215 human/mouse TFs belonging to 27 different TF families passed our QC procedure.

For TFs covered by multiple datasets, only the dataset with the highest R^2 was included in downstream analyses (see Table EV1 for the complete list). As PCA requires only one representative BS sequence for each TF, we separately generated 12-word data using reads from the last round of HT-SELEX, as the last round is expected to be the most specific. We used the top 12-word as the representative BS for each TF. In doing so, as many as 294 TFs were covered in the PCA (Fig 2; see Table EV4 for a complete list of 12-words for the 294 TFs in the PCA).

PCA and linear regression analysis

For each DNA sequence s, the 1-mer, 2-mer, and 3-mer features were encoded into feature vectors ϕ^{1mer}, ϕ^{2mer}, and ϕ^{3mer}, respectively, in a similar way to those used in Zhou *et al* (2015). The ith

nucleotide in s was denoted s^i. Elements of vectors ϕ^{1mer}, ϕ^{2mer}, and ϕ^{3mer} were formulated as follows. For nucleotide position i:

$$\phi_{4*(i-1)+1}^{1mer}(s) = \begin{cases} 0, & \text{if } s^i \neq A \\ 1, & \text{if } s^i = A \end{cases}, i = 1, \ldots, l$$

$$\phi_{4*(i-1)+2}^{1mer}(s) = \begin{cases} 0, & \text{if } s^i \neq C \\ 1, & \text{if } s^i = C \end{cases}, i = 1, \ldots, l$$

$$\phi_{4*(i-1)+3}^{1mer}(s) = \begin{cases} 0, & \text{if } s^i \neq G \\ 1, & \text{if } s^i = G \end{cases}, i = 1, \ldots, l$$

$$\phi_{4*i}^{1mer}(s) = \begin{cases} 0, & \text{if } s^i \neq T \\ 1, & \text{if } s^i = T \end{cases}, i = 1, \ldots, l$$

$$\phi_{16*(i-1)+1}^{2mer}(s) = \begin{cases} 0, & \text{if } s^i s^{i+1} \neq AA \\ 1, & \text{if } s^i s^{i+1} = AA \end{cases}, i = 1, \ldots, l-1$$

$$\phi_{16*(i-1)+2}^{2mer}(s) = \begin{cases} 0, & \text{if } s^i s^{i+1} \neq AC \\ 1, & \text{if } s^i s^{i+1} = AC \end{cases}, i = 1, \ldots, l-1$$

$$\phi_{16*(i-1)+3}^{2mer}(s) = \begin{cases} 0, & \text{if } s^i s^{i+1} \neq AG \\ 1, & \text{if } s^i s^{i+1} = AG \end{cases}, i = 1, \ldots, l-1$$

$\ldots$

$$\phi_{16*i}^{2mer}(s) = \begin{cases} 0, & \text{if } s^i s^{i+1} \neq TT \\ 1, & \text{if } s^i s^{i+1} = TT \end{cases}, i = 1, \ldots, l-1$$

$$\phi_{64*(i-1)+1}^{3mer}(s) = \begin{cases} 0, & \text{if } s^i s^{i+1} s^{i+2} \neq AAA \\ 1, & \text{if } s^i s^{i+1} s^{i+2} = AAA \end{cases}, i = 1, \ldots, l-2$$

$$\phi_{64*(i-1)+2}^{3mer}(s) = \begin{cases} 0, & \text{if } s^i s^{i+1} s^{i+2} \neq AAC \\ 1, & \text{if } s^i s^{i+1} s^{i+2} = AAC \end{cases}, i = 1, \ldots, l-2$$

$$\phi_{64*(i-1)+3}^{3mer}(s) = \begin{cases} 0, & \text{if } s^i s^{i+1} s^{i+2} \neq AAG \\ 1, & \text{if } s^i s^{i+1} s^{i+2} = AAG \end{cases}, i = 1, \ldots, l-2$$

$\ldots$

$$\phi_{64*i}^{3mer}(s) = \begin{cases} 0, & \text{if } s^i s^{i+1} s^{i+2} \neq TTT \\ 1, & \text{if } s^i s^{i+1} s^{i+2} = TTT \end{cases}, i = 1, \ldots, l-2$$

First-order DNA shape features MGW, ProT, Roll, and HelT, denoted ϕ^{MGW}, ϕ^{ProT}, ϕ^{Roll}, and ϕ^{HelT}, respectively, were generated by our DNAshape prediction method (Zhou *et al*, 2013; Chiu *et al*, 2016). For these DNA shape features, the following normalization was performed:

$$\phi_i^{MGW} = (MGW_i - MGW_{min})/MGW_{sd}$$

where MGW_i is the predicted MGW, MGW_{min} is the minimum MGW over all possible pentamers, and MGW_{sd} is the standard deviation of MGW in the data. Similarly:

$$\phi_i^{ProT} = (ProT_i - ProT_{min})/ProT_{sd},$$
$$\phi_i^{Roll} = (Roll_i - Roll_{min})/Roll_{sd},$$
$$\phi_i^{HelT} = (HelT_i - HelT_{min})/HelT_{sd}.$$

Second-order DNA shape features were derived from the first-order features and denoted ϕ^{MGW2}, ϕ^{ProT2}, ϕ^{Roll2}, and ϕ^{HelT2}. These second-order shape features were the product terms of adjacent first-order DNA shape features, normalized by the standard deviation. MGW and ProT were defined for each base pair, and Roll and HelT were defined for each base pair step. Thus, in the feature-selection analysis, DNA shape features at nucleotide position i, denoted as $shape_i$, consisted of ϕ_i^{MGW}, ϕ_i^{ProT}, ϕ_i^{Roll}, ϕ_{i+1}^{Roll}, ϕ_i^{HelT}, ϕ_{i+1}^{HelT}, ϕ_i^{MGW2}, ϕ_{i+1}^{MGW2}, ϕ_i^{ProT2}, ϕ_{i+1}^{ProT2}, ϕ_i^{Roll2}, and ϕ_i^{HelT2}. If the core-motif sequence

was palindromic, then the last step in the feature encoding was to symmetrize the feature vector by averaging it with the feature vector encoding the reverse complementary stand. The DNAshape method predicts shape features based on a pentamer query table that is derived from all-atom Monte Carlo simulations (Zhou *et al*, 2013). As a control, we shuffled the pentamer query table and tested its effects on shape models.

After the feature encoding, L2-regularized MLR and 10-fold cross-validation were performed for each dataset to gauge model performance (Yang *et al*, 2014; Abe *et al*, 2015). L2-regularized MLR was chosen for its simplicity and interpretability. In PCA, the feature vector encoded for the sequence of highest DNA-binding affinity of a TF was used to represent that TF.

3mer and 1mer+2mer+3mer model equivalence in linear regression

The 3mer models and 1mer+2mer+3mer models are equivalently "powerful" in MLR, where the power of a model refers to its descriptive capability. This equivalency can be demonstrated by showing that any solution of a 1mer+2mer+3mer model could be mapped into a 3mer model solution that gives exactly the same prediction of binding affinity for any input DNA sequence, and *vice versa*. Proof for the reverse direction is trivial. We could just keep the learned coefficients, or weights, of the 3-mer features, and set all weights for 1-mer and 2-mer features to be zero. This process results in a 1mer+2mer+3mer model that gives exactly the same prediction of binding affinity for any input DNA sequence as the original 3mer model. Mapping for the other direction is as follows.

Denote a solution to a 1mer+2mer+3mer model as:

$$S_1 = (w_A^1, w_C^1, w_G^1, w_T^1, w_A^2, w_C^2, w_G^2, w_T^2, \ldots,$$
$$w_A^{N-1}, w_C^{N-1}, w_G^{N-1}, w_T^{N-1}, w_A^N, w_C^N, w_G^N, w_T^N,$$
$$w_{AA}^1, w_{AC}^1, w_{AG}^1, w_{AT}^1, w_{CA}^1, w_{CC}^1, w_{CG}^1, w_{CT}^1, \ldots,$$
$$w_{GA}^{N-1}, w_{GC}^{N-1}, w_{GG}^{N-1}, w_{GT}^{N-1}, w_{TA}^{N-1}, w_{TC}^{N-1}, w_{TG}^{N-1}, w_{TT}^{N-1},$$
$$w_{AAA}^1, w_{AAC}^1, w_{AAG}^1, w_{AAT}^1, w_{ACA}^1, w_{ACC}^1, w_{ACG}^1, w_{ACT}^1, \ldots,$$
$$w_{TGA}^{N-2}, w_{TGC}^{N-2}, w_{TGG}^{N-2}, w_{TGT}^{N-2}, w_{TTA}^{N-2}, w_{TTC}^{N-2}, w_{TTG}^{N-2}, w_{TTT}^{N-2}).$$

Denote a solution to the 3mer model as:

$$S_2 = (m_{AAA}^1, m_{AAC}^1, m_{AAG}^1, m_{AAT}^1, m_{ACA}^1, m_{ACC}^1, m_{ACG}^1, m_{ACT}^1, \ldots,$$
$$m_{TGA}^{N-2}, m_{TGC}^{N-2}, m_{TGG}^{N-2}, m_{TGT}^{N-2}, m_{TTA}^{N-2}, m_{TTC}^{N-2}, m_{TTG}^{N-2}, m_{TTT}^{N-2}).$$

Superscript numbers denote nucleotide positions in the DNA sequences. Subscript letters denote what features at those positions the learned weights are for. For any $x, y, z \in \{A, C, G, T\}$, map the weights as follows:

$$m_{xyz}^i = w_{xyz}^i + w_{xy}^i + w_x^i, i = 1, \ldots, N-3$$
$$m_{xyz}^{N-2} = w_{xyz}^{N-2} + w_{xy}^{N-2} + w_x^{N-2} + w_{yz}^{N-1} + w_y^{N-1} + w_z^N.$$

The resulting S_2 will assign the same predicted binding affinity as S_1 to any input DNA sequence.

The equivalency between 3mer and 1mer+2mer+3mer models no longer holds strictly when regularization is added. It is only true if we assume that the training process always ensures that the learned model has the highest generalization accuracy under the MLR framework, that is, the optimal solution. In practice, the solution is not necessarily the optimal one, despite being the goal of the regularization. Thus, 1mer+2mer+3mer models and 3mer models are approximately equivalent in the L2-regularized MLR used here. For this reason, we see that the data points drifted slightly off the diagonal in Appendix Fig S3F.

Generating DNA shape logos

DNA shape logos were generated using the seq2logo program with the PSSM-logo option (Thomsen & Nielsen, 2012). We gauged the importance of each DNA shape feature at each nucleotide position by adding this feature to the baseline 1mer model. We then calculated the ΔR^2 value upon adding this particular feature. These ΔR^2 values were used to construct a position-specific scoring matrix (PSSM), which served as input to the seq2logo program. DNA sequence logos were generated based on PSSMs that were calculated from top 200 M-words for each TF.

Acknowledgements

This work was performed in part while Y.O. and R.S. were visiting the Simons Institute for the Theory of Computing at UC Berkeley. The work was supported by the National Institutes of Health (grants R01GM106056 and U01GM103804 to R.R.) and an Alfred P. Sloan Research Fellowship (to R.R.), the Israel Science Foundation (grant 317/13 to R.S.) and the Raymond and Beverly Sackler Chair in Bioinformatics (to R.S.), and Knut and Alice Wallenberg Foundation and Swedish Research Council grants (to J.T.). L.Y. and Y.O. acknowledge support through Dan David Prize scholarships. Y.O. was partially supported by the Edmond J. Safra Center for Bioinformatics at Tel Aviv University. Open-access charges were defrayed in part through the National Science Foundation (grant MCB-1413539 to R.R.).

Author contributions

LY, YO, RS, and RR conceived and designed the project and wrote the paper. LY and YO developed computational methods and analyzed data. AJ, YY, and JT generated HT-SELEX data. JT directed the HT-SELEX experiments. RS and RR directed the computational study and overall project.

References

Abe N, Dror I, Yang L, Slattery M, Zhou T, Bussemaker HJ, Rohs R, Mann RS (2015) Deconvolving the recognition of DNA shape from sequence. *Cell* 161: 307–318

Barozzi I, Simonatto M, Bonifacio S, Yang L, Rohs R, Ghisletti S, Natoli G (2014) Coregulation of transcription factor binding and nucleosome occupancy through DNA features of mammalian enhancers. *Mol Cell* 54: 844–857

Berger MF, Philippakis AA, Qureshi AM, He FXS, Estep PW, Bulyk ML (2006) Compact, universal DNA microarrays to comprehensively determine transcription-factor binding site specificities. *Nat Biotechnol* 24: 1429–1435

Bulyk ML, Johnson PL, Church GM (2002) Nucleotides of transcription factor binding sites exert interdependent effects on the binding affinities of transcription factors. *Nucleic Acids Res* 30: 1255–1261

Chaney BA, Clark-Baldwin K, Dave V, Ma J, Rance M (2005) Solution structure of the K50 class homeodomain PITX2 bound to DNA and implications for mutations that cause Rieger syndrome. *Biochemistry* 44: 7497–7511

Chiu TP, Comoglio F, Zhou T, Yang L, Paro R, Rohs R (2016) DNAshapeR: an R/Bioconductor package for DNA shape prediction and feature encoding. *Bioinformatics* 32: 1211–1213

Crocker J, Abe N, Rinaldi L, McGregor AP, Frankel N, Wang S, Alsawadi A, Valenti P, Plaza S, Payre F, Mann RS, Stern DL (2015) Low affinity binding site clusters confer hox specificity and regulatory robustness. *Cell* 160: 191–203

Dror I, Zhou T, Mandel-Gutfreund Y, Rohs R (2014) Covariation between homeodomain transcription factors and the shape of their DNA binding sites. *Nucleic Acids Res* 42: 430–441

Farley EK, Olson KM, Zhang W, Brandt AJ, Rokhsar DS, Levine MS (2015) Suboptimization of developmental enhancers. *Science* 350: 325–328

Fernandez-Miranda G, Mendez R (2012) The CPEB-family of proteins, translational control in senescence and cancer. *Ageing Res Rev* 11: 460–472

Gordân R, Shen N, Dror I, Zhou T, Horton J, Rohs R, Bulyk ML (2013) Genomic regions flanking E-box binding sites influence DNA binding specificity of bHLH transcription factors through DNA shape. *Cell Rep* 3: 1093–1104

Jolma A, Kivioja T, Toivonen J, Cheng L, Wei G, Enge M, Taipale M, Vaquerizas JM, Yan J, Sillanpaa MJ, Bonke M, Palin K, Talukder S, Hughes TR, Luscombe NM, Ukkonen E, Taipale J (2010) Multiplexed massively parallel SELEX for characterization of human transcription factor binding specificities. *Genome Res* 20: 861–873

Jolma A, Yan J, Whitington T, Toivonen J, Nitta KR, Rastas P, Morgunova E, Enge M, Taipale M, Wei G, Palin K, Vaquerizas JM, Vincentelli R, Luscombe NM, Hughes TR, Lemaire P, Ukkonen E, Kivioja T, Taipale J (2013) DNA-binding specificities of human transcription factors. *Cell* 152: 327–339

Joshi R, Passner JM, Rohs R, Jain R, Sosinsky A, Crickmore MA, Jacob V, Aggarwal AK, Honig B, Mann RS (2007) Functional specificity of a Hox protein mediated by the recognition of minor groove structure. *Cell* 131: 530–543

Lavery R, Sklenar H (1989) Defining the structure of irregular nucleic acids: conventions and principles. *J Biomol Struct Dyn* 6: 655–667

Levine HA, Nilsen-Hamilton M (2007) A mathematical analysis of SELEX. *Comput Biol Chem* 31: 11–35

Man TK, Stormo GD (2001) Non-independence of Mnt repressor-operator interaction determined by a new quantitative multiple fluorescence relative affinity (QuMFRA) assay. *Nucleic Acids Res* 29: 2471–2478

Mathelier A, Wasserman WW (2013) The next generation of transcription factor binding site prediction. *PLoS Comput Biol* 9: e1003214

Mordelet F, Horton J, Hartemink AJ, Engelhardt BE, Gordân R (2013) Stability selection for regression-based models of transcription factor-DNA binding specificity. *Bioinformatics* 29: i117–i125

Neph S, Vierstra J, Stergachis AB, Reynolds AP, Haugen E, Vernot B, Thurman RE, John S, Sandstrom R, Johnson AK, Maurano MT, Humbert R, Rynes E, Wang H, Vong S, Lee K, Bates D, Diegel M, Roach V, Dunn D *et al* (2012) An expansive human regulatory lexicon encoded in transcription factor footprints. *Nature* 489: 83–90

Nitta KR, Jolma A, Yin Y, Morgunova E, Kivioja T, Akhtar J, Hens K, Toivonen J, Deplancke B, Furlong EE, Taipale J (2015) Conservation of transcription factor binding specificities across 600 million years of bilateria evolution. *Elife* 4: e04837

Rao SSP, Huntley MH, Durand NC, Stamenova EK, Bochkov ID, Robinson JT, Sanborn AL, Machol I, Omer AD, Lander ES, Aiden EL (2015) A 3D map of the human genome at kilobase resolution reveals principles of chromatin looping (vol 159, pg 1665, 2014). *Cell* 162: 687–688

Riley TR, Lazarovici A, Mann RS, Bussemaker HJ (2015) Building accurate sequence-to-affinity models from high-throughput *in vitro* protein-DNA binding data using FeatureREDUCE. *Elife* 4: e06397

Roemer SC, Donham DC, Sherman L, Pon VH, Edwards DP, Churchill ME (2006) Structure of the progesterone receptor-deoxyribonucleic acid complex: novel interactions required for binding to half-site response elements. *Mol Endocrinol* 20: 3042–3052

Rohs R, West SM, Sosinsky A, Liu P, Mann RS, Honig B (2009) The role of DNA shape in protein-DNA recognition. *Nature* 461: 1248–1253

Rohs R, Jin X, West SM, Joshi R, Honig B, Mann RS (2010) Origins of specificity in protein-DNA recognition. *Annu Rev Biochem* 79: 233–269

Slattery M, Riley T, Liu P, Abe N, Gomez-Alcala P, Dror I, Zhou TY, Rohs R, Honig B, Bussemaker HJ, Mann RS (2011) Cofactor binding evokes latent differences in DNA binding specificity between hox proteins. *Cell* 147: 1270–1282

Slattery M, Zhou T, Yang L, Dantas Machado AC, Gordân R, Rohs R (2014) Absence of a simple code: how transcription factors read the genome. *Trends Biochem Sci* 39: 381–399

Stormo GD (2000) DNA binding sites: representation and discovery. *Bioinformatics* 16: 16–23

Tanay A (2006) Extensive low-affinity transcriptional interactions in the yeast genome. *Genome Res* 16: 962–972

Thomsen MC, Nielsen M (2012) Seq2Logo: a method for construction and visualization of amino acid binding motifs and sequence profiles including sequence weighting, pseudo counts and two-sided representation of amino acid enrichment and depletion. *Nucleic Acids Res* 40: W281–W287

Weirauch MT, Hughes TR (2011) A catalogue of eukaryotic transcription factor types, their evolutionary origin, and species distribution. In *A handbook of transcription factors of subcellular biochemistry*, Hughes TR (ed.), Vol. 52, pp 25–73. Netherlands: Springer

Weirauch MT, Cote A, Norel R, Annala M, Zhao Y, Riley TR, Saez-Rodriguez J, Cokelaer T, Vedenko A, Talukder S, Consortium D, Bussemaker HJ, Morris QD, Bulyk ML, Stolovitzky G, Hughes TR (2013) Evaluation of methods for modeling transcription factor sequence specificity. *Nat Biotechnol* 31: 126–134

Whittle CM, Lazakovitch E, Gronostajski RM, Lieb JD (2009) DNA-binding specificity and *in vivo* targets of *Caenorhabditis elegans* nuclear factor I. *Proc Natl Acad Sci USA* 106: 12049–12054

Yang L, Zhou T, Dror I, Mathelier A, Wasserman WW, Gordân R, Rohs R (2014) TFBSshape: a motif database for DNA shape features of transcription factor binding sites. *Nucleic Acids Res* 42: D148–D155

Zhao Y, Granas D, Stormo GD (2009) Inferring binding energies from selected binding sites. *PLoS Comput Biol* 5: e1000590

Zhao Y, Ruan S, Pandey M, Stormo GD (2012) Improved models for transcription factor binding site identification using nonindependent interactions. *Genetics* 191: 781–790

Zhou T, Yang L, Lu Y, Dror I, Dantas Machado AC, Ghane T, Di Felice R, Rohs R (2013) DNAshape: a method for the high-throughput prediction of DNA structural features on a genomic scale. *Nucleic Acids Res* 41: W56–W62

Zhou T, Shen N, Yang L, Abe N, Horton J, Mann RS, Bussemaker HJ, Gordân R, Rohs R (2015) Quantitative modeling of transcription factor binding specificities using DNA shape. *Proc Natl Acad Sci USA* 112: 4654–4659

11

Revisiting biomarker discovery by plasma proteomics

Philipp E Geyer[1,2], Lesca M Holdt[3], Daniel Teupser[3] & Matthias Mann[1,2,*] (iD)

Abstract

Clinical analysis of blood is the most widespread diagnostic procedure in medicine, and blood biomarkers are used to categorize patients and to support treatment decisions. However, existing biomarkers are far from comprehensive and often lack specificity and new ones are being developed at a very slow rate. As described in this review, mass spectrometry (MS)-based proteomics has become a powerful technology in biological research and it is now poised to allow the characterization of the plasma proteome in great depth. Previous "triangular strategies" aimed at discovering single biomarker candidates in small cohorts, followed by classical immunoassays in much larger validation cohorts. We propose a "rectangular" plasma proteome profiling strategy, in which the proteome patterns of large cohorts are correlated with their phenotypes in health and disease. Translating such concepts into clinical practice will require restructuring several aspects of diagnostic decision-making, and we discuss some first steps in this direction.

Keywords biomarkers; diagnostic; mass spectrometry; plasma proteomics; systems medicine

Introduction

The central and integrating role of blood in human physiology implies that it should be a universal reflection of an individual's state or phenotype. Its cellular components are erythrocytes, thrombocytes, and lymphocytes. The liquid portion is called plasma, when all components are retained, and serum, when the coagulation cascade has been activated (blood clotting). For simplicity, we will use the term "plasma" rather than "serum", since most conclusions apply to both.

Concentrations of various plasma components are routinely determined in clinical practice. These include electrolytes, small molecules, drugs, and proteins. The proteins constituting the plasma proteome can be categorized into three different classes (Fig 1A and

B). The first contains abundant proteins with a functional role in blood. These include human serum albumin (HSA, roughly half of total protein mass); apolipoproteins, which have crucial roles in lipid transport and homeostasis; acute phase proteins of the innate immune response; and proteins of the coagulation cascade. The second class are tissue leakage proteins without a dedicated function in the circulation. Examples are enzymes such as aspartate aminotransferase (ASAT) and alanine aminotransferase (ALAT), which are used for the diagnosis of liver diseases, as well as low-level, tissue-specific isoforms of proteins such as cardiac troponins. The third class are signaling molecules like small protein hormones (for instance, insulin) and cytokines, which typically have very low abundances at steady state and are upregulated when needed. Baseline levels of the cytokine interleukin-6 (IL-6) are 5 pg/ml, establishing a minimum 10^{10}-fold dynamic range of the plasma proteome when compared to the concentration of the most abundant protein, HSA, with about 50 mg/ml.

In accepted use, "a biomarker is a defined characteristic that is measured as an indicator of normal biological processes, pathogenic processes, or a response to an exposure or intervention" (FDA-NIH: Biomarker-Working-Group, 2016). For the purpose of this review, we focus specifically on protein or protein modification-based biomarkers. In this sense, there are more than 100 FDA-cleared or FDA-approved clinical plasma or serum tests, mainly in the abundant, functional class (50%), followed by tissue leakage markers (25%), and the rest include receptor ligands, immunoglobulins, and aberrant secretions (Anderson, 2010). Most of these are decades old, and the current introduction rate of novel markers is less than two per year (Anderson *et al*, 2013). A typical test consists of an enzymatic assay or immunoassay against a single target. Clinicians interpret the results in conjunction with other patient information, based on their expert knowledge. Ratios of abundances are only employed in specific cases. Examples are the 60-year-old De Ritis ratio of ASAT/ALAT to differentiate between causes of liver disease (De-Ritis *et al*, 1957) or the more recent sFlt-1/PlGF ratio for diagnosis of preeclampsia (Levine *et al*, 2004).

In contrast to enzymatic and antibody-based methods, mass spectrometry (MS)-based proteomics measures the highly accurate mass and fragmentation spectra of peptides derived from sequence-specific digestion of proteins. Because the masses and sequences of

1 Department of Proteomics and Signal Transduction, Max Planck Institute of Biochemistry, Martinsried, Germany
2 Faculty of Health Sciences, NNF Center for Protein Research, University of Copenhagen, Copenhagen, Denmark
3 Institute of Laboratory Medicine, University Hospital, LMU Munich, Munich, Germany
*Corresponding author. E-mail: mmann@biochem.mpg.de

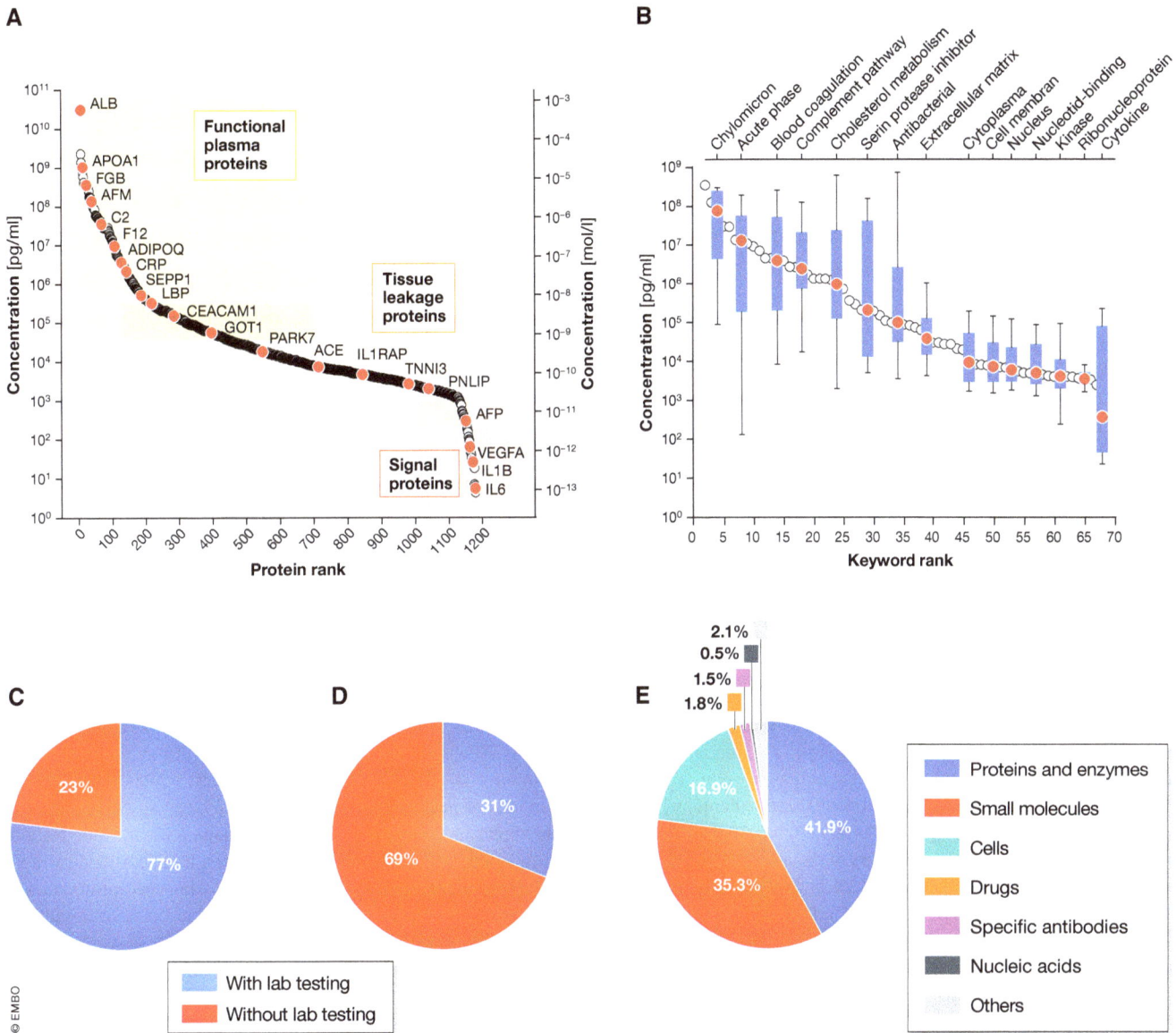

Figure 1. Blood-based laboratory testing in a clinical setting.

(A) Concentration range of plasma proteins with the gene names of several illustrative blood proteins (red dots). Concentrations are in serum or plasma and measured with diverse methods as retrieved from the plasma proteome database in May 2017 (http://www.plasmaproteomedatabase.org/) (Nanjappa *et al*, 2014). (B) Bioinformatic keyword annotation of the plasma proteome database. The blue boxplots with the 10–90% whiskers visualize the range of diverse proteins contributing to distinct functions. (C) Percentage of inpatient admissions receiving blood-based laboratory testing. Numbers are based on 9 million tests performed in the year 2016 at the Institute of Laboratory Medicine, University Hospital Munich. (D) Percentage of outpatient admissions receiving blood-based laboratory testing. (E) Distribution of laboratory tests based on frequency of request. Examples of test for different classes of analytes are as follows: Proteins and enzymes—*liver enzymes, inflammatory proteins, tumor markers*; Small molecules—*electrolytes, substrates, vitamins*; Cells—*red, white blood cells, and platelets*; Drugs—*immunosuppressants, antibiotics, and drugs of abuse*; Specific antibodies—*autoantibodies and antibodies against infectious agents*; and Nucleic acids—*viruses and genetic variants*.

these peptides are unique, proteomics is inherently specific, a constant problem with colorimetric enzyme tests and immunoassays (Wild, 2013). In principle, MS-based proteomics can analyze all the proteins in a system—its proteome—and is in this sense unbiased and hypothesis-free (Aebersold & Mann, 2016). Furthermore, MS methods are ideally suited to discover and quantify post-translational modifications (PTMs) on proteins. These PTMs can also be the basis of diagnostic tests, such as HbA1c levels that serve as a readout of long-term glucose exposure in the context of

diabetes. Nevertheless, none of the routinely performed laboratory tests in plasma is based on proteins that were identified by mass-spectrometric approaches, and in routine analysis, MS is so far only employed for measuring small molecules such as drugs and metabolites (Vogeser & Seger, 2016).

Over the past years, the technology of MS-based proteomics has dramatically improved, and it is now a mainstay of all biological research that involves proteins (Cox & Mann, 2011; Altelaar & Heck, 2012; Richards *et al*, 2015; Zhang *et al*, 2016). In particular, its performance has robustly matured into a sensitivity and dynamic range that makes it interesting for biomarker studies. This review will focus on the prospects of determining proteins in blood by mass spectrometry. We start by empirically assessing the role of proteins in clinical diagnostic today and exhaustively review the literature on previous attempts at finding biomarkers in plasma by MS-based proteomics. So far, proteomics strategies have involved extensive investigations of few samples, to be followed up by targeted approaches in larger cohorts. We discuss how recent advances in technology now enable a new strategy in which deep proteomes are measured for many time points and participants with the prospect to find new biomarkers and biomarker panels. We believe that proteomics will become part of the instrumental routine in the clinical laboratory within the next decade and may even eliminate current technologies in the far future.

The current extent of clinical protein-based diagnostics

Laboratory tests of blood and body fluids aim at disease diagnosis or confirmation, risk prediction, prognosis monitoring, and evaluating treatment effectiveness. It is commonly assumed that 70% of diagnoses are informed by blood testing, even though this number has not been well substantiated. At the Institute of Laboratory Medicine of the University Hospital Munich, laboratory testing is ordered for the vast majority of inpatients at some point during hospitalization (77%; Fig 1C). This fraction is much smaller in patients seen in one of the Hospital's outpatient clinics (31%; Fig 1D). These numbers indicate that hospitalized patients, who are usually sicker, are more likely to receive laboratory tests than ambulatory patients. Based on numbers of requested analyses, clinical routine is dominated by proteins (42% of analyses), followed by small molecules (35%) and cells (17%) (Fig 1E). Thus, already today proteins are the most frequently assayed class of laboratory analytes in clinical practice. We also note that methods suitable for determining plasma proteins have the largest share of the worldwide *in vitro* diagnostics.

Laboratory assays for plasma proteins are based either on classical clinical chemistry, utilizing enzymatic activities of certain plasma proteins, or on antibody-based immunoassays. The costs of enzymatic assays are only in the cent-range, and they run on high-throughput automated analyzers, delivering up to 10,000 test results per hour. In contrast, immunoassays are more expensive (usually several euros/dollars per sample) and throughput of the respective automated analyzers is about 1,000 tests/hour. Large clinical chemistry as well as immunoassay-based analyzers may carry reagents for more than 100 different analytical parameters. Main advantages of immunoassays are a greater degree of flexibility due to the accessibility to plasma proteins devoid of enzymatic activity and a significantly higher sensitivity. Another, clinically relevant issue is the time required per laboratory test. Due to the necessity of immediate decision-making, the majority of enzymatic assays and several immunoassays have to be scaled down to analysis times of < 10 min. In general, immunoassays tend to take longer than enzymatic assays; nevertheless, the vast majority of current automated immunoassays require no more than 30 min.

Systematic review of MS-based plasma proteomics in biomarker research

Plasma proteins had already been investigated by two-dimensional gel electrophoresis in the 1990s, sometimes in combination with MS identification of excised spots. However, these generally identified only a few dozen proteins, and as they preceded MS-based proteomics, they are not discussed in this review. Claims of early cancer detection based on very low-resolution MALDI spectra of plasma that produced patterns but no protein identifications (Petricoin *et al*, 2002) have not been substantiated (Baggerly *et al*, 2004), and these technologies have largely been abandoned today.

To obtain a comprehensive collection of publications dealing with plasma biomarker research and employing MS-based proteomics, we performed an unrestricted PubMed search specifying co-occurrence of the terms "biomarker", "plasma OR serum", "proteome", "proteomics", and "mass spectrometry". This yielded an initial list of 947 publications of which 103 were reviews. We further subtracted studies that did not deal with human subjects or did not involve plasma or serum, leaving 381 original publications (Dataset EV1).

Publications started to appear in 2002 and reached a maximum of 33 per year in 2005, when the special issue on the plasma proteome was released by the Human Proteome Organization (HUPO) (Omenn *et al*, 2005). Two further maxima appeared in 2011 and 2014 with 39 and 43 publications per year, followed by drops in 2013 to 24 and in 2016 to only 20 publications per year (Fig 2A). The observed dynamics contrasts with an ever-expanding community of researchers using proteomics, which is reflected in thousands of publications per year, with a clear upward trend. The ratio of plasma proteome publications to total proteome publications is now < 1% and continues to drop. Given the clear medical need for plasma biomarkers and the success of MS-based proteomics in other areas, this raises the question as to what holds back the field of plasma proteomics.

Of the 381 primary publications, about half dealt with the analytical descriptions of the workflow employed in plasma analysis, whereas the remainder investigated a physiological or pathophysiological question (Fig 2B). About a third of the latter focused on cancer, followed by cardiovascular disease (CVD), topics in human biology, inflammation, diabetes, and infectious diseases (Fig 2B). Clearly, this ordering reflects the interest in the diseases rather than the likelihood of finding relevant changes with the available technology. Only 47% of the studies had any kind of validation of the primary findings (Fig 2C). In half of the cases (24%), these were simple Western blots or ELISAs of candidate proteins performed with the same samples rather than an independent cohort as is usual practice in clinical studies. Only 36 papers used MS-based proteomics to validate potential biomarkers that were proposed independently (Dataset EV1).

The extremely high dynamic range of plasma still makes it difficult to identify more than a few hundred of the most abundant proteins by LC-MS/MS. To partially overcome this challenge, highly abundant plasma proteins are often depleted, generally through columns with immobilized antibodies directed against the top 1 to 20 proteins (Fig 2D). However, these antibodies are never entirely specific and bound proteins—such as HSA—themselves have an affinity for several other proteins (Tu *et al*, 2010; Bellei *et al*, 2011).

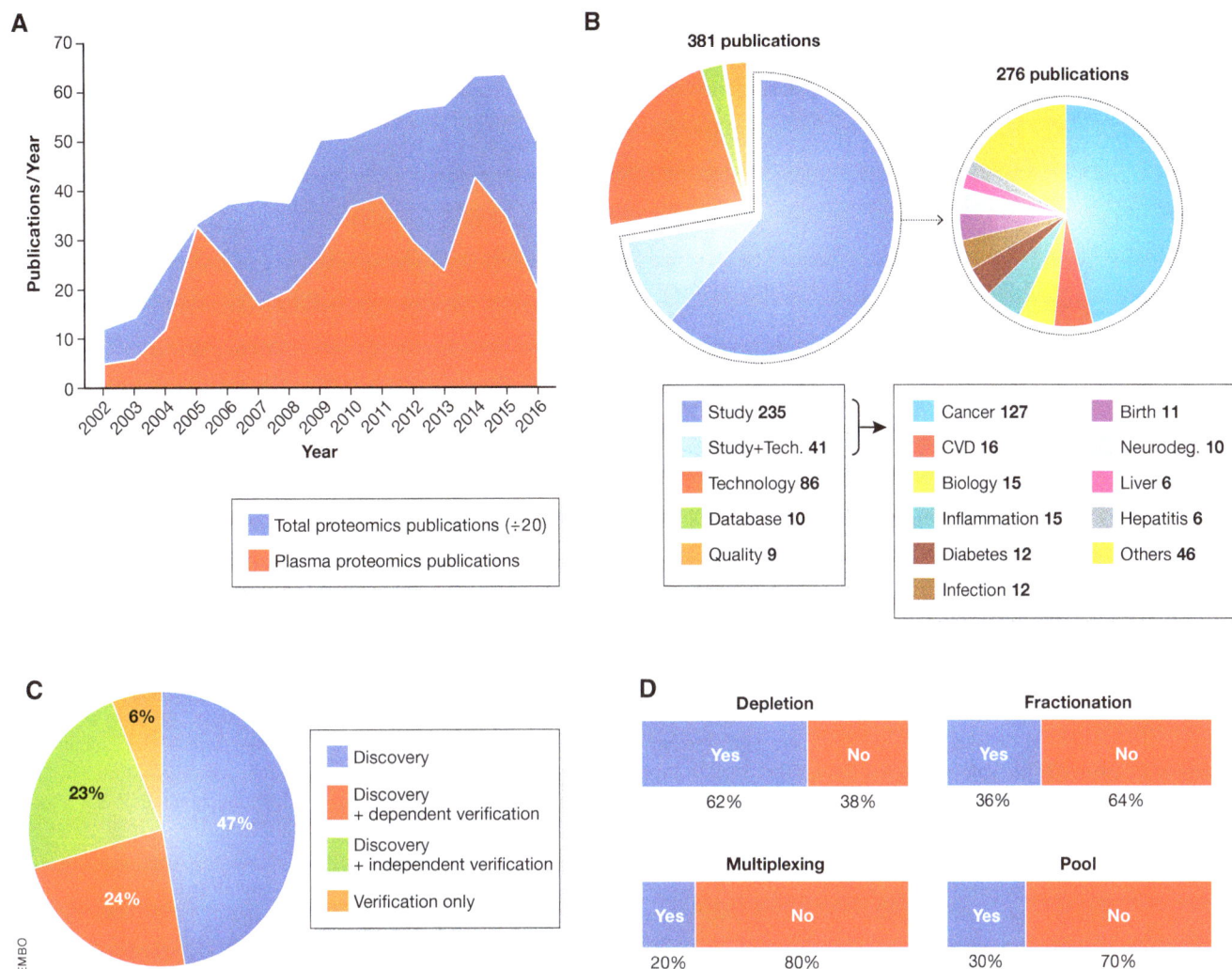

Figure 2. Comprehensive literature review.
(A) Publications using MS-based proteomics in plasma biomarker research (red) compared to the total number of publications in proteomics (blue). (B) Pie charts about the intentions of the investigated studies and proportions of investigated diseases. (C) Overview of the percentage of studies, using discovery and validation phases. (D) Studies using pooled samples, depletion, fractionation, and multiplexing in plasma biomarker research using MS-based proteomics.

Thus, the depleted plasma sample is not a quantitative representation of the original proteome. This is especially true when using "super-depletion" (Qian *et al*, 2008)—a broad mixture of polyclonal antibodies raised against whole plasma—or beads with hexameric peptide mixtures that non-specifically "normalize" the plasma proteome (Thulasiraman *et al*, 2005). Furthermore, these procedures introduce variability and additional expense into the workflow, generally precluding accurate quantification of plasma proteins. Therefore, their use is currently restricted to small discovery projects.

A second strategy to deal with the dynamic range and sensitivity challenge is extensive plasma fractionation, which can be done in various ways at the protein or peptide level. Several studies aiming at in-depth coverage of the plasma proteome by combined depletion and extensive separation (up to hundreds of fractions) identified from several hundred to several thousand proteins (Liu *et al*, 2006; Pan *et al*, 2011; Cao *et al*, 2012; Cole *et al*, 2013; Keshishian *et al*,

2015; Lee *et al*, 2015). Note that many plasma proteome studies continue to use much less stringent statistical identification criteria than the 1% peptide and protein false discovery rates (FDR) that have become standard in MS-based proteomics.

The decrease in throughput implicit in fractionation can partially be recovered by multiplexing. For example, between four and ten samples have been analyzed together using the iTRAQ or TMT strategies, in which samples are labeled with mass neutral tags that give rise to different low mass reporter ions (Kolla *et al*, 2010; Zhou *et al*, 2012; Cominetti *et al*, 2016). Quantification is achieved by fragmenting peptides and quantifying the relative ratios of the reporter ions (Bantscheff *et al*, 2008). Although attractive in principle, these techniques generally suffer from ratio distortion caused by co-isolated peptide species that all contribute to the same reporter ion pattern ("ratio compression"). Regulation of very low-level proteins or those with small but disease-relevant changes may be completely obscured. In shotgun proteomics, eluting peptides are fragmented in order of intensity (data-dependent acquisition), a semi-stochastic process that may lead to missing values across LC-MS/MS runs. Recently introduced data-independent acquisition strategies more consistently identify peptides across runs (Picotti & Aebersold, 2012; Sajic et al, 2015). However, they are incompatible with reporter-ion-based multiplexing because one would quantify the average of groups of peptides.

In about 30% of the studies, plasma samples were pooled to reach a desired plasma proteome coverage within the available measuring time. This approach sacrifices within-group variances and outlier or contaminant proteins in individual samples can skew the whole group, making it all but impossible to assess whether proteins that are different between groups are actually significant on a person-by-person basis.

Partly as a consequence of the demands on instrument time, generally no more than 20–30 samples were analyzed and only few exceeded 500 (Garcia-Bailo et al, 2012; Cominetti et al, 2016; Lee et al, 2017). Considering the large number of measurement points within samples, these are small sample numbers. Accordingly, most studies proposed a few "potential biomarkers", defined as proteins that differ between cases and controls. Furthermore, many of these candidates are unlikely to be specific indicators of the disease in question, because they belong to biological categories that are at best indirectly related to the disease or are likely artifacts of sample preparation (such as keratins and red blood cell proteins). In summary, limitations in proteomics technology and experimental design have prevented the identification of true biomarkers in the published literature to date. To our knowledge, the only possible exception is the OVA1 test, in which the levels of the highly abundant plasma proteins beta-2 macroglobulin, apolipoprotein 1, serum transferrin, and pre-albumin were combined with the previously established ovarian cancer marker CA125 in a narrow, FDA-approved indication (Rai et al, 2002; Zhang et al, 2004).

Triangular MS-based biomarker discovery and validation strategy

The principal advantage of hypothesis-free MS-based proteomics is that no assumptions need to be made regarding the possible nature and number of potential biomarkers, in stark contrast to single protein measurements in classical biomarker research. Conceptually, MS-based proteomics combines all possible hypothesis-driven biomarker studies for each disease into one and furthermore defines the relation of potential biomarkers to each other. In practice, the challenges of plasma proteomics have so far prevented in-depth and quantitative studies on large cohorts. Instead, a stepwise or "triangular" strategy for biomarker discovery has been advocated, with several phases in which the number of individuals increases from a few to many, whereas the number of proteins decreases from hundreds or thousands to just a few (Rifai et al, 2006; Fig 3A).

The typical workflow for hypothesis-free discovery proteomics in plasma is similar to that used in other areas of bottom-up proteomics (Aebersold & Mann, 2016; Altelaar & Heck, 2012; Fig 3B). Briefly, proteins are enzymatically digested into peptides, which are separated by high-pressure liquid chromatography (HPLC) coupled to electrospray ionization. Peptide masses and abundances are measured in the mass spectrometer in full MS scans, whereas a

further step of peptide fragmentation produces MS/MS spectra for peptide identification. Well-established proteomics software platforms automatically and statistically rigorously identify peptides in database searches and quantify them (Cox & Mann, 2008; MacLean et al, 2010; Rost et al, 2014). Furthermore, plasma contains blood components such as lipids that can easily clog HPLC columns, which necessitates dedicated peptide cleanup procedures (Geyer et al, 2016a).

Targeted proteomics for candidate verification is a second phase of the triangular strategy (Fig 3C). A relatively small number of proteins (typically < 10) with differential expression in the discovery phase are tested in a larger and ideally independent cohort. Since immunoassays are often not available, targeted MS methods can be employed. The most widespread of these is "multiple reaction monitoring" (MRM—sometimes also called single or selected reaction monitoring—SRM) (Picotti & Aebersold, 2012; Carr et al, 2014; Ebhardt et al, 2015). For each protein, a set of suitable peptides is selected and their elution and fragmentation behavior is assessed to define an MRM assay. During analysis, the mass spectrometer is programmed to continuously fragment only these peptides as they elute. By monitoring several fragments per peptide, sensitive and specific quantification can be achieved even with low-resolution mass spectrometers. The advantage of MRM over shotgun proteomics for verification is its higher sensitivity and throughput. Inter-laboratory studies have achieved good reproducibility (Addona et al, 2009; Abbatiello et al, 2015), but reported sensitivities typically do not reach the low ng/ml concentration range and practically achieved multiplexing capabilities are limited to dozens of peptides (Percy et al, 2013; Shi et al, 2013; Oberbach et al, 2014; Wu et al, 2015). Nevertheless, two recent studies have reported the targeting of 82 and 192 proteins, respectively (Ozcan et al, 2017; Percy et al, 2017). The sensitivity of MRM can be improved to the low ng/ml or even high pg/ml ranges by more extensive sample preprocessing with depletion or fractionation (Burgess et al, 2014; Kim et al, 2015; Nie et al, 2017).

Absolute and accurate quantification requires internal standards —generally heavy isotope versions of the monitored peptides. Synthesized heavy peptides are added after digestion, creating a source of quantitative inaccuracy since the variability of protein digestion is not taken into account. This can be addressed by embedding the peptide in its original sequence context, for instance, in the SILAC-PrEST strategy, in which a 150- to 250-amino acid stretch of each protein of interest, fused to a quantification tag, is recombinant expressed in a heavy form (Zeiler et al, 2012; Edfors et al, 2014; Geyer et al, 2016a).

Targeted methods can also be combined with immuno-enrichment of proteins or peptides. For instance, in "stable isotope standards and capture by anti-peptide antibodies" (SISCAPA) specific peptides are immunoprecipitated together with their heavy-labeled counterparts, followed by rapid MS-based readout (Anderson et al, 2004; Razavi et al, 2016). This combines the enrichment capabilities of antibodies with the specificity of MS detection; however, development of assays can be difficult and time-consuming—narrowing the advantage compared to purely antibody-based methods.

The final phase in the triangular strategy is the validation with immunoassays, a field that has matured over decades. For maximum specificity, sandwich assays are typically preferred (Fig 3D). While they are costly and laborious to develop, they can achieve

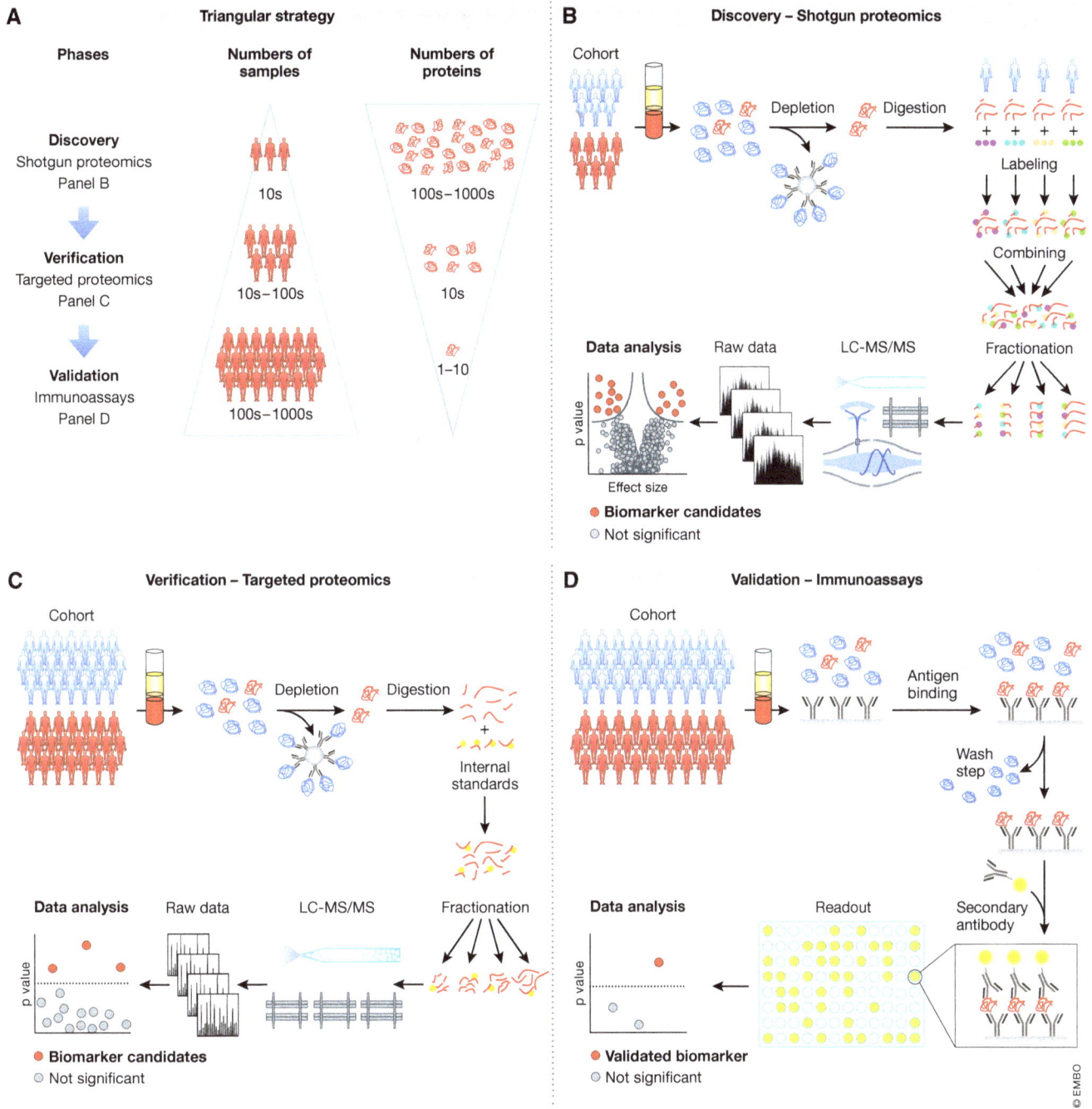

Figure 3. Current paradigms in plasma biomarker research ("triangular approach").

(A) A relatively small number of cases and controls are analyzed by hypothesis-free discovery proteomics in great depth, ideally leading to the quantification of thousands of proteins (top layer in the panel). This may yield tens of candidates with differential expression that are screened by targeted proteomics methods in cohorts of moderate size (middle layer). Finally, for one or a few of the remaining candidates, immunoassays are developed, which are then validates in large cohorts and applied in the clinic (bottom layer). (B) Workflow for hypothesis-free discovery proteomics. (C) Targeted proteomics for candidate verification. (D) Development of immunoassays for clinical validation and application.

high sensitivity and high throughput. Even cohorts with thousands of participants can be tested with this technology, but only for one or a few candidate biomarkers. Such large numbers may be necessary to establish specificity not only against controls but also with respect to other diseases. Standard requirements include insuring

adequate statistical power and replication in an independent population. Today, such clinical studies can be expensive multi-year endeavors, partly explaining the paucity of new biomarkers.

Immunoassays have some inherent limitations, mostly related to antigen-antibody recognition. These include cross-reactivity, interference by background molecules such as triglycerides, and non-linear response ("hook effect") (Hoofnagle & Wener, 2009; Wild, 2013). Furthermore, not all clinically important protein variants are easily recognizable by antibody-based assays. Given these limitations, MS-based methods would be attractive alternatives in at least some large-scale clinical trials, but this requires much more robust, sensitive, and higher throughput technologies than those available today.

Over the last decade, the proteomics community has developed guidelines for proper development of biomarkers that discuss quality standards and emphasize the importance of selecting adequate cohorts that ensure statistical significance of the findings as well as specificity of potential biomarkers and their potential clinical application (Luque-Garcia & Neubert, 2007; Paulovich et al, 2008; Mischak et al, 2010; Surinova et al, 2011; Skates et al, 2013; Parker & Borchers, 2014; Hoofnagle et al, 2016).

Not surprisingly in view of the rigorous requirements of the triangular strategy, there are few, if any, reports in which it has been applied completely and successfully. This may also partly be due to the fact that three different technologies—shotgun proteomics, targeted proteomics, and immunoassay development—are involved. Many publications just describe the first phase or only combine it with immunoassay verification in the same cohort (Dataset EV1).

Among the studies with more than a few participants and with some verification, the majority selected candidates of interest and performed Western blotting, ELISA, or MRM assays. A representative example is the study by Zhang et al (2012) in which depleted plasma of 10 colorectal cancer patients versus controls was labeled with iTRAQ and fractionated, leading to the identification of 72 proteins. Among several up- or downregulated proteins, ORM2 was followed up by ELISAs in 419 individuals. Since this protein is a part of the innate immune system (like the other two upregulated candidates), it is unlikely to be a specific cancer marker. In another study, super-depletion, iTRAQ labeling, and fractionation identified 830 proteins in a discovery cohort of 751 patients with cardiovascular events and controls that had been reduced to 50 pooled samples (Juhasz et al, 2011). The known markers CRP and fibronectin were selected from the list of candidates and found to be significantly upregulated in the original cohort by immunoassays against these proteins. In a heart transplantation study, analysis of depleted and iTRAQ-labeled plasma from 26 patients at five time points before and after surgery identified a total of more than 900 proteins (273 per individual; Cohen et al, 2013). MRM assays and ELISAs against five medium-abundant proteins in a partially independent follow-up cohort of 43 individuals served to develop a computational pipeline for risk markers for organ rejection. In an approach of potential clinical utility, depleted plasma from a mouse model of breast cancer allowed the identification of more than 1,000 plasma proteins from which 88 were selected for MRM assays in an independent verification cohort of 80 animals (Whiteaker et al, 2011).

Rectangular biomarker strategy and plasma proteome profiling

In the last few years, the community has substantially improved all aspects of the workflow of MS-based proteomics. In sample

preparation, laborious, multi-stage preparation workflows have been replaced by robust, single-vial processing with a minimum of manipulation steps. This also helps with automation and increases throughput. The sensitivity and sequencing speed of MS instruments have improved severalfold. The entire LC-MS/MS system has become much more robust, although this is still far from what will be needed for routine clinical application. Finally, bioinformatic analysis of the results is now statistically sound and straightforward to use and increasingly enables correlation of MS results with a wide range of other classical clinical and additional "omics" data. Illustrating the power of cutting edge MS-based proteomics, cell lines can now routinely be quantified to a depth of more than 10,000 different proteins in a relatively short time, sometimes even without any fractionation (Mann et al, 2013; Richards et al, 2015; Sharma et al, 2015; Bekker-Jensen et al, 2017).

Given this technological progress of proteomics in cell line and tissue samples, we asked whether one could also develop a fast and automated workflow that would quantify the plasma proteome in depth in a large number of samples (Geyer et al, 2016a). We reasoned that this would then enable a "rectangular strategy" in which as many proteins as possible are measured for as many individuals and conditions as possible. In contrast to the triangular workflow, the initial discovery cohort would be much larger, ideally encompassing hundreds or thousands of participants, resulting in a greater likelihood to reveal any patterns that might differentiate the investigated groups or conditions. These larger initial numbers of plasma proteomes would allow the discovery of statistically significant, but small differences and changes associated with a group of proteins. In the proposed rectangular strategy, discovery and validation cohorts would both be measured by shotgun proteomics in great depth. This removes the dependency of validation on discovery, meaning that both cohorts can be analyzed together (Fig 4A). Moreover, having separate cohorts allows unmasking study-specific confounders. A further advantage of the rectangular strategy is its ability to discover and validate protein patterns that are characteristic of particular health or disease states, in addition to single biomarker candidates, something that is unattainable with the triangular approach.

Interestingly, an analogous change of concept has already happened a number of years ago for genome-wide association studies (GWAS). Researchers in this field found that joint analysis of as many samples as possible was superior to a sequential pipeline (Skol et al, 2006). In proteomics, the obvious challenge is achieving sufficient proteomics depth in a short time, ideally without depletion and in a robust workflow. This goal has not been achieved at the time of writing, but the current rate of technological improvements promises to make it feasible in the near future. Below, we discuss four examples of this emerging approach.

The first of these investigated a cohort of 36 monozygotic and 22 dizygotic twin pairs to determine the influence of genetic background on the levels of plasma proteins (Liu et al, 2015). The authors established a spectral library using depleted, fractionated, and pooled samples and measured their samples with data-independent acquisition (DIA). A total of 232 plasma samples were then measured with 35-min gradients in a data-independent mode, leading to the consistent quantification of 1,904 peptides and 342 proteins. Interestingly, protein levels were often relatively stable within individuals as compared to between individuals.

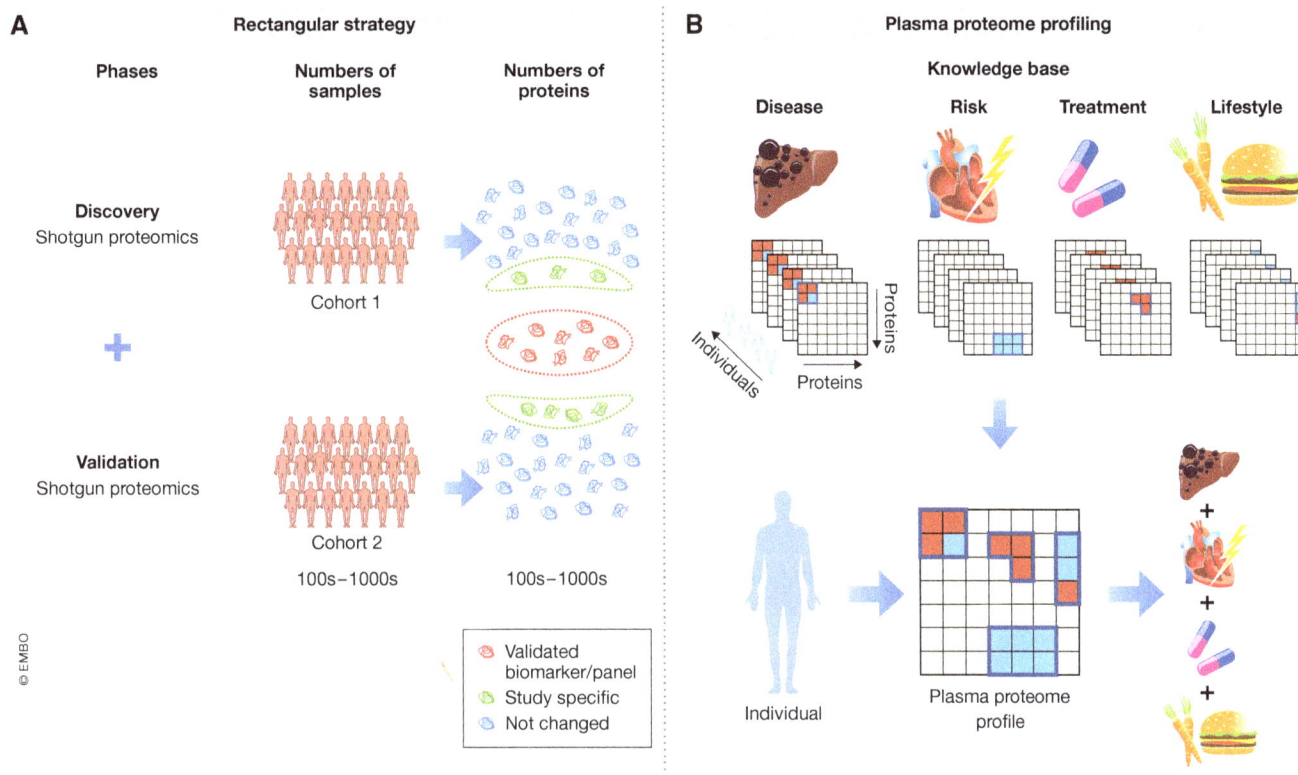

Figure 4. Rectangular workflow.

(A) A large cohort is investigated in the discovery phase with as much proteome coverage as possible. In the validation phase, another cohort is analyzed to confirm the biomarker candidates, but it uses the same technology and similar cohort size. Both cohorts can be analyzed in parallel, but only the proteins that are statistically significantly different in both studies (orange as opposed to green circle in the right-hand part of panel A) are validated biomarkers. (B) Plasma proteome profiling of diverse lifestyle, disease, treatment, or other relevant alterations will over time build up a knowledge base that connects plasma protein changes to perturbations in a general manner (upper panel). The plasma proteome profile of a given individual can then be deconvoluted using the information and algorithms associated with the knowledge base (lower panel).

Furthermore, there were clear indications for the levels of some proteins to be under genetic control. For instance, processes connected to "immune response" and "blood coagulation" tended to be heritable, whereas those associated with "hormone response" did not. Although a pioneering study, the number of plasma proteomes analyzed was relatively small in view of the generality of the research question posed. Generally, genetics studies routinely investigate thousands of participants to tease out subtle heritable effects, illustrating the need for much higher throughput in clinical proteomics.

Malmström *et al* (2016) induced sepsis in mice by injecting *S. pyogenes* and followed their plasma proteomes through three time points on non-depleted, non-fractionated samples. A library of diverse mouse tissues was employed to support data-independent identifications as well as to determine the origin of tissue damage proteins. In this way, 2-h runs quantified an average of 786 mouse proteins, although it should be noted that proper FDR criteria for inferring peptide identities in the complex DIA MS/MS spectra are still being discussed (Nesvizhskii *et al*, 2007; Bruderer *et al*, 2017; Rosenberger *et al*, 2017). Several expected categories of plasma proteins increased during sepsis, as well as some markers associated with damage to the vascular system. Some of the changes were related to mobilization of the immune system against the pathogen, and others appeared to be correlated with necrosis in severely affected animals.

In a workflow termed "plasma proteome profiling", we focused on the rapid and robust analysis of only 1 μl of undepleted plasma from a single fingerpick (Geyer *et al*, 2016a). Total gradient time was only 20 min, enabling extensive investigation of analytical, intra-assay, intra-individual, and inter-individual variation of the plasma proteome. Based on the quantification of 300 plasma proteins, about 50 FDA-approved biomarkers were covered with label-free quantification (CV < 20%). Rapid analysis of a wide range of samples also revealed different sets of quality markers that clearly classified samples with evidence of red blood cell lysis, those with partial activation of the coagulation cascade due to inappropriate sample handling, and those with exogenous contaminations such as keratins. Even though this study provided a useful overview of the information content of the plasma proteome, the depth of coverage was not yet sufficient to address low-level, regulatory plasma proteins. A single step of fractionation yielded a quantitative plasma proteome of about 1,000 proteins, including 183 proteins with a reported concentration of < 10 ng/ml, however at the cost of longer measurement times per sample.

An improved version of the plasma proteome profiling workflow allowed the robotic preparation and measurement of nearly 1,300 plasma proteome samples in a weight loss study (Geyer *et al*, 2016b). Quadruplicate analysis of individuals captured the dynamics of an average of 437 proteins upon losing weight and over a year of weight maintenance. Weight loss itself had a broad effect on the human plasma proteome with 93 significantly changed proteins. Quantitative differences were often small but physiologically meaningful, such as a 16% reduction of the adipocyte-secreted factor SERPINF1. The longitudinal study design in which the individuals sustained an average 12% weight loss for 1 year allowed capturing the long-term dynamics of the plasma proteome and categorizing it into proteins stable within versus between individuals. Multi-protein patterns reflected the lipid homeostasis system (apolipoprotein family), low-level inflammation, and insulin resistance. These patterns quantified the benefits of weight loss at the level of the individual, potentially opening up for individualized treatment and lifestyle recommendations.

Together, these studies also highlight the advantages of longitudinal over cross-sectional study designs, because the plasma proteome tends to be much more constant within an individual over time than between different individuals. Furthermore, they are similar in that they use less bias-prone undepleted plasma, and identify many proteins in a given analysis time (up to 20 proteins/min).

Regarding the question of how many proteins should be covered, we found that a proteomic depth of more than 1,500 proteins in undepleted plasma allows the coverage of tissue leakage proteins such as liver-based lipoprotein receptors and is within reach of technological capabilities that are currently being developed. Among the first 300 highest abundant proteins, every fourth protein is a biomarker, whereas in the next 1,200 proteins, it is only every 25th protein (Fig 5). As there is no a priori reason that biomarkers should have a skewed abundance distribution, this suggests that many biomarkers are still to be found. We believe that the real promise of plasma proteome profiling using the rectangular strategy is that it can discover proteins and protein patterns that have not been considered as biomarkers yet. The exponential increase in the underlying LC-MS/MS technology will stimulate a matching increase in the number of plasma proteome datasets recorded in laboratories around the world. This will create an extensive database of plasma proteomes and their dynamics, involving many clinical studies and individuals. Such data could then be aggregated to build up a knowledge base that connects proteome states to a wide diversity of "perturbations", including diseases, risks, treatments, and lifestyles. At a minimum, this approach will reveal all the different conditions in which a given set of biomarkers is involved, in addition to the specific context where they were discovered. Proteome overlap between disease conditions could reveal commonalities between them (Fig 4B, upper panel). An individual's plasma proteome profile and its dynamics could then be interpreted by comparing it to the global knowledge base. This could be used to deconvolute co-morbidities and to guide treatment and monitor effectiveness (Fig 4B, lower panel).

Standardization of the proteomic biomarker discovery pipeline

It has been suggested that the current lack of biomarkers making their way into the market may be the result of various technical, scientific, and political aspects including undervaluation, resulting from inconsistent regulatory standards, and lack of evidence for analytical validity and clinical utility (Hayes et al, 2013). To overcome these challenges, systematic pipelines for biomarker

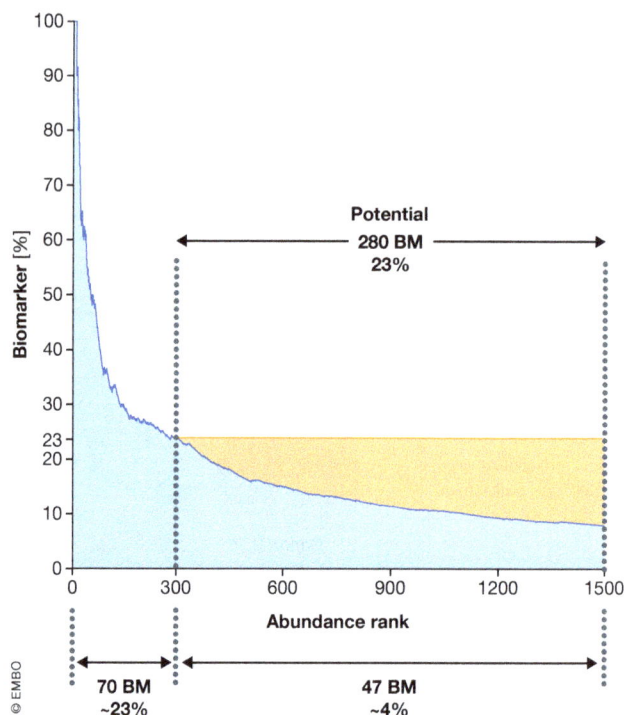

Figure 5. Biomarker distribution across the abundance range.
The blue area illustrates the percentage of biomarker (BM) as a function of increasing depth of the plasma proteome. Within the 300 most abundant proteins, 23% are already known biomarkers. The top of the yellow region extrapolates this proportion to the remainder of the plasma proteome. If the portion of biomarkers remained as high as it is in the 300 most abundant proteins, there are at least 233 potential biomarkers to be discovered (yellow area of the figure).

development have been advocated (Pavlou et al, 2013; Duffy et al, 2015). In the context of moving from a triangular to a rectangular strategy of biomarker discovery, it will be particularly important to consider the following principles.

(1) Analytical performance characteristics: Analytical validity is the capacity of a test to provide an accurate and reliable measurement of a biomarker. Establishment of analytical validity of the plasma proteomics methodology will be key, because the same method will often be carried on from discovery to application. Detailed standards to determine analytical validity have been developed by the Clinical and Laboratory Standards Institute (CLSI) (www.clsi.org). An overview can be found in Grant and Hoofnagle (2014) and Jennings et al (2009). Some of these standards have been recognized by the U.S. Food and Drug Administration (FDA) and are accepted for bringing in vitro diagnostic test to the market (https://www.accessdata.fda.gov/scripts/cdrh/cfdocs/cfstandards/search.cfm). Even though starting off with a full analytical validation conforming to FDA standards might be prohibitive in biomarker discovery, at least some of the key criteria, such as carryover, accuracy, precision, analytical sensitivity, analytical specificity, and limit of quantification, should be tested early on. This is in line with what we advocate in the context of the rectangular strategy and is also in the interest of saving resources, because the step following biomarker discovery is biomarker validation, where analytical validity will be mandatory.

(2) Clinical performance characteristics: Clinical validity relates to the associated diseases and clinical conditions of patients and is different from analytical validity, which focuses on the correct measurement of analytes targeted by the assay. According to International Standard Organization (ISO) 15189 and ISO 17025, validation is the "confirmation, through the provision of objective evidence, that the requirements for a specific intended use or application have been fulfilled". Therefore, establishing clinical performance is the main goal in the validation phase of a biomarker. Clinical performance characteristics include (i) defining normal reference ranges by measuring cohorts of apparently healthy individuals, (ii) determining clinical sensitivity, which is defined as the proportion of individuals who have the disease and are tested positive, and (iii) determining clinical specificity, which is defined as the proportion of disease-free individuals who are tested negative. Derived statistics such as receiver operating characteristic (ROC) plots are particularly helpful in assessing the clinical performance of biomarkers (Zweig & Campbell, 1993; Obuchowski et al, 2004).

(3) Study design and pre-analytics: Careful study design and well-controlled pre-analytical conditions are key requirements at any time during a biomarker study. With respect to study design, it is mandatory to clearly define the clinical question and the medical need that should be addressed by the biomarker. A common problem in biomarker studies is that samples from cases and controls have been collected independently and are mismatched for age, ethnicity, sex, and other factors that may or may not lead to unintentional bias (Duffy et al, 2015). Methods against bias include proper study design as well as precise and deep clinical phenotyping of participants, using systematic classifications such as the International Statistical Classification of Diseases (http://apps.who.int/classifications/icd10/browse/2016/en) or the human phenome ontology (Kohler et al, 2017). In this way, if a person has multiple disease conditions, this can be properly accounted for. Sample collection is important as well, and it is imperative that all samples (including cases and controls) are treated equally from blood drawing to the analytical phase. Another critical step in many biomarker studies is biobanking. When employing ELISAs, we have found that storage of protein-based biomarkers for 3 months requires temperatures of −80°C or below (Zander et al, 2014). Sample stability for longer periods is only poorly investigated. However, in our experience, shotgun proteomics has a high tolerance for variation in sample history, because there are no protein epitopes that need to be preserved and even partial protein degradation may be tolerable as long as the majority of subsequently generated proteolytic peptides remain unaltered.

The road to clinical application

The current progress in plasma proteomics opens exciting novel avenues for research and the clinic. How likely is it, given all the aforementioned precautions that the outlined approaches will lead to the discovery of novel protein-based biomarkers? And what will the proteomic biomarker of the future look like? A key theme in this context is the discriminative power of a biomarker to distinguish between the presence and absence of a particular disease state or risk, in other words its clinical performance. Examples of currently used biomarkers with high specificity and high sensitivity are cardiac troponins, which are structural proteins specifically expressed in cardiomyocytes and therefore highly specific for myocardial damage. For this reason, cardiac troponins have even been incorporated into the universal definition of myocardial infarction (Roffi et al, 2016).

It is likely that proteomics approaches will succeed in the identification of additional biomarkers with similar performance, at least for certain diseases. In fact, we need to be aware that most biomarkers used today are either highly abundant or originate from a known pathophysiological context. As a thought experiment, we have extrapolated the ratio of the number of biomarkers relative to the number of proteins in the high abundance range to lower abundance protein range, which indicates the potential for several hundred novel biomarkers, which might be accessible with appropriate technology (Fig 5). In analogy to GWAS, where a significant number of hits turned out to be related to previously unknown pathophysiology of the investigated disease (Holdt & Teupser, 2013; Manolio, 2013), it is quite likely that new markers, which have hidden below the radar of previous strategies, will be identified by novel systematic proteomics approaches. These biomarkers may also have the potential to improve our understanding of disease pathophysiology not only in diagnostics but also for therapy. Note, however, that the identified biomarkers might not always be directly involved in the disease pathophysiology but may only be associated with it.

The human genome encodes for about 20,000 protein coding genes, which is opposed to more than 14,500 diseases classified by an ICD code. This makes it even conceptually difficult to imagine that one gene or protein is associated with each disease condition, as is often implied in current efforts to find biomarkers. In contrast, the rectangular strategy, allowing to screen large cohorts for multiple markers, holds great promise to discover and validate protein patterns that are characteristic of particular health or disease states. Indeed, multi-marker combinations may achieve higher specificity and sensitivity compared to single markers and first tools for selecting accurate marker combinations out of omics data have been developed (Mazzara et al, 2017). However, a common problem with new biomarkers combined with existing ones is that they frequently only lead to minor classification improvements, in particular when added to well-performing ones (Pencina et al, 2010). Contrary to common and intuitive assumptions, it has been shown that correlation (especially negative correlation) between predictors can be beneficial for discrimination (Demler et al, 2013). More research in this area is clearly warranted, and new proteomics technologies will provide the data required for the validation of appropriate statistical methods.

Finally, how will these markers be applicable in a clinical setting? We favor in-depth measurement of the entire plasma proteome regardless of the occasion, as this provides the most complete information. Over time, it adds to the longitudinal plasma proteome profile that could usefully be obtained even of healthy subjects. As mentioned above, plasma protein levels tend to generally be stable but person-specific, allowing individual-specific interpretation instead of population-based cutoff values. Furthermore, co-morbidities are the rule rather than the exception in many patient groups. These are much more easily and economically addressed by a generic diagnostic test such as plasma proteomic profiling rather than a succession of individual ELISA tests. Nevertheless, there would clearly be many situations in which a universal test will not be appropriate because it may inadvertently uncover other conditions. Similar issues arise with other technologies such as genome sequencing or imagining techniques, where individuals may not want to learn about predispositions that they can do little about. In these cases and generally to avoid the risk of overdiagnosis (Hofmann & Welch, 2017), clinicians may prefer plasma proteomics tests of a more directed nature that focuses on a particular disease context. This could be accomplished by the abovementioned MS techniques targeting a panel of proteins, rather than the entire proteome.

For either whole-proteome diagnostic tests or panel-based tests, the question arises how doctors would deal with the resulting multidimensional data. Figure 6A shows the current single/oligo biomarker diagnostics, which is integrated into decision-making

largely based on clinical knowledge and intuition. New biomarkers clearly hold the promise of better informed clinical decisions, but also imply the risk of generating patterns exceeding the human cognitive capacity of interpretation (Fig 6B). A solution to this problem might be the algorithmic combination of multiple biomarkers into a quantitative panel, possibly combined with clinical metadata, which might substantially aid clinical decision-making (Fig 6C). Given rapid developments in "deep learning" and "big data", it will be very interesting to see whether this combination can provide powerful and unprecedented associations. We note that there are already multi-parameter scores in clinical practice today. For instance, the Child–Pugh score and the Framingham Risk Score have each combined several blood values with patient data, to aid

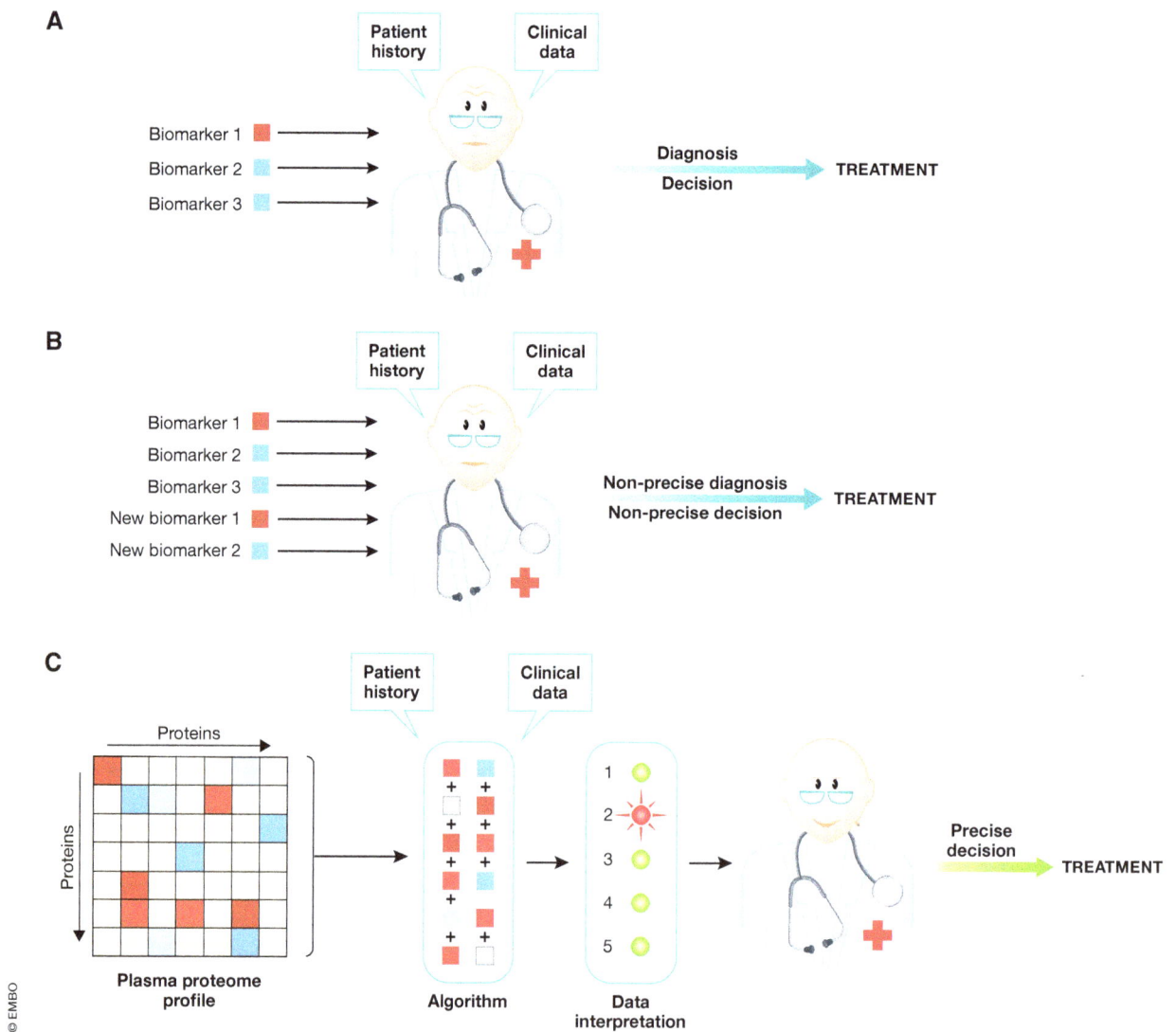

Figure 6. Implementation of proteomic data in clinical decisions.
(A) Currently, physicians make treatment decisions on the basis of a few plasma biomarker tests, combined with patient history and clinical data (upper panel). (B) Adding new biomarkers would quickly overwhelm the current paradigm—leading to suboptimal clinical decisions. (C) Multi-protein panels and the data from past studies (the knowledge base in Fig 4B) are combined algorithmically. This will aid the physician in making more precise recommendations for treatment, while still taking patient history and other clinical data into account.

clinician's decision in treating liver disease and cardiovascular treatment, respectively, for decades. This also suggests a way how plasma proteomics could be accepted into evidence-based medical practice, a huge challenge given the many parameters and parameter combinations involved, which clearly cannot all be validated with separate clinical trials. A pragmatic alternative might be to devise trials in which doctors randomly obtain the proteomic information and associated decision support. It would then be straightforward to determine whether there is a significant benefit in patient outcomes.

Conclusions

Staking stock of the current practice in laboratory medicine shows that the majority of treatment decisions are made on the basis of blood tests and that protein measurements are even today the most prominent among them. Despite successfully being carried out by the millions every year, these assays are almost always directed against individual proteins and the pace of introduction of new protein tests has slowed to a trickle.

MS-based proteomics clearly has the potential for multiplexed and highly specific measurements, in which protein patterns rather than single biomarkers could be the relevant readout. Our review of the literature revealed that past efforts were held back by the great analytical challenges of the plasma proteome, something that is only now giving way to exciting technological developments. We argue that the analysis of large numbers of conditions and participants in all stages of the discovery and validation process has the potential to produce biomarker panels that are likely to be of clinical value. When coupled to large knowledge bases of changes in protein patterns in defined conditions, such a plasma proteome profiling strategy could in principle exploit the entire information contents of this body fluid.

To make this vision a reality, further improvements in throughput, depth of proteome coverage, robustness, and accessibility of the underlying workflow are crucial. Furthermore, plasma proteomics can also be extended to the analysis of post-translation modifications. Likewise, plasma metabolomics also uses MS-based workflows and could routinely be integrated with plasma proteomics in the future. We are confident that the required technological developments can and will all be achieved over time. At least as much of a challenge will be conceptual and "political", as the proteomic information deluge needs to be turned into actionable data for the physician and the healthcare system. This will require a dedicated and untiring commitment from all partners involved. We believe that the promise of much more precise and specific diagnostics will amply reward such efforts.

Acknowledgements

We thank all members of the Proteomics and Signal Transduction and the Clinical Proteomics groups for help and discussions, in particular Peter V. Treit for assistance with the literature search and Sophia Doll, Lili Niu, and Atul Deshmukh for helpful comments. The work carried out in this project was partially supported by the Max Planck Society for the Advancement of Science and by the Novo Nordisk Foundation (grant NNF15CC0001).

References

Abbatiello SE, Schilling B, Mani DR, Zimmerman LJ, Hall SC, MacLean B, Albertolle M, Allen S, Burgess M, Cusack MP, Gosh M, Hedrick V, Held JM, Inerowicz HD, Jackson A, Keshishian H, Kinsinger CR, Lyssand J, Makowski L, Mesri M et al (2015) Large-scale interlaboratory study to develop, analytically validate and apply highly multiplexed, quantitative peptide assays to measure cancer-relevant proteins in plasma. Mol Cell Proteomics 14: 2357–2374

Addona TA, Abbatiello SE, Schilling B, Skates SJ, Mani DR, Bunk DM, Spiegelman CH, Zimmerman LJ, Ham AJ, Keshishian H, Hall SC, Allen S, Blackman RK, Borchers CH, Buck C, Cardasis HL, Cusack MP, Dodder NG, Gibson BW, Held JM et al (2009) Multi-site assessment of the precision and reproducibility of multiple reaction monitoring-based measurements of proteins in plasma. Nat Biotechnol 27: 633–641

Aebersold R, Mann M (2016) Mass-spectrometric exploration of proteome structure and function. Nature 537: 347–355

Altelaar AF, Heck AJ (2012) Trends in ultrasensitive proteomics. Curr Opin Chem Biol 16: 206–213

Anderson NL, Anderson NG, Haines LR, Hardie DB, Olafson RW, Pearson TW (2004) Mass spectrometric quantitation of peptides and proteins using Stable Isotope Standards and Capture by Anti-Peptide Antibodies (SISCAPA). J Proteome Res 3: 235–244

Anderson NL (2010) The clinical plasma proteome: a survey of clinical assays for proteins in plasma and serum. Clin Chem 56: 177–185

Anderson NL, Ptolemy AS, Rifai N (2013) The riddle of protein diagnostics: future bleak or bright? Clin Chem 59: 194–197

Baggerly KA, Morris JS, Coombes KR (2004) Reproducibility of SELDI-TOF protein patterns in serum: comparing datasets from different experiments. Bioinformatics 20: 777–785

Bantscheff M, Boesche M, Eberhard D, Matthieson T, Sweetman G, Kuster B (2008) Robust and sensitive iTRAQ quantification on an LTQ Orbitrap mass spectrometer. Mol Cell Proteomics 7: 1702–1713

Bekker-Jensen DB, Kelstrup CD, Batth TS, Larsen SC, Haldrup C, Bramsen JB, Sørensen KD, Høyer S, Ørntoft TF, Andersen CL, Nielsen ML, Olsen JV (2017) An optimized shotgun strategy for the rapid generation of comprehensive human proteomes. Cell Syst 4: 587–599.e4

Bellei E, Bergamini S, Monari E, Fantoni LI, Cuoghi A, Ozben T, Tomasi A (2011) High-abundance proteins depletion for serum proteomic analysis: concomitant removal of non-targeted proteins. Amino Acids 40: 145–156

Bruderer R, Bernhardt OM, Gandhi T, Xuan Y, Sondermann J, Schmidt M, Gomez-Varela D, Reiter L (2017) Heralds of parallel MS: data-independent acquisition surpassing sequential identification of data dependent acquisition in proteomics. Mol Cell Proteomics https://doi.org/10.1074/mcp.M116.065730

Burgess MW, Keshishian H, Mani DR, Gillette MA, Carr SA (2014) Simplified and efficient quantification of low-abundance proteins at very high multiplex via targeted mass spectrometry. Mol Cell Proteomics 13: 1137–1149

Cao Z, Tang HY, Wang H, Liu Q, Speicher DW (2012) Systematic comparison of fractionation methods for in-depth analysis of plasma proteomes. J Proteome Res 11: 3090–3100

Carr SA, Abbatiello SE, Ackermann BL, Borchers C, Domon B, Deutsch EW, Grant RP, Hoofnagle AN, Huttenhain R, Koomen JM, Liebler DC, Liu T, MacLean B, Mani DR, Mansfield E, Neubert H, Paulovich AG, Reiter L, Vitek O, Aebersold R et al (2014) Targeted peptide measurements in biology and medicine: best practices for mass spectrometry-based assay development using a fit-for-purpose approach. Mol Cell Proteomics 13: 907–917

Cohen Freue GV, Meredith A, Smith D, Bergman A, Sasaki M, Lam KK,

Hollander Z, Opushneva N, Takhar M, Lin D, Wilson-McManus J, Balshaw R, Keown PA, Borchers CH, McManus B, Ng RT, McMaster WR, Biomarkers in T, the NCECPoOFCoET (2013) Computational biomarker pipeline from discovery to clinical implementation: plasma proteomic biomarkers for cardiac transplantation. *PLoS Comput Biol* 9: e1002963

Cole RN, Ruczinski I, Schulze K, Christian P, Herbrich S, Wu L, Devine LR, O'Meally RN, Shrestha S, Boronina TN, Yager JD, Groopman J, West KP Jr (2013) The plasma proteome identifies expected and novel proteins correlated with micronutrient status in undernourished Nepalese children. *J Nutr* 143: 1540–1548

Cominetti O, Nunez Galindo A, Corthesy J, Oller Moreno S, Irincheeva I, Valsesia A, Astrup A, Saris WH, Hager J, Kussmann M, Dayon L (2016) Proteomic biomarker discovery in 1000 human plasma samples with mass spectrometry. *J Proteome Res* 15: 389–399

Cox J, Mann M (2008) MaxQuant enables high peptide identification rates, individualized p.p.b.-range mass accuracies and proteome-wide protein quantification. *Nat Biotechnol* 26: 1367–1372

Cox J, Mann M (2011) Quantitative, high-resolution proteomics for data-driven systems biology. *Annu Rev Biochem* 80: 273–299

Demler OV, Pencina MJ, D'Agostino RB Sr (2013) Impact of correlation on predictive ability of biomarkers. *Stat Med* 32: 4196–4210

De-Ritis F, Coltorti M, Giusti G (1957) An enzymic test for the diagnosis of viral hepatitis - the transaminase serum activities. *Clin Chim Acta* 2: 70–74

Duffy MJ, Sturgeon CM, Soletormos G, Barak V, Molina R, Hayes DF, Diamandis EP, Bossuyt PM (2015) Validation of new cancer biomarkers: a position statement from the European group on tumor markers. *Clin Chem* 61: 809–820

Ebhardt HA, Root A, Sander C, Aebersold R (2015) Applications of targeted proteomics in systems biology and translational medicine. *Proteomics* 15: 3193–3208

Edfors F, Bostrom T, Forsstrom B, Zeiler M, Johansson H, Lundberg E, Hober S, Lehtio J, Mann M, Uhlen M (2014) Immunoproteomics using polyclonal antibodies and stable isotope-labeled affinity-purified recombinant proteins. *Mol Cell Proteomics* 13: 1611–1624

FDA-NIH: Biomarker-Working-Group (2016) *BEST (Biomarkers, EndpointS, and other Tools) resource.* Maryland: Silver Spring (MD): Food and Drug Administration (US); Bethesda (MD): National Institutes of Health (US)

Garcia-Bailo B, Brenner DR, Nielsen D, Lee HJ, Domanski D, Kuzyk M, Borchers CH, Badawi A, Karmali MA, El-Sohemy A (2012) Dietary patterns and ethnicity are associated with distinct plasma proteomic groups. *Am J Clin Nutr* 95: 352–361

Geyer PE, Kulak NA, Pichler G, Holdt LM, Teupser D, Mann M (2016a) Plasma proteome profiling to assess human health and disease. *Cell Syst* 2: 185–195

Geyer PE, Wewer Albrechtsen NJ, Tyanova S, Grassl N, Iepsen EW, Lundgren J, Madsbad S, Holst JJ, Torekov SS, Mann M (2016b) Proteomics reveals the effects of sustained weight loss on the human plasma proteome. *Mol Syst Biol* 12: 901

Grant RP, Hoofnagle AN (2014) From lost in translation to paradise found: enabling protein biomarker method transfer by mass spectrometry. *Clin Chem* 60: 941–944

Hayes DF, Allen J, Compton C, Gustavsen G, Leonard DG, McCormack R, Newcomer L, Pothier K, Ransohoff D, Schilsky RL, Sigal E, Taube SE, Tunis SR (2013) Breaking a vicious cycle. *Sci Transl Med* 5: 196cm196

Hofmann B, Welch HG (2017) New diagnostic tests: more harm than good. *BMJ* 358: j3314

Holdt LM, Teupser D (2013) From genotype to phenotype in human atherosclerosis—recent findings. *Curr Opin Lipidol* 24: 410–418

Hoofnagle AN, Wener MH (2009) The fundamental flaws of immunoassays and potential solutions using tandem mass spectrometry. *J Immunol Methods* 347: 3–11

Hoofnagle AN, Whiteaker JR, Carr SA, Kuhn E, Liu T, Massoni SA, Thomas SN, Townsend RR, Zimmerman LJ, Boja E, Chen J, Crimmins DL, Davies SR, Gao Y, Hiltke TR, Ketchum KA, Kinsinger CR, Mesri M, Meyer MR, Qian WJ et al (2016) Recommendations for the generation, quantification, storage, and handling of peptides used for mass spectrometry-based assays. *Clin Chem* 62: 48–69

Jennings L, Van Deerlin VM, Gulley ML, College of American Pathologists Molecular Pathology Resource C (2009) Recommended principles and practices for validating clinical molecular pathology tests. *Arch Pathol Lab Med* 133: 743–755

Juhasz P, Lynch M, Sethuraman M, Campbell J, Hines W, Paniagua M, Song L, Kulkarni M, Adourian A, Guo Y, Li X, Martin S, Gordon N (2011) Semi-targeted plasma proteomics discovery workflow utilizing two-stage protein depletion and off-line LC-MALDI MS/MS. *J Proteome Res* 10: 34–45

Keshishian H, Burgess MW, Gillette MA, Mertins P, Clauser KR, Mani DR, Kuhn EW, Farrell LA, Gersztzen RE, Carr SA (2015) Multiplexed, quantitative workflow for sensitive biomarker discovery in plasma yields novel candidates for early myocardial injury. *Mol Cell Proteomics* 14: 2375–2393

Kim YJ, Sertamo K, Pierrard MA, Mesmin C, Kim SY, Schlesser M, Berchem G, Domon B (2015) Verification of the biomarker candidates for non-small-cell lung cancer using a targeted proteomics approach. *J Proteome Res* 14: 1412–1419

Kohler S, Vasilevsky NA, Engelstad M, Foster E, McMurry J, Ayme S, Baynam G, Bello SM, Boerkoel CF, Boycott KM, Brudno M, Buske OJ, Chinnery PF, Cipriani V, Connell LE, Dawkins HJ, DeMare LE, Devereau AD, de Vries BB, Firth HV et al (2017) The human phenotype ontology in 2017. *Nucleic Acids Res* 45: D865–D876

Kolla V, Jeno P, Moes S, Tercanli S, Lapaire O, Choolani M, Hahn S (2010) Quantitative proteomics analysis of maternal plasma in Down syndrome pregnancies using isobaric tagging reagent (iTRAQ). *J Biomed Biotechnol* 2010: 952047

Lee SE, West KP Jr, Cole RN, Schulze KJ, Christian P, Wu LS, Yager JD, Groopman J, Ruczinski I (2015) Plasma proteome biomarkers of inflammation in school aged children in Nepal. *PLoS ONE* 10: e0144279

Lee SE, Stewart CP, Schulze KJ, Cole RN, Wu LS, Yager JD, Groopman JD, Khatry SK, Adhikari RK, Christian P, West KP Jr (2017) The plasma proteome is associated with anthropometric status of undernourished nepalese school-aged children. *J Nutr* 147: 304–313

Levine RJ, Maynard SE, Qian C, Lim KH, England LJ, Yu KF, Schisterman EF, Thadhani R, Sachs BP, Epstein FH, Sibai BM, Sukhatme VP, Karumanchi SA (2004) Circulating angiogenic factors and the risk of preeclampsia. *N Engl J Med* 350: 672–683

Liu T, Qian WJ, Gritsenko MA, Xiao W, Moldawer LL, Kaushal A, Monroe ME, Varnum SM, Moore RJ, Purvine SO, Maier RV, Davis RW, Tompkins RG, Camp DG II, Smith RD, Inflammation, the Host Response to Injury Large Scale Collaborative Research P (2006) High dynamic range characterization of the trauma patient plasma proteome. *Mol Cell Proteomics* 5: 1899–1913

Liu Y, Buil A, Collins BC, Gillet LC, Blum LC, Cheng LY, Vitek O, Mouritsen J, Lachance G, Spector TD, Dermitzakis ET, Aebersold R (2015) Quantitative variability of 342 plasma proteins in a human twin population. *Mol Syst Biol* 11: 786

Luque-Garcia JL, Neubert TA (2007) Sample preparation for serum/plasma profiling and biomarker identification by mass spectrometry. *J Chromatogr*

MacLean B, Tomazela DM, Shulman N, Chambers M, Finney GL, Frewen B, Kern R, Tabb DL, Liebler DC, MacCoss MJ (2010) Skyline: an open source document editor for creating and analyzing targeted proteomics experiments. *Bioinformatics* 26: 966–968

Malmström E, Kilsgard O, Hauri S, Smeds E, Herwald H, Malmstrom L, Malmstrom J (2016) Large-scale inference of protein tissue origin in gram-positive sepsis plasma using quantitative targeted proteomics. *Nat Commun* 7: 10261

Mann M, Kulak NA, Nagaraj N, Cox J (2013) The coming age of complete, accurate, and ubiquitous proteomes. *Mol Cell* 49: 583–590

Manolio TA (2013) Bringing genome-wide association findings into clinical use. *Nat Rev Genet* 14: 549–558

Mazzara S, Rossi RL, Grifantini R, Donizetti S, Abrignani S, Bombaci M (2017) CombiROC: an interactive web tool for selecting accurate marker combinations of omics data. *Sci Rep* 7: 45477

Mischak H, Allmaier G, Apweiler R, Attwood T, Baumann M, Benigni A, Bennett SE, Bischoff R, Bongcam-Rudloff E, Capasso G, Coon JJ, D'Haese P, Dominiczak AF, Dakna M, Dihazi H, Ehrich JH, Fernandez-Llama P, Fliser D, Frokiaer J, Garin J et al (2010) Recommendations for biomarker identification and qualification in clinical proteomics. *Sci Transl Med* 2: 46ps42

Nanjappa V, Thomas JK, Marimuthu A, Muthusamy B, Radhakrishnan A, Sharma R, Ahmad Khan A, Balakrishnan L, Sahasrabuddhe NA, Kumar S, Jhaveri BN, Sheth KV, Kumar Khatana R, Shaw PG, Srikanth SM, Mathur PP, Shankar S, Nagaraja D, Christopher R, Mathivanan S et al (2014) Plasma Proteome Database as a resource for proteomics research: 2014 update. *Nucleic Acids Res* 42: D959–D965

Nesvizhskii AI, Vitek O, Aebersold R (2007) Analysis and validation of proteomic data generated by tandem mass spectrometry. *Nat Methods* 4: 787–797

Nie S, Shi T, Fillmore TL, Schepmoes AA, Brewer H, Gao Y, Song E, Wang H, Rodland KD, Qian WJ, Smith RD, Liu T (2017) Deep-dive targeted quantification for ultrasensitive analysis of proteins in non-depleted human blood plasma/serum and tissues. *Anal Chem* 89: 9139–9146

Oberbach A, Schlichting N, Neuhaus J, Kullnick Y, Lehmann S, Heinrich M, Dietrich A, Mohr FW, von Bergen M, Baumann S (2014) Establishing a reliable multiple reaction monitoring-based method for the quantification of obesity-associated comorbidities in serum and adipose tissue requires intensive clinical validation. *J Proteome Res* 13: 5784–5800

Obuchowski NA, Lieber ML, Wians FH Jr (2004) ROC curves in clinical chemistry: uses, misuses, and possible solutions. *Clin Chem* 50: 1118–1125

Omenn GS, States DJ, Adamski M, Blackwell TW, Menon R, Hermjakob H, Apweiler R, Haab BB, Simpson RJ, Eddes JS, Kapp EA, Moritz RL, Chan DW, Rai AJ, Admon A, Aebersold R, Eng J, Hancock WS, Hefta SA, Meyer H et al (2005) Overview of the HUPO Plasma Proteome Project: results from the pilot phase with 35 collaborating laboratories and multiple analytical groups, generating a core dataset of 3020 proteins and a publicly-available database. *Proteomics* 5: 3226–3245

Ozcan S, Cooper JD, Lago SG, Kenny D, Rustogi N, Stocki P, Bahn S (2017) Towards reproducible MRM based biomarker discovery using dried blood spots. *Sci Rep* 7: 45178

Pan S, Chen R, Crispin DA, May D, Stevens T, McIntosh MW, Bronner MP, Ziogas A, Anton-Culver H, Brentnall TA (2011) Protein alterations associated with pancreatic cancer and chronic pancreatitis found in human plasma using global quantitative proteomics profiling. *J Proteome Res* 10: 2359–2376

Parker CE, Borchers CH (2014) Mass spectrometry based biomarker discovery, verification, and validation–quality assurance and control of protein

Paulovich AG, Whiteaker JR, Hoofnagle AN, Wang P (2008) The interface between biomarker discovery and clinical validation: the tar pit of the protein biomarker pipeline. *Proteomics Clin Appl* 2: 1386–1402

Pavlou MP, Diamandis EP, Blasutig IM (2013) The long journey of cancer biomarkers from the bench to the clinic. *Clin Chem* 59: 147–157

Pencina MJ, D'Agostino RB, Vasan RS (2010) Statistical methods for assessment of added usefulness of new biomarkers. *Clin Chem Lab Med* 48: 1703–1711

Percy AJ, Chambers AG, Yang J, Borchers CH (2013) Multiplexed MRM-based quantitation of candidate cancer biomarker proteins in undepleted and non-enriched human plasma. *Proteomics* 13: 2202–2215

Percy AJ, Michaud SA, Jardim A, Sinclair NJ, Zhang S, Mohammed Y, Palmer AL, Hardie DB, Yang J, LeBlanc AM, Borchers CH (2017) Multiplexed MRM-based assays for the quantitation of proteins in mouse plasma and heart tissue. *Proteomics* https://doi.org/10.1002/pmic.201600097

Petricoin EF, Ardekani AM, Hitt BA, Levine PJ, Fusaro VA, Steinberg SM, Mills GB, Simone C, Fishman DA, Kohn EC, Liotta LA (2002) Use of proteomic patterns in serum to identify ovarian cancer. *Lancet* 359: 572–577

Picotti P, Aebersold R (2012) Selected reaction monitoring-based proteomics: workflows, potential, pitfalls and future directions. *Nat Methods* 9: 555–566

Qian WJ, Kaleta DT, Petritis BO, Jiang H, Liu T, Zhang X, Mottaz HM, Varnum SM, Camp DG II, Huang L, Fang X, Zhang WW, Smith RD (2008) Enhanced detection of low abundance human plasma proteins using a tandem IgY12-SuperMix immunoaffinity separation strategy. *Mol Cell Proteomics* 7: 1963–1973

Rai AJ, Zhang Z, Rosenzweig J, Shih I, Pham T, Fung ET, Sokoll LJ, Chan DW (2002) Proteomic approaches to tumor marker discovery - Identification of biomarkers for ovarian cancer. *Arch Pathol Lab Med* 126: 1518–1526

Razavi M, Leigh Anderson N, Pope ME, Yip R, Pearson TW (2016) High precision quantification of human plasma proteins using the automated SISCAPA Immuno-MS workflow. *N Biotechnol* 33: 494–502

Richards AL, Merrill AE, Coon JJ (2015) Proteome sequencing goes deep. *Curr Opin Chem Biol* 24: 11–17

Rifai N, Gillette MA, Carr SA (2006) Protein biomarker discovery and validation: the long and uncertain path to clinical utility. *Nat Biotechnol* 24: 971–983

Roffi M, Patrono C, Collet JP, Mueller C, Valgimigli M, Andreotti F, Bax JJ, Borger MA, Brotons C, Chew DP, Gencer B, Hasenfuss G, Kjeldsen K, Lancellotti P, Landmesser U, Mehilli J, Mukherjee D, Storey RF, Windecker S (2016) 2015 ESC Guidelines for the management of acute coronary syndromes in patients presenting without persistent ST-segment elevation. Task Force for the Management of Acute Coronary Syndromes in Patients Presenting without Persistent ST-Segment Elevation of the European Society of Cardiology (ESC). *G Ital Cardiol* 17: 831–872

Rosenberger G, Bludau I, Schmitt U, Heusel M, Hunter CL, Liu Y, MacCoss MJ, MacLean BX, Nesvizhskii AI, Pedrioli PGA, Reiter L, Röst HL, Tate S, Ting YS, Collins BC, Aebersold R (2017) Statistical control of peptide and protein error rates in large-scale targeted data-independent acquisition analyses. *Nat Methods* 14: 921–927

Rost HL, Rosenberger G, Navarro P, Gillet L, Miladinovic SM, Schubert OT, Wolski W, Collins BC, Malmstrom J, Malmstrom L, Aebersold R (2014) OpenSWATH enables automated, targeted analysis of data-independent acquisition MS data. *Nat Biotechnol* 32: 219–223

Sajic T, Liu Y, Aebersold R (2015) Using data-independent, high-resolution mass spectrometry in protein biomarker research: perspectives and clinical applications. *Proteomics Clin Appl* 9: 307–321

Sharma K, Schmitt S, Bergner CG, Tyanova S, Kannaiyan N, Manrique-Hoyos

N, Kongi K, Cantuti L, Hanisch UK, Philips MA, Rossner MJ, Mann M, Simons M (2015) Cell type- and brain region-resolved mouse brain proteome. *Nat Neurosci* 18: 1819–1831

Shi T, Fillmore TL, Gao Y, Zhao R, He J, Schepmoes AA, Nicora CD, Wu C, Chambers JL, Moore RJ, Kagan J, Srivastava S, Liu AY, Rodland KD, Liu T, Camp DG II, Smith RD, Qian WJ (2013) Long-gradient separations coupled with selected reaction monitoring for highly sensitive, large scale targeted protein quantification in a single analysis. *Anal Chem* 85: 9196–9203

Skates SJ, Gillette MA, LaBaer J, Carr SA, Anderson L, Liebler DC, Ransohoff D, Rifai N, Kondratovich M, Tezak Z, Mansfield E, Oberg AL, Wright I, Barnes G, Gail M, Mesri M, Kinsinger CR, Rodriguez H, Boja ES (2013) Statistical design for biospecimen cohort size in proteomics-based biomarker discovery and verification studies. *J Proteome Res* 12: 5383–5394

Skol AD, Scott LJ, Abecasis GR, Boehnke M (2006) Joint analysis is more efficient than replication-based analysis for two-stage genome-wide association studies. *Nat Genet* 38: 209–213

Surinova S, Schiess R, Huttenhain R, Cerciello F, Wollscheid B, Aebersold R (2011) On the development of plasma protein biomarkers. *J Proteome Res* 10: 5–16

Thulasiraman V, Lin S, Gheorghiu L, Lathrop J, Lomas L, Hammond D, Boschetti E (2005) Reduction of the concentration difference of proteins in biological liquids using a library of combinatorial ligands. *Electrophoresis* 26: 3561–3571

Tu C, Rudnick PA, Martinez MY, Cheek KL, Stein SE, Slebos RJ, Liebler DC (2010) Depletion of abundant plasma proteins and limitations of plasma proteomics. *J Proteome Res* 9: 4982–4991

Vogeser M, Seger C (2016) Mass spectrometry methods in clinical diagnostics - state of the art and perspectives. *Trac-Trends Analyt Chem* 84: 1–4

Whiteaker JR, Lin C, Kennedy J, Hou L, Trute M, Sokal I, Yan P, Schoenherr RM, Zhao L, Voytovich UJ, Kelly-Spratt KS, Krasnoselsky A, Gafken PR, Hogan JM, Jones LA, Wang P, Amon L, Chodosh LA, Nelson PS, McIntosh MW et al (2011) A targeted proteomics-based pipeline for verification of biomarkers in plasma. *Nat Biotechnol* 29: 625–634

Wild D (2013) *The immunoassay handbook: theory and applications of ligand binding, ELISA, and related techniques*, 4th edn. Oxford; Waltham, MA: Elsevier

Wu HY, Goan YG, Chang YH, Yang YF, Chang HJ, Cheng PN, Wu CC, Zgoda VG, Chen YJ, Liao PC (2015) Qualification and verification of serological biomarker candidates for lung adenocarcinoma by targeted mass spectrometry. *J Proteome Res* 14: 3039–3050

Zander J, Bruegel M, Kleinhempel A, Becker S, Petros S, Kortz L, Dorow J, Kratzsch J, Baber R, Ceglarek U, Thiery J, Teupser D (2014) Effect of biobanking conditions on short-term stability of biomarkers in human serum and plasma. *Clin Chem Lab Med* 52: 629–639

Zeiler M, Straube WL, Lundberg E, Uhlen M, Mann M (2012) A Protein Epitope Signature Tag (PrEST) library allows SILAC-based absolute quantification and multiplexed determination of protein copy numbers in cell lines. *Mol Cell Proteomics* 11: O111.009613

Zhang Z, Bast RC Jr, Yu Y, Li J, Sokoll LJ, Rai AJ, Rosenzweig JM, Cameron B, Wang YY, Meng XY, Berchuck A, Van Haaften-Day C, Hacker NF, de Bruijn HW, van der Zee AG, Jacobs IJ, Fung ET, Chan DW (2004) Three biomarkers identified from serum proteomic analysis for the detection of early stage ovarian cancer. *Cancer Res* 64: 5882–5890

Zhang X, Xiao Z, Liu X, Du L, Wang L, Wang S, Zheng N, Zheng G, Li W, Zhang X, Dong Z, Zhuang X, Wang C (2012) The potential role of ORM2 in the development of colorectal cancer. *PLoS ONE* 7: e31868

Zhang X, Ning Z, Mayne J, Moore JI, Li J, Butcher J, Deeke SA, Chen R, Chiang CK, Wen M, Mack D, Stintzi A, Figeys D (2016) MetaPro-IQ: a universal metaproteomic approach to studying human and mouse gut microbiota. *Microbiome* 4: 31

Zhou C, Simpson KL, Lancashire LJ, Walker MJ, Dawson MJ, Unwin RD, Rembielak A, Price P, West C, Dive C, Whetton AD (2012) Statistical considerations of optimal study design for human plasma proteomics and biomarker discovery. *J Proteome Res* 11: 2103–2113

Zweig MH, Campbell G (1993) Receiver-operating characteristic (ROC) plots: a fundamental evaluation tool in clinical medicine. *Clin Chem* 39: 561–577

12

In situ genotyping of a pooled strain library after characterizing complex phenotypes

Michael J Lawson[†], Daniel Camsund[†], Jimmy Larsson, Özden Baltekin, David Fange & Johan Elf[*] ⓘ

Abstract

In this work, we present a proof-of-principle experiment that extends advanced live cell microscopy to the scale of pool-generated strain libraries. We achieve this by identifying the genotypes for individual cells *in situ* after a detailed characterization of the phenotype. The principle is demonstrated by single-molecule fluorescence time-lapse imaging of *Escherichia coli* strains harboring barcoded plasmids that express a sgRNA which suppresses different genes in the *E. coli* genome through dCas9 interference. In general, the method solves the problem of characterizing complex dynamic phenotypes for diverse genetic libraries of cell strains. For example, it allows screens of how changes in regulatory or coding sequences impact the temporal expression, location, or function of a gene product, or how the altered expression of a set of genes impacts the intracellular dynamics of a labeled reporter.

Keywords DuMPLING; live cell; microfluidic; single cell; strain libraries
Subject Categories Methods & Resources; Quantitative Biology & Dynamical Systems

Introduction

Recent years have seen a rapid development in genome engineering, which, in combination with decreased costs for DNA oligonucleotide synthesis, have made it possible to design and produce pool-generated cell libraries with overwhelming genetic diversity (Wang *et al*, 2009; Dixit *et al*, 2016; Jaitin *et al*, 2016; Peters *et al*, 2016; Garst *et al*, 2017; Otoupal *et al*, 2017). A similarly impressive development in microscopy enables the investigation of complex phenotypes at high temporal resolution and spatial precision in living cells (Liu *et al*, 2015; Balzarotti *et al*, 2017). Biological imaging has benefited greatly from developments in microfluidics which have enabled well-controlled single-cell observations of individual strains over many generations (Wang *et al*, 2010; Uphoff *et al*, 2016; Wallden *et al*, 2016). Despite the rapid technological progress within these areas, there is currently no efficient technique for mapping phenotypes related to intracellular

dynamics or localization to their corresponding genotype for pool-generated libraries of genetically different cell strains. Recent work observing multiple bacterial strains on agarose pads allows for sensitive microscopy (Kuwada *et al*, 2015; Shi *et al*, 2017), but the genetic diversity is capped since the strain production and handling is not pooled. On the other end, droplet fluidics allows working with large genetic diversity (Dixit *et al*, 2016) but cannot be used to characterize phenotypes that require sensitive time-lapse imaging.

Here, we present a method that solves the problem by *in situ* genotyping the library of strains after the phenotypes have been studied in time-lapse microscopy. Thus, the genotype of the cell is not known at the time of phenotyping but revealed through the spatial position of the cell after fixation and *in situ* genotyping.

Results

The DuMPLING approach

We refer to our solution of the library phenotyping problem as DuMPLING—**d**ynamic **u**-fluidic **m**icroscopy-based **p**henotyping of a library before *in situ* **g**enotyping. DuMPLING is composed of three key components: strain generation, live cell phenotyping, and *in situ* genotyping (schematically outlined in Fig 1). All three components can be made in different ways, but in the current study, we have selected this implementation:

1   *Pool-generated strain library:* We have constructed a library of CRISPRi/dCas9 knockdowns. We generated a recipient strain harboring chromosomal inducibly expressed dCas9 and T7 polymerase. We used Golden Gate assembly to generate a small plasmid-expressed library of sgRNA spacers (to direct the dCas9 chromosomal binding and create knockdowns) and neighboring barcode sequences (for later genetic identification) (Figs 2, EV1, and EV2). Note that in 167 nt, we fit the variable regions (i.e., the barcode sequence and sgRNA spacer sequence), the constant elements between the variable regions and the constant regions on the ends for PCR and assembly (see Supplement for sequence design details). This length of oligo is easily procured from companies, and much larger libraries have been built following this approach with purchased oligo pools (Dixit *et al*, 2016), making it clear that

Department of Cell and Molecular Biology, Science for Life Laboratory, Uppsala University, Uppsala, Sweden
  *Corresponding author. E-mail: johan.elf@icm.uu.se
  [†]These authors contributed equally to this work

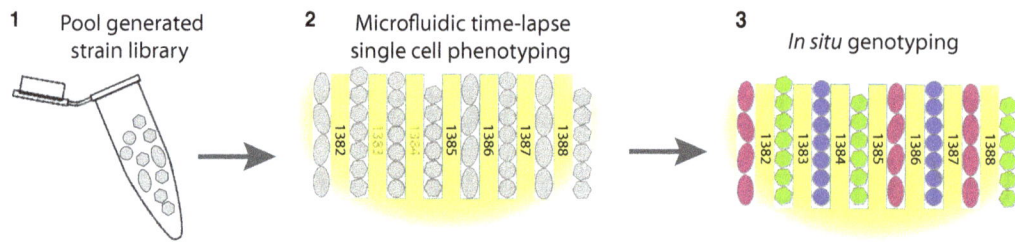

Figure 1. The DuMPLING strategy.

(1) Pooled strain library generation. (2) Live single cell phenotyping using microscopy. (3) Genotypes recovered by *in situ* genotyping.

this strategy of library construction can be extended to a genomewide knockdown library.

2   *Live cell phenotyping in a microfluidic device where each strain occupies a defined position:* The mixed strains are loaded into a microfluidic chip which harbors 4,000 cell channels, sustains continuous exponential growth, and allows single-cell imaging for days (Fig 3A, Movies EV1 and EV2). After only a few generations, all cells in a channel are the progeny of the cell at the back of the channel and thus share the same genotype. The chip design is similar to the mother machine (Wang *et al*, 2010), but we have introduced a 300 nm opening in the back of each cell channel such that media and reagents can be passed over the cells. This redesign facilitates cell loading and is essential for genotyping.

3   *In situ genotyping to identify which strain is in which position:* As mentioned above, each plasmid expresses a unique RNA-based barcode that allows genotype identification. The barcode is expressed from a T7 promoter, and the T7 polymerase is under control of an inducible arabinose promoter. The orthogonal and inducible nature of this system prevents it from interfering with cell physiology during phenotyping. After induction of the barcode RNA expression, the cells are fixed *in situ* with formaldehyde and permeabilized in 70% EtOH before sequential fluorescent *in situ* hybridization (FISH). The individual barcodes are identified by sequential hybridization of fluorescent 37-nt-long oligonucleotides (probes). The multiplexed process of designing and producing the probe library is described in the Materials and Methods section. The templates

for probe synthesis are procured in the same array format as the barcoded sgRNA templates. Here, we use probes of two different colors in two sequential rounds of probing, which is sufficient for identifying the three genotypes in this study.

In general, C^N genotypes can be identified where C is the number of colors and N is the number of rounds of probing. Thus, genotyping can straightforwardly be extended to more strains by using more colors or rounds of probing. For example, a recent publication (Shah *et al*, 2016) showed four rounds of single-molecule FISH probing in five colors (i.e., 625 genotypes), and they observed a miss-call rate of ~1%. We would however expect a lower error rate than this as we are imaging ~6 cells of the same genotype, each containing many RNA rather than individual RNA molecules. To demonstrate that it is possible to reprobe many times, we perform $N = 6$ consecutive rounds (Fig 3B) of probing in each position. It is however likely that more rounds are possible without loss of specificity. For example, in a recent study, Chen *et al* were able to successfully probe single RNA molecules 16 rounds (Chen *et al*, 2015).

Proof-of-concept demonstration

To exemplify the use of DuMPLING, we performed targeted knockdowns of different components of the *lac* operon in *Escherichia coli* using a set of sgRNA-expressing plasmids that repressed *lacY*, an unrelated gene or *lacI* (Fig 2A–C). As described above, the plasmids are made from pooled oligos including the sgRNA and its unique barcode. The pooled approach has previously been used to generate libraries of thousands of genotypes (Dixit *et al*, 2016; Jaitin *et al*,

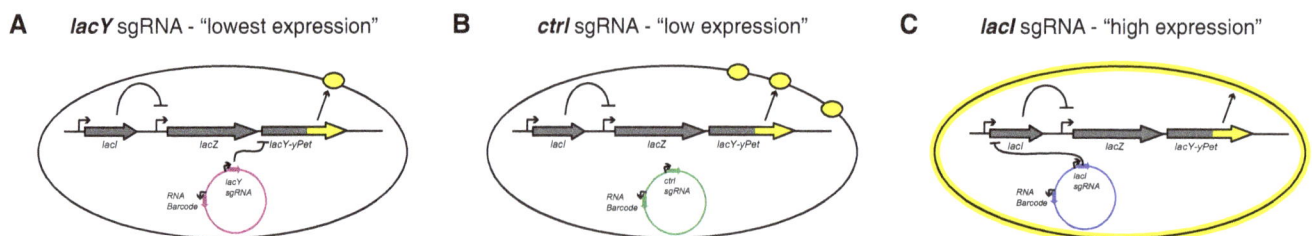

Figure 2. Three strain *lac* operon knockdown library: Repression network for the three different plasmids used.

A   *lacY* knockdown (lowest LacY-YPet expression, purple).

B   No knockdown (low LacY-YPet expression, green).

C   *lacI* knockdown (high LacY-YPet expression, blue).

Data information: Color scheme holds throughout this paper.

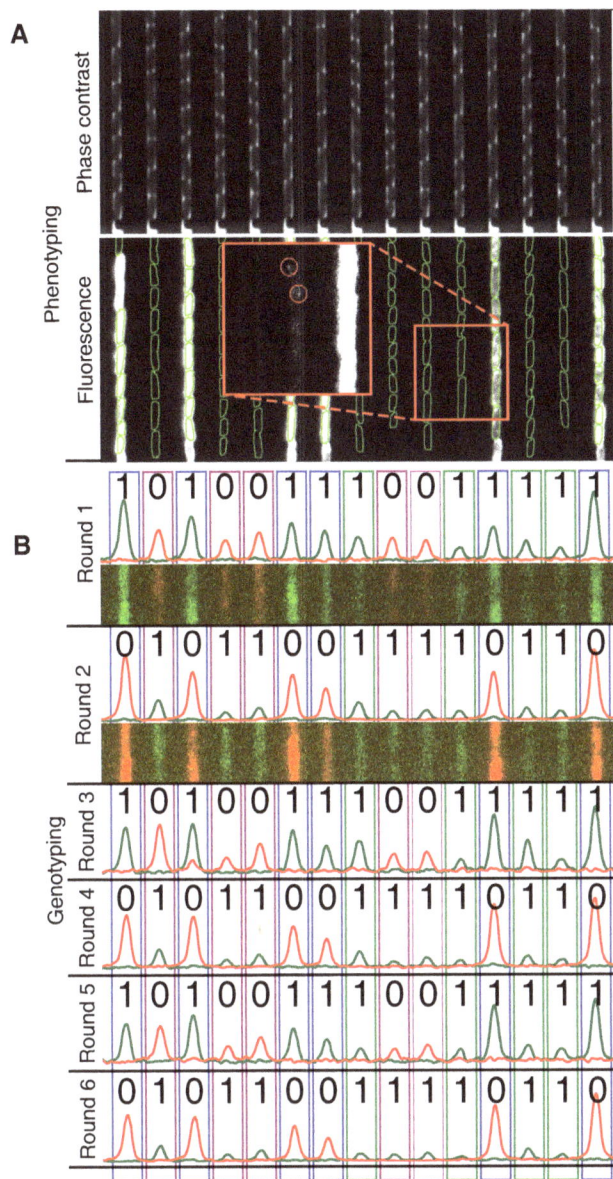

Figure 3. Mapping phenotypes to genotypes.

A Examples of channels and cells in the custom-made microfluidic device which are imaged in both phase contrast (top) and fluorescence microscopy (bottom). Phase contrast is used to segment the cells (green outlines), and single-molecule fluorescence microscopy is used to detect gene expression (red circles in red inset box, which is a blow up of the figure as indicated by the smaller red square and has a change of levels to allow visualization of single molecules) from the *lac* operon.

B *In situ* genotyping with six sequential rounds of FISH probe hybridization and stripping. Cropped images of two cells that are representative of all cells in the trap are shown for the first two rounds (overlay of Cy3 (green) and Cy5 (red) images). The genotype is called by summing the signal in the channel: 0 is assigned for Cy5 (red) and 1 for Cy3 (green). Rectangles indicate assigned genotype (10: lacI knockdown; 01: lacY knockdown; 11: no knockdown).

regulated chromosomal promoter (the promoter is tightly regulated to prevent bias in growth before loading and induction, Fig EV3). Furthermore, the *lacY* gene is fused with the gene for the fluorescent protein YPet to obtain a detectable single-molecule phenotype.

In our experiments, 233 channels are imaged every 60 s using phase contrast and every 13 min using single-molecule-sensitive wide-field fluorescence for a total of 272 min. Phase contrast images are used for cell detection and lineage tracking. Individual LacY-YPet molecules, detected using wide-field epifluorescence, are over-laid on the phase contrast images to allow assignment of individual molecules to individual cells.

We were able to track a cell lineage over the full time-course of the experiment (six generations) and quantify the growth curves of each member of the family tree (see example in Fig 4C). In addition, the long time course of single-cell/single-molecule microscopy allowed us to reproducibly measure mean expression of less than one YPet molecule per generation and distinguish a < 3× change at this expression level (compare distribution of single-molecule counts per cell in Figs 4B and EV4). This type of phenotyping is not possible in most other settings (e.g., flow cytometry) and would not scale to hundreds of strains in those where it is possible (e.g., agarose pads mounted on a microscope).

While the phenotypic difference between the two low-expression strains can only be resolved with extensive single-molecule time-lapse imaging, we also included the *lacI* knockdown phenotype, which is trivial to identify, to test for correct genotype to phenotype assignments. All 74 channels with cells that express a high level of LacY-YPet (Fig 3A) have been correctly found to express the barcode RNA associated with the sgRNA against *lacI* (blue boxes in Fig 3B and blue bars in Fig 4A), and all channels with cells with the barcode RNA associated with the sgRNA against *lacI* express high levels of LacY-YPet. The observed sensitivity and specificity for identifying the genotype in this experiment is therefore 100%. If we also consider the limited sample size and the redundant genotyping as independent, the sensitivity is > 97.5% and the specificity > 99% (see Materials and Methods section for details).

Discussion

This paper describes a proof-of-principle application of the DuMPLING concept, that is, the possibility to use advanced microscopy to phenotype a pool-generated library of live cells and then genotype *in situ*. The advantage of our method compared to the state of the art is the combination of pooled handling of library generation and characterization of complex phenotypes based on dynamic changes in single cells. We have used a microfluidic device to both phenotype the cells in a constant growth environment for an extended period of time and perform the subsequent genotyping.

We note that each of the components (strain library generation, phenotyping, and genotyping) can be performed in different ways depending on the specific question. For example, one can make pooled dCas9 libraries based on plasmids harboring both a genotype-identifying barcode and a sgRNA gene (Dixit *et al*, 2016; Jaitin *et al*, 2016; Peters *et al*, 2016; Garst *et al*, 2017; Otoupal *et al*, 2017) for labeling genetic loci (Chen *et al*, 2013) or knocking down/activating genes throughout the chromosome. Alternatively, pooled

2016), but here, we limit to three variants to be able to precisely evaluate the accuracy of each step. The mixed plasmids are electroporated into an *E. coli* strain, where dCas9 is expressed from a

A

B

C

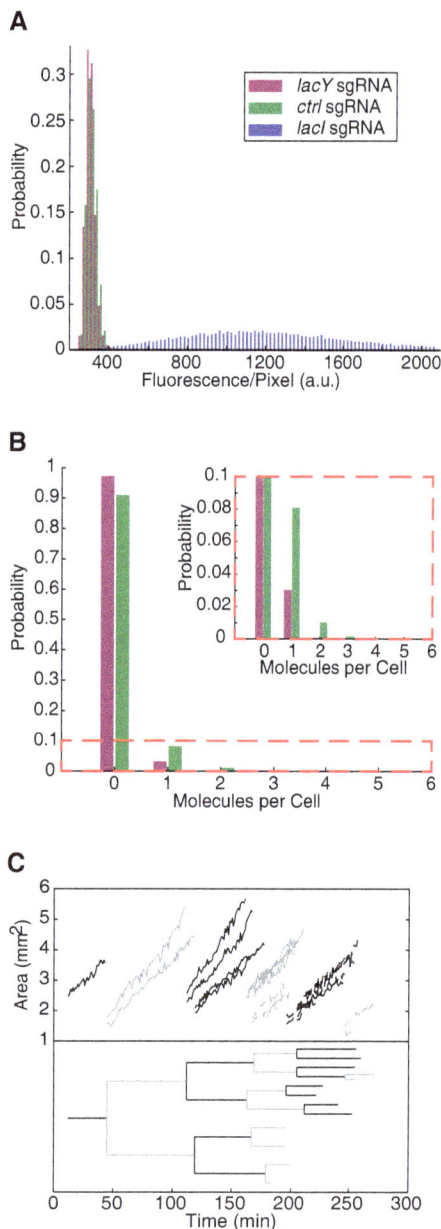

Figure 4. Phenotype data.

A Gene expression categorized by assigned genotype.

B Single-molecule counting of expression from the two low-expression genotypes.

C Top: Growth curves for one cell lineage (from one channel). Dashed lines indicate the end of detection of a branch. Bottom: Corresponding lineage tree.

chromosomal libraries with variants of promoters, ribosome binding sites (RBS), or coding sequences (Wang *et al*, 2009; Keren *et al*, 2016) can be made. Furthermore, it is in general not necessary to introduce the barcode in direct proximity to the genetic alterations as long as the barcode can be connected to the genotype in some other way than through the oligo synthesis. For example, long sequence reads can connect random barcodes to the genetic alteration that causes a phenotype.

Similarly, sensitive single-cell time-lapse imaging can be used to characterize a bewildering diversity of cell phenotypes than are not accessible with snapshot measurement as obtained in FACS or droplet fluidics (Norman *et al*, 2013; Hammar *et al*, 2014; Taheri-Araghi *et al*, 2015; Potvin-Trottier *et al*, 2016; Wallden *et al*, 2016). Depending on the cell types and the experiment, it may also be more convenient to use an open culture dish instead of the fluidic device.

Also, the method for identifying the barcode can be implemented in different ways such as *in situ* sequencing (Ke *et al*, 2013; Lee *et al*, 2014). One advantage of direct *in situ* sequencing is that the genotype may be identified directly without the use of a barcode.

In short, while we have presented a CRISP-FISH-DuMPLING, the DuMPLING can have many other fillings.

Materials and Methods

Design and construction of the DuMPLING screening strain

The dCas9 expression cassette from Qi *et al* (2013), which includes the TetR repressor and the bidirectional PRPA promoter that regulates both *tetR* and *dcas9*, was introduced into the chromosome of *E. coli* and optimized for low leakage of dCas9 under non-induced conditions.

Briefly, the dCas9 expression cassette (plasmid pdcas9 Addgene 44249) and a spectinomycin resistance (SpecR) cassette were separately amplified using Phusion polymerase (Thermo Scientific, all PCRs were performed with Phusion unless otherwise specified) (primers: revL3S2P11-tetR-f1, olp-rrnBTwP(SpR)-r1; olp-rrnBTwP (SpR)-f1, L3S2P55-SpR-r1) and fused together via overlap PCR. Fragments from *intC* were separately amplified and fused together via overlap PCR to enable chromosomal recombination (primers revIntC-f1, olp-smaI-revIntC-r1; olp-smaI-revIntC-f2, revIntC-r2). The fusion was digested with SmaI, and the dCas9-SpecR cassette was inserted [cloning steps were performed in the pGEM-T easy (Promega) vector and confirmed with Sanger sequencing (Eurofins Genomics)].

The construct was subcloned into the NotI site of pKO3 (Link *et al*, 1997), inserted into *intC* of *E. coli* BW25993 using double recombination as previously described (Link *et al*, 1997), and sequence-verified. Finally, the complete, integrated construct was transferred by P1 phage generalized transduction to a BW25993 strain carrying a translational fusion of *ypet* to *lacY* in the native *lacZYA* operon, generating strain BW25993 *intC::tetR-dcas9-aadA lacY::ypet-cat*.

The TetR and dCas9 promoters were optimized to minimize non-induction leakage

First, the PRPA bidirectional promoter, driving the expression of both TetR and dCas9, was replaced with two separate promoters driving each gene. The PLtetO-1 promoter (Lutz & Bujard, 1997) and a strong synthetic RBS designed using the RBS calculator (Salis *et al*, 2009; Espah Borujeni *et al*, 2014) were used for regulating the expression of dCas9. For driving the expression of TetR, the combined promoter and RBS sequence element PN25 was used (Lutz & Bujard, 1997). However, the dCas9 leakage levels were not sufficiently low, so PN25 was switched out for the stronger proB TetR expression elements (Rogers *et al*, 2015) in a second step. The dCas9 and TetR promoter engineering was made in the BW25993 *intC::tetR-dcas9-aadA lacY:: ypet-cat* strain with λ-RED recombination using the pKD46 plasmid (Datsenko & Wanner, 2000). The PRPA promoter region was exchanged with a kanamycin resistance (kanR)-sacB cassette produced using Phusion polymerase and primers URStetR-kanRsacR-f1 and DRSdcas9-kanRsacR-r1. For producing the PN25-PLtetO-1 recombination fragment, oligonucleotides PN25-SpOLP and SpOLP-PLtet-strRBS were fused using PrimeSTAR polymerase (Takara) overlap extension, and the product was used in a PrimeSTAR PCR with primers URStetR-PN25-f1 and DRSdcas9-strRBS-r1. The integrated kanR-sacB cassette was exchanged with the PN25-PLtetO-1 recombination fragment. To engineer stronger TetR expression, the PN25 expression region was first exchanged with a kanR-sacB cassette amplified using primers URStetR-kanRsacR-f1 and DRS-spacer-kanRsacR-r1. The proB recombination fragment was generated by fusing the two oligonucleotides tetR-Rogers-spac-oligo1 and tetR-Rogers-spac-oligo2 using PrimeSTAR. Finally, the integrated kanR-sacB cassette was exchanged with the proB recombination fragment to produce the final dCas9 construct, and all modified regions were confirmed by sequencing. The completed dCas9 construct was transduced to a fresh *lacY::ypet-cat* strain.

To complete the DuMPLING screening strain, the T7 RNA polymerase gene regulated by the *araBAD* promoter was P1 phage transduced from the BL21 AI strain (Invitrogen), producing the final BW25993 *intC::tetR-dcas9-aadA lacY::ypet-cat araB::T7 RNAP-tetA ΔaraB* strain with the new name EL101. All oligonucleotides used for cloning are available in Table EV1.

Design and construction of the CRISPRi/RNA barcode plasmid library

To directly connect each unique sgRNA spacer with a specific barcode, the sgRNA and the barcode were placed in close proximity and expressed from divergent promoters (Fig EV1). This makes it possible to fit both the variable part of the sgRNA and the barcode within the current commercial synthesis limit. The sgRNA and the barcode RNA are driven by a constitutive sigma70 promoter and a T7 promoter, respectively. The two expression units are separated by a spacer that positions the UP-element of the sgRNA promoter in the constant region of the RNA barcode promoter. The final construct, including flanking priming regions to facilitate cloning, came to 167 nt.

Each of the three library oligonucleotides (probe1-lacY-spacer (P1-lacY), probe2-lacI-spacer (P2-lacI), and probe3-control-spacer (P3-control)) consists of a unique FISH RNA barcode sequence paired with a unique sgRNA spacer targeting *lacY*, *lacI*, and a control spacer with five mismatches toward *ypet*, respectively. To avoid the formation of chimeras, emulsion PCR, with library oligos sufficiently diluted to ensure that each droplet at most contains one template, was used (Williams *et al*, 2006; Shao *et al*, 2011; Fig EV1). The PCR was performed with DreamTaq on templates in a 1:5 ratio with the expected number of emulsion droplets (primers library-fw and library-rv). The emulsion PCR product was recovered (Shao *et al*, 2011) and then purified using a commercial kit (Purelink Quick PCR Purification Kit, Invitrogen). The pGuide backbone was PCR-amplified to introduce GG adapters with Q5 DNA polymerase (primers forward pGuide3-early and BpiI-pG7-d0-rv). The PCR product was DpnI-treated and gel-purified (Purelink Quick Gel Extraction Kit,

Invitrogen). The Golden Gate assembly was carried out using BpiI and an approximately 1:1 molecular ratio of amplified pooled library DNA and pGuide backbone product (Engler & Marillonnet, 2013). The assembled pGuide plasmid library was purified (PCR Purification Kit, Invitrogen) and electroporated into the DuMPLING screening strain. After recovery, the cell library was either selected in liquid media (LB + 50 μg/ml kanamycin at 37°C for 3 h) to make cryostocks or plated on LB agar + 50 μg/ml kanamycin plates for estimating library construction accuracy and diversity.

Library colony PCR was carried out using DreamTaq polymerase and the seq-pguide-f1 and r1 primers. Sequence verification of 24 colonies confirmed the absence of library chimeras.

Golden Gate assembly (Engler & Marillonnet, 2013) was used to combine the variable RNA barcode and sgRNA spacer sequences (flanked by GG priming sequences) with the plasmid backbone. For specific amplification of library subpools, the Golden Gate adaptors are flanked by 20-nt primer binding sequences (see upper section of Fig EV1). The RNA barcode is expressed from the strong T7 promoter and transcriptionally fused to the 5′ end of the stable structural d0 RNA (Delebecque *et al*, 2011). The first two guanines of the consensus T7 transcript were kept fixed to ensure strong expression (Imburgio *et al*, 2000). The expression of the sgRNA is driven by the synthetic constitutive promoter J23101 (iGEM Registry of Standard Biological Parts). The putative transcriptional start site of J23101 was kept fixed with an adenine, which was found to be favored (Vvedenskaya *et al*, 2015). The pGuide plasmid backbone, which provides kanamycin resistance (kanR) and contains a high copy number pUC origin of replication, was designed to contain the minimal sequences required for selecting and replicating the dual RNA expression cassette (Fig EV2).

Bulk growth rate and CRISPRi repression assay

To investigate the bulk growth and CRISPRi characteristics of the DuMPLING proof-of-principle system, the P1-lacY, P2-lacI, and P3-control pGuide plasmids in the BW25993 *intC::tetR-dcas9-aadA lacY::ypet-cat araB::T7 RNAP-tetA ΔaraB* screening strain were assayed for growth (OD$_{600}$) and YPet fluorescence using an Infinite M200 plate reader (Tecan).

Cultures
Overnight cultures of the wild-type BW25993 strain with the empty pGuide plasmid, the DuMPLING screening strain with the empty pGuide plasmid, the P1-lacY, P2-lacI, and P3-control pGuide plasmids were grown in LB + 50 μg/ml kanamycin at 37°C shaking at 200 rpm.

Pre-plate
In the morning, overnight cultures were diluted 1:400 into 200 μl supplemented M9 medium [100 μM CaCl$_2$, 2 mM MgSO$_4$, 1× M9 salts, 0.8% v/v glycerol, 1× RPMI amino acid mix (Sigma)] + 50 μg/ml kanamycin + 0.85 g/l Pluronic F108 in a transparent 96-well plate with lid (Costar Assay Plate, REF 3370, Corning). LacY-Ypet and dCas9 were induced by adding isopropyl β-D-1-thiogalactopyranoside (IPTG) (1 mM final concentration) and anhydrotetracycline (aTc) (1 ng/μl final concentration), respectively. To control for the EtOH in the aTc stock, 100 ppm EtOH was added to media without aTc. Plate reader cultures were grown at 37°C, with shaking (1 min,

4.5 mm amplitude) and measurements (OD_{600} and fluorescence with 510 ± 9 nm excitation and 540 ± 20 nm emission) every 5 min.

Experiment run

The pre-plate cultures were diluted 1:200 once they hit exponential phase and run for 20 h as described above (Costar Assay Plate, REF 3904, Corning).

Analysis

The raw data were analyzed using custom MATLAB scripts. The maximum growth rates were converted to minimum doubling times (Fig EV3A). After subtracting the medium background absorption and fluorescence, the fluorescence was normalized with OD_{600} (Fig EV3B) and these values were used for calculating CRISPRi repression ratios.

Results

P1-lacY sgRNA: LacY-YPet was repressed 19.8-fold upon dCas9 induction. The leakage repression was negligible compared with the empty vector control. *P2-lacI sgRNA:* LacY-YPet expression was activated to 24.1-fold over the cell background level and 0.43-fold of the maximal IPTG induction levels in the empty vector control, due to suppression of LacI expression. *P3-control strain:* The ratio of LacY-YPet expression with the empty vector culture was close to 1 (1.04 for induction with both IPTG and aTc, 0.90 for just aTc).

These data illustrate the low leakage of dCas9 expression in the DuMPLING screening strain, which is important to avoid biasing the screening population before phenotyping.

The microfluidic chip

The microfluidic chip is a PDMS (Polydimethylsiloxane)–glass hybrid disposable device where the flow is driven by pressure. We describe the microfluidic chip design, production, and operation in Baltekin *et al* (2017). The chip is designed to rapidly capture individual bacterial cells from liquid growth cultures and exchange the liquid media around the cells effectively while keeping the captured cells in place throughout the experiment. Here, the chip design enables effective delivery and exchange of different media, probes, and buffers during the genotyping.

Microscope setup

All imaging were carried out using a Nikon Ti-E setup for both phase contrast and epifluorescence microscopy. The microscope was equipped with 100× CPI Plan Apo Lambda (Nikon). Phase contrast images were acquired using a dmk23u274 (The Imaging Source). Bright-field and fluorescence images were acquired using a Zyla 4.2 PLUS sCMOS (Andor).

For wide-field epifluorescence-based phenotyping, a 300 ms excitation [shuttered using an AOTFnC (AA Opto Electronics)] from a 514-nm CW-laser at 415 W/cm^2 (Fandango, Cobolt) was used. The laser light was reflected on a zt514.5rdc (Chroma) dichroic before hitting the sample. The Ypet emitted light was transmitted through the above dichroic and filtered through a BrightLine Fluorescence 542/27 (Semrock) before hitting the sCMOS camera. The genotyping and DAPI imaging were carried out using LED white light source (Sola, Lumencore) together with the appropriate filter cubes. Filter cube for Cy3 detection: excitation filter: FF01-543/22 (Semrock), dichroic mirror: FF562-Di03 (Semrock), emission filter: FF01-586/20 (Semrock). Filter cube for Cy5 detection: excitation filter: FF01-635/18 (Semrock), dichroic mirror: FF652-Di01 (Semrock), emission filter: FF01-680/42 (Semrock).

Loading cells into the microfluidic chip

Overnight cultures of the strains to be loaded were grown in LB + 50 µg/ml kanamycin at 37°C shaking at 200 rpm. In the morning, cells were diluted 1:200 in M9 + 0.2% Glucose + 1× RPMI + 50 µg/ml kanamycin + 0.85 g/l Pluronic F108 and grown for 2 h at 37°C shaking at 200 rpm, at which point cells were flown into the chip and into the cell channels where they are caught by the 300 nm constriction at the end of the cell channels [as described in (Baltekin *et al*, 2017)]. The cells were grown in the chip overnight in M9 + 0.2% Glucose + 1× RPMI + 50 µg/ml kanamycin + 0.85 g/l Pluronic + 0.1 ng/µl aTc at 30°C, and then imaged.

Imaging phenotypes

Cells were imaged for 272 min in the same conditions as overnight growth. Phase contrast images were taken every minute. Bright-field images and epifluorescence images were taken every 13 min. Microscope and accessory equipment were controlled using micromanager (version 1.4.20) (Edelstein *et al*, 2010).

Genotyping by sequential FISH

After phenotype imaging was complete, the media was switched to LB + 20% arabinose + 50 µg/ml kanamycin + 0.85 g/l Pluronic, and the cells were grown 3 h further at 30°C. After arabinose induction, the cells were fixed in a solution of 1× PBS + 4% formaldehyde for 10 min at room temperature (all steps from this point forward were carried out at room temperature). The cells were then washed with 1× PBS + Ribolock (Thermo Scientific). The cells were then permeabilized with 70% EtOH for 45 min. The 70% EtOH was washed away with 50% EtOH, then 25% EtOH, and finally with 1× PBS + Ribolock.

For each round of FISH, the appropriate probe pool was flowed into the chip [30 µl hybridization probes + 7.5 µl Ribolock (Thermo Scientific) + 30 µl *E. coli* tRNA (0.65 mg/ml) + 233 µl (0.05 g/ml Dextran sulfate sodium salt, 20% formamide and 2× SSC)]. Hybridization was allowed to proceed overnight (~16 h). The excess probes were washed away with PBS + DAPI stain + Ribolock and then imaged in DAPI, Cy3, and Cy5 using the white light source (SOLA). After imaging, the cells were incubated in a solution of 90% formamide + 2× SSC for 1 h to wash away bound probes and then washed again in PBS + DAPI stain + Ribolock to remove the previous reagents. The cells were again imaged as before to ensure that the probes were fully removed. This was the completion of one round of probing, and at this point, the next pool of probes was flowed into the chip.

Fluorescent *in situ* hybridization probe production

The steps for probe production, adapted from Beliveau *et al* and Chen *et al* (Beliveau *et al*, 2012; Chen *et al*, 2015), are seen in

Fig EV5. Sequences can be found in Table EV2. Templates for FISH probe elongation rounds 1 and 2 (P1 R1 E0, P2 R1 E1, P3 R1 E0 or P1 R2 E1, P2 R2 E0, P3 R2 E0) were pooled separately and PCR-amplified with DreamTaq polymerase using phosphorylated forward primers and phosphorothioate-modified reverse primers (R1 FWD and R1 REV or R2 FWD and R2 REV). The PCR product was purified using the PureLink quick PCR purification kit (Invitrogen). The phosphorylated strand was selectively digested by lambda exonuclease (Thermo Scientific) treatment for 30 min at 37°C followed by heat inactivation at 80°C for 10 min. The ssDNA was purified using the MinElute PCR purification kit (Qiagen). The ssDNA template was elongated by hybridization of the corresponding phosphorothioate-modified Cy3 or Cy5 elongation probes (E0 Cy3 and E1 Cy5) at 55°C for 5 min after an initial heating step at 96°C for 3 min. Elongation was performed with DreamTaq polymerase and dNTP in DreamTaq buffer at 72°C for 15 min. The elongated product was purified using the PureLink quick PCR purification kit (Invitrogen) and cleaved by the SchI FD enzyme for 30 min at 37°C. After this step, lambda exonuclease was added directly to the SchI digestion for an additional 30 min at 37°C. The processed FISH probes were purified using phenol/chloroform/isoamylalcohol (VWR), washed with chloroform (Sigma-Aldrich), and extracted by means of centrifugation after precipitation with EtOH and sodium acetate. The DNA pellet was washed once in 70% EtOH and dried at room temperature before being resolved in water. To remove any additional undesirable DNA, the probe mixture was purified on a 4% agarose gel. The expected DNA band was excised from the gel, sliced in small pieces, and incubated overnight in water. The extracted probe was phenol/chloroform/iso-amylalcohol-purified, washed, and extracted as previously described followed by filtration in Ultrafree-MC microcentrifuge filters (Sigma-Aldrich) before being used in the microfluidic experiment.

Polyacrylamide gel analysis of produced probes

Samples were collected throughout the probe production protocol, mixed with 10× FD green buffer, and loaded on a 10% polyacrylamide gel (Bio-Rad). As size references, Cy3 and Cy5 39-nt ssDNA probes with two phosphodiester bonds, and also the Cy3 and Cy5 19-nt probes used for elongation, were loaded onto the gel. The gel was run in 1× TBE buffer in a Mini-PROTEAN system (Bio-Rad) and analyzed with a Chemidoc system (Bio-Rad) (Fig EV6).

Image analysis

Phenotyping

Cell outlines were identified using cell segmentation (Ranefall et al, 2016) of phase contrast images. Before segmentation, the image of a trap designed to be without cells was deducted from all traps imaged in phase contrast as described in Baltekin et al (2017). Using the detected cell outlines, lineages were constructed using the Baxter algorithm (Magnusson et al, 2015) where Jaccard indices between consecutive cell outlines were used to score migration and division events. The division event scores were calculated to require binary fission. Cell lineages from Baxter were filtered based on the following criteria: (i) Cell outlines where the size transiently dropped or increased by large amounts were deemed as missegmentation and not used in further analysis. (ii) Lineages from one cell generation with large shifts in size (non-transient) were excluded

from further analysis. (iii) Lineages from one cell generation with large center of mass movements of the cell outlines were excluded from further analysis. Finally (iv), lineages from one cell generation with very short life span were excluded from further analysis unless they contained both a mother and two daughter cells. Fluorescently labeled LacY-YPet molecules were localized using the dot detection algorithm suggested by Loy and Zelinsky (Loy & Zelinsky, 2003). Given that phase contrast and fluorescence images were acquired using different cameras, a transformation was required to place dots inside segmented cell outlines. This transformation was estimated before the start of the experiment using landmarks in images captured on the two different cameras.

Genotyping

DAPI, Cy3, and Cy5 images were summed vertically (see Fig 3B). The locations of the cell traps were determined using the vertically summed DAPI signal. The log ratio of vertically summed Cy3 and Cy5 signals was used to call a 1 or 0. The genotype of the trap was then associated with all cells in that trap.

Sensitivity and specificity

The bright fluorescent phenotype is easy to identify, which makes it possible to use this as a reference when calculating sensitivity and specificity for the genotyping in our experiment. There were 74 traps with bright cells and 159 with the other strains. One of the 1,398 attempts to read a barcode gave the wrong answer, which would lead to a misclassification of a bright trap as a non-bright trap. However, this could be corrected since only two rounds of probing are needed to call the genotype and we probed six rounds. This implies that each genotype has been determined $6/2 = 3$ times.

In terms of sensitivity and specificity, the true-positive identifications of the bright genotype was made $(74 \times 3) - 1 = 224$ times. The true-negative identifications of the bright genotype was made $159 \times 3 = 477$ times. There is one false negative and 0 false positive. Based on this, we calculated the 95% Clopper–Pearson confidence intervals for the sensitivity to be 97.55–99.99% and for specificity to be 99.23–100.00% (Clopper & Pearson, 1934).

If this experiment is used as a proxy for a library that requires six rounds of probing to identify each genotype, then the true-positive identification of the bright genotype was made $74 - 1 = 73$ times. The true-negative identification of the bright genotype was made 159 times. There is one false negative and 0 false positives. Based on this, we calculated the 95% confidence intervals for the sensitivity to be 92.7–99.97% and for specificity to be 97.71% to 100%.

Note on chemicals and reagents

All chemicals were acquired from Sigma-Aldrich unless otherwise stated. All synthetic DNAs are from Integrated DNA Technologies, and unlabeled DNA oligonucleotides above 100 nt were bought as Ultramers. DreamTaq DNA polymerase (Thermo Scientific) was used for colony PCRs, PCRs for sequencing reactions, preparative PCRs of small fragments (< 200 bp), and emulsion PCR. Phusion DNA polymerase (Thermo Scientific) was used for preparative PCRs. For difficult preparative PCRs, PrimeSTAR DNA polymerase (Takara) was used. For preparative PCRs requiring extra high accuracy, Q5 RNA polymerase was used (NEB). Restriction enzymes, ligases, and other cloning-related enzymes were procured from Thermo Scientific unless otherwise stated.

Acknowledgements

E. Gullberg is gratefully acknowledged for providing the DA25184 strain that carries the spectinomycin resistance cassette and Prune Leroy for providing the LacY-YPet strain. The authors would like to thank George Church and Mats Nilsson for helpful discussions. In addition, the authors would like to acknowledge funding from the Knut and Alice Wallenberg foundation, the Swedish Research Council and the European Research Council.

Author contributions

JE conceived the concept and coordinated the project. MJL developed the phenotyping and genotyping protocols and carried out the corresponding experiments. DC designed the strains and made them. JL and MJL developed the probe synthesis method. ÖB developed the microfluidic device. DF and MJL developed the microscopy and analysis methods. JE, MJL, DF, DC, and JL wrote the manuscript. However, the authors worked closely on the whole project and made substantial contributions in each other's main areas.

References

Baltekin Ö, Boucharin A, Tano E, Andersson DI, Elf J (2017) Antibiotic susceptibility testing in less than 30 min using direct single-cell imaging. *Proc Natl Acad Sci USA* 114: 9170−9175

Balzarotti F, Eilers Y, Gwosch KC, Gynnå AH, Westphal V, Stefani FD, Elf J, Hell SW (2017) Nanometer resolution imaging and tracking of fluorescent molecules with minimal photon fluxes. *Science* 355: 606−612

Beliveau BJ, Joyce EF, Apostolopoulos N, Yilmaz F, Fonseka CY, McCole RB, Chang Y, Li JB, Senaratne TN, Williams BR, Rouillard J-M, Wu C-T (2012) Versatile design and synthesis platform for visualizing genomes with oligopaint FISH probes. *Proc Natl Acad Sci USA* 109: 21301−21306

Chen B, Gilbert LA, Cimini BA, Schnitzbauer J, Zhang W, Li G-W, Park J, Blackburn EH, Weissman JS, Qi LS, Huang B (2013) Dynamic imaging of genomic loci in living human cells by an optimized CRISPR/Cas system. *Cell* 155: 1479−1491

Chen KH, Boettiger AN, Moffitt JR, Wang S, Zhuang X (2015) RNA imaging. Spatially resolved, highly multiplexed RNA profiling in single cells. *Science* 348: aaa6090

Clopper CJ, Pearson ES (1934) The use of confidence or fiducial limits illustrated in the case of the binomial. *Biometrika* 26: 404−413

Datsenko KA, Wanner BL (2000) One-step inactivation of chromosomal genes in *Escherichia coli* K-12 using PCR products. *Proc Natl Acad Sci USA* 97: 6640−6645

Delebecque CJ, Lindner AB, Silver PA, Aldaye FA (2011) Organization of intracellular reactions with rationally designed RNA assemblies. *Science* 333: 470−474

Dixit A, Parnas O, Li B, Chen J, Fulco CP, Jerby-Arnon L, Marjanovic ND, Dionne D, Burks T, Raychowdhury R, Adamson B, Norman TM, Lander ES, Weissman JS, Friedman N, Regev A (2016) Perturb-Seq: dissecting molecular circuits with scalable single-cell RNA profiling of pooled genetic screens. *Cell* 167: 1853−1866.e17

Edelstein A, Amodaj N, Hoover K, Vale R, Stuurman N (2010) Computer control of microscopes using μManager. *Curr Protoc Mol Biol* 92: 14.20.1−14.20.17

Engler C, Marillonnet S (2013) Combinatorial DNA assembly using golden gate cloning. *Methods Mol Biol* 1073: 141−156

Espah Borujeni A, Channarasappa AS, Salis HM (2014) Translation rate is controlled by coupled trade-offs between site accessibility, selective RNA unfolding and sliding at upstream standby sites. *Nucleic Acids Res* 42: 2646−2659

Garst AD, Bassalo MC, Pines G, Lynch SA, Halweg-Edwards AL, Liu R, Liang L, Wang Z, Zeitoun R, Alexander WG, Gill RT (2017) Genome-wide mapping of mutations at single-nucleotide resolution for protein, metabolic and genome engineering. *Nat Biotechnol* 35: 48−55

Hammar P, Walldén M, Fange D, Persson F, Baltekin O, Ullman G, Leroy P, Elf J (2014) Direct measurement of transcription factor dissociation excludes a simple operator occupancy model for gene regulation. *Nat Genet* 46: 405−408

Imburgio D, Rong M, Ma K, McAllister WT (2000) Studies of promoter recognition and start site selection by T7 RNA polymerase using a comprehensive collection of promoter variants. *Biochemistry* 39: 10419−10430

Jaitin DA, Weiner A, Yofe I, Lara-Astiaso D, Keren-Shaul H, David E, Salame TM, Tanay A, van Oudenaarden A, Amit I (2016) Dissecting immune circuits by linking CRISPR-pooled screens with single-cell RNA-seq. *Cell* 167: 1883−1896.e15

Ke R, Mignardi M, Pacureanu A, Svedlund J, Botling J, Wählby C, Nilsson M (2013) In situ sequencing for RNA analysis in preserved tissue and cells. *Nat Methods* 10: 857−860

Keren L, Hausser J, Lotan-Pompan M , IV S, Alisar H, Kaminski S, Weinberger A, Alon U, Milo R, Segal E (2016) Massively parallel interrogation of the effects of gene expression levels on fitness. *Cell* 166: 1282−1294.e18

Kuwada NJ, Traxler B, Wiggins PA (2015) Genome-scale quantitative characterization of bacterial protein localization dynamics throughout the cell cycle. *Mol Microbiol* 95: 64−79

Lee JH, Daugharthy ER, Scheiman J, Kalhor R, Yang JL, Ferrante TC, Terry R, Jeanty SSF, Li C, Amamoto R, Peters DT, Turczyk BM, Marblestone AH, Inverso SA, Bernard A, Mali P, Rios X, Aach J, Church GM (2014) Highly multiplexed subcellular RNA sequencing in situ. *Science* 343: 1360−1363

Link AJ, Phillips D, Church GM (1997) Methods for generating precise deletions and insertions in the genome of wild-type *Escherichia coli*: application to open reading frame characterization. *J Bacteriol* 179: 6228−6237

Liu Z, Lavis LD, Betzig E (2015) Imaging live-cell dynamics and structure at the single-molecule level. *Mol Cell* 58: 644−659

Loy G, Zelinsky A (2003) Fast radial symmetry for detecting points of interest. *IEEE Trans Pattern Anal Mach Intell* 25: 959−973

Lutz R, Bujard H (1997) Independent and tight regulation of transcriptional units in *Escherichia coli* via the LacR/O, the TetR/O and AraC/I1-I2 regulatory elements. *Nucleic Acids Res* 25: 1203−1210

Magnusson KEG, Jalden J, Gilbert PM, Blau HM (2015) Global linking of cell tracks using the viterbi algorithm. *IEEE Trans Med Imaging* 34: 911−929

Norman TM, Lord ND, Paulsson J, Losick R (2013) Memory and modularity in cell-fate decision making. *Nature* 503: 481−486

Otoupal PB, Erickson KE, Escalas-Bordoy A, Chatterjee A (2017) CRISPR perturbation of gene expression alters bacterial fitness under stress and reveals underlying epistatic constraints. *ACS Synth Biol* 6: 94−107

Peters JM, Colavin A, Shi H, Czarny TL, Larson MH, Wong S, Hawkins JS, Lu CHS, Koo B-M, Marta E, Shiver AL, Whitehead EH, Weissman JS, Brown ED, Qi LS, Huang KC, Gross CA (2016) A comprehensive, CRISPR-based functional analysis of essential genes in bacteria. *Cell* 165:

Potvin-Trottier L, Lord ND, Vinnicombe G, Paulsson J (2016) Synchronous long-term oscillations in a synthetic gene circuit. *Nature* 538: 514–517

Qi LS, Larson MH, Gilbert LA, Doudna JA, Weissman JS, Arkin AP, Lim WA (2013) Repurposing CRISPR as an RNA-guided platform for sequence-specific control of gene expression. *Cell* 152: 1173–1183

Ranefall P, Sadanandan SK, Wählby C (2016) Fast adaptive local thresholding based on ellipse fit. In *2016 IEEE 13th International Symposium on Biomedical Imaging (ISBI)*, 205–208.

Rogers JK, Guzman CD, Taylor ND, Raman S, Anderson K, Church GM (2015) Synthetic biosensors for precise gene control and real-time monitoring of metabolites. *Nucleic Acids Res* 43: 7648–7660

Salis HM, Mirsky EA, Voigt CA (2009) Automated design of synthetic ribosome binding sites to control protein expression. *Nat Biotechnol* 27: 946–950

Shah S, Lubeck E, Zhou W, Cai L (2016) *In situ* transcription profiling of single cells reveals spatial organization of cells in the mouse hippocampus. *Neuron* 92: 342–357

Shao K, Ding W, Wang F, Li H, Ma D, Wang H (2011) Emulsion PCR: a high efficient way of PCR amplification of random DNA libraries in aptamer selection. *PLoS One* 6: e24910

Shi H, Colavin A, Lee TK, Huang KC (2017) Strain library imaging protocol for high-throughput, automated single-cell microscopy of large bacterial collections arrayed on multiwell plates. *Nat Protoc* 12: 429–438

Taheri-Araghi S, Bradde S, Sauls JT, Hill NS, Levin PA, Paulsson J, Vergassola M, Jun S (2015) Cell-size control and homeostasis in bacteria. *Curr Biol* 25: 385–391

Uphoff S, Lord ND, Okumus B, Potvin-Trottier L, Sherratt DJ, Paulsson J (2016) Stochastic activation of a DNA damage response causes cell-to-cell mutation rate variation. *Science* 351: 1094–1097

Vvedenskaya IO, Zhang Y, Goldman SR, Valenti A, Visone V, Taylor DM, Ebright RH, Nickels BE (2015) Massively systematic transcript end readout, "MASTER": transcription start site selection, transcriptional slippage, and transcript yields. *Mol Cell* 60: 953–965

Wallden M, Fange D, Lundius EG, Baltekin Ö, Elf J (2016) The synchronization of replication and division cycles in individual *E. coli* cells. *Cell* 166: 729–739

Wang HH, Isaacs FJ, Carr PA, Sun ZZ, George X, Forest CR, Church GM (2009) Programming cells by multiplex genome engineering and accelerated evolution. *Nature* 460: 894–898

Wang P, Robert L, Pelletier J, Dang WL, Taddei F, Wright A, Jun S (2010) Robust growth of *Escherichia coli. Curr Biol* 20: 1099–1103

Williams R, Peisajovich SG, Miller OJ, Magdassi S, Tawfik DS, Griffiths AD (2006) Amplification of complex gene libraries by emulsion PCR. *Nat Methods* 3: 545–550

13

Single-cell sequencing maps gene expression to mutational phylogenies in PDGF- and EGF-driven gliomas

Sören Müller[1,2,†], Siyuan John Liu[1,2,†], Elizabeth Di Lullo[2,3], Martina Malatesta[1,2], Alex A Pollen[2,3], Tomasz J Nowakowski[2,3], Gary Kohanbash[1], Manish Aghi[1], Arnold R Kriegstein[2,3], Daniel A Lim[1,2,4,*] & Aaron Diaz[1,2,**]

Abstract

Glioblastoma multiforme (GBM) is the most common and aggressive type of primary brain tumor. Epidermal growth factor (EGF) and platelet-derived growth factor (PDGF) receptors are frequently amplified and/or possess gain-of-function mutations in GBM. However, clinical trials of tyrosine-kinase inhibitors have shown disappointing efficacy, in part due to intra-tumor heterogeneity. To assess the effect of clonal heterogeneity on gene expression, we derived an approach to map single-cell expression profiles to sequentially acquired mutations identified from exome sequencing. Using 288 single cells, we constructed high-resolution phylogenies of EGF-driven and PDGF-driven GBMs, modeling transcriptional kinetics during tumor evolution. Descending the phylogenetic tree of a PDGF-driven tumor corresponded to a progressive induction of an oligodendrocyte progenitor-like cell type, expressing pro-angiogenic factors. In contrast, phylogenetic analysis of an *EGFR*-amplified tumor showed an up-regulation of pro-invasive genes. An in-frame deletion in a specific dimerization domain of PDGF receptor correlates with an up-regulation of growth pathways in a proneural GBM and enhances proliferation when ectopically expressed in glioma cell lines. In-frame deletions in this domain are frequent in public GBM data.

Keywords copy-number variation; glioblastoma; PDGFRA; single-cell RNA-sequencing; tumor phylogeny

Subject Categories Cancer; Chromatin, Epigenetics, Genomics & Functional Genomics; Genome-Scale & Integrative Biology

Introduction

Glioblastoma multiforme (GBM) is an extremely aggressive type of brain tumor, characterized by a high degree of intra-tumor heterogeneity (Patel *et al*, 2014). Amplifications and gain-of-function mutations in receptor-tyrosine kinases (RTKs) are common in GBM. However, these mutations are typically regional and mosaic (Szerlip *et al*, 2012), and combinatorial application of RTK inhibitors is required to achieve a complete treatment *in vitro* (Stommel *et al*, 2007). Clinical trials of RTK inhibitors have shown only minimal efficacy, and these limitations may be in part due to intra-tumor heterogeneity (Prados *et al*, 2015). More broadly, developing treatments that circumvent specific, regional genotypic differences is challenging, since the number of biopsies per tumor is generally limited, and bulk-sequencing methods reduce such regional variation to population averages.

There is strong evidence that treatment itself can drive clonal evolution. Temozolomide chemotherapy is part of the current standard of care for newly diagnosed GBM. But, for some glioma patients, temozolomide treatment can also drive a hyper-mutated phenotype, as has been demonstrated by phylogenetic analysis of exome-sequencing (exome-seq) data (Johnson *et al*, 2014). Phylogenetic analyses of bulk exome-seq and methylation array data in glioma cohorts have also been used to identify recurrent events in tumor evolution (Johnson *et al*, 2014; Mazor *et al*, 2015). Recent advances in single-cell sequencing have enabled fine mapping of *EGFR* variant heterogeneity (Francis *et al*, 2014), as well as studies of the evolutionary history of individual tumors at unprecedented resolution (Navin *et al*, 2011; Garvin *et al*, 2014). Furthermore, large-scale copy-number variations (CNVs) have been inferred from single-cell RNA sequencing (RNA-seq) in GBM (Patel *et al*, 2014). In this study, we called CNVs from bulk exome-seq and quantified them in single-cell

---
1 Department of Neurological Surgery, University of California San Francisco, San Francisco, CA, USA
2 Eli and Edythe Broad Center of Regeneration Medicine and Stem Cell Research, University of California San Francisco, San Francisco, CA, USA
3 Department of Neurology, University of California San Francisco, San Francisco, CA, USA
4 Veterans Affairs Medical Center, San Francisco, CA, USA
  *Corresponding author. E-mail: daniel.lim@ucsf.edu
  **Corresponding author. E-mail: aaron.diaz@ucsf.edu
  †These authors contributed equally to this work

RNA-seq from the same tumor sample. Based on this, we produced a clonal ordering of individual cells that we used to infer transcriptional kinetics during tumor evolution and to perform inter-clone differential transcriptomics. We used this approach to contrast EGF-driven and PDGF-driven GBMs and identified pathways that show a dose–response to EGF- and PDGF-receptor copy number.

Results

Primary GBMs contain heterogeneous mixtures of cell types with recurrent transcriptional signatures

We collected fresh tissue from three cases of primary, untreated GBM directly from the operating room (SF10282, SF10345, and SF10360) and subjected these biopsies to both single-cell RNA-seq and bulk exome-seq (Fig 1A). We also performed bulk exome-seq on a separate blood sample from each patient. We characterized the landscape of genomic, somatic mutations for each patient using a robust exome-seq pipeline (Johnson *et al*, 2014), this analysis included identifying single nucleotide variants (SNVs), small insertions/deletions (indels), and copy-number variants (CNVs; Materials and Methods).

All cases demonstrated an amplification of growth factor genes. We found *EGFR* to be highly amplified in SF10345 (122 copies) and *PDGFRA* to be amplified in SF10282 (12 copies). Deletion and putative loss-of-function mutations in tumor suppressors were also common events (Datasets EV1–EV6). For example, all cases had non-synonymous point mutations in *PTEN* (with variant allele frequencies (VAFs) from 41 to 89%). A copy of chromosome 10 was lost in SF10345 and SF10360. Furthermore, these two cases harbored a deletion in chromosome 9, in a region encoding tumor suppressor genes *KLHL9* and *CDKN2A/B*. *KLHL9* deletions are correlated with the mesenchymal GBM subtype and poor prognosis (Chen *et al*, 2014). In our data, *KLHL9* is not expressed in either SF10345 or SF10360, and both samples classify as mesenchymal/classical. SF10360 and SF10282 share other mutations, such as a loss of 13q14 that contains the tumor suppressive micro-RNA cluster miR-15a/16 (Aqeilan *et al*, 2010; Afonso-Grunz & Müller, 2015). Between 5 and 35, small indels were detected per sample, for example, *TP53* (SF10282, frame-shift deletion), *NF1* (SF10360, frame-shift deletion), and *PLAGL1* (SF10345, frame-shift deletion).

Prior to the analysis of single-cell RNA-seq libraries, low-complexity and low-coverage libraries were filtered (Fig EV1A and B), and stromal/non-malignant cells were identified (Materials and Methods). This workflow left 61, 66, and 63 tumor cells from SF10282, SF10345, and SF10360, respectively. Consistent with previous reports (Patel *et al*, 2014), classification of single cells according to the Verhaak subtypes (Verhaak *et al*, 2010) identified heterogeneous mixtures of distinct subtypes within the same tumor (Fig 1C). SF10345 and SF10360 are classical/mesenchymal and predominantly EGFR driven. SF10282 is predominantly pro-neural, up-regulates PDGF-pathway genes, and markers of oligodendrocyte progenitor cells (OPCs) are broadly expressed. Yet SF10282 contains a subpopulation of cells with a neural stem cell (NSC)-like expression profile (Fig 1C). These cells classify as

mesenchymal/classical in the Verhaak scheme. We sought to infer the relative ordering of the NSC and OPC-like cells in the tumor's phylogeny and more generally to establish cellular phylogenies for all samples.

Phylogenies of copy-number alterations map gene expression to clonal structure

We chose to focus on large, somatic CNVs of 100 exons or more (Materials and Methods). The median size of CNVs exceeding this threshold was 18–21 mega base-pairs, comprising 300–400 genes. This size is much larger than the size of CNVs previously observed to occur frequently in the germline (Sudmant *et al*, 2015), which had a median size of 36 kilo base-pairs. We found that GBM to normal-brain control single-cell expression ratios correlated with CNV status (Appendix Fig S1), motivating us to quantify these CNVs in individual cells (Materials and Methods). Briefly, for each CNV identified in the exome-seq, the 5% significance level of the distribution of normal-brain read counts covering that locus was used as a threshold to assign CNV presence/absence calls to individual cells (Fig 2A). This triage was unaffected by the application of a wavelet-smoothing filter to the single-cell data. Furthermore, exome-seq read histograms showed excellent agreement with single-cell trend-lines (Fig 2B and C). This indicated that our approach was robust to the stochastic expression of individual genes. We validated the error rate of this classifier using 10-fold cross-validation, as well as empirical testing on a control dataset (Pollen *et al*, 2015; Appendix Fig S2, Materials and Methods). Using Jaccard distance between CNV genotypes to assess inter-cell similarity, we fit phylogenies to each of the tumor samples using the Fitch–Margoliash method (Materials and Methods). The amplifications of chromosomes 7 and 19p13.3, which were shared across cases in the exome-seq, occurred early in all three of our single-cell phylogenies. In SF10360, chromosome 7 gain was a founding event, occurring together with a loss of chromosome 10 (Fig 3A). Intriguingly, a loss of chromosome 13 arose independently in two distinct subclones of SF10360. Since 13q14 harbors the miR-15a/16 micro-RNA cluster, a known tumor suppressor in prostate cancer (Bonci *et al*, 2008) and chronic lymphocytic leukemia (Pekarsky & Croce, 2015), this loss may convey a survival advantage here as well.

MiR-15a/16 mutant cells up-regulate downstream adhesion and Aurora B kinase pathway genes, enriched in the leading edge and infiltrating tumor

We compared gene expression in chromosome 13 wild-type cells to cells harboring the deletion in SF10360 (Dataset EV7) and scanned the promoters of differentially expressed genes ($P < 0.05$) for conserved transcription factor recognition motifs (Materials and Methods). 16% of genes up-regulated upon chromosome 13 deletion were validated direct targets of miR-15a/16; and an additional 78% were expressed from loci enriched for motifs of transcription factors targeted by miR-15a/16 (Fig 3B; Chou *et al*, 2015). The most significant differentially expressed genes (adjusted $P < 0.1$) showed distinct patterning across tumor anatomical structures, when cross-referenced with the Ivy Glioblastoma Atlas

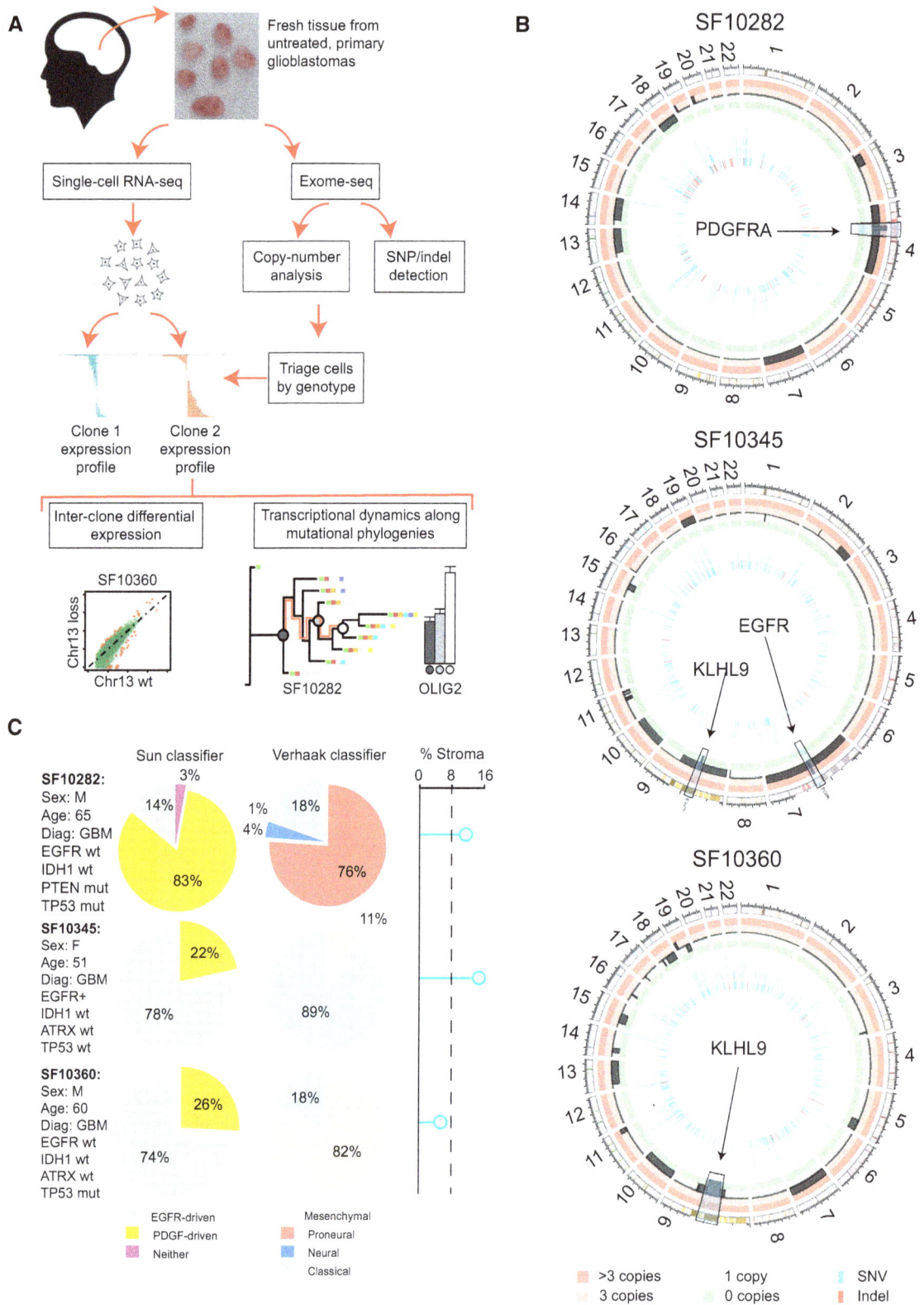

Figure 1.   Experimental design and pipeline, pathology, and genomic and transcriptomic signatures for three primary GBMs.

A   The sample acquisition and processing pipeline.

B   Circos plot of somatic, genomic alterations detected in the bulk DNA of each patient using ADTEx. Copy-number alterations are highlighted in the outer circle by thick
    black bars, and SNVs (MuTect) and small indels (Pindel) by the vertical lines in the inner circle. Regions with strong amplifications/losses are highlighted.

C   Summaries of the patient's sex, age, pathology report, sample's stromal infiltration, and molecular classification.

A

B

C

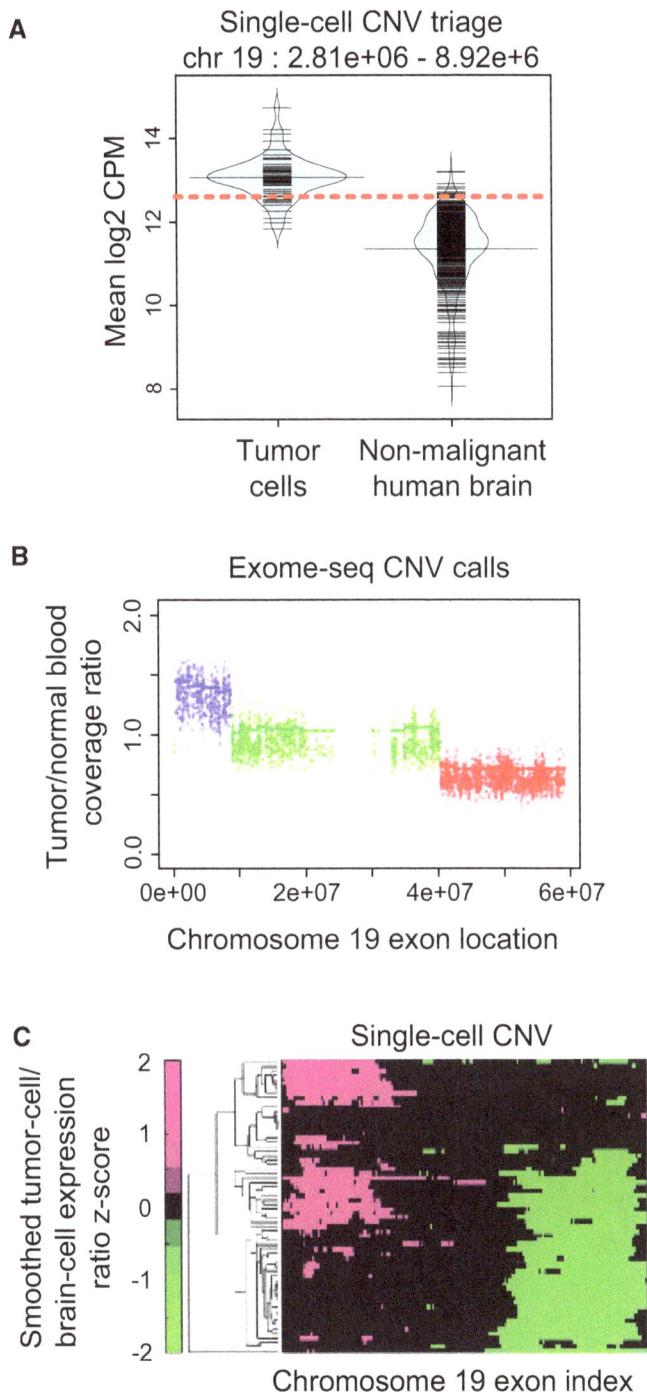

Figure 2. Presence/absence assignments in individual cells of CNVs called from bulk exome-seq.

A   Read-count distributions in a locus of copy-number gain on 19p13.3, comparing cells in SF10360 to a human, normal-brain control. The 5th percentile of the normal-cell distribution is indicated by a red line.

B   Per-exon, normalized bulk exome-seq read-count computed by ADTEx, compared between SF10360 and a blood control. Regions with a read-count ratio > 1.3 are putative DNA copy-number gains (blue), and regions with fold-changes < 0.7 are putative losses (red).

C   A heatmap visualizing the z-score of ratios, of wavelet-smoothed read counts, compared between single-cell RNA-seq expression in tumor cell (rows) and the median expression in single-cell RNA-seq of normal-brain tissue.

(glioblastoma.alleninstitute.org). Genes up-regulated in the wild-type cells were enriched in the peri-vascular region, and genes up-regulated upon chromosome 13 deletion were enriched in the leading edge and infiltrating tumor (Figs 3C and EV2). Consistent with an infiltrating phenotype, cell-adhesion molecules were over-represented (q = 0.06; Materials and Methods). This included junction-adhesion molecules, integrins, disintegrins, and cell-surface receptors implicated in invasion (Fig 3D; Nath *et al*, 2000; Sloan, 2005; Tenan *et al*, 2010; Reyes *et al*, 2013; Sarkar *et al*, 2015; Venkatesh *et al*, 2015). 76% of these cell-adhesion pathway genes were enriched for NF-κB/REL recognition motifs (Fig 3B). Taken together with the EGFR-driven/mesenchymal classification of the case, we speculated that the loss of the miR-15a/16 cluster enhances growth factor-stimulated cell invasion here, as has been described in other cancers (Bonci *et al*, 2008). Among the direct targets of miR-15a/16 that were up-regulated, Aurora B kinase, survivin, and genes that complex with them were overrepresented (q = 0.03; Fig 3E). Survivin is a well-studied inhibitor of apoptosis. Aurora B kinase overexpression increases genomic instability (Ota *et al*, 2002), resulting in multinuclearity and aneuploidy (Tatsuka *et al*, 1998).

Dose–response analysis correlates an in-frame deletion in *PDGFRA* to a pro-growth signature *in vivo*

We found that the PDGF-receptor alpha encoding gene, *PDGFRA*, was amplified in SF10282 (12 copies as estimated by exome-seq). We also detected a small deletion in exon 7 that was broadly expressed (Fig 4A). This mutant transcript, which we denote as *PDGFRA*Δ^7, is found in 98% of SF10282's cells that express *PDGFRA* (69% of cells overall). Since the deletion is in-frame, broadly expressed, and affects an immunoglobulin-like fold involved in receptor dimerization, we reasoned that *PDGFRA*Δ^7 might enhance PDGF-receptor signaling. We sorted cells by *PDGFRA*Δ^7 expression and identified genes that showed a strong rank correlation with *PDGFRA*Δ^7 levels (Materials and Methods). Positively correlated genes were enriched for the PDGF-receptor signaling network and cell cycle, when compared to the Pathway Commons and DAVID databases (Huang *et al*, 2009; Cerami *et al*, 2011). Negatively correlating genes were enriched for oxidative phosphorylation (Fig 4B). Genes correlating with increasing *PDGFRA*Δ^7 were scanned for overrepresented transcription factor binding motifs. Genetic and physical interaction databases were queried against significant transcription factors (Fig 4C), implicating STAT1 and NF-κB as downstream effectors of *PDGFRA*Δ^7. By comparison, an analogous dose–response analysis of *EGFR* in SF10345 identified an increasing gene set of cell cycle genes, as well as genes related to chromatin modification and cell motion. Inference of mediating transcription factors implicated STAT signaling, as in SF10282. Additionally, SOX2 [a pluripotency factor highly expressed in embryonic, neural and glioma stem cells (Suvà *et al*, 2014)] and c-Jun [an anti-apoptotic factor involved in glioma-genesis (Yoon *et al*, 2012)] targets correlated to *EGFR* dose (Fig 5).

PDGFRAΔ^7 confers a growth advantage and stimulates wild-type _PDGFRA_ expression *in vitro*

We expressed *PDGFRA*Δ^7, wild-type *PDGFRA* and a GFP control from lentivirus, in two patient-derived cell lines that we had cultured as monolayers (Fig 6A). One we derived from SF10360 (described

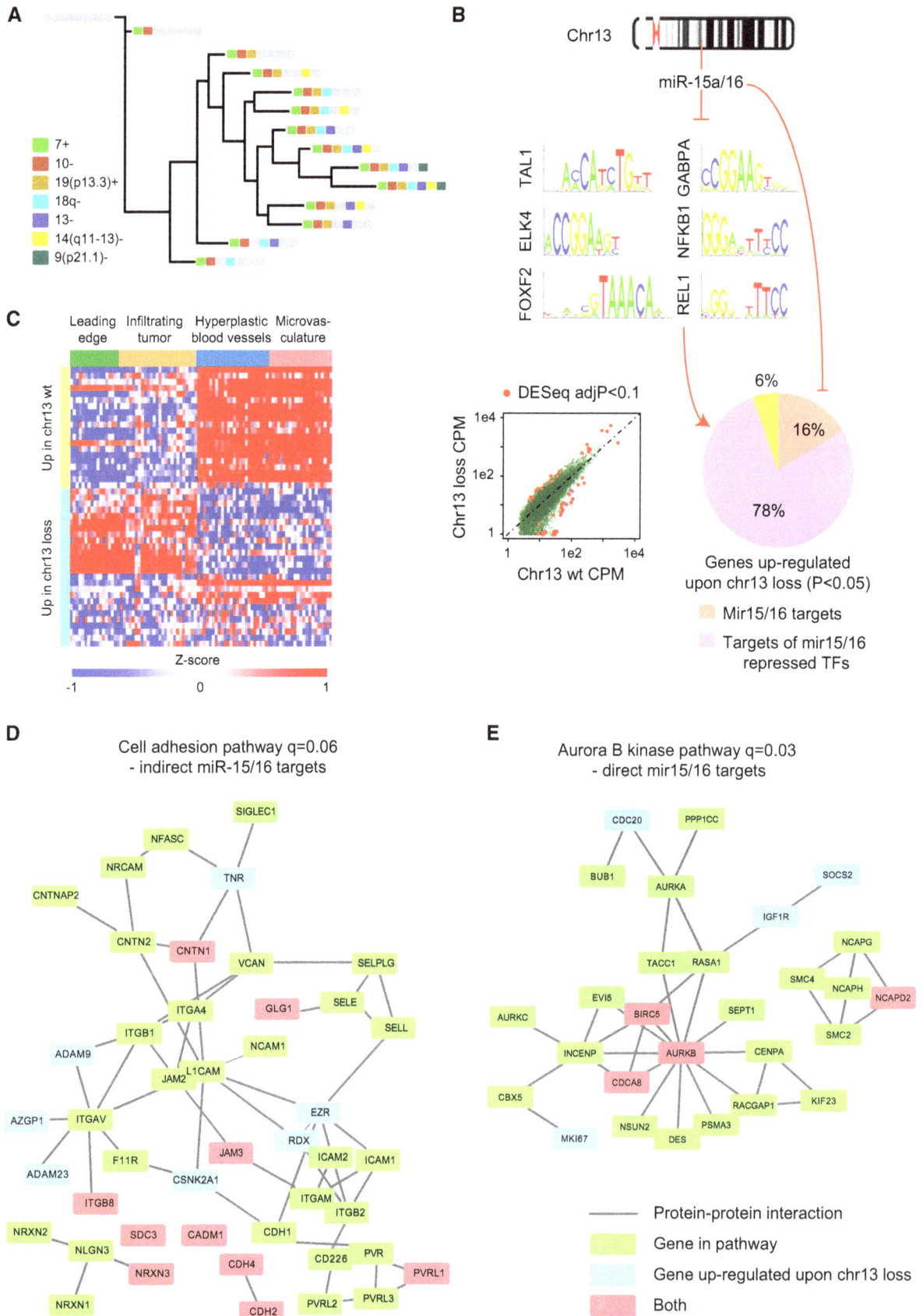

Figure 3.

Figure 3. Analysis of gene sets differentially expressed between subclones of SF10360.

A A phylogeny of CNV cellular genotypes identified in SF10360. Each leaf corresponds to a genotype, defined by a set of CNV presence/absence calls shared between a group of cells.

B Overrepresented transcription factor recognition motifs in genes up-regulated (DESeq, $P < 0.05$) upon chromosome 13 loss. 16% of up-regulated genes are direct, validated miR-15a/16 targets. Another 78% of up-regulated genes are targets of transcription factors that are repressed by miR-15a/16.

C Heatmap of the most significantly, differentially expressed genes (DESeq, FDR < 0.1) upon chromosome 13 loss in the Ivy Glioblastoma Atlas. Each row is a gene, each column is an RNA-seq from an anatomically defined tumor compartment, micro-dissected from an untreated GBM biopsy.

D Protein interaction network of genes differentially expressed upon chromosome 13 loss, which are targets of a miR-15a/16 regulated transcription factor, with an overlay of protein interactions in the cell-adhesion pathway.

E Protein interaction network of genes differentially expressed upon chromosome 13 loss, which are direct, validated targets of miR-15a/16, with an overlay of protein interactions in the Aurora B kinase.

Data information: Significance of network overlaps in (D, E) was computed via JEPETTO ($q < 0.1$).

here), and the second was from a primary GBM: SF10281. These cell lines do not strongly express *PDGFRA* endogenously, but we detected robust expression of *PDGFRA* and *PDGFRA*Δ^7 mRNA in the respective cultures where they were ectopically expressed (Fig 6B). To specifically quantify the expression of wild-type *PDGFRA*, we designed an RT-qPCR assay with a probe targeted to the deleted region. Intriguingly, we found that endogenous, wild-type *PDGFRA* was induced in both cell lines upon ectopic expression of *PDGFRA*Δ^7 (Fig 6C). When we identified genes that were differentially, recurrently expressed in both *PDGFRA*Δ^7 cultures compared to wild-type *PDGFRA* and GFP, we found that these genes enriched for gene-ontology molecular functions associated with PDGF binding and the binding of other growth factors (Fig 6D). In particular, we saw an up-regulation of the epiregulin encoding mRNA (*EREG*), an epidermal growth factor family member which ligates EGFR and most members of the v-erb-b2 oncogene homolog (ERBB) family. We saw an up-regulation of the notch-receptor ligand jagged 1 (*JAG1*) and the master regulator of angiogenesis, *VEGFA*. Additionally, we saw an induction of regulators of inflammation *IL1B* and *IL6*, as well as *COX-2* and colony-stimulating factor 3 (*CSF3*). *VEGFA*, *COX-2*, and *CSF3* all encode chemotactic factors for MDSC (Lechner *et al*, 2010; Fujita *et al*, 2011; Cao *et al*, 2014; Fig 6E). Additionally, we performed an MTT colorimetric assay and found that ectopic *PDGFRA*Δ^7 expression significantly enhanced cell growth *in vitro*, compared to over-expression of wild-type *PDGFRA* or GFP control (Fig 6F).

In-frame deletions in the PFGFRA dimerization domain are frequent events in The Cancer Genome Atlas's GBM data

We then processed exome-seq data from 389 GBM patients and corresponding blood controls available from The Cancer Genome Atlas (TCGA) and quantified the frequency of in-frame deletions in *PDGFRA* (Materials and Methods). An in-frame deletion resulting in the loss of exons 8 and 9 (*PDGFRA*$\Delta^{8,9}$) had been previously cloned from a GBM biopsy (Kumabe *et al*, 1992). *PDGFRA*$\Delta^{8,9}$ affects the same dimerization domain as *PDGFRA*Δ^7 and has been shown to be transforming (Clarke & Dirks, 2003). A more recent TCGA study showed that *PDGFRA* mRNA lacking exons 8 and 9 was expressed in 17.8% of GBMs; however, *PDGFRA*$\Delta^{8,9}$ prevalence was not interrogated at the DNA level (Brennan *et al*, 2014). We compared the distributions of reads mapping exons 8 and 9 in *PDGFRA* between tumor samples and the blood controls in TCGA data. The tumor distribution is clearly bimodal (Fig 7A), and this second mode

corresponds to a set of samples depleted of reads mapping exons 8 and 9. By thresholding at the 10% level of the blood distribution as a control, we estimate *PDGFRA*$\Delta^{8,9}$ occurs in 16% of cases in our dataset ($n = 389$), after Benjamini–Hochberg correction for multiple hypothesis testing. This is remarkably close to the 17.8% of cases where *PDGFRA*$\Delta^{8,9}$-consistent mRNA was observed in the TCGA study ($n = 206$). Additionally, we found a second family of deletions in exon 7, occurring in 1.8% of cases we analyzed (including SF10282). The frequency of *PDGFRA* amplification was 13.6%, but we did not find a strong correlation between *PDGFRA*$\Delta^{8,9}$ and *PDGFRA* amplification. On the other hand, all of the small deletions occurred in *PDGFRA* amplified cases (Fig 7B). Both *PDGFRA*Δ^7, *PDGFRA*$\Delta^{8,9}$, and the other small in-frame deletions target immunoglobulin I-set sub-domains of the extracellular domain of PDGFRA (Fig 7C). These domains are involved in receptor dimerization (Chen *et al*, 2012).

Accumulating mutations correlates with the acquisition of an OPC signature in a proneural GBM and an invasion signature in a classical/mesenchymal case

We performed differential gene expression analysis between SF10282 and SF10345, and as expected, *EGFR* was up-regulated in SF10345 (Dataset EV8). Additionally, there was an overrepresentation of cell-adhesion molecules and genes mediating motility in SF10345's differentially expressed genes (Fig 8A). For example, *CD44*, encoding an adhesion molecule that mediates stem cell homing (Pietras *et al*, 2014), was over 14-fold enriched in SF10345 (Fig 8B). Intriguingly, transcripts coding for chemotactic factors for myeloid-derived suppressor cells (MDSC) were differentially expressed in SF10345. C3 convertase is a core component of the complement cascade, mediating inflammation, and the innate immune response. Complement pathway cytokines attract MDSC and induce their expression of reactive-oxygen species, contributing to a tumor-supportive microenvironment (Markiewski *et al*, 2008). Periostin (*POSTN*) is secreted by glioma stem cells, recruiting tumor-associated macrophages that enhance tumor growth (Zhou *et al*, 2015). Both *C3* and *POSTN* were up-regulated in SF10345 by several hundred fold (Fig 8B).

On the other hand, the neuron-differentiation pathway was significant in genes up-regulated in SF10282 (Fig 8A). Upon inspection, however, these genes were factors predominantly expressed by OPCs during development: *PDGFRA*, *NKX2-2*, *SOX10*, *SEMA5A*, *LINGO1*, *S100B*, *MAP2* (Shafit-Zagardo *et al*, 2000; Deloulme *et al*,

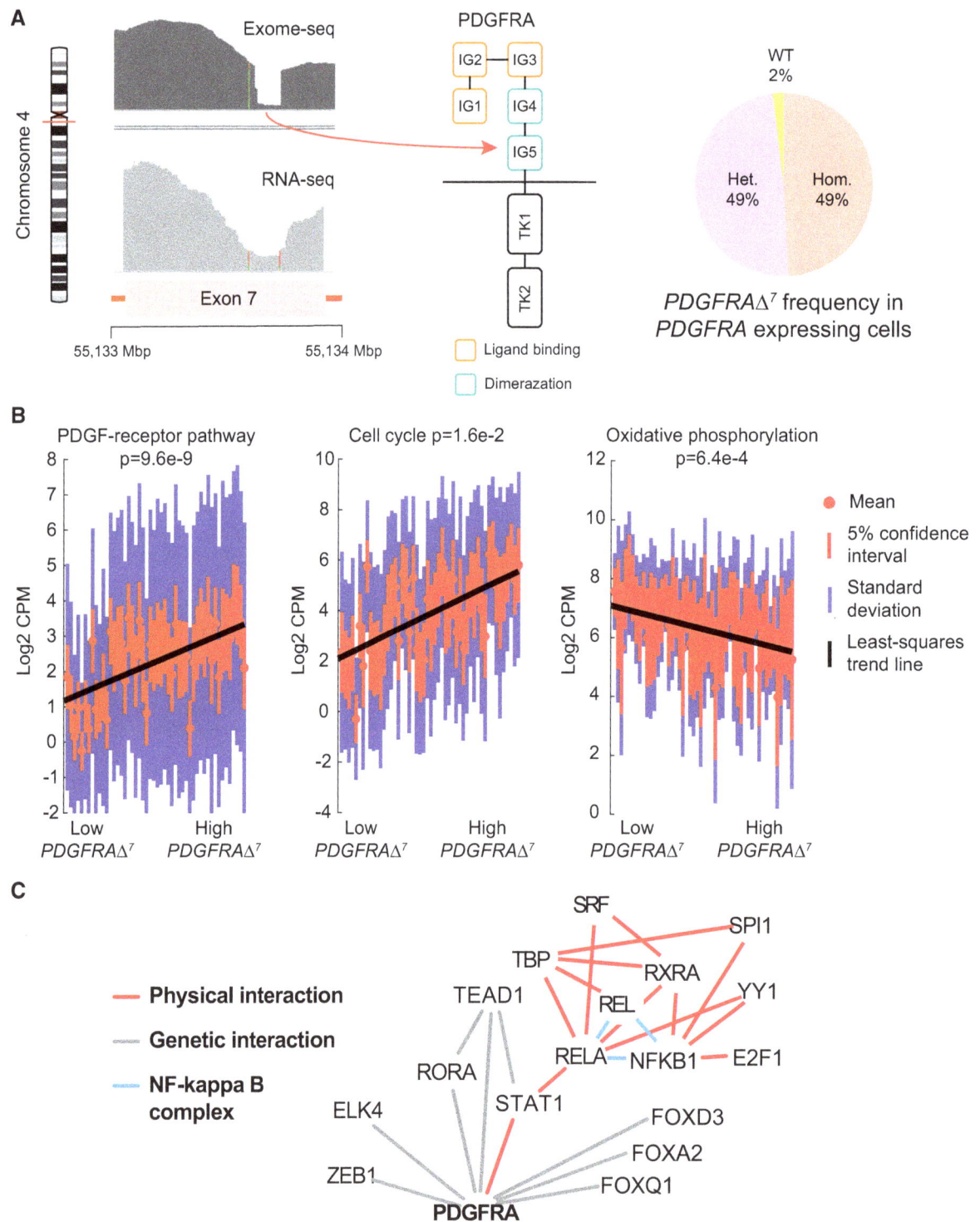

Figure 4. Dose–response analysis of a *PDGFR* mutant.

A   Coverage of exome-seq (top left) and RNA-seq reads (bottom left) in exon 7 of the *PDGFRA* gene. The deletion targets the immunoglobulin-like domain IG5 of the PDFG receptor (center). 49% of *PDGFRA* expressing cells express *PDGFRAΔ⁷* homozygously, another 49% express it heterozygously (right).

B   Enriched gene sets (WEBGESTALT, DAVID, adj. *P*-value < 0.05) correlated to *PDGFRAΔ⁷*. Distributions of in-pathway genes in individual cells, sorted from low *PDGFRAΔ⁷* to high *PDGFRAΔ⁷*.

C   An interaction network (generated via geneMANIA) of physical and genetic interactions of transcription factors, whose recognition motifs are overrepresented (OPOSSUM, z-score ≥ 10, Fisher score ≥ 7) in correlated genes. Physical interactions are interactions between the protein products, identified from proteomics experiments. Genetic interactions are changes in gene expression that occur when another gene is suppressed in a knockdown experiment.

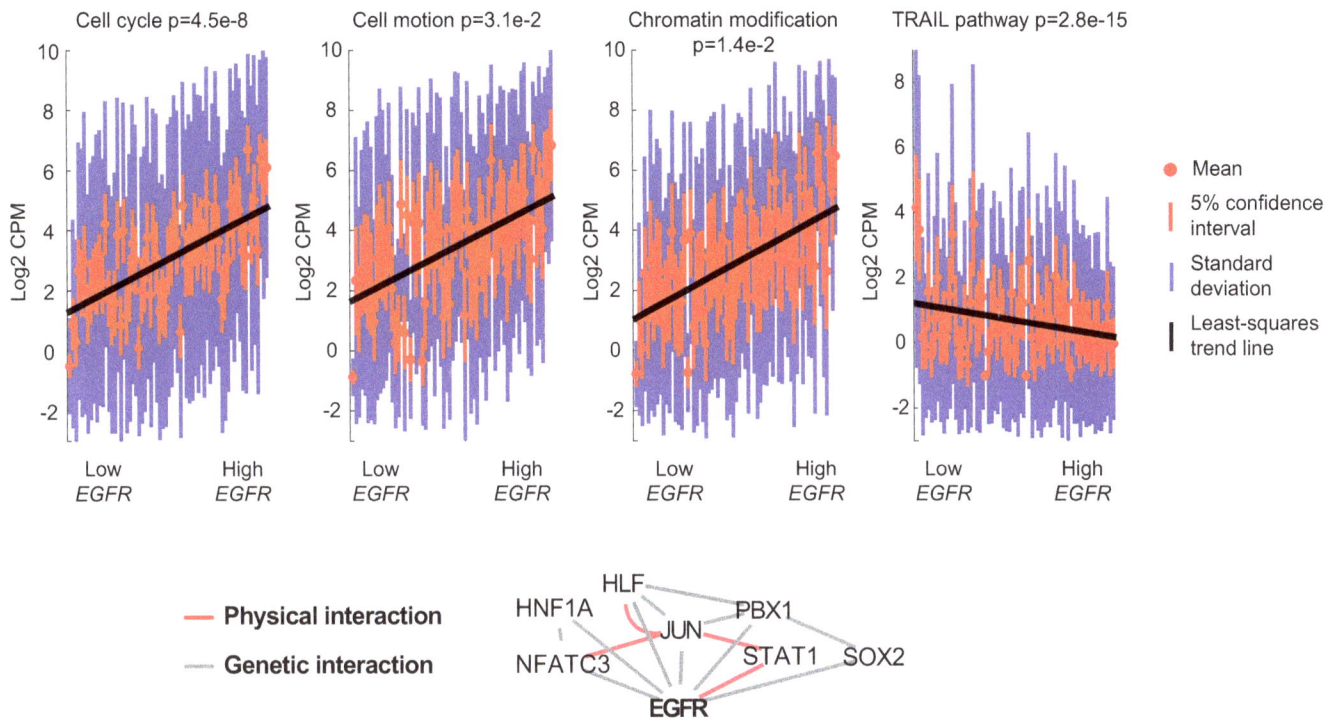

Figure 5. Dose–response analysis of an *EGFR*-amplified case.

Enriched gene sets (WEBGESTALT, DAVID, adj. *P*-value < 0.05) correlated to *EGFR*. Distributions of in-pathway genes in individual cells, sorted from low *EGFR* to high *EGFR* (top panel). An interaction network (generated via geneMANIA) of physical and genetic interactions of transcription factors, whose recognition motifs are overrepresented (OPOSSUM, *z*-score ≥ 10, Fisher score ≥ 7) in correlated genes. Physical interactions are interactions between the protein product, identified from proteomics experiments. Genetic interactions are changes in gene expression that occur when another gene is suppressed in a knockdown experiment.

2004; Goldberg, 2004; Petryniak *et al*, 2007; Jepson *et al*, 2012). Our initial inspection of SF10282 had revealed a NSC-like subpopulation (Fig 1C), which occurred early in our phylogeny of SF10282. Analysis of NSC and OPC gene expression in our phylogeny showed constitutively high expression of *PAX6*, *SOX2*, and *TNC*, but a gradual increase in the OPC genes *OLIG2*, *ASCL1*, *NKX2-2,* and *SOX10* along pseudo-time. PI3K/AKT pathway genes and genes implicated in angiogenesis also increased concomitantly (Fig 8C). By comparison, SF10345 showed a progressive up-regulation of genes encoding extracellular matrix and transmembrane proteins associated with glioma motility and invasion, such as tenacin-C, neurocan, and integrin (Cuddapah *et al*, 2014). AKT pathway genes *AKT2* and *AKT3*, which contribute to glioma invasiveness and malignancy

(Chautard *et al*, 2014), and class II myosins, required for glioma invasion and neural stem cell migration (Beadle *et al*, 2008; Ostrem *et al*, 2014), were also progressively up-regulated as one moves along the backbone of the SF10345 phylogeny (Fig 8D).

Discussion

Despite standard of care treatment, GBM has an extremely high recurrence rate, approximately 90% (Weller *et al*, 2013). There is an urgent need for combinatorial strategies to address residual disease (Prados *et al*, 2015). In particular, intra-tumor receptor heterogeneity is a confounder for tyrosine-kinase inhibitor therapy

Figure 6. *In vitro* analysis of *PDGFRAΔ7*.

A Schematic representation of the *in vitro* experiment.

B Reads mapped to exon 7 of *PDGFRA* in *PDGFRAΔ7* over-expressing, wild-type *PDGFRA* over-expressing, and GFP control cultures.

C Quantitative PCR with a probe targeted to the region deleted in *PDGFRAΔ7*. Results (mean ± SD) comparing wild-type *PDGFRA* expression between GFP control and *PDGFRAΔ7* expressing cells from SF10360. The asterisk indicates *P* < 0.05 (*t*-test).

D Top: Volcano plot of gene expression between *PDGFRAΔ7* and *PDGFRA* wild-type expressing cells from SF10281 (left) and SF10360 (right). Differentially expressed genes (adjusted *P*-value < 0.05, ANODEV test from DEGSeq2 package) are indicated in red. Bottom: Gene-ontology enrichment of genes differentially expressed between *PDGFRAΔ7* and *PDGFRA* wild type in both cell lines (right).

E Bar plots of mean gene expression (± SD) across duplicates in GFP, wild-type PDGFRA, and expressing cells from SF10281 (left) and SF10360 (right).

F WST-1 assay (*n* = 3) comparing proliferation (mean ± SD) of SF10360c cells expressing GFP, PDGFRA WT or PDGFRA delta7. The asterisk indicates *P* < 0.05 (*t*-test).

G Genetic interactions and physical interactions (via Genemania) of transcription factors (via OPOSSUM) whose motifs are enriched in the promoters of genes correlating with *PDGFRAΔ7 in vivo*. This is compared to the genetic interactions between upstream transcription factors of genes which are differentially expressed in the *PDGFRAΔ7* over-expression experiment.

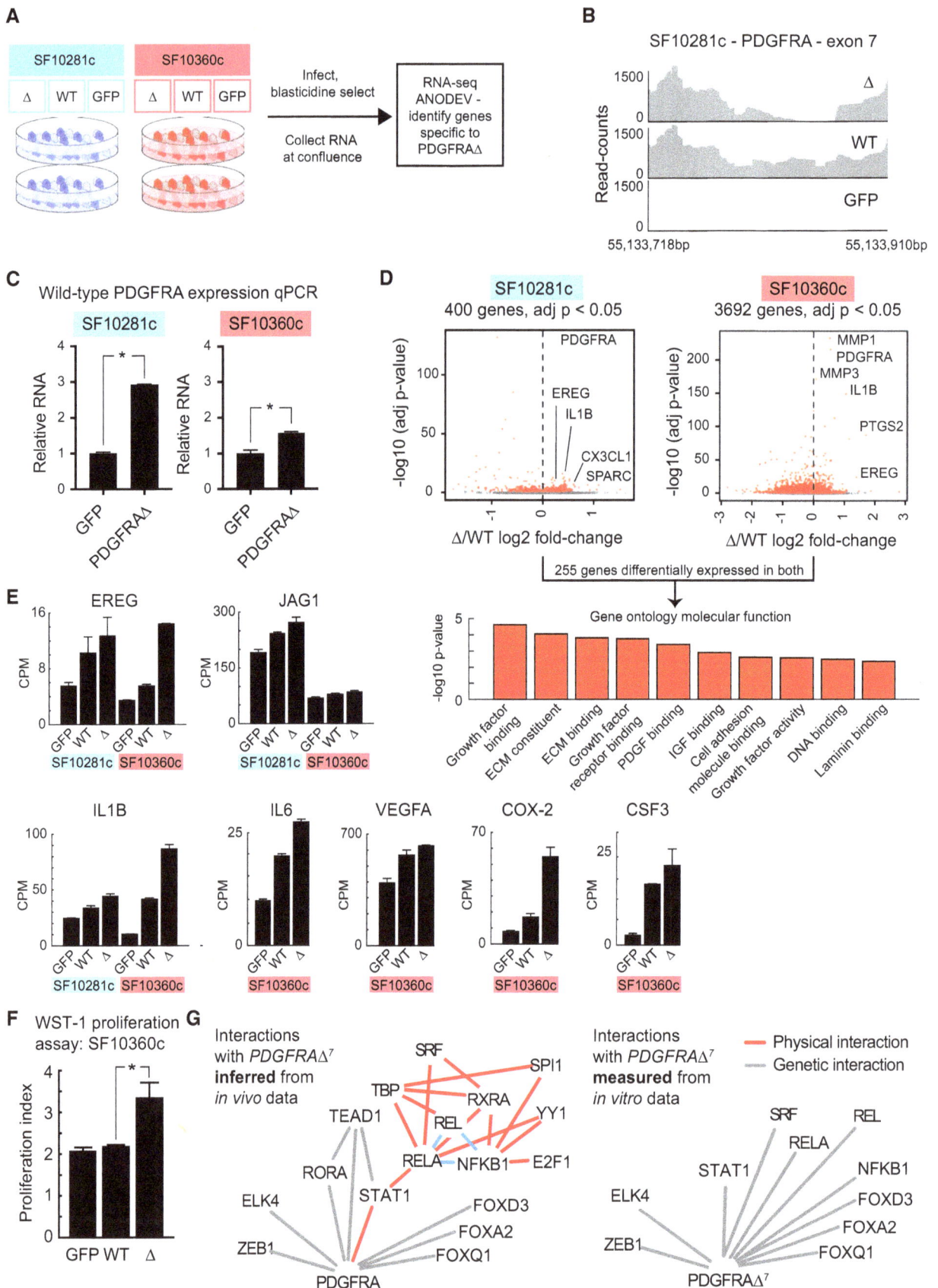

Figure 6.

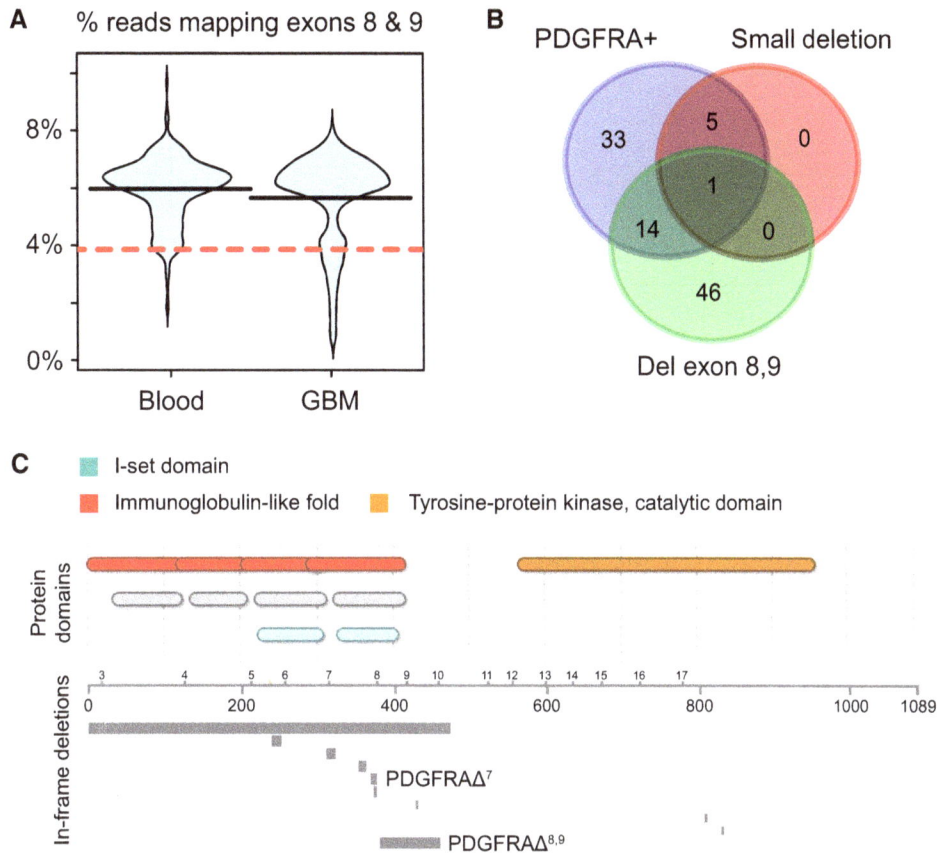

Figure 7. Analysis of TCGA data reveals a family of *PDGFRA* mutations affecting the dimerization domain.

A   The fraction of reads assigned to exons 8 and 9 from all exome-seq reads mapping to *PDGFRA*, compared between blood control and GBM samples. The 10th percentile of the blood control distribution is indicated by a red line.

B   Venn diagram of GBM cases harboring an amplification, a deletion of exons 8 and 9, or a small deletion affecting the dimerization domain of PDGFRA.

C   Visualization of deletions detected in *PDGFRA* from TCGA data. Domains are indicated by color (top), exons and protein residues by number (middle), and deleted regions by bars (bottom).

(De Witt Hamer, 2010). Recent advances in single-cell analysis have enabled high-resolution estimates of clonal heterogeneity (Francis *et al*, 2014; Patel *et al*, 2014; Meyer *et al*, 2015). In this study, we combined whole exome and single-cell mRNA sequencing to map transcriptional signatures to mutational phylogenies in EGF- and PDGF-driven GBMs. These data implicate in-frame deletions in the dimerization domain of PDGFRA as potential therapeutic targets. And, they identify a cell-type hierarchy, which occurs in early brain development, as being recapitulated during tumor evolution.

In the developing forebrain, OPCs arise from neuroepithelial stem cells in sequential waves of oligodendrocyte production (Kessaris *et al*, 2006; Menn *et al*, 2006). ASCL1 and OLIG2 are interacting transcription factors required for oligodendrogenesis and highly expressed in OPCs (Zhou & Anderson, 2002; Petryniak *et al*, 2007; Nakatani *et al*, 2013). OLIG2 drives SOX10 transcription in OPCs (Kuspert *et al*, 2011), which in turn regulates myelin expression (Stolt *et al*, 2002), critical for oligodendrocyte function. Consistent with aberrant activation of this developmental program in GBM, SF10282 progressively up-regulated the above OPC-specific genes in its model of tumor evolution. Early cells expressed *PAX6*, *SOX2*, and other markers of neural stem cells but produced daughters with a more OPC-like profile (Fig 8C). *PDGFRA* expression is characteristic of OPCs in non-malignant brain and was amplified in SF10282.

Figure 8. Differences in gene expression between an EGF- and a PDGF-driven GBM.

A   A scatterplot of log2 mean expression between SF10345 and SF10282. Adjusted *P*-values of biological process GO terms, enriched in differentially expressed genes (SCDE), with an adjusted *P*-value < 0.05 (on a −log10 scale).

B   Single-cell gene expression estimates (colored lines), their SF10345/SF10282 log2 fold-changes, and their joint posteriors for select genes. A 95% confidence interval is represented by dotted lines.

C, D   Phylogenetic trees for SF10282 and SF10345. Each leaf corresponds to a unique set of cells with the same CNV genotype. Bar plots show mean (± SD) expression of select genes across sets of cells that progressively gain CNVs.

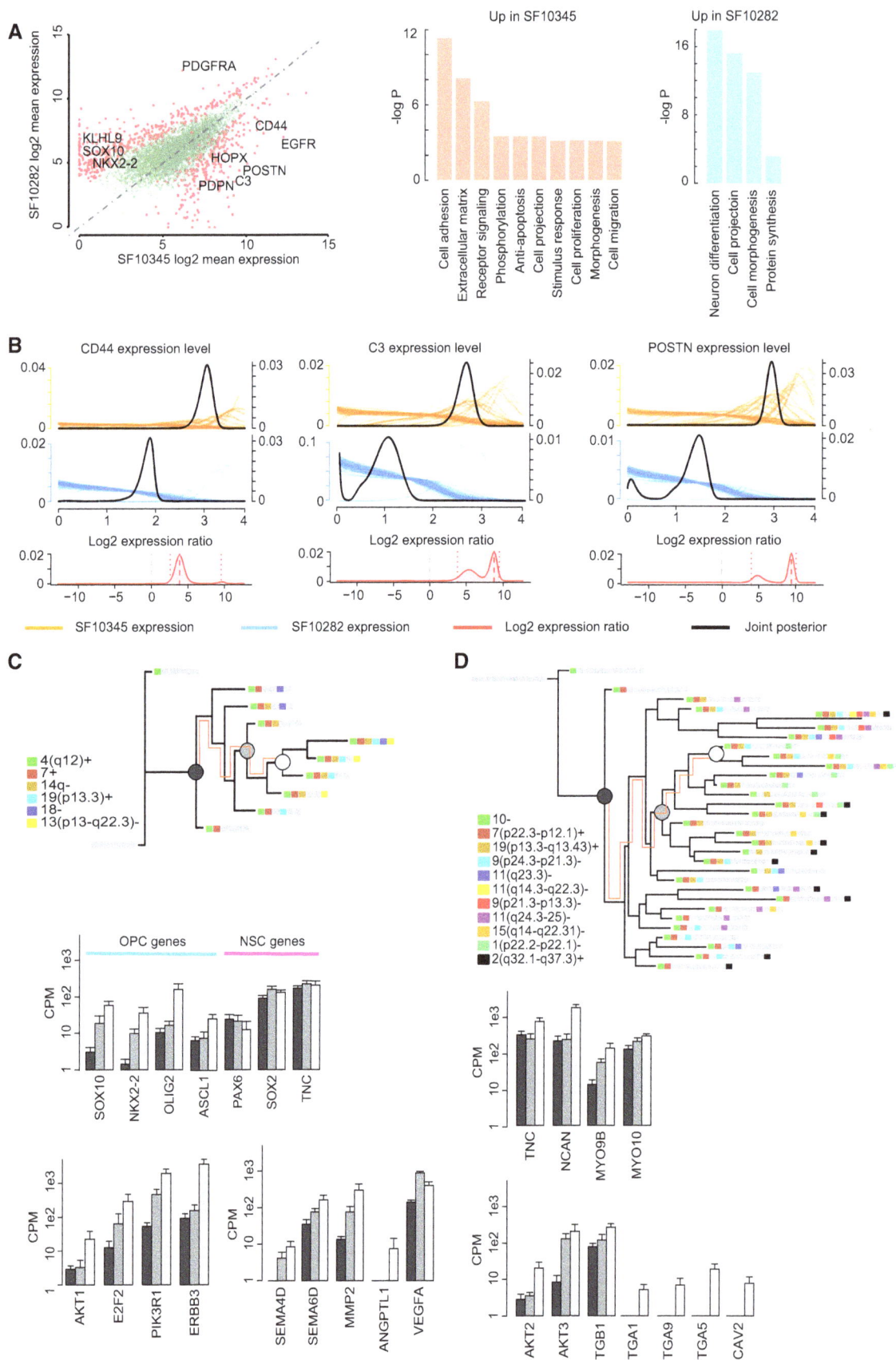

Figure 8.

However, 68% of tumor cells in SF10282 express an in-frame deletion mutant, $PDGFRA\Delta^7$. We found that $PDGFRA\Delta^7$ is part of a family of in-frame deletions in $PDGFRA$'s dimerization domain that occur in GBM. The most common of these mutations, $PDGFRA\Delta^{8,9}$, occurs with 16% frequency in the TCGA DNA sequencing data we analyzed (n = 389). Our finding is consistent with a recent TCGA paper that found $PDGFRA$ mRNA lacking exons 8 and 9 in 17.8% of their samples (n = 206; Brennan et al, 2014). $PDGFRA\Delta^{8,9}$ expression has been shown to induce constitutive signaling of the PDGF receptor and is sufficient for malignant transformation (Clarke & Dirks, 2003). In-frame deletions in $PDGFRA$'s dimerization domain have also been observed in pediatric high grade gliomas, and those studies implicate constitutive receptor signaling too (Paugh et al, 2013). Our single-cell data show that increasing $PDGFRA\Delta^7$ dosage correlates in $vivo$ with an activation of genes downstream of the PDGF-receptor, in particular, targets of rapid-acting transcription factors (e.g. STAT, NF-κB). When ectopically over-expressed, in patient-derived glioma cell lines, $PDGFRA\Delta^7$ enhances proliferation compared to wild-type $PDGFRA$ over-expression. $PDGFRA\Delta^7$ induces wild-type $PDGFRA$ expression in $vitro$, along with tyrosine-kinase receptor ligands and pro-angiogenesis genes.

The deletions we observed aggregate in the I-set domains of $PDGFRA$'s immunoglobulin-like folds. These highly conserved domains are involved with receptor dimerization and in principle can interact with a variety of different domain types; however, these domains are typically glycosylated, which protects against promiscuous receptor interaction in the absence of ligand (Barclay, 2003). All of the deletions we have observed are either on or near predicted glycosylation sites (Fig EV3).

Single-cell RNA-seq has enabled composition assessments and lineage reconstruction in the highly dynamic, highly heterogeneous context of the developing brain (Darmanis et al, 2015; Pollen et al, 2015; Diaz et al, 2016; Liu et al, 2016). Single-cell RNA-seq of tumor biopsies provide a similar snapshot of tumor evolution. By using a proven approach such as bulk exome-seq to identify mutations, single-cell RNA-seq libraries can then be triaged to assign a transcriptional signature to a mutational profile or phylogeny. By cross-referencing public atlases such as the Ivy Glioblastoma Atlas, we further related this information to tumor anatomical structure. When we cross-referenced the Ivy Glioblastoma Atlas with the chromosome 13 deletion subclone of SF10360, we identified a spatial segregation of differentially expressed genes, aggregating in the perivascular niche and the leading edge, respectively. Cells harboring the chromosome 13 deletion were also characterized by an up-regulation of pro-invasion, adhesion genes. However, whether the deletion event preceded or succeeded any physical separation of these two clones could not be inferred from these data. Targeted resections that record relative biopsy position can resolve spatial evolution more accurately (Sottoriva et al, 2013).

Materials and Methods

Sample acquisition and processing

Fresh tumor tissue was acquired from patients undergoing resection for glioblastoma. De-identified samples were obtained from the Neurosurgery Tissue Bank at the University of California San

Francisco (UCSF). Sample use was approved by the Committee on Human Research at UCSF. The experiments conformed to the principles set out in the WMA Declaration of Helsinki and the Department of Health and Human Services Belmont Report. All patients provided informed written consent. Tissues were minced in collection media (Leibovitz's L-15 medium, 4 mg/ml glucose, 100 μg/ml penicillin, 100 μg/ml streptomycin) with a scalpel. Samples were further dissociated in a mixture of papain (Worthington Biochem. Corp) and 2,000 units/ml of DNase I freshly diluted in EBSS and incubated at 37°C for 30 min. The suspension was centrifuged for 5 min at 300 g and re-suspended in PBS. Suspensions were triturated by pipetting up and down 10 times and passed through a 70-μm strainer cap (BD Falcon), followed by centrifugation for 5 min at 300 g. Pellets were re-suspended in PBS and passed through a 40-μm strainer cap (BD Falcon), followed by centrifugation for 5 min at 300 g. Dissociated, single cells were then re-suspended in GNS (Neurocult NS-A (Stem Cell Tech.), 2 mM L-glutamine, 100 U/ml penicillin, 100 μg/ml streptomycin, N2/B27 supplement (Invitrogen), sodium pyruvate).

Single-cell capture and cDNA generation was performed using the Fluidigm C1 Single-Cell Integrated Fluidic Circuit (IFC) and SMARTer Ultra Low RNA Kit. cDNA was quantified using Agilent High Sensitivity DNA Kits and diluted to 0.15–0.30 ng/μl. Dual indexing and amplification were performed using the Nextera XT DNA Library Prep Kit (Illumina) according to the Fluidigm C1 protocol. 96 single-cell RNA-seq libraries were generated from each tumor sample and were pooled for 96-plex sequencing. Amplified and pooled cDNA was purified and size selected twice using 0.9× volume of Agencourt AMPure XP beads (Beckman Coulter). Final cDNA libraries were quantified using High Sensitivity DNA Kits (Agilent) and sequenced on a HiSeq 2500 (Illumina), using the paired-end 100 base pair (bp) protocol.

Exome-sequencing and genomic mutation identification

Exome capture was done using NimbleGen SeqCap EZ Human Exome Kit v3.0 (Roche) exome capture kits on a tumor sample and a blood control sample from each patient. Sequencing was carried out with an Illumina-HiSeq 2500 machine acquiring 100-bp paired-end reads. Reads were aligned to the human genome (hg19) using BWA (Li & Durbin, 2009), whereas only uniquely matched paired reads were retained. PicardTools (http://broadinstitute.github.io/picard/) and the GATK toolkit (McKenna et al, 2010) were used for quality score re-calibration, duplicate-removal, and re-alignment around indels. The resulting BAM files were sorted by genomic coordinates. Subsequently, the percent contamination was assessed with ContEst (SF10345: 0.1%, SF10360: 0.1%, SF10282: 0.2%). OxoG metrics were calculated with PicardTools' CollectOxoGMetrics (Dataset EV9). After these control steps, single nucleotide variants (SNVs), short indels (< 50 bps), and large CNVs comprising more than three exons were detected. Somatic SNVs were inferred with MuTect (https://www.broadinstitute.org/cancer/cga/mutect) for each tumor/control pair and annotated with. SNVs with < 10% variant frequency in the tumor, with more than five variant reads in the patient-matched normal, or > 10% variant frequency in the patient-matched normal were excluded from further analysis. Small indels were detected with Pindel (Ye et al, 2009) and those with fewer than six supporting reads in the tumor, any supporting reads or

< 14 total reads in the patient-matched normal, and replacements for which the deletion and non-template inserted sequence were of the same length were excluded. All indels and SNVs were annotated for their mutational context and effect using the Annovar software package (Wang *et al*, 2010). Only protein-coding or splice-site mutations were retained for further analysis. ADTEx (Amarasinghe *et al*, 2014) was used for detection of large somatic CNVs. Only CNVs comprising more than 100 exons were retained for downstream analysis. Proximal (< 1 Mbp) somatic CNVs were merged in the output file to maximize CNV regions.

Single-cell RNA-sequencing data preprocessing, quality control, and GBM subtype classification

Reads were trimmed for quality and Nextera adapters removed with TrimGalore! (http://www.bioinformatics.babraham.ac.uk/projects/trim_galore/), and paired-end reads were mapped to the human genome (hg19) with tophat2 (Kim *et al*, 2013) using the –prefilter-multihits option and a GENCODE V19 transcriptome-guided alignment. Quantification of GENCODE genes was carried out with featureCounts (Liao *et al*, 2014). Only fragments corresponding to uniquely mapped, correctly paired reads were kept. Expression values were normalized to CPM in each cell. We filtered samples whose background fraction is significantly high, via a threshold on the (Benjamini–Hochberg corrected) q-value of a Lorenz statistic on the samples' cumulative densities, described previously (Diaz *et al*, 2016). In our tests, samples that have a small q-value have low complexity, as measured by Gini-Simpson index, and they have low coverage, as estimated by the Good–Turing statistic (Fig EV1).

To compare genotypes from the single-cell RNA-seq data across patients and to the exome-seq data, we identified 17, 21, and 15 single nucleotide variants (SNVs) in the single-cell RNA-seq that are patient specific in SF10282, SF10345, and SF10360, respectively. These SNVs are likewise detected in, and only in, their respective blood and tumor-derived exome-seq datasets. The median difference in RNA-seq to tumor exome-seq variant allele frequency (VAF) is 0.056. By comparison, the median difference in tumor exome-seq to blood exome-seq VAF is 0.044 (see Appendix Fig S3 and Appendix Table S1).

Infiltrating stromal/non-malignant cells were identified as those cells which contained none of the CNVs, none of the indels or point mutations found in the exome-seq data, and clustered away from tumor cells in a hierarchical clustering. These putative stromal cells formed two clusters: one that is comprised of cells from all samples that expresses markers of immune cells (particularly those of macrophages/microglia) and a second cluster comprised of cells from SF10282 that express oligodendrocyte markers (Fig EV2C). We classified all of the remaining cells according to the Verhaak *et al* (2010) and Sun *et al* (2014) molecular subtypes. To perform the Verhaak classification in individual cells, we fit a linear regression model using the four centroids in the original Verhaak clustering as predictors and the gene expression profile of the tumor cell to be classified as a response. We restricted expression profiles in individual cells to those genes used in the original Verhaak clustering and represented expression by standardized, log-transformed CPMs. The cell to be classified is assigned to the subtype whose corresponding centroid has the largest regression coefficient in magnitude. Sun *et al* determined their subtypes by identifying gene modules that

co-expressed with *PDGFRA* or *EGFR* in a large cohort of adult diffuse gliomas. In each cell to be classified, we averaged gene expression (measured by log-CPM) across both of these two gene modules. If either module's average was more than twofold higher than the other, then we assigned the cell to the corresponding subtype.

Single-cell CNV presence/absence calls

Based on the premise that copy-number changes are reflected in RNA-seq read counts when averaged over large, adjacent genomic regions (Patel *et al*, 2014), we examined loci that were called for somatic CNV in our ADTEx pipeline. For each CNV candidate region (CNVCR), we sum the library-size normalized read counts across genes in that region, for each cell in our tumor sample. We do the same for each cell in a non-malignant, human brain control (Darmanis *et al*, 2015). We use the distribution across cells in the control for each CNVCR, to assess the sum in a given tumor cell. We use the 5% significance level of the control distribution as our threshold for making a CNV call (Fig 2A), and control for multiple hypothesis testing using Benjamini–Hochberg correction. This results in a genotype assignment to each cell, determined by which CNVs called in the exome-seq data are present in that cell. To estimate the false discovery rate (FDR) of this classification procedure, we performed 10-fold cross-validation using the normal-brain control cells. For each patient, we randomly selected tranches of 10% as test and 90% as training data. We estimated the FDR as (# positive CNV calls)/(# total CNV calls), for each of the 10 folds. We found the FDR to be < 0.01 for all tests (Appendix Fig S2A). As a second estimator of the FDR, we classified the presence of CNVs on a dataset comprised of non-malignant, fetal-brain cells (Pollen *et al*, 2015), and estimated the FDR as above. We found these FDR estimates to all be < 0.06 (Appendix Fig S2B).

Phylogenetic trees

We measure pairwise distance between individual cells using Jaccard distance between CNV genotypes. This measures the number of shared CNV calls, as a fraction of the number of unique calls in either cell. To obtain a phylogenetic tree of tumor cells based on this distance metric, we use the Fitch–Margoliash method (Fitch & Margoliash, 1967) as implemented in the phylip R package (Revell & Chamberlain, 2014), adding a "normal" genotype with no CNVs as an out-group to root the tree. We identified 5–6 cells per sample that did not harbor the CNV with the highest frequency (chromosome 7 gain in SF10345, chromosome 4q12 gain in SF10282 and chromosome 10 loss in SF10360). Since these mutations are nearly ubiquitous, they represent founding mutations for the dominant clones sampled in our cells. This is consistent with the fact that they affect cancer driver genes *EGFR*, *PDGFRA*, and *PTEN*, respectively. These rare cells lacking founding mutations may be technical outliers, or members of a lineage under-sampled in the biopsy.

PDGFR *in vitro* overexpression

Primary dissociated tumors were plated for cell culture in DMEM-F12 with 10% FBS. Expression vectors driving wild-type *PDGFR*, *PDGFRΔ⁷*, or GFP (pLV[Exp]-EGFP/Bsd-EF1A) were generated and packaged into third generation lentivirus particles (VectorBuilder, Cyagen Biosciences). Triplicate cell cultures were infected with lentiviruses at equal titers at a MOI of ~1.0 and with 0.5 µg/ml polybrene. Two days following infection, cells were selected with 33.3 µg/ml blasticidine for 3 days. For RNA collection, cells were grown for one additional day with no drug selection and then harvested. For proliferation assays, cells were grown continually in the presence of 33.3 µg/ml blasticidine. Vector expression was confirmed using fluorescence microscopy and RNA-seq.

Bulk RNA-seq sample processing

RNA was harvested using TRIzol reagent, followed by Direct-zol MiniPrep RNA purification kits (Zymo Research) with the on-column DNase digestion step. RNA integrity was confirmed using the Agilent 2200 RNA ScreenTape (Agilent Technologies). RNA-seq libraries were generated using TruSeq Stranded mRNA kit according to manufacturer's protocol (Illumina). cDNA was validated using the Agilent 2200 DNA 1000 ScreenTape, Qubit 2.0 Fluorometer (Life Technologies), and ddPCR (Bio-Rad). Cluster generation and sequencing was performed on a HiSeq 4000, using the single-end 50 read protocol.

Reads were mapped to the human genome (hg19) with tophat2 (Kim *et al*, 2013). Quantification of GENCODE V19 genes was carried out with featureCounts (Liao *et al*, 2014) using only uniquely mapped reads. Differential expression analyses were performed via DESeq2, using the likelihood ratio test applied against the wild-type *PDGFR*, *PDGFRΔ7*, and GFP triplicate samples as a three-level factor.

Dose–response analysis

In SF10282, cells were sorted by *PDGFRAΔ7* expression, in SF10345 cells were sorted by *EGFR* expression. Genes which had a Spearman's rank correlation in the top 5% with *PDGFRAΔ7* were considered for further analysis. Pathway analysis was done using Pathway Commons (Cerami *et al*, 2011) via WEBGESTALT (Wang *et al*, 2013), and DAVID (Huang *et al*, 2009), transcription factor motif enrichment analysis was done via OPOSSOM (Sui Ho *et al*, 2007). Network interactions were computed via GeneMANIA (Montojo *et al*, 2010).

In vitro proliferation and wild-type *PDGFRA* quantification assays

For the WST-1 proliferation assay, 1×10^4 cells were cultured in a 96-well plate for 24 h in 100 µl of complete media. Then, 10 µl of WST-1 reagent (Roche) was added to each well. Cells were incubated at 37°C, 5% CO_2 for 4 h, and placed on a shaker for 1 min. The plates were then read on a microplate reader with a wavelength of 420 nm and a reference at 620 nm. For the TaqMan gene expression assay, cDNA was synthesized from 50 ng of glioma cell line-derived RNA with qScript™ XLT cDNA SuperMix (Quanta Biosciences, Gaithersburg, MD). 2 µl of converted cDNA was then used in the qRT–PCR reaction with PerfeCta® FastMix® II (Quanta Biosciences), according to the manufacturer's protocol. A custom TaqMan assay specific to wild-type, but not mutant, *PDGFRA* was designed by Life Technologies. The fluorescent probe was targeted to the region deleted in *PDGFRAΔ7*, but present in wild-type PDGFRA. Real-time

PCR and data analysis were performed using the StepOnePlus Real-Time PCR system (Life Technologies, Carlsbad, CA). GAPDH (Assay ID: Hs02758991_g1) expression was used as the housekeeping gene and relative expression was determined using the $2^{ΔΔC_T}$ method.

Differential expression and time-series analysis

To identify genes differentially expressed between SF10345 and SF10282 and assess inter-cellular heterogeneity, we used the scde R package (Kharchenko *et al*, 2014). We used DESeq to compare chromosome 13 loss to wild-type cells in SF10360 and treated each cell as a replicate. Genes that were expressed in more than 80% of cells at < 1 CPM were filtered prior to each analysis. Transcription factor motif enrichment analysis was done via OPOSSOM (Sui Ho *et al*, 2007). JEPETTO (Winterhalter *et al*, 2014), run via Cytoscape (Shannon *et al*, 2003), was used to compute pathways having significant overlap with genes up-regulated upon chromosome 13 loss, and their protein interactions. For the time-series analysis, we chose paths in our phylogenetic trees corresponding to the most frequent mutation at each level. Our goal was to identify an ordering of the dominant clones in the sample. But, in principle, expression along an arbitrary path can be measured. We grouped cells with identical copy-number profiles into three intermediate points along each branch: early, mid, and late. We subjected monotonically increasing genes to gene-ontology analysis via DAVID (Huang *et al*, 2009).

TCGA data analysis

Alignments for all GBM exome-seq with available paired blood controls from TCGA (http://cancergenome.nih.gov/) were retrieved from CGHub (https://cghub.ucsc.edu/) in BAM format. Reads in FASTQ format were extracted with bedtools from these alignments (Quinlan & Hall, 2010). We detected the quality encoding with a custom perl script and clipped adapters/low quality based with Trimmomatic (Bolger *et al*, 2014). Next, we mapped reads to the human genome (hg19) with HISAT2 (Kim *et al*, 2015). Small Indels were detected with pindel (Ye *et al*, 2009) for each tumor/control pair. If multiple sequencing libraries existed for one patient, the most recently published library was used. For the detection of loss of exons 8/9, we calculated the fraction of reads mapping to these exons from all reads aligned to the PDGFRA in each GBM and each blood sample. We used the distribution of all blood controls to assign significance values to each GBM sample. *P*-values were corrected for multiple testing with Benjamini–Hochberg (Benjamini & Hochberg, 1995) and an adjusted *P*-value of < 0.25 was considered significant, which is in agreement with the significance threshold for CNVs used by the TCGA consortium (Brennan *et al*, 2014).

Acknowledgements

We would like to thank Joanna Phillips and Anny Shai of the UCSF Neurosurgery Tissue Core, who facilitated tissue acquisition, and Pamela Paris (UCSF) for consultation on genomic arrays. This work has been supported by a Shurl and Kay Curci Foundation Research Grant, a UCSF Brain Tumor SPORE Career Development Award (P50-CA097257-13:7017), and a gift from the Dabbiere Family to A.D., a Damon Runyon Cancer Research Foundation postdoctoral fellowship (DRG-2166-13) to A.P., NIH award 1R01NS091544-01A1, VA award 5I01 BX000252, and gifts from the Hana Jabsheh Initiative and the

Dabbiere Family to D.A.L., and by NIH awards U01 MH105989 and R01NS075998 to A.R.K.

Author contributions

AD conceived of and designed the study. SM developed the algorithms for inferring tumor evolution. ED, GK, SJL, MM, TJN, and AAP performed the experiments under the supervision of AD, ARK, and DAL. MA provided the surgical specimens via the UCSF Tissue Core. AD and SM performed the bioinformatics analysis of single-cell and exome-seq data. AD and SJL performed the bioinformatics analysis of the *in vitro* RNA-seq data. AD and SM wrote the manuscript with input from all authors. All authors read and approved the final manuscript.

References

Afonso-Grunz F, Müller S (2015) Principles of miRNA-mRNA interactions: beyond sequence complementarity. *Cell Mol Life Sci* 72: 3127–3141

Amarasinghe KC, Li J, Hunter SM, Ryland GL, Cowin PA, Campbell IG, Halgamuge SK (2014) Inferring copy number and genotype in tumour exome data. *BMC Genom* 15: 732

Aqeilan RI, Calin GA, Croce CM (2010) miR-15a and miR-16-1 in cancer: discovery, function and future perspectives. *Cell Death Differ* 17: 215–220

Barclay AN (2003) Membrane proteins with immunoglobulin-like domains - a master superfamily of interaction molecules. *Semin Immunol* 15: 215–223

Beadle C, Assanah MC, Monzo P, Vallee R, Rosenfeld SS, Canoll P (2008) The role of myosin II in glioma invasion of the brain. *Mol Biol Cell* 19: 3357–3368

Benjamini Y, Hochberg Y, Benjamini Y, Hochberg Y (1995) Controlling the false discovery rate: a practical and powerful approach to multiple testing. *J R Stat Soc B* 57: 289–300

Bolger AM, Lohse M, Usadel B (2014) Trimmomatic: a flexible trimmer for Illumina sequence data. *Bioinformatics* 30: 2114–2120

Bonci D, Coppola V, Musumeci M, Addario A, Giuffrida R, Memeo L, D'Urso L, Pagliuca A, Biffoni M, Labbaye C, Bartucci M, Muto G, Peschle C, De Maria R (2008) The miR-15a–miR-16-1 cluster controls prostate cancer by targeting multiple oncogenic activities. *Nat Med* 14: 1271–1277

Brennan CW, Verhaak RGW, McKenna A, Campos B, Noushmehr H, Salama SR, Zheng S, Chakravarty D, Sanborn Z, Berman SH, Beroukhim R, Bernard B, Chin L (2014) The somatic genomic landscape of glioblastoma. *Cell* 155: 462–477

Cao Y, Slaney CY, Bidwell BN, Parker BS, Johnstone CN, Rautela J, Eckhardt BL, Anderson RL (2014) BMP4 inhibits breast cancer metastasis by blocking myeloid-derived suppressor cell activity. *Cancer Res* 74: 5091–5102

Cerami EG, Gross BE, Demir E, Rodchenkov I, Babur O, Anwar N, Schultz N, Bader GD, Sander C (2011) Pathway commons, a web resource for biological pathway data. *Nucleic Acids Res* 39: D685–D690

Chautard E, Ouédraogo ZG, Biau J, Verrelle P (2014) Role of Akt in human malignant glioma: from oncogenesis to tumor aggressiveness. *J Neurooncol* 117: 205–215

Chen P-HH, Chen X, He X (2012) Platelet-derived growth factors and their receptors: structural and functional perspectives. *Biochim Biophys Acta* 1834: 2176–2186

Chen JCC, Alvarez MJJ, Talos F, Dhruv H, Rieckhof GEE, Iyer A, Diefes KLL, Aldape K, Berens M, Shen MMM, Califano A (2014) Identification of causal genetic drivers of human disease through systems-level analysis of regulatory networks. *Cell* 159: 402–414

Chou C-H, Chang N-W, Shrestha S, Hsu S-D, Lin Y-L, Lee W-H, Yang C-D, Hong H-C, Wei T-Y, Tu S-J, Tsai T-R, Ho S-Y, Jian T-Y, Wu H-Y, Chen P-R,

Lin N-C, Huang H-T, Yang T-L, Pai C-Y, Tai C-S *et al* (2015) miRTarBase 2016: updates to the experimentally validated miRNA-target interactions database. *Nucleic Acids Res* 44: D239–D247

Clarke ID, Dirks PB (2003) A human brain tumor-derived PDGFR-α deletion mutant is transforming. *Oncogene* 22: 722–733

Cuddapah VA, Robel S, Watkins S, Sontheimer H (2014) A neurocentric perspective on glioma invasion. *Nat Rev Neurosci* 15: 455–465

Darmanis S, Sloan SA, Zhang Y, Enge M, Caneda C, Shuer LM, Hayden Gephart MG, Barres BA, Quake SR (2015) A survey of human brain transcriptome diversity at the single cell level. *Proc Natl Acad Sci USA* 112: 7285–7290

De Witt Hamer PC (2010) Small molecule kinase inhibitors in glioblastoma: a systematic review of clinical studies. *Neuro Oncol* 12: 304–316

Deloulme JC, Raponi E, Gentil BJ, Bertacchi N, Marks A, Labourdette G, Baudier J (2004) Nuclear expression of S100B in oligodendrocyte progenitor cells correlates with differentiation toward the oligodendroglial lineage and modulates oligodendrocytes maturation. *Mol Cell Neurosci* 27: 453–465

Diaz A, Liu SJ, Sandoval C, Pollen A, Nowakowski TJ, Lim DA, Kriegstein A (2016) SCell: integrated analysis of single-cell RNA-seq data. *Bioinformatics* 32: 2219–2220

Fitch WM, Margoliash E (1967) Construction of phylogenetic trees. *Science* 155: 279–284

Francis JM, Zhang CZ, Maire CL, Jung J, Manzo VE, Adalsteinsson VA, Homer H, Haidar S, Blumenstiel B, Pedamallu CS, Ligon AH, Love JC, Meyerson M, Ligon KL (2014) EGFR variant heterogeneity in glioblastoma resolved through single-nucleus sequencing. *Cancer Discov* 4: 956–971

Fujita M, Kohanbash G, Fellows-Mayle W, Hamilton RL, Komohara Y, Decker SA, Ohlfest JR, Okada H (2011) COX-2 blockade suppresses gliomagenesis by inhibiting myeloid-derived suppressor cells. *Cancer Res* 71: 2664–2674

Garvin T, Aboukhalil R, Kendall J, Baslan T, Atwal GS, Hicks J, Wigler M, Schatz M (2014) Interactive analysis and quality assessment of single-cell copy-number variations. *bioRxiv* 12: 11346

Goldberg JL (2004) An oligodendrocyte lineage-specific semaphorin, sema5A, inhibits axon growth by retinal ganglion cells. *J Neurosci* 24: 4989–4999

Huang DW, Sherman BT, Lempicki RA (2009) Systematic and integrative analysis of large gene lists using DAVID bioinformatics resources. *Nat Protoc* 4: 44–57

Jepson S, Vought B, Gross CH, Gan L, Austen D, Frantz JD, Zwahlen J, Lowe D, Markland W, Krauss R (2012) LINGO-1, a transmembrane signaling protein, inhibits oligodendrocyte differentiation and myelination through intercellular self-interactions. *J Biol Chem* 287: 22184–22195

Johnson BE, Mazor T, Hong C, Barnes M, Aihara K, McLean CY, Fouse SD, Yamamoto S, Ueda H, Tatsuno K, Asthana S, Jalbert LE, Nelson SJ, Bollen AW, Gustafson WC, Charron E, Weiss WA, Smirnov IV, Song JS, Olshen AB *et al* (2014) Mutational analysis reveals the origin and therapy-driven evolution of recurrent glioma. *Science* 343: 189–193

Kessaris N, Fogarty M, Iannarelli P, Grist M, Wegner M, Richardson WD (2006) Competing waves of oligodendrocytes in the forebrain and postnatal elimination of an embryonic lineage. *Nat Neurosci* 9: 173–179

Kharchenko PV, Silberstein L, Scadden DT (2014) Bayesian approach to single-cell differential expression analysis. *Nat Methods* 11: 740–742

Kim D, Pertea G, Trapnell C, Pimentel H, Kelley R, Salzberg SL (2013) TopHat2: accurate alignment of transcriptomes in the presence of insertions, deletions and gene fusions. *Genome Biol* 14: R36

Kim D, Langmead B, Salzberg SL (2015) HISAT: a fast spliced aligner with low memory requirements. *Nat Methods* 12: 357–360

Kumabe T, Sohma Y, Kayama T, Yoshimoto T, Yamamoto T (1992) Amplification of alpha-platelet-derived growth factor receptor gene lacking an exon coding for a portion of the extracellular region in a primary brain tumor of glial origin. *Oncogene* 7: 627–633

Kuspert M, Hammer A, Bosl MR, Wegner M (2011) Olig2 regulates Sox10 expression in oligodendrocyte precursors through an evolutionary conserved distal enhancer. *Nucleic Acids Res* 39: 1280–1293

Lechner MG, Liebertz DJ, Epstein AL (2010) Characterization of cytokine-induced myeloid-derived suppressor cells from normal human peripheral blood mononuclear cells. *J Immunol* 185: 2273–2284

Li H, Durbin R (2009) Fast and accurate short read alignment with Burrows-Wheeler transform. *Bioinformatics* 25: 1754–1760

Liao Y, Smyth GK, Shi W (2014) FeatureCounts: an efficient general purpose program for assigning sequence reads to genomic features. *Bioinformatics* 30: 923–930

Liu SJ, Nowakowski TJ, Pollen AA, Lui JH, Horlbeck MA, Attenello FJ, He D, Weissman JS, Kriegstein AR, Diaz AA, Lim DA (2016) Single-cell analysis of long non-coding RNAs in the developing human neocortex. *Genome Biol* 17: 67

Markiewski MM, DeAngelis RA, Benencia F, Ricklin-Lichtsteiner SK, Koutoulaki A, Gerard C, Coukos G, Lambris JD (2008) Modulation of the antitumor immune response by complement. *Nat Immunol* 9: 1225–1235

Mazor T, Pankov A, Brett E, Chang SM, Jun S, Costello JF (2015) DNA methylation and somatic mutations converge on the cell cycle and define similar evolutionary histories in brain tumors article DNA methylation and somatic mutations converge on the cell cycle and define similar evolutionary histories in brain. *Cancer Cell* 28: 307–317

McKenna A, Hanna M, Banks E, Sivachenko A, Cibulskis K, Kernytsky A, Garimella K, Altshuler D, Gabriel S, Daly M, DePristo MA (2010) The genome analysis toolkit: a MapReduce framework for analyzing next-generation DNA sequencing data. *Genome Res* 20: 1297–1303

Menn B, Garcia-Verdugo JM, Yaschine C, Gonzalez-Perez O, Rowitch D, Alvarez-Buylla A (2006) Origin of oligodendrocytes in the subventricular zone of the adult brain. *J Neurosci* 26: 7907–7918

Meyer M, Reimand J, Lan X, Head R, Zhu X, Kushida M, Bayani J, Pressey JC, Lionel AC, Clarke ID, Cusimano M, Squire JA, Scherer SW, Bernstein M, Woodin MA, Bader GD, Dirks PB (2015) Single cell-derived clonal analysis of human glioblastoma links functional and genomic heterogeneity. *Proc Natl Acad Sci USA* 112: 851–856

Montojo J, Zuberi K, Rodriguez H, Kazi F, Wright G, Donaldson SL, Morris Q, Bader GD (2010) GeneMANIA Cytoscape plugin: fast gene function predictions on the desktop. *Bioinformatics* 26: 2927–2928

Nakatani H, Martin E, Hassani H, Clavairoly A, Maire CL, Viadieu A, Kerninon C, Delmasure A, Frah M, Weber M, Nakafuku M, Zalc B, Thomas J-L, Guillemot F, Nait-Oumesmar B, Parras C (2013) Ascl1/Mash1 promotes brain oligodendrogenesis during myelination and remyelination. *J Neurosci* 33: 9752–9768

Nath D, Slocombe PM, Webster A, Stephens PE, Docherty a, Murphy G (2000) Meltrin gamma(ADAM-9) mediates cellular adhesion through alpha(6)beta(1)integrin, leading to a marked induction of fibroblast cell motility. *J Cell Sci* 113(Pt 1): 2319–2328

Navin N, Kendall J, Troge J, Andrews P, Rodgers L, McIndoo J, Cook K, Stepansky A, Levy D, Esposito D, Muthuswamy L, Krasnitz A, McCombie WR, Hicks J, Wigler M (2011) Tumour evolution inferred by single-cell sequencing. *Nature* 472: 90–94

Ostrem BEL, Lui JH, Gertz CC, Kriegstein AR (2014) Control of outer radial glial stem cell mitosis in the human brain. *Cell Rep* 8: 656–664

Ota T, Suto S, Katayama H, Han Z, Suzuki F, Maeda M, Tanino M, Terada Y, Tatsuka M (2002) Increased mitotic phosphorylation of histone H3

number instability. *Cancer Res* 62: 5168–5177

Patel AP, Tirosh I, Trombetta JJ, Shalek AK, Gillespie SM, Wakimoto H, Cahill DP, Nahed BV, Curry WT, Martuza RL, Louis DN, Rozenblatt-Rosen O, Suvà ML, Regev A, Bernstein BE (2014) Single-cell RNA-seq highlights intratumoral heterogeneity in primary glioblastoma. *Science* 344: 1396–1401

Paugh BS, Zhu X, Qu C, Endersby R, Diaz AK, Zhang J, Bax DA, Carvalho D, Reis RM, Onar-Thomas A, Broniscer A, Wetmore C, Zhang J, Jones C, Ellison DW, Baker SJ (2013) Novel oncogenic PDGFRA mutations in pediatric high-grade gliomas. *Cancer Res* 73: 6219–6229

Pekarsky Y, Croce CM (2015) Role of miR-15/16 in CLL. *Cell Death Differ* 22: 6–11

Petryniak MA, Potter GB, Rowitch DH, Rubenstein JLR (2007) Dlx1 and Dlx2 control neuronal versus oligodendroglial cell fate acquisition in the developing forebrain. *Neuron* 55: 417–433

Pietras A, Katz AM, Ekström EJ, Wee B, Halliday JJ, Pitter KL, Werbeck JL, Amankulor NM, Huse JT, Holland EC (2014) Osteopontin-CD44 signaling in the glioma perivascular niche enhances cancer stem cell phenotypes and promotes aggressive tumor growth. *Cell Stem Cell* 14: 357–369

Pollen AA, Nowakowski TJ, Chen J, Retallack H, Sandoval-Espinosa C, Nicholas CR, Shuga J, Liu SJ, Oldham MC, Diaz A, Lim DA, Leyrat AA, West JA, Kriegstein AR (2015) Molecular identity of human outer radial glia during cortical development. *Cell* 163: 55–67

Prados MD, Byron SA, Tran NL, Phillips JJ, Molinaro AM, Ligon KL, Wen PY, Kuhn JG, Mellinghoff IK, de Groot JF, Colman H, Cloughesy TF, Chang SM, Ryken TC, Tembe WD, Kiefer JA, Berens ME, Craig DW, Carpten JD, Trent JM (2015) Toward precision medicine in glioblastoma: the promise and the challenges. *Neuro Oncol* 17: 1051–1063

Quinlan AR, Hall IM (2010) BEDTools: a flexible suite of utilities for comparing genomic features. *Bioinformatics* 26: 841–842

Revell LJ, Chamberlain SA (2014) Rphylip: an R interface for PHYLIP. *Methods Ecol Evol* 5: 976–981

Reyes SB, Narayanan AS, Lee HS, Tchaicha JH, Aldape KD, Lang FF, Tolias KF, McCarty JH (2013) avb8 integrin interacts with RhoGDI1 to regulate Rac1 and Cdc42 activation and drive glioblastoma cell invasion. *Mol Biol Cell* 24: 474–482

Sarkar S, Zemp FJ, Senger D, Robbins SM, Yong VW (2015) ADAM-9 is a novel mediator of tenascin-C-stimulated invasiveness of brain tumor-initiating cells. *Neuro Oncol* 17: 1095–1105

Shafit-Zagardo B, Davies P, Rockwood J, Kress Y, Lee SC (2000) Novel microtubule-associated protein-2 isoform is expressed early in human oligodendrocyte maturation. *Glia* 29: 233–245

Shannon P, Markiel A, Ozier O, Baliga NS, Wang JT, Ramage D, Amin N, Schwikowski B, Ideker T (2003) Cytoscape: a software environment for integrated models of biomolecular interaction networks. *Genome Res* 13: 2498–2504

Sloan KE (2005) CD155/PVR enhances glioma cell dispersal by regulating adhesion signaling and focal adhesion dynamics. *Cancer Res* 65: 10930–10937

Sottoriva A, Spiteri I, Piccirillo SGM, Touloumis A, Collins VP, Marioni JC, Curtis C, Watts C, Tavaré S (2013) Intratumor heterogeneity in human glioblastoma reflects cancer evolutionary dynamics. *Proc Natl Acad Sci USA* 110: 4009–4014

Stolt CC, Rehberg S, Ader M, Lommes P, Riethmacher D, Schachner M, Bartsch U, Wegner M (2002) Terminal differentiation of myelin-forming oligodendrocytes depends on the transcription factor Sox10. *Genes Dev* 16: 165–170

Stommel JM, Kimmelman AC, Ying H, Nabioullin R, Ponugoti AH, Wiedemeyer R, Stegh AH, Bradner JE, Ligon KL, Brennan C, Chin L, DePinho R (2007) Coactivation of receptor tyrosine kinases affects the response of tumor cells to targeted therapies. *Science* 318: 287–290

Sudmant PH, Rausch T, Gardner EJ, Handsaker RE, Abyzov A, Huddleston J, Zhang Y, Ye K, Jun G, Hsi-Yang Fritz M, Konkel MK, Malhotra A, Stütz AM, Shi X, Paolo Casale F, Chen J, Hormozdiari F, Dayama G, Chen K, Malig M *et al* (2015) An integrated map of structural variation in 2,504 human genomes. *Nature* 526: 75–81

Sui Ho SJ, Fulton DL, Arenillas DJ, Kwon AT, Wasserman WW (2007) oPOSSUM: integrated tools for analysis of regulatory motif over-representation. *Nucleic Acids Res* 35: 245–252

Sun Y, Zhang W, Chen D, Lv Y, Zheng J, Lilljebjörn H, Ran L, Bao Z, Soneson C, Sjögren HO, Salford LG, Ji J, French PJ, Fioretos T, Jiang T, Fan X (2014) A glioma classification scheme based on coexpression modules of EGFR and PDGFRA. *Proc Natl Acad Sci USA* 111: 3538–3543

Suvà ML, Rheinbay E, Gillespie SM, Patel AP, Wakimoto H, Rabkin SD, Riggi N, Chi AS, Cahill DP, Nahed BV, Curry WT, Martuza RL, Rivera MN, Rossetti N, Kasif S, Beik S, Kadri S, Tirosh I, Wortman I, Shalek AK *et al* (2014) Reconstructing and reprogramming the tumor-propagating potential of glioblastoma stem-like cells. *Cell* 157: 580–594

Szerlip NJ, Pedraza A, Chakravarty D, Azim M, McGuire J, Fang Y, Ozawa T, Holland EC, Huse JT, Jhanwar S, Leversha MA, Mikkelsen T, Brennan CW (2012) Intratumoral heterogeneity of receptor tyrosine kinases EGFR and PDGFRA amplification in glioblastoma defines subpopulations with distinct growth factor response. *Proc Natl Acad Sci USA* 109: 3041–3046

Tatsuka M, Katayama H, Ota T, Tanaka T, Odashima S, Suzuki F, Terada Y (1998) Multinuclearity and increased ploidy caused by overexpression of the Aurora- and Ipl1-like midbody-associated protein mitotic kinase in human cancer cells. *Cancer Res* 58: 4811–4816

Tenan M, Aurrand-Lions M, Widmer V, Alimenti A, Burkhardt K, Lazeyras F, Belkouch M-C, Hammel P, Walker PR, Duchosal MA, Imhof BA, Dietrich P-Y (2010) Cooperative expression of junctional adhesion molecule-C and -B supports growth and invasion of glioma. *Glia* 58: 524–537

Venkatesh HS, Johung TB, Caretti V, Noll A, Tang Y, Nagaraja S, Gibson EM, Mount CW, Polepalli J, Mitra SS, Woo PJ, Malenka RC, Vogel H, Bredel M, Mallick P, Monje M (2015) Neuronal activity promotes glioma growth through neuroligin-3 secretion. *Cell* 161: 803–816

Verhaak RGW, Hoadley KA, Purdom E, Wang V, Qi Y, Wilkerson MD, Miller CR, Ding L, Golub T, Mesirov JP, Alexe G, Lawrence M, O'Kelly M, Tamayo P, Weir BA, Gabriel S, Winckler W, Gupta S, Jakkula L, Feiler HS *et al* (2010) Integrated genomic analysis identifies clinically relevant subtypes of glioblastoma characterized by abnormalities in PDGFRA, IDH1, EGFR, and NF1. *Cancer Cell* 17: 98–110

Wang K, Li M, Hakonarson H (2010) ANNOVAR: functional annotation of genetic variants from high-throughput sequencing data. *Nucleic Acids Res* 38: 1–7

Wang J, Duncan D, Shi Z, Zhang B (2013) WEB-based GEne SeT AnaLysis Toolkit (WebGestalt): update 2013. *Nucleic Acids Res* 41: W77–W83

Weller M, Cloughesy T, Perry JR, Wick W (2013) Standards of care for treatment of recurrent glioblastoma–are we there yet? *Neuro Oncol* 15: 4–27

Winterhalter C, Widera P, Krasnogor N (2014) JEPETTO: a cytoscape plugin for gene set enrichment and topological analysis based on interaction networks. *Bioinformatics* 30: 1029–1030

Ye K, Schulz MH, Long Q, Apweiler R, Ning Z (2009) Pindel: a pattern growth approach to detect break points of large deletions and medium sized insertions from paired-end short reads. *Bioinformatics* 25: 2865–2871

Yoon C-H, Kim M-J, Kim R-K, Lim E-J, Choi K-S, An S, Hwang S-G, Kang S-G, Suh Y, Park M-J, Lee S-J (2012) c-Jun N-terminal kinase has a pivotal role in the maintenance of self-renewal and tumorigenicity in glioma stem-like cells. *Oncogene* 31: 4655–4666

Zhou Q, Anderson DJ (2002) The bHLH transcription factors OLIG2 and OLIG1 couple neuronal and glial subtype specification. *Cell* 109: 61–73

Zhou W, Ke SQ, Huang Z, Flavahan W, Fang X, Paul J, Wu L, Sloan AE, McLendon RE, Li X, Rich JN, Bao S (2015) Periostin secreted by glioblastoma stem cells recruits M2 tumour-associated macrophages and promotes malignant growth. *Nat Cell Biol* 17: 170–182

Permissions

List of Contributors

Nicholas A Graham
Crump Institute for Molecular Imaging, David Geffen School of Medicine, University of California, Los Angeles, CA, USA
Department of Molecular & Medical Pharmacology, David Geffen School of Medicine, University of California, Los Angeles, CA, USA
Mork Family Department of Chemical Engineering and Materials Science, University of Southern California, Los Angeles, CA, USA

Rong Qiao
Department of Molecular & Medical Pharmacology, David Geffen School of Medicine, University of California, Los Angeles, CA, USA

Aspram Minasyan, Anastasia Lomova, Ashley Cass, Nikolas G Balanis, Michael Friedman, Shawna Chan, Sophie Zhao, Adrian Delgado, James Go, Lillie Beck, Johanna ten Hoeve and Nicolaos Palaskas
Crump Institute for Molecular Imaging, David Geffen School of Medicine, University of California, Los Angeles, CA, USA
Department of Molecular & Medical Pharmacology, David Geffen School of Medicine, University of California, Los Angeles, CA, USA

Christian Hurtz and Carina Ng
Department of Laboratory Medicine, University of California, San Francisco, CA, USA

Hong Wu
Department of Molecular & Medical Pharmacology, David Geffen School of Medicine, University of California, Los Angeles, CA, USA
Jonsson Comprehensive Cancer Center, David Geffen School of Medicine, University of California, Los Angeles, CA, USA
School of Life Sciences & Peking-Tsinghua Center for Life Sciences, Peking University, Beijing, China

Markus Müschen
Department of Laboratory Medicine, University of California, San Francisco, CA, USA
Department of Haematology, University of Cambridge, Cambridge, UK

Asha S Multani
Department of Genetics, M. D. Anderson Cancer Center, The University of Texas, Houston, TX, USA

Elisa Port
Department of Surgery, Icahn School of Medicine at Mount Sinai, New York, NY, USA

Steven M Larson
Department of Radiology, Memorial Sloan Kettering Cancer Center, New York, NY, USA

Nikolaus Schultz
Marie-Josée and Henry R. Kravis Center for Molecular Oncology, Memorial Sloan Kettering Cancer Center, New York, NY, USA
Human Oncology and Pathogenesis Program, Memorial Sloan Kettering Cancer Center, New York, NY, USA

Daniel Braas
Crump Institute for Molecular Imaging, David Geffen School of Medicine, University of California, Los Angeles, CA, USA
Department of Molecular & Medical Pharmacology, David Geffen School of Medicine, University of California, Los Angeles, CA, USA
UCLA Metabolomics Center, David Geffen School of Medicine, University of California, Los Angeles, CA, USA

Heather R Christofk
Department of Molecular & Medical Pharmacology, David Geffen School of Medicine, University of California, Los Angeles, CA, USA
Jonsson Comprehensive Cancer Center, David Geffen School of Medicine, University of California, Los Angeles, CA, USA
UCLA Metabolomics Center, David Geffen School of Medicine, University of California, Los Angeles, CA, USA

Ingo K Mellinghoff
Human Oncology and Pathogenesis Program, Memorial Sloan Kettering Cancer Center, New York, NY, USA
Department of Neurology, Memorial Sloan Kettering Cancer Center, New York, NY, USA
Department of Pharmacology, Weill Cornell Medical College, New York, NY, USA
Department of Neurology, Weill Cornell Medical College, New York, NY, USA

Thomas G Graeber
Crump Institute for Molecular Imaging, David Geffen School of Medicine, University of California, Los Angeles, CA, USA
Department of Molecular & Medical Pharmacology, David Geffen School of Medicine, University of California, Los Angeles, CA, USA
Jonsson Comprehensive Cancer Center, David Geffen School of Medicine, University of California, Los Angeles, CA, USA
UCLA Metabolomics Center, David Geffen School of Medicine, University of California, Los Angeles, CA, USA
California NanoSystems Institute, David Geffen School of Medicine, University of California, Los Angeles, CA, USA

Thomas E Gorochowski, Amin Espah Borujeni, Yongjin Park, Alec AK Nielsen, Jing Zhang and Bryan S Der
1 Synthetic Biology Center, Department of Biological Engineering, Massachusetts Institute of Technology, Cambridge, MA, USA

D Benjamin Gordon and Christopher A Voigt
Synthetic Biology Center, Department of Biological Engineering, Massachusetts Institute of Technology, Cambridge, MA, USA
Broad Institute of MIT and Harvard, Cambridge, MA, USA

Sunjae Lee, Cheng Zhang, Zhengtao Liu, Natasha Sikanic, Sumit Deshmukh, Azadeh M Harzandi Tim Kuijpers and Mathias Uhlen
Science for Life Laboratory, KTH – Royal Institute of Technology, Stockholm, Sweden

Martina Klevstig, Marcus Ståhlman, Mattias Bergentall, Fredrik Bäckhed, Ulf Smith and Jan Boren
Department of Molecular and Clinical Medicine, University of Gothenburg and Sahlgrenska University Hospital, Gothenburg, Sweden

Bani Mukhopadhyay, Resat Cinar, Joshua K Park and George Kunos
Laboratory of Physiologic Studies, National Institute on Alcohol Abuse and Alcoholism, National Institutes of Health, Bethesda, MD, USA

Morten Grøtli
Department of Chemistry and Molecular Biology, University of Gothenburg, Gothenburg, Sweden

Simon J Elsässer
Department of Medical Biochemistry and Biophysics, Karolinska Institutet, Stockholm, Sweden

Brian D Piening and Michael Snyder
Department of Genetics, Stanford University, Stanford, CA, USA

Jens Nielsen and Adil Mardinoglu
Science for Life Laboratory, KTH – Royal Institute of Technology, Stockholm, Sweden
Department of Biology and Biological Engineering, Chalmers University of Technology, Gothenburg, Sweden

Mary Lee and Eric Puttock
Department of Mathematics, University of California, Irvine, Irvine, CA, USA

George T Chen
Department of Microbiology and Molecular Genetics, University of California, Irvine, Irvine, CA, USA

Kehui Wang and Robert A Edwards
Department of Pathology, School of Medicine, University of California, Irvine, Irvine, CA, USA
Chao Family Comprehensive Cancer Center, University of California, Irvine, Irvine, CA, USA

Marian L Waterman
Department of Microbiology and Molecular Genetics, University of California, Irvine, Irvine, CA, USA
Chao Family Comprehensive Cancer Center, University of California, Irvine, Irvine, CA, USA
Center for Complex Biological Systems, University of California, Irvine, Irvine, CA, USA

John Lowengrub
Department of Mathematics, University of California, Irvine, Irvine, CA, USA
Chao Family Comprehensive Cancer Center, University of California, Irvine, Irvine, CA, USA
Center for Complex Biological Systems, University of California, Irvine, Irvine, CA, USA
Department of Biomedical Engineering, University of California, Irvine, Irvine, CA, USA

Kasper Karlsson
Stanford Cancer Institute, Stanford University School of Medicine, Stanford, CA, USA

Peter Lönnerberg and Sten Linnarsson
Laboratory for Molecular Neurobiology, Department of Medical Biochemistry and Biophysics, Karolinska Institutet, Stockholm, Sweden]

Bernhard Schmierer, Sandeep K Botla and Jilin Zhang
Department of Medical Biochemistry and Biophysics, Karolinska Institutet, Stockholm, Sweden

Mikko Turunen and Teemu Kivioja
Genome-Scale Biology Research Program, Faculty of Medicine, University of Helsinki, Helsinki, Finland

Jussi Taipale
Department of Medical Biochemistry and Biophysics, Karolinska Institutet, Stockholm, Sweden
Genome-Scale Biology Research Program, Faculty of Medicine, University of Helsinki, Helsinki, Finland

Kate Sokolina, Saranya Kittanakom, Jamie Snider, Victoria Wong, Javier Menendez and Zhong Yao
Donnelly Centre, University of Toronto, Toronto, ON, Canada

Max Kotlyar and David Otasek
Princess Margaret Cancer Centre, University Health Network, University of Toronto, Toronto, ON, Canada

Pascal Maurice
Inserm, U1016, Institut Cochin, Paris, France
CNRS UMR 8104, Paris, France
Sorbonne Paris Cité, University of Paris Descartes, Paris, France
UMR CNRS 7369 Matrice Extracellulaire et Dynamique Cellulaire (MEDyC), Université de Reims Champagne Ardenne (URCA), UFR Sciences Exactes et Naturelles, Reims, France

Jorge Gandía Xavier Morató and Francisco Ciruela
Unitat de Farmacologia, Departament de Patologia i Terapèutica Experimental, Facultat de Medicina, IDIBELL, Universitat de Barcelona, L'Hospitalet de Llobregat, Barcelona, Spain
Institut de Neurociències, Universitat de Barcelona, Barcelona, Spain

Abla Benleulmi-Chaachoua, Atsuro Oishi, Ralf Jockers and Kenjiro Tadagaki
Inserm, U1016, Institut Cochin, Paris, France
CNRS UMR 8104, Paris, France
Sorbonne Paris Cité, University of Paris Descartes, Paris, France

Ramy H Malty,Viktor Deineko, Hiroyuki Aoki, Shahreen Amin, and Mohan Babu
Department of Biochemistry, Research and Innovation Centre, University of Regina, Regina, SK, Canada

Igor Jurisica
Princess Margaret Cancer Centre, University Health Network, University of Toronto, Toronto, ON, Canada
Departments of Medical Biophysics and Computer Science, University of Toronto, Toronto, ON, Canada
Institute of Neuroimmunology, Slovak Academy of Sciences, Bratislava, Slovakia

Igor Stagljar
Donnelly Centre, University of Toronto, Toronto, ON, Canada
Department of Molecular Genetics, University of Toronto, Toronto, ON, Canada
Department of Biochemistry, University of Toronto, Toronto, ON, Canada

Hiroyuki Kobayashi and Michel Bouvier
Department of Biochemistry, Institute for Research in Immunology & Cancer, Université de Montréal, Montréal, QC, Canada

Daniel Auerbach
Dualsystems Biotech AG, Schlieren, Switzerland

Stephane Angers
Department of Pharmaceutical Sciences, Leslie Dan Faculty of Pharmacy and Department of Biochemistry, Faculty of Medicine, University of Toronto, Toronto, ON, Canada

Natasa Przulj
Department of Computing, University College London, London, UK

Jaydeep K Srimani and Allison J Lopatkin
Department of Biomedical Engineering, Duke University, Durham, NC,USA

Shuqiang Huang
Center for Synthetic Biology Engineering Research, Shenzhen Institutes of Advanced Technology, Chinese Academy of Sciences, Shenzhen, China

Lingchong You
Department of Biomedical Engineering, Duke University, Durham, NC,USA
Center for Genomic and Computational Biology, Duke University, Durham, NC, USA
Department of Molecular Genetics and Microbiology, Duke University School of Medicine, Durham, NC, USA

Panagiotis L Kastritis, Marie-Therese Mackmull, Thomas Bock, Yuanyue Li, Matt Z Rogon, Wim J Hagen, Martin Beck, Anne-Claude Gavin, Peer Bork, Katarzyna Buczak, Natalie Romanov and Marco L Hennrich
European Molecular Biology Laboratory, Structural and Computational Biology Unit, Heidelberg, Germany

Francis J O'Reilly
European Molecular Biology Laboratory, Structural and Computational Biology Unit, Heidelberg, Germany
Chair of Bioanalytics, Institute of Biotechnology, Technische Universität Berlin, Berlin, Germany

Matthew J Betts and Robert B Russell
Cell Networks, Bioquant & Biochemie Zentrum Heidelberg, Heidelberg University, Heidelberg, Germany

Khanh Huy Bui
European Molecular Biology Laboratory, Structural and Computational Biology Unit, Heidelberg, Germany
Department of Anatomy and Cell Biology, McGill University, Montreal, QC, Canada

Juri Rappsilber
Chair of Bioanalytics, Institute of Biotechnology, Technische Universität Berlin, Berlin, Germany
Wellcome Trust Centre for Cell Biology, School of Biological Sciences, University of Edinburgh, Edinburgh, UK

Lin Yang and Remo Rohs
Molecular and Computational Biology Program, Departments of Biological Sciences, Chemistry, Physics & Astronomy, and Computer Science, University of Southern California, Los Angeles, CA, USA

Yaron Orenstein and Ron Shamir
Blavatnik School of Computer Science, Tel Aviv University, Tel Aviv, Israel

Arttu Jolma, Yimeng Yin and Jussi Taipale
Division of Functional Genomics and Systems Biology, Department of Medical Biochemistry and Biophysics, Karolinska Institutet, Stockholm, Sweden

Philipp E Geyer and Matthias Mann
Department of Proteomics and Signal Transduction, Max Planck Institute of Biochemistry, Martinsried, Germany
Faculty of Health Sciences, NNF Center for Protein Research, University of Copenhagen, Copenhagen, Denmark

Lesca M Holdt and Daniel Teupser
Institute of Laboratory Medicine, University Hospital, LMU Munich, Munich, Germany

Michael J Lawson, Daniel Camsund, Jimmy Larsson, Özden Baltekiń, David Fange and Johan Elf
Department of Cell and Molecular Biology, Science for Life Laboratory, Uppsala University, Uppsala, Sweden

Gary Kohanbash and Manish Aghi
Department of Neurological Surgery, University of California San Francisco, San Francisco, CA, USA

Sören Müller, Siyuan John Liu, Martina Malatesta and Aaron Diaz
Department of Neurological Surgery, University of California San Francisco, San Francisco, CA, USA
Eli and Edythe Broad Center of Regeneration Medicine and Stem Cell Research, University of California San Francisco, San Francisco, CA, USA

Elizabeth Di Lullo, Alex A Pollen, Tomasz J Nowakowski and Arnold R Kriegstein
Eli and Edythe Broad Center of Regeneration Medicine and Stem Cell Research, University of California San Francisco, San Francisco, CA, USA
Department of Neurology, University of California San Francisco, San Francisco, CA, USA

Daniel A Lim
Department of Neurological Surgery, University of California San Francisco, San Francisco, CA, USA
Eli and Edythe Broad Center of Regeneration Medicine and Stem Cell Research, University of California San Francisco, San Francisco, CA, USA
Veterans Affairs Medical Center, San Francisco, CA, USA

Index

www.ingramcontent.com/pod-product-compliance
Lightning Source LLC
Chambersburg PA
CBHW080258230326
41458CB00097B/5113